James Joseph Sylvester

∽James Joseph Sylvester∽

Jewish Mathematician
in a Victorian World

KAREN HUNGER PARSHALL

The Johns Hopkins University Press
Baltimore

2 4 6 8 9 7 5 3 1

The Johns Hopkins University Press
2715 North Charles Street
Baltimore, Maryland 21218-4363
www.press.jhu.edu

Library of Congress Cataloging-in-Publication Data
Parshall, Karen Hunger, 1955–
    James Joseph Sylvester : Jewish mathematician in a Victorian world / Karen Hunger Parshall.
        p. cm.
    Includes bibliographical references and index.
    ISBN 0-8018-8291-5 (alk. paper)
        1. Sylvester, James Joseph, 1814–1897. 2. Mathematicians—Great Britain—History—19th
    century—Biography. 3. Jews—Great Britain—History—19th century—Biography. 4. Great
    Britain—History—19th century—Biography.   I. Title.
    QA29.S96P375 2006
    510′.92—dc22[B]   2005052035

A catalog record for this book is available from the British Library.

To my adviser, Allen G. Debus,

to my parents, "Mike" and Maurice Hunger,

and to my husband, Brian Parshall

Contents

Acknowledgments

Much of my work over the past two decades has centered in one way or another on the English mathematician James Joseph Sylvester, on the world in which he lived, on the mathematics he helped to create. From the beginning, I adopted a two-pronged research agenda. First, I confronted the extant archival record. That resulted in 1998 in *James Joseph Sylvester: Life and Work in Letters*, a selection of Sylvester's sizeable *Nachlaß* with extensive histor ical and mathematical commentary. Second, I used those archives as a frame upon which to weave a biography of a man who, to me at least, was a fascinating figure in the history of nineteenth-century science. With the publication of the present book, this research agenda is now complete.

In the course of this project, I have incurred more debts, made more friends, and established more professional relationships than I could ever adequately acknowledge. As always, it is the greatest of pleasures to thank those who have helped me and made me welcome in far-flung archives on my numerous research trips in search of Sylvesteriana: the librarians of St. John's College, Cambridge; Cynthia Requardt, Joan Grattan, and the staff in the archives at the Johns Hopkins University; Caroline Dalton and her staff in the library at New College, Oxford; Gillian Furlong and her staff in Special Collections at University College London; Brian O'Donnell, formerly at Trinity College Dublin; the librarians, John Rogers and Vin Roper, of the Athenæum of Liverpool; the staff at the Liverpool Record Office; Maureen Watry and her staff at the archives of the University of Liverpool; David Raymont at the library of the Institute of Actuaries and Claire Milne at the archives of the Inner Temple, both in London; Helmut Rohlfing and his staff at the Handschriftenabteilung of the Niedersächsische Staats- und Universitätsbibliothek in Göttingen; the staffs at the Paris Académie des Sciences and at the Rare Book and Manuscript Library of Columbia University in New York; and Paul Theerman, Marc Rothenberg, and both the Joseph Henry Papers Project and the Smithsonian Institution Archives.

For permission to quote here from the archives held in their collections, I gratefully acknowledge the Master and Fellows of St. John's College, Cambridge; Special Collections, the Milton S. Eisenhower Library, Johns

Hopkins University; the Warden and Fellows of New College, Oxford; University College London, Library Service; the London Mathematical Society; Special Collections, University of Virginia Library; the Paris Académie des Sciences; the British Library; the Syndics of Cambridge University Library; the Women's Library, London Metropolitan University; the Inner Temple Archives; the Institut Mittag-Leffler, Djursholm, Sweden; the Royal Institution Archives; the Royal Society of London; the Smithsonian Institution Archives; the Staatsbibliothek Berlin; and Trinity College Dublin. Other archives cited here appeared with permission in *James Joseph Sylvester: Life and Work in Letters.*

Most of the travel and much of the research that have gone into this book would have been impossible without the generous financial support I have received for my work from the National Science Foundation (especially grants SES 8509795 and DIR 9011625) and from the John Simon Guggenheim Foundation (in 1996–97). Moreover, the first phase of my Sylvester project was completed while I held a Sesquicentennial Fellowship from the University of Virginia in 1996–97; I finished this second and final phase during another Sesquicentennial Fellowship granted for the spring semester of 2004 followed by a University Summer Research Fellowship. The University of Virginia has consistently provided me with the research support and intellectual stimulation needed to pursue my research. In particular, I thank my departmental chairs during the 2003–2004 academic year, Chuck McCurdy in History and Jim Howland in Mathematics, for all their help and encouragement during the final stages of my work on this biography.

Friends and colleagues at home and abroad have also (seemingly) never tired of talking with me and answering questions about Sylvester and his times. Here in Charlottesville, my colleagues in History, Paul Halliday and Mark Thomas, read my manuscript with an eye to the Victorian British historical scene, while Joe Kett provided advice on everything from sources on the history of American higher education to how best to craft a "word picture" to how to streamline and sharpen arguments. Joe, Elizabeth Meyer, and John Stagg also cheerfully sat through lunch after lunch of conversations punctuated by more than they probably ever wanted to hear about Sylvester, his exploits, his accomplishments, and certainly his mathematics. Holly Shulman, the editor of the digital edition of the letters of Dolley Madison, also stepped outside of her usual research bailiwick to share with me not only her deep insights into Jewish identity but also to discuss with me at length the problems that I encountered in trying to understand Sylvester as a nineteenth-century English Jew. Holly also introduced me to Justin Brummer in London, who, aided by his knowledge of Hebrew, explored for me many of the Jewish archives in London in search of Sylvester's Jewish roots. In the Mathematics Department, my graduate students, Sloan Despeaux, Laura Martini, and Deborah Kent, listened to and provided critical feedback on numerous talks on "research in progress" during our biweekly research seminar as did my colleagues in the Commonwealth Consortium for the History of Mathematics, Adrian Rice and Della Fenster. Also from the Mathematics Department, Holly Carley secured Sylvester's correspondence with the Swedish mathematician Gösta Mittag-Leffler for me during her research sojourn as a graduate student at the Institut Mittag-Leffler in Djursholm, Sweden. Finally, Julie Riddleberger, fondly known as the Department's "TeX goddess," good-naturedly solved even the most complicated typesetting questions.

Farther from home, my French colleague, Catherine Goldstein, alerted me to some interesting correspondence relating to Sylvester that she had unearthed during the course of her research in the Dirichlet *Nachlaß* in Berlin. Albert Lewis of the Charles S. Peirce Papers Project at the Indiana University–Purdue University of Indiana shared his insights both on Peirce and on Sylvester's student, George Bruce Halsted, while Dan Silver of the University of South Alabama forced me finally to come to terms with Sylvester's Jewish identity when he invited me to give not only a mathematical colloquium to his department but also a lecture to the greater Mobile Jewish community. The talk that I prepared for the latter occasion in some sense came to shape the entire argument of the biography I have written.

In England, several friends and colleagues opened their homes and their pantries to me and shared their numerous insights. In London, June Barrow-Green provided a homebase from which I retraced Sylvester's steps around Bloomsbury, the Inns of Court, and Mayfair. In Bengeo, Ivor and Enid Grattan-Guinness welcomed me on so many occasions that I have lost count. Ivor has also always been a keen and most valued critic of my work on Sylvester, reading numerous chapters in draft, pointing me to new sources, grilling me on my arguments, sharing his vast knowledge of the entire nineteenth-century, European mathematicophysical scene. In Liverpool, Carole and Arnold Lewis introduced me to Liverpool's Jewish community, helped me navigate the local archives, and made it possible for me to meet and talk with Joseph Wolfman about his work on Jews in nineteenth-century Liverpool. Their hospitality, their knowledge of the local historical literature, and their untiring efforts to help me trace Sylvester's English ancestral roots resulted in many of the glimpses of Sylvester's early life that so enhanced the picture I was able to present in chapter 1. Finally, Tony Crilly always shared new archives as he encountered them in his research on Sylvester's best friend and mathematical alter ego, Arthur Cayley, while Hugh Stewart and Paul Garcia put me on to sources that I never would have encountered on my own.

I close by offering my greatest thanks to my adviser, Allen G. Debus, my parents, "Mike" and Maurice Hunger, and my husband, Brian Parshall. This book is dedicated to the four of you, since I could never have written it without your continuous support and encouragement.

James Joseph Sylvester

James Joseph Sylvester:
The Myth, the Mathematician, the Man

*Young Americans could hardly realize that the great Sylvester, who
with Cayley outranks all English-speaking mathematicians, was
actually at work in our land. All young men who felt within
themselves the divine longing of creative power hastened to
Baltimore, made at once by this Euclid a new Alexandria.*
—George Bruce Halsted to Florian Cajori, 25 December 1888

In Baltimore, 22 February 1877 was a day of celebration. Bright, cloudless,
uncharacteristically springlike, the birthday of America's first president was
feted in fine style. Bunting hung from the balconies; salutes were fired from
nearby Fort McHenry; large crowds of men, women, and children, both black
and white, filled the streets and the parks "where the signs of reviving nature
[were] beginning to appear in freshly bursting buds and pale green shoots after
the long sleep of winter."[1]

Downtown at the city's Johns Hopkins University, it was also a day of
celebration; just a year earlier on this day, the university had been officially
inaugurated. In the intervening twelve months, buildings had been acquired, a
faculty had been assembled, students had been admitted, and classes had begun.
The university rejoiced not only in Washington's birthday but also in its own
initial success with a by-invitation-only event. Hopkins Hall was resplendent.
A grand memorial tablet to founder Johns Hopkins had been mounted on the
front wall and ringed with greenery, while the dais bloomed with flowers from
the late Hopkins's estate. The dais was also adorned with some of Baltimore's
finest: its mayor, Ferdinand Latrobe; Hopkins president, Daniel Coit Gilman;
the university's trustees; and its professors.

The third speaker to take the podium that day was a man of singular
appearance. Short and stout, he had sharp gray eyes, a full gray beard, and

an enormous, domed pate fringed with unruly gray hair. He seemed to one who saw him "a sort of subconscious adjunct to his head."[2] This was James Joseph Sylvester, the professor of mathematics, a sixty-two-year-old Englishman internationally renowned for his researches in the theory of invariants, a mathematician deemed by many in his audience "one of the historic figures of all time, one of the immortals."[3]

Sylvester eloquently delivered an "elaborately prepared" speech that ranged, like his mind, swiftly from topic to topic.[4] Mathematics, how best to study and teach it, the role of the Johns Hopkins University in nurturing it in the United States, the new university's place in the pantheon of higher education internationally—Sylvester voiced his thoughts on all these things on Commemoration Day, but one topic that particularly exercised him, and toward which his remarks effectively built, was sectarian influence in higher education. He abhorred it. He had personally suffered its consequences as a Jew in England in an educational system that was little more, in his words, than "the monopoly of a party and the appanage of a sect," and he was thankful that there would be none of that kind of bigotry at the Johns Hopkins.[5] "Happy the young men gathered under our wing," he said, "who, unfettered and untrammeled by any other test than that of diligence and attainments, have here afforded to them an opportunity of filling up a complete scheme of education, such as a Milton or a Locke would have deemed adequate to their ideal" (83).

Sylvester's audience was riveted, and no one more than the American poet, James Russell Lowell, who shared the dais with him and who would soon follow him at the podium. One onlooker wrote that he would "never forget the emotion and astonishment exhibited by . . . Lowell while listening to this unexpected climax" of Sylvester's address.[6] At the end of the day, many more than the few students in his Hopkins classroom had come to the realization "that the great Sylvester . . . was actually at work in our land."[7]

≈ ≈ ≈

WHO WAS THIS IMPASSIONED ORATOR? For all that was written about him at the time of his death in 1897 and in the decades leading up to and following the publication in 1937 of Eric Temple Bell's *Men of Mathematics*, Sylvester remained little more than a bare set of mostly consistent biographical facts upon which was imposed—more often than not—a caricature of the absent-minded mathematician, the fiery and often difficult personality, the true eccentric.[8] This, however, was the myth that subsequently developed, not the man that the Hopkins audience had come to hear in February 1877. They had come to see a living mathematical legend, the "Euclid" who would make Baltimore a "new Alexandria."[9]

By 1877, Sylvester had made his international reputation during the course of a career that had already spanned four decades. He had finished as second wrangler on the Mathematical Tripos at Cambridge University in 1837 and, that same year, had published his first mathematical paper. Forty years later, he had done seminal research characterized by a fundamentally algebraic point of view at a time when many, especially in England, worked in a geometrical or applied mathematical vein. Sylvester's first algebraic foray was in the theory of elimination,[10] next in the new theory of invariants that he developed in the 1850s hand-in-hand with his friend and mathematical confidant, Arthur Cayley, next in the theory of partitions, and on and off in the theory of numbers. He had also succeeded—where none other than Isaac Newton had failed—in providing a proof, over the years 1864 and 1865, of Newton's Rule for isolating pairs of imaginary roots of a polynomial equation. His mathematical output—over 1,300 quarto pages by 1877—had won him numerous honors and awards: the Royal Medal of the Royal Society of London in 1860, the position of foreign corresponding member of the Paris Académie des Sciences in 1863, the presidency of Section A (Mathematics and Physics) of the British Association for the Advancement of Science (BAAS) in 1869, and recognition by foreign academies and societies in Germany, Russia, Italy, and the United States. His results had also appeared in journals at home and abroad, notably in France and Italy. In fact, Sylvester had consciously worked to establish his mathematical reputation both in England and on the Continent. He, unlike many of his countrymen, had an international outlook and saw mathematics as a universal endeavor in which researchers, regardless of their nationalities, built on and extended a common and ever-expanding body of knowledge. All the better if the work built on and extended was his own! Moreover, in his view, England was a lesser mathematical stage in the mid-nineteenth century, behind France and Germany. It was important to establish a reputation at home, but real mathematical reputation was conferred by those across the English Channel.

If this provides a sketch of how the Hopkins Hall audience had come by its sense of an internationally renowned mathematician in its midst, what did they know of the man behind the mathematics? As he captivated them from the podium, Sylvester allowed them several glimpses into his inner self. He was humorous in his false self-deprecation. "An eloquent mathematician," he told them within moments of opening his remarks, "must, from the nature of things, ever remain as rare a phenomenon as a talking fish, and it is certain that the more anyone gives himself up to the study of oratorical effect the less will he find himself in a fit state of mind to mathematize."[11] He was colorful and genuine in his gratitude, admitting that "I should be heartless, indeed, and more callous than the oyster, who, twin-soul to the mathematician, working in silence and seclusion between the folding doors of his mansion, elaborates the pearl that may, hereafter, deck an empress's brow, could I be insensible to the many proofs of kind and generous feeling which, both within and without the walls of this University, have been so widely and unequivocally accorded to me" (73). He was passionate in his beliefs and in his sense of injustice, likening sectarian influence to "a black drop of gall, a taint of congenital rancour and animosity, which infects all it comes in contact with, more indelible, more difficult to wring out or efface, than that dread smear on Lady Macbeth's hand, which could 'the multitudinous sea incaradine'" (82). He had a spirit of generosity that infused his praise for his new Hopkins colleagues. "How rejoiced should I be," he announced, "were I of less ripe years and under less peremptory obligations as to the disposal of my time, branching out from mathematics as my natural mental centre of gravity, to diverge into the physical and chemical studies which lie so near to it, and which there are here such ample means accorded of studying under the most competent instructors" (84–85). Sylvester was a living, breathing, feeling man as well as a mathematician, but what forces had actually shaped him?

SYLVESTER HAD BEEN BORN A JEW IN ENGLAND IN 1814 at a time when Jews and other dissenters from the Church of England, although tolerated, did not enjoy the same rights and privileges as their Anglican brethren. In 1831, the year he matriculated at St. John's College, Cambridge, a required dissolution of the government on the ascension to the throne of William IV marked the beginning of a decades-long period of reform that ultimately came to affect everything from the rights of the working man, to the rights of Jews, to the way in which the military was staffed, to education at all levels, to the right to vote, to the judiciary, to the civil service. These changes—most of which were associated with Victoria's rule from 1837 to 1901—had significant consequences, not the least of which was the firm establishment of a middle class and of a new social category: the professional man. Time, place, and religion, these factors very much shaped Sylvester's life.

Sylvester himself, while perhaps too intimately connected to time and place to recognize their effects on his life's trajectory, was nevertheless so fully aware of the consequences his religion had had in his life that at the age of sixty-two he singled out precisely this factor in the very public venue of his Commemoration Day address in Baltimore. As he alluded there, the fact that he was Jewish and unwilling to renounce his religion and his heritage had meant that, although he could attend Cambridge, he could neither actually take his degree nor compete for a fellowship there. One obvious avenue for a mathematically talented high wrangler was thus closed to him, forcing him to seek his livelihood elsewhere, to shape his own career rather than merely to fall into a comfortable, well-defined niche.

Sylvester wanted a career in academe that would allow him to pursue his mathematical researches. He wanted to be a professional mathematician, a mathematician, in his view, who might earn a living through teaching but who had the time and the drive to produce and publish original research. Incredibly, the chair of natural philosophy at University College, London, the nonsectarian university that had opened in 1828, fell vacant just as Sylvester began this task of self-definition. As a Cambridge-trained mathematician, he had been exposed to and had mastered quite a bit of physics, although mostly from a theoretical, mathematical point of view, so this position, although not as comfortable for him as a mathematics professorship might have been, was nonetheless a viable option.[12] As he soon learned, however, he was far from a natural in the laboratory. In 1841, he left what was effectively the only university-level appointment open to him in England to take the professorship of mathematics at Thomas Jefferson's University of Virginia on the other side of the Atlantic. It was a daring move made in search of the kind of career he envisioned for himself, but it ended in failure. After four-and-a-half months, he quit over a matter of principle regarding academic freedom and failed in his subsequent efforts to find academic employment in the United States.

On his return to England late in 1843, he again cast about for a suitable position. He had to earn a living. In 1844, he took the post of actuary and subsequently secretary at the Equity and Law Life Assurance Society in London. Although a position in which he could put some of his mathematical expertise to use, it was not a job in academe. Sylvester ultimately persevered at Equity and Law for more than ten years before winning in 1855 the type of prize he had been seeking, the professorship of mathematics at the Royal Military Academy in Woolwich. After fifteen years on the job there, in which he fought with only some success to realize his conception of the professional mathematician, he was forcibly retired by a change in the law that prohibited anyone over the age of fifty-five to hold an Academy professorship. This seemed the end of a career that had never quite met its crafter's expectations. Despite the volume of research he had produced, despite the recognition and reputation he had secured among his scientific confrères both at home and abroad, he had not succeeded in effecting his ideal of a professional mathematician. That opportunity would only present itself in 1876 when he came out of retirement to accept the professorship of mathematics at the Johns Hopkins University in Baltimore.

This equally daring transatlantic move not only brought him seven and a half productive and rewarding years of research, teaching, and training future researchers but also strengthened his resolve that his vision of the professional mathematician was realizable and right for England. The only problem was that he was no longer at home; he was in the United States. In 1883, however, the Savilian professorship of geometry at Oxford became vacant, and he, thanks to reforms enacted in the early 1870s, was actually eligible for it despite his Jewish heritage. Sylvester applied. He was in his late sixties and mentally wearied by the weight of his responsibilities in Baltimore. In December 1883, he received the word that he had almost dared not to expect. He had been appointed to the Oxford chair. The first Jew to hold a professorship at Oxbridge, he had finally broken the barrier that had forced him down the convoluted career path he had had to carve step by step. In principle, he had finally won one major life battle over the forces of religion, but as he soon came to realize, Oxford, unlike Baltimore, was not yet ready to support his ideal of the professional mathematician. It was not yet quite the time or the place.

In addition to his own personal quest to define a new category of professional mathematician within the growing group of professional men in England, Sylvester had long worked to encourage and promote the accoutrements needed to support such a category. He participated in the Royal Society of London and in the BAAS, but more particularly in the BAAS's specialized Section A and, after its founding in 1865, in the London Mathematical Society. He edited key specialized journals: the *Quarterly Journal of Pure and Applied Mathematics* in England and the *American Journal of Mathematics* in the United States. He refereed for and submitted his original research to journals in Great Britain, on the Continent, and on the other side of the Atlantic. He recognized, in short, that the professional mathematician could only come about through the establishment of an effective system of gatekeeping. Mathematics was a meritocracy, open to any and all with the requisite talent, but as such, the standards of merit had to be set, met, and maintained. Only in this way would England (and the United States) begin to catch up with mathematicians in Europe, and especially in France and Germany.

Sylvester's commitment to the ideal of the professional mathematician was, in fact, only one manifestation of evolutionary changes in nineteenth-century England that were creating the broader category of the professional man. As early as the 1840s, Sylvester had participated in the development of the Institute of Actuaries as a means for legitimizing the growing industry of which he was then a part as well as for certifying the mathematical competence of its practitioners. The field, under attack in the 1840s and 1850s by the citizenry and by the government on account of the fraudulent business practices of some fly-by-night firms, required oversight by those of recognized talent and unquestionable ethics. What better way to ensure such oversight than by making provisions for it freely from within rather than having it imposed punitively from without?

Similarly, Sylvester's support of initiatives to remove all remaining disabilities for Jews was only one of the ways in which the Victorian spirit of reform shaped his life. In the 1850s, he offered his services to the government on more than one occasion to help affect fiscal and other reforms under consideration in Parliament. In the 1850s, 1860s, and 1870s, he was committed to the movement for workingmen's education, examining for the Society of Arts and giving Penny Readings in the evenings to working class men, women, and children who sought to better themselves through literature, poetry, art, and general cultural literacy. In the 1870s, he also sought to bring his educational values to bear on the issue of primary and secondary education by running (although unsuccessfully) for a seat on the London School Board.

If Sylvester, the man, responded in key ways to these political and social forces of time and place, he was also moved profoundly by more timeless forces of European culture: literature, poetry, music, and travel. An avid reader throughout his life, he could carry on knowledgeable conversations about Greek, Roman, English, French, and German literature. Beginning in the 1840s and almost literally until the day he died, he composed his own original verses in English, Latin, and Italian and, in 1870, published a book on what he deemed to be *The Laws of Verse*.[13] A capable tenor, he also reportedly studied voice for a time with the French composer Charles Gounod and sang both in private at musical gatherings with friends and in public at Penny Readings.[14] From at least the late 1830s and throughout his life, he traveled widely—within the British Isles, to Europe, to the United States, to Russia, to Scandinavia, to North Africa—taking in the scenery and the culture, making connections with the people and with those of scientific bent, learning that there was a world beyond England to consider and in which to participate.

Sylvester the Victorian, Sylvester the Jew, Sylvester the mathematician—these were not three separate entities to be understood in three distinct ways. They were one man who, simultaneously shaped by his religion, his time, and his place, crafted a life for himself as a mathematician in the Victorian era while laying foundations—both mathematical and professional—upon which others would later build.

THIS BOOK AIMS TO TELL, for the first time, the complex story of Sylvester's life by situating that life as fully as possible within the political, religious, mathematical, and social currents of nineteenth-century England. It aims to demythologize the man by placing him in his milieux at the same time that it demystifies his mathematics by revealing it as a very human endeavor. It aims to show how the man lived his life, what choices he made and why, how the world in which he lived affected him, and how he affected that world.

This approach—reminiscent of what Adrian Desmond has termed "contextual biography"—makes no apologies for centering on one particular life in its effort both to illuminate that life *per se* and to provide key insights into a range of broader issues.[15] In the case of the nineteenth-century, Anglo-Jewish mathematician, Sylvester, those broader issues range, as indicated above, from the intellectual history of mathematics, to the social history of nineteenth-century science, to the history of science in Britain, to the history of professionalization, and to Anglo-Jewish history.[16] Biography may be case study, but it is case study with "legs." The story of Sylvester's life, for example, also sheds light on the evolution of mathematical thought, the ways in which mathematics may be done, and what factors may shape the mathematician's ideas. It further illuminates historical contingencies of professionalization, of nineteenth-century mathematical culture, of the development of what would come to be recognized as "modern algebra," of nationalism, internationalism, and internationalization in creating scientific communities and bodies of scientific knowledge. And, it highlights the very human side—shaped by factors as diverse as religion, ego, and depression—of what many view as that most inhuman and otherworldly of intellectual endeavors: mathematics.

In 1998, I published *James Joseph Sylvester: Life and Work in Letters*.[17] That book paved the way for the present study by providing a representative sample of and extensive commentary on the surviving archival record related to Sylvester's life and work. It followed Sylvester's life from 1837—his twenty-third year and the year of the first surviving letter I was able to find in his far-flung *Nachlaß*—through to its end in 1897. It was most successful in documenting Sylvester's mathematical thought in the making; that is what Sylvester wrote about most, or, at least, that is the primary subject matter of the correspondence that survives. The book sought both to illuminate as fully as possible Sylvester's nineteenth-century mathematical ideas to a modern audience and to highlight the hints peppered throughout of the man behind those ideas. It was a necessary prerequisite for this, its narrative companion, yet much of what is covered here was only barely hinted at in the letters.

Sylvester may have written voluminously to his correspondents about his mathematical ideas, but he almost never mentioned or even alluded to his Jewish heritage. Sylvester the Jew is virtually invisible in the correspondence. The young Sylvester is also absent. No letters exist (apparently) from the boy away at boarding school or from the young man away at Cambridge to provide a first-person account of his experiences, his hopes, and his dreams. Sylvester also did not write (again apparently) about his life as an actuary, or (much) about his career as a professor of mathematics to mostly teenaged boys training for careers as officers in the British Army, or about his first retirement in the early 1870s, or about his final retirement in 1894. Although much about Sylvester does come through in his surviving letters, then, much does not. This biography thus aims to fill in as many of the gaps left by the archival record as possible by providing the broader contextualization of a life that must, of necessity, always elude the one actually living it.

James Joseph Sylvester could be an absent-minded, fiery-tempered eccentric. There is almost always truth behind a myth. But Sylvester was so much more. It is the many dimensions of the life of this nineteenth-century, Anglo-Jewish mathematician that the biography to follow exposes and explores.

CHAPTER 1

Born to "the Faith in Which the Founder of Christianity Was Educated"

*I feel what irreparable loss of facilities for domestic and foreign
study, for full mental development and the growth of productive
power, I have suffered, what opportunities for usefulness
been cut off from, under the effect of this oppressive monopoly
[the Church of England], this baneful system of protection
of such old standing and inveterate tenacity of existence.*
—J. J. Sylvester, 22 February 1877

James Joseph Sylvester, the future mathematician, was born in London on Saturday, 3 September 1814, into the large and prosperous Jewish family of Abraham and Miriam Joseph. England was at a critical juncture geopolitically that September. Anglo-Jewry did not then exist as a well-defined social and cultural category. Essential tensions were nevertheless in evidence that would soon effect key changes.

In April 1814, just five months before Sylvester's birth, Napoleon had signed the Treaty of Paris bringing an end both to years of conflict in Europe and to a war with Britain that had effectively begun in February 1793. The tide had clearly been turning in England's favor for some months. Her naval ships had long blockaded France's coasts, and her privateers—who had, in part, contributed to the Joseph family's wealth—had made Atlantic waters unsafe for French ships. On 3 September, the *Times* noted that "the opening of the Congress in Vienna," which aimed to reestablish political stability in Europe in the wake of Napoleon's aggression, was "looked to with the most lively impatience."[1] "Well-informed persons," it added, were "convinced" that the Congress would indeed "have a happy issue." From a British point of view, their convictions proved well founded. In fact, a new world order, in which Britain was a major force, resulted from the friction between the self-conscious nation building and the outward-looking imperialism that became so characteristic of the post-Napoleonic era.

One of the manifestations of that evolving world order could already be seen in Paris, a city that would provide, in its royally funded Académie des Sciences, a key locus for the mature Sylvester's international scientific activities. There, the Bourbon regime had been returned to the throne in the person of Louis XVIII, and the *Times* described his triumphal return to Paris in vivid terms.[2] The day had begun with "discharges of artillery." By early in the afternoon, the Champs Élysées had become the site of all manner of diversions, and "here and there were seen fountains of wine, and booths filled with provisions, which were distributed among the people." When at five o'clock the king himself appeared in the royal procession of carriages, "incessant acclamations and shouts of joy accompanied [him] as he passed." "Thus after so many years of calamity," the *Times* intoned, "the city of Paris has at last been permitted to express to its King the true sentiments with which all its inhabitants are animated."

England's war with France may have appeared to be over and the Bourbon dynasty restored, but war with the United States—another country that would strongly affect Sylvester's life and career—had been going on since 1812 with no end in sight. News from that front trickled in. Writing from Bermuda on 14 July, "an Officer" in the British Navy expressed his frustration that "we were long since to have left this for the coast, with about 1000 marines, which are too few to commence any serious operation, and too many to fritter away without effecting any object."[3] "My fears," he confessed to his readers back home, "are that the country is looking up to the navy to perform extraordinary services . . . Who will be to blame," he asked accusingly, "if the severity of the weather should oblige the expedition to be reembarked almost as soon as it lands, because the navy cannot keep the coast?" On 3 September, the outcome of the War of 1812 may have been far from certain, but the persistence and drive of that nation on the other side of the Atlantic pointed to the existence of another player on the newly internationalizing stage.

Although these and other external affairs fundamentally affected nineteenth-century England and the life Sylvester would lead there, forces operating within the island nation would also prove of critical influence. Although Jews had been expelled there by royal decree in 1290, their resettlement had begun informally in 1656, and their peaceful coexistence in and adaptation to a country dominated by the Church of England had made them by the early nineteenth century a recognized, if not fully accepted, component of the broader English cultural landscape.[4] Jews unquestionably faced disabilities in Britain, but disabilities of a social and political rather than of a financial nature. Jews could not vote. Their legal right to own land was questioned. Anti-Semitism was part of the English social fabric. Still, Jews could live, work, and worship freely in England, privileges not enjoyed by many of their coreligionists on the Continent. For the outward-looking and enterprising eighteenth-century European Jew, England had represented a land of opportunity, but one of fresh cultural challenges. By the first half of the nineteenth century, however, second- and third-generation Jews found themselves confronted with the question of how best to be—or even *to be*—Jews in an Anglican land. Tensions between traditional Jewish and English ideals, traditional Jewish and English ways of life, had resulted in confrontations in the political arena and in the definition of Anglo-Jewry as a religiously informed way of life compatible with and resonant of Anglican, English mores. The Joseph family and the mathematical son it would rear in Sylvester were equally part of this evolving Anglo-Jewish world.

Time, place, religion. Sylvester could not choose them, but these circumstances of his birth, and the tensions they created, profoundly shaped his life.

The Josephs of Liverpool

The Joseph family had taken root in England in 1754. In London that year, the thirty-two-year-old German emigré, Simon Joseph, wed Zipporah, a young Hamburg-born woman three years his junior.[5] As Ashkenazim, or descendants of Jews from Germany or central Europe (as opposed to Sephardim descended from Iberian Jews), these newlyweds were part of a small but well-defined Jewish migration from the German-speaking states to England in search of a brighter economic future and, perhaps, of a life less restricted by the mores of traditional Judaism.[6] Most of these emigrés settled permanently in London, but some, like Simon and Zipporah, made their ways to provincial cities farther afield.

By 1756 when their first son, Elias, was born, the young Joseph family had settled in the Yorkshire town of Wakefield just south of Leeds. Four more sons, Samuel, Abraham, Joseph, and Ellis, and a daughter, Sarah, followed in that order at two-year intervals. In the increasingly industrial West Riding, Wakefield was a center for the manufacture of woolens and worsteds, a major conduit of goods passing from the east of England to the West Riding and to the growing industrial regions of south Lancashire, and a seat of local government. Leeds, however, had already surpassed Wakefield in economic importance late in the seventeenth century, making the town a less-than-natural choice for the financially enterprising.[7] Wakefield was also not one of the English provincial towns that would come to support an established Jewish community in the last half of the eighteenth century, although at least two other Jewish heads of household lived there in the 1750s and 1760s, Simon's brother, Samuel Joseph, and his brother-in-law, Jonas Israel.[8] Exactly what convinced the extended Joseph family that Wakefield was where it would best make its way in England thus remains unclear, but Simon Joseph and his family unquestionably prospered there. Judging by local property records, Simon, who most likely earned his living at this time as a jeweler and watchmaker, already owned a considerable property in Wakefield by 1761 and continued to hold it in 1771.[9] By local standards of the day, the family was comfortably well off.

Sometime in the 1770s, however, the Simon Joseph family left Wakefield to settle in the bustling port city of Liverpool on England's northwest coast.[10] Liverpool at this moment was a city on the rise. There, in 1715, England's first commercial wet dock had opened, establishing the city at the mouth of the Mersey River as a major port. By the end of the century, the growth of the Atlantic slave trade, trade "in Irish wool, linen, and beef, American cotton, colonial tobacco and sugar, Midlands metals and pottery, Cheshire salt, and Lancashire coal and textiles," as well as successful privateering during decades of war had swelled Liverpool's population to some 90,000, making it second in England only to London with a population of about 1,000,000.[11] The western port's commercial success had also made it an outward-looking world city poised, in retrospect, to serve as the gateway to empire. From an economic point of view, then, Liverpool was most definitely a city of opportunities for a jeweler and watchmaker, even if, with its narrow and congested streets running up from the docks, its constant pall of smoke from the saltworks, its open-fronted slaughterhouses, and its sailors, stevedors, and other rough-and-tumble inhabitants, it was a far cry from Wakefield as a place to raise a family.[12]

Liverpool also had the advantage over Wakefield of having a distinct Jewish presence. As early as 1753, there were already enough Jews in Liverpool—perhaps mainly of Sephardi origin—to found a small synagogue and burial ground off Stanley Street in the congested business district between the town hall and the docks. This location was necessarily near the businesses and often business-residences of the congregants, most of whom made their living through trade with the seamen who ebbed and flowed through the port. Although the continuity of this original community has been debated, a second burial ground was acquired in 1773, and an Ashkenazi congregation had firmly established itself by 1775 also in the center of Liverpool's business district but in a small house in Turton Court (a minimum of ten was required by Jewish law to form a *minyan* or quorum for public worship).[13] Just three years later in 1778, the congregation had grown sufficiently to warrant yet another move to a somewhat larger property on Frederick Street adjacent to the new burial ground. That house afforded space not only for worship but also for a *mikvah* or ritual bath, one of the ancillaries critical in establishing a sense of authenticity and permanence for a Jewish community.[14] The Jews of Liverpool were thus laying down their roots ever more deeply in the port city and with them were Simon Joseph and his family. In fact, the Joseph family patriarch has been called the late-eighteenth-century "leader" of Liverpool Jewry.[15]

By 1790, that leader ran his wholesale watchmaking business from 29 Pool Lane, a bustling and congested commercial street that ran up from the docks to the more fashionable Castle Street. His eldest son, Elias, had a silversmith shop a few doors away at 16 Pool Lane, and his second son, Samuel, was a silversmith and jeweler at 35 Castle Street.[16] Daughter, Sarah, who had married Morris Lewin Mozley of Portsmouth in 1785, had returned with her husband to Liverpool, where Mozley, not unlike his inlaws, earned a living as a silversmith and watchmaker. His store was at 28 Castle Street just doors from his brother-in-law's business.[17] By 1796, sons Joseph and Abraham had also set up shop in Liverpool, with Joseph a goldsmith, jeweler, and privateers' agent at 1 Pool Lane and Abraham a silversmith, jeweler, and naval agent operating like his father out of 29 Pool Lane.[18]

Abraham, Simon and Zipporah's third son and the father of the future mathematician, had been born in 1760 and had married Miriam Woolf of Portsmouth in 1790. Although he undoubtedly learned the silversmith's trade from his father, he may well have engaged as a naval agent under the influence of his father-in-law in the key Royal Navy port of Portsmouth.[19] The possibility of pursuing such a dual career—and of garnering the profits it promised—had arisen from the political circumstances of nearly a half-century of war.

Naval agents essentially served as money lenders to and financial representatives of seamen—officers were represented by so-called prize agents—at a time when the Royal Navy paid its sailors only irregularly and when it institutionalized, but only marginally supervised, the payment of supplementary prize money based on the sale of ships and cargo captured as spoils of war.[20] Naval agents thus lent money at interest—perhaps 5%—to seamen in anticipation of future prize money. When a prize was captured, the agent then secured for the seaman he represented—and for an additional fee—the seaman's financial share of it, once the Admiralty had sold it off and announced the total cash value; every member of the crew from the captain to the lowliest swabbie received a cut of the prize.[21] A given seaman's share would then be used to pay off his debts to his agent with any remaining funds representing a financial windfall to the sailor. In the absence of a more direct system of payment administered by the Admiralty itself, these agents, many of whom were Jewish, thus provided a valuable service, especially in light of the fact that naval officers and seamen were often out of port when prizes were announced and arrangements for collection had to be made.[22] Abraham's father-in-law, Gershon Woolf, supplemented his income as a silversmith by serving as a naval agent to various of those seamen based in Portsmouth.

Obvious abuses were inherent in a system such as the one in which Gershon Woolf engaged, however; unfortunately, he got caught up in one in February 1795 when he and others were found to be in possession of forged official stamps from a number of ports and were sent to London for prosecution.[23] A year later, with Abraham and his wife Miriam set up in the commercial as opposed to naval port of Liverpool, Abraham's designation of himself as a "naval agent" in *Gore's Liverpool Directory* may have been directly related to former naval clients of his father-in-law then residing in or otherwise attached to Liverpool. Abraham's brother Joseph functioned in a similar capacity but as a privateers' agent for private merchantmen homebased in Liverpool.[24] The Josephs and the Woolfs thus engaged in and capitalized on an occupation born of the politics of war and of England's ever-increasing dominance of the seas. In particular, the Josephs, like other of their coreligionists in Liverpool, became comfortably well off in the bargain.

Indicative of this personal prosperity, by 1800, Abraham Joseph and his family made their private residence on Liverpool's fashionable Rodney Street. Well away from the town center, Rodney Street had been laid out beginning in 1783 by, among others, William Roscoe, banker, antislavery campaigner, botanist, poet, historian, and Liverpool's leading intellectual and cultural light. It had quickly become *the* address for Liverpool's turn-of-the-century élite.[25] The stately, four-story, brick townhouses—most, like the Joseph's, with three bays and handsome doorcases—visually attested to wealth and station in society. And that society was English.

Although they were prominent members of Liverpool's growing Jewish community, the Josephs were also quite self-consciously English and fully acculturated—unlike some of their coreligionists—to the English way of life. In Georgian England, Jews, especially those in the middle and upper classes, increasingly conformed—like the Josephs—to English customs and to the English way of life. Unlike in Germany or Central Europe, they neither lived in closed societies governed by communal institutions that monitored personal behavior nor were they required to belong to any sort of legally recognized Jewish communal organization.[26] They moved out of, in other words, a Jewish world governed by Jewish law (*Halachah*) and into an English world governed by secular law. These facts had led over the course of the eighteenth century to what might be perceived as a certain religious laxity that manifested itself in several ways. Many Anglo-Jews of this period tended either to ignore or to follow only sporadically the Jewish dietary laws. They also tended neither to observe the Sabbath nor to attend synagogue regularly. As historian Todd Endelman has noted in regard specifically to England by 1830, "Judaism ceased to be an all-encompassing civilization, a way of life, a point of orientation by which to measure all other values and activities. It became, instead, a religion— something that affected the individual English Jew in a much less definitive way than it had his ancestors. Judaism as a religion became one concern, one element, one interest among many in the lives of most English Jews" (289). Nevertheless, this secularization and acculturation did not necessarily bring with it social integration or full assimilation in the sense of intermarriage or outright conversion. Jews—again like the Josephs—may have increasingly "left behind the cultural and social isolation of the traditional community," in Endelman's words, but they firmly maintained their sense of cultural identity as members of a Jewish people.[27] That this characterization of Anglo-Jewry most probably fits the Abraham Joseph family seems borne out not only by its Liverpool address but also by some of the values it embraced.

In 1797, several of Liverpool's most distinguished and enlightened citizens— William Roscoe, physician James Currie, banker William Clarke, among others—resolved that a library and newsroom to be called the Athenæum should be established in Liverpool to provide its members with access to a full range of books and newspapers in English and in other modern languages.[28] It was a time of war on the Continent and of troubles in Ireland, and educated men needed to keep abreast of these and other world events as well as of events and policy decisions closer to home. The organizers thus formulated a scheme for getting the venture off the ground that involved selling a limited number of shares to individuals who would then have the right to full use of the facility. The campaign was a success. The initial 250 shares offered in November 1797 sold quickly, and by July 1798 another fifty were offered and also quickly sold. On 1 January 1799, the newsroom opened in Church Street, an easy walk from the homes of most of the shareholders, with the library following a year-and-a-half later on 1 May 1800. From the beginning, the Athenæum subscribed to some forty-five different papers, magazines, and reviews from London, the English provinces, and abroad, and by 1802 its library boasted a collection of 6000 volumes on topics as diverse as theology, ethics, metaphysics, Greek and Roman writers, belles lettres and criticism, biography, fine arts, history, geography, natural history, arts and sciences, mathematics and natural philosophy, chemistry, and medicine. The members who assembled there—mostly in the evenings after a long day of business transactions to read of current events by candlelight—were a diverse lot, yet, according to the group's first honorary treasurer, John Rutter, "the establishment of the Athenæum, . . . had the effect of bringing into active co-operation, for a common object, a number of gentlemen, whose opinions on political subjects widely differed; and who greatly to their honour, laid aside all such differences, and acted together with utmost harmony."[29] Among the first 400 founding members of this enlightened society were only four Jews—Abraham Joseph, his brothers Samuel and Elias, and his brother-in-law Morris Mozley. Their presence at the Athenæum underscored not only their membership in Liverpool's intelligensia but also the extent of their acculturation into English society.[30]

Like the Joseph family, Liverpool's Jewish community as a whole had prospered and become a more established part of the fabric of Liverpudlian society over the course of the Napoleonic era. By 1806, enough money had been raised by subscription—in the form of an initial deposit and then eight equal installments—for the design and construction of a new, purpose-built synagogue to be located in Seel Street and to replace the converted residence on Frederick Street. Abraham's eldest brother, Elias, was the driving force behind the new Seel Street synagogue, and according to the "Subscription and Accompt Book for the Intended New Synagogue," begun in October 1804, the Joseph family pledged the most of any members of the Liverpool Jewish community for the new place of worship. Elias Joseph, treasurer of the fund drive, gave £20 down and an additional £80 in eight £10 installments for a total of £100, and his brothers Samuel and Joseph gave £60 and £36, respectively. Abraham Joseph also pledged £36 but ultimately made no down payment or installments.[31]

The synagogue's construction coincided with a period of change in the life of the Abraham Joseph family. Abraham was listed in *Gore's Liverpool Directory* in 1807—the year his daughter Fanny was born in Liverpool—but not in the next edition of 1810; he *is* listed in London's *Post Office Annual Directory* of 1810. These facts, together with the fact that the synagogue was only consecrated in September 1808, suggest that Abraham and his family had moved to London some time between 1807 and 1810 and probably in 1808 or 1809 and that their plans to relocate may have begun to take shape as early as late in 1804, hence the unfulfilled pledge.[32] Abraham Joseph may ultimately have needed all his resources to embark on new business ventures planned for London. Whatever the reason for the unrealized pledge, the family most likely left Liverpool at the time the new Seel Street synagogue was redefining the center of the city's Jewish life.[33] Abraham, Miriam, and their family of eight (two other children had died in infancy) left Liverpool—like Abraham's brother-in-law and sometimes business partner, Morris Mozley—to seek their fortunes in the capital city.

The Abraham Joseph Family of London

In 1810, when the then fifty-year-old Abraham Joseph first appeared in the *Post Office Annual Directory*, he listed his occupation as "merchant and shipowner" and his address as 7 Magdalen Row, Prescott Street. This address, in the traditionally Jewish area of Goodman's Fields on the southeastern border of the City of London proper, was located just to the north and east of the Tower of London and just east of the heavily Jewish Aldgate Ward, site of both the Sephardi Bevis Marks Synagogue and the Ashkenazi Great Synagogue.[34] Styled as one of "four streets of elegant houses surrounding Goodman's Fields," Prescott, in 1810, would have been a highly desirable and fashionable address for a Jewish family newly transplanted to London. For the Josephs, it was perhaps only a business and not a residential address, and they were only on the Prescott Street end of Magdalen Row and not actually on Prescott Street itself.[35]

As for Abraham Joseph's designation as "merchant and shipowner," it certainly departed both from the occupation of "silversmith" that had been a constant in his listings in *Gore's Liverpool Directory* and from the occupation of "naval agent" that had appeared in his listings of 1796 and 1800. In entries from

1800, 1803, and 1804, however, he was also listed, in addition to "silversmith, jeweler, and naval agent" as "merchant" and as "Joseph, Abraham, Mozley and Co. Merchants." Whether he singly or he and his brother-in-law together dealt as "merchants" in silver or perhaps in watches (Abraham Joseph had listed his occupation as "watchmaker" in 1787, and Morris Mozley was well-known in that craft) or whether they dealt in some other sorts of goods, Abraham no longer styled himself a silversmith after his move to London. Moreover, although he continued to be listed as a "merchant" through 1827, "shipowner" had already dropped from his entry by 1811. That side of his new career in the capital city must have been either speculative or unprofitable.

One clear constant from Abraham Joseph's former life in Liverpool, however, was his connection to the broader Jewish community and to the Jewish faith. He set up his business and possibly his family as well in Goodman's Fields among coreligionists, and he is listed on the membership rolls of the orthodox Great Synagogue.[36] In 1810, the family included, in addition to parents, Abraham and Miriam, eight children ranging in age from eighteen-year-old daughter Ellen to three-year-old daughter Fanny. In between came Sarah (aged fifteen), Nathaniel (thirteen), Sylvester (twelve), Henry (ten), Elias (nine), and Elizabeth (seven). Two more children followed after the move to London; a son, Frederick, was born in 1812 and another son, James, the future mathematician, was born two years later in 1814.[37] This large family would have made quite an impression on those occasions when it may have walked from its home to the Great Synagogue on the Sabbath.

Abraham Joseph's formal synagogue membership aside, it is difficult to know precisely what role Judaism may have played in the actual day-to-day life of the Joseph family either before or after its move to London. That it was fully acculturated to English life may suggest that it was less observant of Jewish rites than the Simon Joseph or Gershon Woolf families had been a generation earlier, but a relaxation of the old orthodoxy did not necessarily imply a retreat from the Jewish fold. It may simply have implied a loosening and stretching of the rules governing the day-to-day, not a renunciation in any way of the family's Jewish identity. The latter interpretation seems consonant with the family's decision regarding young James's education.

In 1820, when James was six years old, Abraham and Miriam sent him to England's first private boarding school for Jewish boys. The school, founded in fashionable Highgate to the northwest of central London by the Polish-born Hyman Hurwitz in 1799, was run by him until his retirement in 1821 and catered specifically to the children of the Anglo-Jewish élite.[38] Housed in what has been described as a "fine, eighteenth-century mansion" at 10 South Grove in Highgate, the school enrolled some 100 boys and had a *mikvah* in addition to a synagogue where worship services were regularly conducted.[39] It was thus a school that maintained and upheld Jewish rituals and customs and, as such, provided an educational alternative—for the sons of Jewish families who could afford it—either to local, non-Jewish education or to home-schooling.[40]

The curriculum taught there first by Hurwitz, the master in young James Joseph's first year in Highgate, and then by Leopold Neumegen, the boy's teacher from the ages of seven to twelve, was by no means that of the yeshiva, however. It had both traditional and secular elements. On the traditional side, Hebrew and biblical studies were taught and, to the boys whose families desired it, Talmud, that is, the written tradition of rabbinical commentaries and interpretations. On the secular side, studies included Latin, Greek, and mathematics.[41] It was, in short, an education that prepared boys not only for their lives as Jews but also to some extent for their lives in English society, because their secular studies were similar to those of the non-Jewish, English boys with whom they would ultimately interact as adults.

That Abraham and Miriam Joseph exercised the option of sending James to Highgate reflects both their sense of the religious and cultural values in which they felt their son should be raised and their refusal to deny their religious and cultural identity at a time when some British Jews were choosing apostasy in an effort better to acculturate themselves into English society.[42] The Abraham Joseph family was unapologetically Jewish in a country thoroughly dominated by the Church of England, but they were also English, and that meant that they equally embraced and participated in England's cultural, historical, and intellectual heritage. They thus exposed their son to both kinds of learning by sending him to the Highgate school.

James seemingly took his studies—both traditional and secular—seriously. His exposure to Hebrew reflected itself in mathematical papers he published much later in life and in which he used letters of the Hebrew alphabet as a not-so-subtle testament to his Jewish heritage.[43] He also may have been one of those boys who studied Talmud as well as Torah, because, again, much later in life, he reported having attended a lecture on the Talmud at London's Royal Institution and having "some ladies" approach him and "implore" him "to tell them what they should do to get up the Talmud."[44] His early attainments in mathematics, however, are easier to document. By the age of eleven, he had so impressed his teacher Neumegen with his mathematical abilities that the latter arranged for Olinthus Gregory, the Professor of Mathematics at the Royal Military Academy in Woolwich, to examine the boy in algebra.[45] Gregory "pronounced him to be possessed of great talents, and recommended his mathematical tutor to pay great attention to his instruction." Some two years later in March 1827, the Woolwich professor was still following young James's progress, for he asked Abraham Joseph "to drop me a line as soon as your son returns home that I may endeavour to fix a day in which I may have the pleasure of seeing him here, and tracing his progress since I saw him before."

In the interim, James had left Neumegen's school for a year-and-a-half at a school in Islington run by one Mr. Daniell and where he had come under the mathematical tutelage of Mr. Downes.[46] James's brief time in Islington coincided, moreover, with two other key events in his life: the departure of his brother for the United States in 1826 and his thirteenth birthday in September 1827.

In 1826, his elder brother, the twenty-eight-year-old, Sylvester Joseph, left England to seek his fortunes on the other side of the Atlantic.[47] The 1820s, in fact, witnessed a significant Jewish immigration not only to the United States but also to Africa and to Asia in pursuit of more favorable economic conditions and opportunities.[48] This appears to have motivated Sylvester Joseph, who established himself, under the name of Sylvester Joseph Sylvester, as a broker and lottery agent in New York City.

The adoption of the surname "Sylvester" by the sons of Abraham Joseph, in fact, appears to have begun as early as 1821. In that year, the eldest son, Nathaniel, was listed in the newly renamed *Post Office London Directory* as "Sylvester, N. Joseph. Solicitor. 28, St. Swithin's Lane," while the directory for 1822–1826 carried the entry "Sylvester, S. J. Merchant. 7 Magdalen Row, Great Prescott Street," the same address as his father, who maintained the surname Joseph throughout his life.[49] The younger sons, Frederick and James eventually followed their brothers' lead. Although the reason for this name change is unclear, it may have represented a conscious decision on their parts to distance themselves from their parents' sense of being a Jew in late Georgian England. Religion, after all, need not have attention called to it unnecessarily by a surname, even if the surname Joseph was not exclusively Jewish.[50] A certain precedent had been set, moreover, by a number of English Jews in this regard in the opening decades of the nineteenth century.[51] The Joseph brothers may have wanted to acculturate themselves further than their parents into English society; they may have wanted an unequivocally non-Jewish surname as their calling card in life. They certainly would not have been alone in their efforts to blend even more seamlessly into English society, although the effects that their efforts may have had on their parents unfortunately remain unknown. At least they apparently did not cause a rift within the family, because Sylvester Joseph Sylvester was in business in Magdalen Row with his father, Abraham Joseph, before his emigration. As for the youngest son, James, he was only twelve in 1826 when his brother Sylvester set off for the United States. He only reached the age of thirteen, the age of religious duty and responsibility recognized and celebrated in the *bar mitzvah*, in September of the following year. His decisions on the issues of Anglo-Jewry his elder brothers faced still lay in the future.

Other decisions, however, were more imminent. Both Neumegen and Gregory had recognized real mathematical talent in James's work. This, in fact, may have been one reason for the boy's move to Daniell's school where he could have more targeted instruction from a mathematical tutor. Moreover, as Gregory well knew, moves were afoot in London for the founding of a new university that would be, unlike Oxbridge, free from sectarian influences and restrictions. In April 1827, just a month after writing to Abraham Joseph about his son, Gregory himself had been one of those whose name was inscribed on the foundation stone laid for the building that would become a year later the new London University.[52] He may well have not only had his eye on Sylvester as a potential student for the new university but also encouraged the family to enroll him there. The fourteen-year-old James was, indeed, in the school's first entering class in the fall of 1828. In providing their son with this level of secular education, the Josephs represented the exception rather than the rule. At this point of time, Ashkenazi families tended to view a few years in a private school like Neumegen's as sufficient for their sons' education.[53] The Josephs, apparently in recognition of James's mathematical abilities and in appreciation of secular learning and scholarship, wanted to give him the opportunity to nurture his talent.

London University, later called University College London, was located on Gower Street in Bloomsbury on London's increasingly—but not yet completely—fashionable westside and was within a reasonable walk from the family's residence (in 1829) on Gloucester Street leading into unquestionably fashionable Queen Square.[54] Support for the university had come from several influential men, among them, the liberal and politically influential member of the House of Lords, Sir Henry Brougham, and the successful Jewish financier, (later Sir) Isaac Goldsmid. These men and others like Gregory deplored the fact that England's ancient universities were so firmly attached to the Church of England. In fact, before 1828 and the repeal of the Test and Corporation Acts—acts that had been in place since the seventeenth century that debarred dissenters from the Church of England from holding public office—various Anglican oaths had to be sworn for admission to Oxford and for taking a degree at Cambridge. After 1828, however, only one oath remained and that was "on the true faith of a Christian." Thus, Jews were still effectively debarred from both institutions. The men behind the new university in London, the so-called "godless institution on Gower Street," wanted religion to play no role in their educational enterprise.[55]

In mathematics, after a protracted search that had begun as early as 1826, the University finally hired Augustus De Morgan as its Professor of Mathematics in February 1828. Just twenty-one years old at the time of his appointment, De Morgan had entered Trinity College, Cambridge, in 1823, had come out fourth wrangler on the Mathematical Tripos of 1827, and had proceeded to law studies at Lincoln's Inn in London. An untried talent, he had nevertheless won the London University post over a number of highly qualified competitors, and he had set about immediately to craft a suitable curriculum for those students who would be in his junior and senior classes come the fall.[56]

James enrolled—as James Joseph Sylvester—in the university in November 1828.[57] Entering students took three courses: ten hours of Latin and ten hours of Greek each week in addition to six hours of mathematics.[58] Sylvester's first hurdle, with respect to mathematics at least, was an interview with De Morgan designed so "that the Professor may ascertain the state of preparation of the student . . . [and] so that he may adapt his instruction accordingly."[59] De Morgan, like Neumegen and Gregory before him, was apparently impressed because he placed Sylvester directly into the higher division of his senior class.

De Morgan opened the term with an introductory lecture on 5 November; the curriculum that he had prepared for his students was extensive. Material for the senior class, in particular, included conic sections; transcendental algebra, or the theories of continued fractions, equations, approximation of solutions of equations, and indeterminate analysis; the theory of projections; trigonometrical analysis; algebraic geometry, that is, the geometrical construction of algebraic quantities, the classification of curves, the construction of the solutions of cubic and biquadratic equations, and the analysis of three dimensions; differential calculus; integral calculus; and the theory of probabilities.[60] In his inaugural lecture, however, De Morgan made it clear that he intended this list as a guideline, a list of possibilities, rather than a hard and fast curriculum. As he put it, "I shall not consider myself bound to carry the class through the whole of what is contained in it if it shall appear that their interest will be more effectually consulted by my confining myself to the more prominent parts of it."[61]

Exactly what De Morgan taught that first year remains unknown, but that Sylvester thoroughly mastered all that De Morgan brought his way was clear. Writing some years later, De Morgan stated categorically that Sylvester "became by far the first pupil in" that most advanced course, adding that he had, in fact, "never, before or since, [seen] mathematical talent so strongly marked in a boy of that age."[62] In his view, Sylvester had distinguished himself "by the facility with which he acquired a knowledge of the higher branches of Mathematics and the singularity of his power to apply them."[63] Unfortunately, Sylvester had only a very short time in which to profit from De Morgan's instruction.

Storm clouds were visible on the horizon by January 1829. Sylvester's family had just received the second monthly report on his conduct and attendances, and the news was not good. "I very much regret the necessity you are under of thus complaining of his inattention to his duties," wrote brother Elias, "and have severely admonished him which I trust will be the means of producing a better report for the future."[64] "I am most anxious for his Welfare and interest," Elias continued, "and shall spare neither expense nor pains to promote both." That admonishment apparently did not suffice, for Sylvester allegedly took "a table-knife from the refectory with the intention of sticking it into a fellow student who had incurred his displeasure."[65] On 25 February, Elias Sylvester once again wrote to the authorities, this time withdrawing his brother from the university. "I have this day heard from his Family," Elias explained, "who are of opinion and by whom I am directed to inform you that owing to the extreme youth of my Brother and the fact of his requiring constant control and attention they deem it advisable that he should for the present withdraw from the London University and reside with his family in Lancashire until he attain a more mature age when I doubt not he will resume his studies at the University with additional ardor."[66] Sylvester's immaturity and his apparent inability to comport himself appropriately deprived him at the age of fourteen of the chance to continue under De Morgan's mathematical tutelage. He was sent to Liverpool to live under the wings of his elder sisters Ellen, Sarah, and perhaps Elisabeth at the home at 8 Great George Street, where, by 1832, they ran a ladies' seminary.[67] He was enrolled at Liverpool's nearby Royal Institution School.

The Royal Institution School, Liverpool

The Royal Institution had opened its doors in a mansion on Colquitt Street in Liverpool in November 1817, the result of sustained efforts by, among others, the same William Roscoe who had been so instrumental in founding the Athenæum twenty years earlier. The new initiative was to be a society "for promoting the increase and diffusion of Literature, Science and the Arts."[68] Among the ways in which it aimed to achieve these goals were the establishment not only of a series of public lectures but also of academic schools for the intellectual training of teenaged boys. By February 1819, special lectures had already been given in subjects ranging from physiology and the elements of natural history to ancient and modern poetry, and two such schools had been founded, one for classical and one for mathematical learning (13–20). The schools, in particular, emulated the curriculum at the ancient universities, were staffed by a number of Cambridge graduates, and sought to prepare their students for entry into Oxbridge. When Sylvester enrolled in both of these schools in February 1829, ten years after their opening, Thomas Williamson Peile, 1828 graduate of Trinity College, Cambridge, and a bracketed second in classics and eighteenth wrangler in mathematics, was the head master and teacher of classics, and William Marrat, a self-taught writer on physics, astronomy, and mathematics, was the mathematical master.

Peile, although only twenty-two at the time of his appointment as head master, had strong views not only on how to run the school but also on how the curriculum should be organized. In the upper school of which Sylvester was a member, five hours a week were allocated to mathematics with one hour only spent on history, English composition, and geography. The rest of the time was devoted to the ancient classics, with the modern language instructors permitted to arrange classes, for those interested, only outside the regular instructional day.[69] Peile's insistence on the primacy of the classics was not universally appreciated. Several members of the school's governing committee quit in protest of what they viewed as his overemphasis of ancient relative to modern knowledge, and Marrat waged his own private battle to prove that "mathematics and not classics were the key to knowledge." This clash apparently took the public form of an open competition between Peile and Marrat as to who could produce the best scholar among the students. Peile's special student was one John Charles Beadey, who had entered the school in July 1827; Marrat's was Sylvester. Beadey died in 1830 at the age of seventeen, an unproven talent; that same year, Sylvester "crushed all his rivals" and effectively won the competition for Marrat. Peile was so disheartened by his experiences at the school that he tendered his resignation in August 1830, eight months after Beadey's death and two months after Sylvester had finished his studies, although he was reappointed in January 1831 and ultimately stayed on in his post until December 1833.[70]

Sylvester—a strong student in both classics and mathematics, who in February 1830 won the first prize in the first class of mathematics and the second prize in the first class of classics—found himself caught in the middle of the power play between his two masters.[71] This only added to what had apparently already been a difficult situation at his new school. "I have more than once seen him hunted by his schoolfellows, in the open street, for no worse reason than that he was a Jew, and very much cleverer, especially in mathematics, than they were," reported one Royal Institution School student years later.[72] Another, William Leece Drinkwater, recalled that one classmate "fought a long battle with Sylvester, and punished poor Sylvester a good deal, but did not succeed in making him give in" despite the fact that "we had to separate them."[73] "Of course, as a Jew," Drinkwater noted, "he had to bear a good many schoolboy remarks respecting his creed; but he knew that I did not share in the disrespectful observations, and perhaps on that account, though we occasionally quarreled, we always made it up."[74]

And Sylvester's troubles were not just with his fellow students. "He was fully aware when at school of his great knowledge of mathematics," Drinkwater recounted, "and used to puzzle our mathematical master, who was not a great authority, with questions which the latter was unable to answer." Sylvester also crossed swords with the English composition master on at least one occasion. When his class was asked to write on the topic of despotism, "he wrote a very good theme, giving various instances, high and low, of the abuses of it, but reflecting unmistakably upon the case of himself and the under master (who was a respectable clergyman) and ending, 'Thus we see that power begets tyranny, whether in the case of the mightiest monarch or of the paltry usher of a school or institution.'" His sense of justice—as well as his intellectual impertinence—resulted on this occasion in a caning before his peers. That it was a public humiliation of the worst sort is borne out by Drinkwater's recollection of it decades later. Sylvester "wore very tight clothes at the time," Drinkwater recalled, "and his antics while under the punishment were something remarkable; but he bore it with great courage."

These instances with both his fellow students and his teachers paint a portrait of an irascible, quick-witted, sensitive Sylvester who had trouble fitting in in his mid teens on account of his religion, his intellectual abilities, his sense of injustice, and his temper. At one point during his brief time in Liverpool, in fact, he felt so frustrated that he tried to escape his various torments by running away. With reportedly only a few shillings in his pocket, he went down to the Liverpool docks and hopped a ship for Dublin. On his arrival in Ireland, he happened to encounter Richard Keatinge, Judge of the Prerogative Court of Ireland and husband of Sylvester's first cousin.[75] The Keatinges entertained the young runaway and promptly put him back on a ship to Liverpool, where he simply had to learn to deal with his unhappy situation at school.[76]

If there had been troubles in Liverpool, there were also some triumphal moments. From his lottery and brokerage house in New York City, James's brother Sylvester wrote to Liverpool to relate a combinatorial problem to his younger sibling that had been posed to the American contractors of lotteries.[77] Lotteries had become big business in the United States after 1790, being used to fund everything from internal improvements like roads and bridges to churches and schools to the liquidation of personal debt. Perhaps the most celebrated case of the latter type was that of former President of the United States, Thomas Jefferson, in 1826. Deep in the red and approaching the end of his life, Jefferson made a personal appeal to the legislature of the Commonwealth of Virginia for permission to launch a lottery to raise sufficient funds to allow him to save his home, Monticello.[78] Sylvester Joseph Sylvester, as a lottery ticket broker, earned his living by buying large blocks of tickets at a discount once a lottery—like the one eventually launched for Jefferson—was announced and then by selling them to individuals who were willing to make a small gamble in the hopes of winning one of the prizes. Contractors played another key role in this system. In exchange for a fee from the grantors of a given lottery, they set the prices of tickets and generally oversaw the event, contracting with particular brokers for the sale of the tickets (82). Although the exact nature of the contractors' problem exchanged between the Sylvester brothers is unknown, the young mathematical prodigy back in Liverpool apparently solved it and received a $500 prize for his efforts from the grateful contractors.[79]

On balance, however, Sylvester's year and a half at Liverpool's Royal Institution School had been a trial. Sylvester, an English boy, had acquired an English education and had excelled in his studies. He had received the early training requisite for the English gentleman and, so, for the members of the new and rising English middle class who wanted to emulate and be accepted in the society of gentlemen. Yet, Sylvester also neither denied nor abandoned his religious heritage at a time when, as Endelman has put it, "young men and women whose ambitions outstripped those of their parents often took themselves to the baptismal font in the belief that this would ease their way in the world."[80] That his "way in the world" had indeed not been smooth in Liverpool is evidenced by the fact that his religious heritage was well known to his schoolmates, some of whom persecuted him for those beliefs. Even at this young age, Sylvester seemed to view religion as a private matter that should have nothing whatsoever to do with one's ability to participate fully in English society. He had, after all, fought on more than one occasion for his right as a fellow fee-paying student to go about his studies unharrassed. He was a Jew in an Anglican world, who sensed, already in his teens, "what irreparable loss of facilities . . . for full mental development and the growth of productive power" he would suffer "under the effect of this oppressive monopoly, this baneful system of protection of such old standing and inveterate tenacity of existence" (see epigraph note).

The skirmishes that Sylvester had had at the Royal Institution School mirrored, in fact, larger battles that were being waged at just the same moment in the wider political arena. In Liverpool, Sylvester's uncle-in-law and his father's sometime business partner, Morris Lewin Mozley, was active in the nationwide agitation that had begun in 1828 for the right of Jews to hold seats in Parliament. In 1829 and 1830, for example, he had written in support of the movement to its London leader Isaac Goldsmid, the same Isaac Goldsmid who had been so instrumental in founding London University in 1828.[81] In the view of Mozley, Goldsmid, and other Jews and non-Jews alike, it was time for the end of social and political disabilities in England based on religion.[82] Sylvester, who had been born just as the Napoleonic wars were coming to an end on the international geopolitical scene, entered into adulthood as this new conflict surrounding his religion began to escalate at home. He soon found himself on one of its key battlegrounds, Cambridge University.

CHAPTER 2

A Price of Dissent

*Here you have a youth, it will be said, who has satisfied all your own
tests of merit to the uttermost—his talents are rare, his industry
unwearied, his acquirements vast, his character spotless; but because
in addition to these his many virtues, he has the misfortune to have a
conscience and will not strain it by subscription, you expel him from
your body as an unclean thing, and bid him
seek his fortune as he can.*

When Sylvester was entered onto the rolls at St. John's College, Cambridge, in the summer of 1831, significant changes were underway in England. The much disliked George IV had died in 1830, prompting the *Times* to proclaim that "there never was an individual less regretted by his fellow creatures than this deceased King. What eye has wept for him? What heart has heaved one sob of unmercenary sorrow?"[1] His brother, William IV, succeeded him, and although he was esteemed only somewhat more highly, the change in monarch required a dissolution of the Parliament and a new general election. Decades of conservative Tory rule ended, and what would be more than a quarter-century of almost unbroken, more liberal- and reform-minded Whig control of the government began.[2]

The first major legislation to issue from this new political climate, the so-called Great Reform Bill of 1832, fundamentally altered the nation's voting structure by disenfranchising or partially disenfranchising a number of small boroughs, giving representation to many large ones that had previously not had a voice, and standardizing the voting criteria for property holders. The bill did not pass without a fight, and it also did not materialize out of thin air in 1832.

Agitation in the 1820s had already resulted in the repeal in 1828 of the Test and Corporation Acts, a law passed during the reign (1660–1685) of Charles II that debarred dissenters from the Church of England from holding public office, as well as in the passage of the Catholic Emancipation Bill in 1829 permitting Catholics to hold, most significantly, seats in Parliament. Ironically, this reform legislation did not extend to Jews. The repeal of 1828 no longer required those assuming public office to take the Sacrament as prescribed by the Church of England or to take a variety of oaths, but, as passed (although not as originally proposed), it did require a swearing-in declaration that included the words "on the true faith of a Christian." From the late 1820s through the 1830s, this ongoing disability for Jews received extensive coverage in printed pamphlets, in the press, and in Parliament.[3] The spirit of reform that spurred these changes and debates deeply penetrated English society, affecting even its most tradition-bound institutions, at precisely the moment Sylvester entered upon his university studies at Cambridge.

Cambridge Debates Dissent

At Cambridge, a bastion of mathematical and classical learning, this spirit of reform manifested itself in at least two key ways. First, curricular reforms had, by the 1820s, resulted in the introduction of Continental mathematics, especially the calculus in its Leibnizian and Lagrangian forms (as opposed to the Newtonian version), into the previously Euclid-dominated course of study.[4] Second, statutory reforms were hotly debated that would remove restrictions on those who did not adhere to Anglican teachings. (At Cambridge, students had to subscribe to the Thirty-nine Articles of faith of the Church of England to take their degrees; fellows and professors were subject to this test as well.) As early as 1818, a move to allow the President of the Linnean Society, the dissenting Sir James Edward Smith, to give a course of lectures on botany in the University had been squelched by a vocal group of tutors.[5] The call to admit nonconformists into University society was again sounded in 1827 by George Dyer in his book, *Academic Unity*, but the issue came dramatically to the fore beginning in 1833 in the wake of the Reform Bill's success in Parliament.[6]

In an opening salvo, Robert Beverley published a no-holes-barred "Letter to His Royal Highness the Duke of Gloucester, on the Present Corrupt State of the University of Cambridge" in November 1833. An 1821 graduate of Trinity, Beverley had joined the dissenting ranks after leaving Cambridge and had come to view as a sham the supposed religious training of the Cambridge undergraduate. His vivid, but exaggerated, pamphlet purported to describe faithfully his own observations of undergraduate life at Cambridge. In painting a bacchanalian picture of habitual drunkenness, uncontrolled gambling, and outright profligacy, Beverley aimed to counter the University's claim that it trained its students "in the principles and practices of the English Church," and thereby to undermine the main argument against the complete participation of dissenters in University life.[7]

Also in 1833, the Regius Professor of Physic and fellow of St. John's College, John Haviland, M.D., published "A Letter to the Members of the Senate on the Subject of the Subscription Required of Medical Graduates in the University of Cambridge," in which he, too, called for the removal of religious disabilities, but only for candidates for medical degrees. St. John's countered by strongly opposing *all* efforts to remove the test requirement,[8] and when Haviland's colleague, the Downing Professor of Medicine, Cornelius Hewett, brought the question officially before the governing Caput of the University Senate in February 1834, it was roundly defeated there as well.[9] Advocates of the measure then decided to take matters into their own hands. As astronomer George Biddell Airy put it in his diary, this resulted in a "furious discussion about the admission of dissenters into the university."[10]

In addition to Hewett, Adam Sedgwick, the highly respected Woodwardian Professor of Geology and fellow of Trinity, spearheaded a move to circumvent the Senate by directly petitioning Parliament to lift the test requirement. In mid-March 1834, the group quickly gathered the signatures of sixty-three members of the University Senate—including Airy, Charles Babbage, and the Masters of Corpus Christi and Caius—and had the petition brought before both the House of Commons and the Lords. After a heated month-long debate that received lengthy reportage in the conservative *Quarterly Review*, the measure passed the Commons but, failing to pass in the Lords, was defeated.[11] This outcome, however, silenced neither the opponents of nor the proponents for change. Thomas Turton, the Regius Professor of Divinity, and Connop Thirlwall, assistant tutor at Trinity, carried on the debate in heated broadsides later in 1834, with Turton against and Thirlwall for the admission of dissenters.[12]

External agitation followed in 1835 when a bill was introduced in Parliament calling for the abolition of religious tests at both Cambridge and Oxford. (At Oxford, students had to subscribe to the Thirty-nine Articles on matriculation, and so dissenters were effectively barred from attending there.) This bill failed as well, but the wheels had been set in motion for the reform of statutes long in place at both of the ancient universities.[13]

The Cambridge of the 1830s was thus not insulated from broader issues under debate in the nation at large. On the one hand, long-standing—but slowly evolving—academic traditions there continued to aim both at establishing lifelong social connections and at providing a liberal education for English gentlemen. On the other hand, a spirit of reform centered precisely on the issue of religious tolerance came to pervade the cloistered colleges. It was into this environment that Sylvester, a sensitive, seventeen-year-old, Jewish, mathematical prodigy, came to take up residence on 6 October 1831.

Within the Walls of St. John's

The two largest and best endowed colleges at Cambridge were Trinity and St. John's, the former perceived as Cambridge's nineteenth-century citadel of classical scholarship, the latter as its stronghold of mathematical learning. In his Cambridge memoirs, American and Trinity student (from 1840 to 1845), Charles Astor Bristed, conveyed the sense of rivalry between these two neighboring colleges and their respective senses of disciplinary superiority. When William Whewell became the Master of Trinity in 1842, Bristed reported, he pushed for the introduction of more mathematics into the competitions for Trinity scholarships and fellowships. As a result, "even the Mathematicians [i.e., mathematically inclined students and fellows] did not all agree with the master, their College pride getting the better of their professional and scientific *amour propre*. Not believing that Trinity could be brought up to the Mathematical point of Johns, they feared he would only endanger its Classical superiority by his experiments."[14] In fact, from 1824 when the Classical Tripos was instituted to 1845 when Bristed left Cambridge, 19 St. John's students had come out either senior or second wrangler[15] on the Mathematical Tripos to Trinity's 12. On the Classical Tripos, Trinity outran St. John's in senior and second wranglers 32 to 11 over the same twelve-year period.[16] In light of these numbers and perceptions, the student of the 1830s and 1840s primarily interested in the classics, like Bristed, might naturally have gravitated to Trinity, whereas the student drawn to mathematics, like Sylvester, might have slightly favored St. John's.[17]

Of course, general reputations only partly accounted for a particular student's association with a particular college. Family traditions and connections, college traditions and statutes, the availability of scholarships, size, and facilities, these and other factors also contributed to the makeup of a given college's student body. In Sylvester's case, the decision to enroll at St. John's may have been influenced by his connection with a college fellow and by the college's ability to offer financial support to a relatively large number of students.

While at Liverpool's Royal Institution School, Sylvester had received private tuition in mathematics from Johnian Richard Wilson.[18] Wilson, who had earned his B.A. in 1824, held one of his college's fellowships from 1826 to 1831 and received the M.A. there in 1827. Following Sylvester's graduation from the Liverpool school in June 1831, Wilson wrote his pupil's certificate of recommendation to the tutors of St. John's.[19] The latter, a statement of qualification for admission that was reviewed and judged before a candidate's name was officially entered on a college's rolls,[20] must have sufficed. Sylvester was admitted on 7 July 1831 and took up residence in October.[21] Wilson, the fifteenth wrangler in 1824 and the author of *A System of Plane and Spherical Trigonometry*,[22] had been well qualified to guide the young Sylvester along a Cambridge-directed mathematical path, and although the details of his influence on his student are unknown, it is not inconceivable that Sylvester's decision to attend St. John's may have owed, at least in part, to the example and advice of this teacher who felt "powerfully interested" in his charge's welfare.[23]

The size and wealth of St. John's may also have played a role in Sylvester's attendance there. In the 1830s, Cambridge University consisted of 17 colleges, 16 of which had been founded before 1600 and one, Downing College, that dated from 1800. Of the more ancient colleges, the smallest was Trinity Hall. Its foundation supported a Master, 12 fellows, and 18 scholars, that is, students who, based on their academic merit after matriculation and on their performance in various examinations and contests, were given financial relief often in the form of rentfree rooms and monetary awards. Gonville and Caius was of roughly the average size, funding the Master, 29 fellows, and 50 scholars. Trinity and St. John's were by far the largest. In addition to the Master, Trinity had 60 fellows and 69 scholars, while St. John's had a Master and the same number of fellows but a whopping 107 scholars.[24] The gifted student hopeful of some financial relief for his education would thus have had more chances for a scholarship at St. John's, although the overall size of the college's student body would have stiffened the competition.[25]

A college's relative wealth also made another sort of financial support possible: the sizarship. In the nineteenth century and before, Cambridge students fell into four categories. Fellow-commoners, often members of the nobility and always wealthy, fully paid their way and enjoyed additional privileges such as dining with the college fellows and wearing distinctive academic garb with gold-tassled caps. Pensioners also paid full fees and tended to be sons of the gentry or of clergymen. Scholars, as noted above, received some college support, although it rarely sufficed to cover the actual costs of a year in residence. Finally, sizars tended to be students of lesser means who received free lodging as well as additional financial breaks.[26] Unlike the smaller and poorer colleges, St. John's had the financial means to offer not only a large number of scholarships but also more than 50 sizarships by midcentury.[27]

Entering students competed for the sizarships on the basis of their performance on a set examination taken in the October before their official matriculation. This test covered elementary geometry, arithmetic, algebra, and knowledge of a number of works by classical authors announced the preceding June. Although Sylvester came from a middle class family of comfortable means, he nevertheless sat for the sizarship examination in October 1831 and won one of the places.[28] This gave him free board and allowed him to pay 15 shillings per quarter of tuition instead of the £2.10 required of pensioners. It also afforded him a discount in the caution money that each student had to pay to his assigned college tutor at the time of admission as a hedge against future debts.[29] With this examination behind him, Sylvester matriculated on 14 November and officially embarked on his undergraduate career.

As a facility, the St. John's Sylvester entered in 1831 differed dramatically from that of even two or three years earlier. In 1831, the massive New Court with its more than 100 additional sets of rooms—the largest building erected by a Cambridge college up to that time—was completed at a sobering cost of £78,000.[30] This elegant structure, designed in a perpendicular Gothic style, was the first new construction at the college since the Third Court was completed in 1673. On the other side of the River Cam from First, Second, and Third Courts, it was reached through the Venetian-like "Bridge of Sighs" and looked out onto spacious grounds landscaped in the early 1820s.[31]

As a sizar, Sylvester would not have had his set of rooms—a study, a bedroom, and possibly a small pantry—in this new part of the college. As late as the 1850s, the scholars and sizars tended to be housed in the Labyrinth, formerly the college infirmary and located on the north side of First Court until it and the old chapel were razed in the 1860s to make way for the present college chapel. With rooms described by midcentury Johnian and author Samuel Butler as "dingy" and "tumble-down,"[32] the Labyrinth effectively delineated the students of St. John's into those perceived as "haves" and those seen as "have nots" within the cloistered society. These perceptions were only reinforced in the college hall.

John William Colenso, an accomplished student of mathematics and later the Bishop of Natal and an authority on the Zulu language, entered St. John's as did Sylvester in 1831. As serious students and as sizars, the two young men were in the same social class within the walls of the college. They soon struck up a long-lasting friendship.[33] The sizar's circumstances Colenso described in 1832 thus applied equally well to his friend, Sylvester. "We get our meat from the fellow's table," he wrote, "and though the joints are not very hot when they reach us, yet we manage to make very capital dinners, for they take care to get the best of meat etc."[34] With the sizars sequestered in hall at a separate table of their own, this transfer of dishes—day in, day out—from the fellows' table presented a strong image and reminder of differences in status among the students. These differences also extended to the more social activities within the college.

Another of the young men who matriculated with Sylvester in 1831 was William Henry Bateson. Bateson, who later served as the master of St. John's and as an active, liberal, reform-minded member of the University community, was admitted as a pensioner and earned one of the foundation scholarships prior to taking his degree as third wrangler on the Classical Tripos in 1836. As a student, his liberal leanings were already in evidence when he fought to change the rules that forbade sizars from participating in the college's two sports clubs, the cricket club and the Lady Margaret Boat Club (founded in 1825). Membership in both of these groups was by subscription only,[35] so the thinking was presumably and patronizingly that sizars, as scholarship students, should be prohibited from spending money on such extras. Bateson apparently saw the matter differently, finding it offensive that his classmates like Colenso,

Sylvester, and others could not take part in such central aspects of the college's social life if they so chose.

Bateson's efforts aside, it would not have been that likely for a sizar to have wanted to participate in activities that took precious time from his studies. Almost by definition, sizars were "reading men," that is, students who came to Cambridge to work hard, to do well on the college and university examinations, and to earn one of the college fellowships after taking the B.A. They and other reading men contrasted sharply with the nonreading or "rowing" men, students there for the society who studied little, socialized much, and were content to graduate as "poll men" with ordinary, nonhonors degrees. Still, even the reading men occasionally entertained their friends at the local pubs or in their rooms with dinners and wine parties,[36] and although sizars generally did not offer such lavish hospitality, they certainly got together with friends for evenings of conversation or for afternoon strolls around Cambridge.

Sizars were thus doubly categorized—or stigmatized—within a college's social structure. Among the numerous nonreading men, the sizars were apt to be viewed as uninteresting, closeted bookworms, and the social snobs looked down on them simply because they may not have shared all the trappings of wealth. These perceptions thus tended to restrict the society of the sizars within and without college walls. Butler gave testimony to the persistence of this classism at St. John's when he reported that even at midcentury the sizars "were rarely seen except in hall or chapel or at lecture, where their manners of feeding, praying and studying were alike objectionable."[37]

Couple these general attitudes with the fact that Sylvester was Jewish, and the following anecdote recounted by Sylvester's schoolmate from the Royal Institution School and fellow Johnian, William Leece Drinkwater, gives an indication of the social pressures Sylvester may have felt while a Cambridge student. Writing in 1897, Drinkwater first offered the opinion that Sylvester "was very much respected by those whose respect was worth having" at St. John's, yet he clearly implied that not all of the society fell into this category. Drinkwater, two years Sylvester's senior, "heard that on one occasion at the Hall dinner some remark was made, or thought by him to be made, I think, about his eating ham, and he lost his temper and threw a plate or dish at the offending person's head."[38] Whether fact or merely hearsay, this story—and Drinkwater's telling of it— implies that Sylvester was known at Cambridge as he had been at the Royal Institution School to be of Jewish heritage, that such a remark would have rung true for the St. John's of the early 1830s, and that it would likely have provoked some reaction from Sylvester. Like the anecdotes from his student days at University College and in Liverpool, this further suggests a youthful Sylvester who was sensitive—even painfully so—to perceived insults, slights, and prejudices.

Sylvester's life as a reading man within the walls of St. John's was only partly defined by the college's social context, however. Intellectual pursuits occupied the majority of his time with the big prize being a high place on the Tripos examination prior to the end of twelve terms of residence. Numerous preliminary trials of various sorts would help him hone his skills along the way and build his academic reputation.

When Sylvester entered, the master was James Wood, a Johnian since he had entered as a sizar in 1778. Wood took his B.A. in 1782 as senior wrangler and first Smith's Prizeman[39] and proceeded directly to the College fellowship he held until he was named to the mastership in 1815.[40] As a fellow, Wood authored several of the mathematical textbooks used by a generation of Cambridge undergraduates—among them Sylvester—and strongly upheld the University's mathematical traditions. As master, the politically and academically conservative Wood strove to maintain the status quo during an almost quarter-century term in office that found St. John's opposed to Catholic emancipation in the late 1820s as well as to the move in the 1830s to abolish subscription to the Thirty-nine Articles. At the same time, however, he oversaw key changes in the college's statutes and its expansion into the buildings of New Court.[41]

Sylvester would have had few direct dealings with the master, although Wood's policies and administration did set the tone of the College throughout the young man's tenure there. Of much more immediate concern to Sylvester was his assigned tutor, Richard Gwatkin. The forty-year-old Gwatkin had entered St. John's in 1810 and, like Wood, had graduated in four years as senior wrangler and first Smith's Prizeman and had promptly become a fellow. As tutor, Gwatkin was responsible not only for doing his share of college teaching but also for managing the financial dealings of each of his students. In Bristed's words, the tutor "is the medium of all the student's pecuniary relations with the college. He sends in their accounts every term, and receives the money through his banker; nay more, he takes in the bills of their tradesmen, and settles them also."[42] The tutor thus acted *in loco parentis* for the young men in his care.

The assignment of a tutor at St. John's also meant assignment to one of two "teams" or "sides." Because of the college's relatively large size, the entering student body was divided into two separate groups of roughly forty-five taught by two different sets of instructors. This set up a competitive collegiate atmosphere that had no doubt contributed to the Johnian dominance of the Mathematical Tripos by the late 1830s. Sylvester's tutorial lecturers during his first two years in residence were, in addition to Gwatkin, Henry Hunter Hughes in mathematics and Edward Bushby in classics. As a member of Gwatkin's side, Sylvester attended college lectures exclusively with his half of his St. John's entering class and would have seen those in the other half regularly only in the college hall or at the eight mandatory chapel attendances each week, although he might have gotten to know a few socially.[43]

A typical day began with a half-hour-long chapel service at 7:00 A.M. followed perhaps by a brief promenade around the college grounds and by a breakfast of tea and rolls in one's rooms. College lectures occupied the rest of the morning. As a freshman, Sylvester had two hour-long lectures: one in mathematics, the other in classics. The rest of the day was spent in study and in preparation of the material assigned in the morning's lectures, with time out perhaps for some sort of exercise, generally a brisk walk with friends in the Cambridge environs. A dinner typically of meat, potatoes, cabbage, and college ale followed in hall at 4:00 P.M., with evening chapel at 6:00 P.M. (if one had not attended chapel that morning or if one chose), and more studying from 7:00 P.M. until bedtime, usually between 11:00 P.M. and midnight. During these evenings of study, it was not uncommon for students to host small study groups in their rooms, pooling intellectual resources and creating bonds of friendship.[44]

During his first year, Sylvester attended mathematical lectures principally by Hughes on Euclid, algebra, and trigonometry in addition to Bushby's classical lectures covering, for example, a Greek tragedy and the works of both a Greek and a Roman historian or orator. Describing a Caius in the 1850s that would have differed little from the St. John's of the 1830s, John Venn noted that "there was . . . no selection or discrimination" in the college lecture rooms. "The brilliant scholar from the best public school, and the young man who had given up business and was beginning his Greek letters with a view to taking Orders; the destined high wrangler who had read his conic sections as a schoolboy, and the youth to whom Euclid and his mysterious pictures were a daily puzzle, sat side by side on the benches of our lecture room, and tried to make the most of the lecture, or at any rate of the time during which the lecture was delivered."[45] The mathematics lectures consisted mainly of questions posed by the lecturer and answers given by students as called upon, with the rest of the class scribbling furiously. The classics lectures involved much sight translation, again, with the lecturer randomly choosing his victims. Both lecturers also recommended volumes of outside reading and assigned regular homework on the material covered in class. In mathematics, the latter tended to take the form of "deductions," that is, statements to prove that were different, but derivable, from those actually found in the text.[46] The textbook statements themselves were meticulously copied and memorized, the so-called "book-work," for the purposes of regurgitation on later examinations. In classics, it involved the translation and memorization of passages. And so the days passed until May.

All the freshman year's work aimed at the college's May examination, a four-day written ordeal in both mathematics and the classics that took place in the college hall. Bristed vividly described the general scene and conveyed the frenzy of the experience:

> At nine next morning, the Hall doors were thrown open to us. . . . The tables were decked with green baize instead of white linen, and the goodly joints of beef and mutton and dishes of smoking potatoes were replaced by a profusion of stationery.

Even the dais shared the general fate. . . . I was now doomed . . . to write against the time for four mortal hours. . . .

At one, "close your papers, gentlemen," says the examiner, who has been solemnly pacing up and down all the time. (The examiner is never your college lecturer or tutor, and of course never your private tutor.) At two the hall assumes its more legitimate and welcome guise, dinner being thrown back two hours; at four the grinding begins again and lasts until eight . . . .

Thus passed four days; eight hours a day thinking and writing together at full speed; two or three hours of cramming in the intervals (for though the principle and theory is never to look at a book during an examination, or indeed for two or three days before, that your mind may be fresh and vigorous, few men are cool enough to put this into practice); and long lounges at night, very different from the ordinary constitutional.[47]

When Sylvester went to see the posted results of his May examination in 1832, he found his name at the top of the rank-ordered list. He had come out St. John's first Freshman Prizeman. He had also won one of the college's nine so-called proper sizarships, a distinction conferred by the master and the seniors upon those sizars who had both excelled academically and set an example by their conduct.[48] Academically, it had been a very rewarding year.

The second or "junior soph" year followed a similar pattern, except that an additional examination, the university-wide Previous Examination colloquially termed the "Little Go," took place toward the end of the second—the so-called Lent—term of the year.[49] The "Little Go" had been instituted in 1822 as part of the university reforms aimed at holding the undergraduates more accountable academically. It and the Tripos, taken near the beginning of the Lent Term of the fourth year in residence, were the only university examinations at Cambridge, all others being conducted within the walls of the individual colleges. Still, as its nickname suggested, the "Little Go" was almost a joke. It was a pass/fail examination (the candidates were not rank-ordered according to performance) that covered one Greek and one Latin author in addition to one of the four Gospels and William Paley's *Natural Theology; or, Evidences of the Existence and Attributes of the Deity* (1802). Bristed minced no words in characterizing the test's intellectual level. "There is nothing in the *Little Go* to occupy a good schoolboy of fifteen more than three or four months," he wrote. "And for a Second-year Cantab of good standing, there is really nothing to prepare except the Paley; he might without danger trust to the light of nature for his Classics, or if scrupulous to run no risk, read them up sufficiently for practical purposes in three days, and the same time properly applied would make him master of his Evidences."[50] Given his performance in his freshman year, Sylvester probably found little that challenged him in the "Little Go." Things would be different come May.

All year long, the junior sophs had been reading their classical authors—Homer, Herodotus, Livy, Tacitus, Plato—and studying such mathematical topics as algebra from the works of Silvestre Lacroix and Jean Guillaume Garnier, differential and integral calculus, conic sections, spherical trigonometry, and the first three sections of Newton's *Principia*.[51] The college examination in May was principally mathematical in nature, although it did include questions on a Platonic dialogue, on one of the Gospels, and on moral philosophy from Paley, Dugald Stewart, and others. In general, the reading men did not exert themselves that much for this test; the real trial would come in another five terms when they sat for the Tripos. Still, for the best mathematicians, even this May examination was important. It served as a predictor of things to come. If those who scored well in their first year also made a strong showing here, they were good bets for the high wranglerships later on.[52] In this light, Sylvester's tie for fifth place on the junior soph May examination must have been a real disappointment for him. The top scorer, Henry Cotterill, did, indeed, go on to come out senior wrangler in 1835. Hugh William Smith came in second and went on to a twentieth place finish. William Chatterley Bishop and George Buckmaster Gibbons placed third and fourth, respectively, yet failed to secure wranglerships, ranking in the second tier of senior optimes. Sylvester tied with fellow sizar and future divine, William Edward Scudamore, 1835's ninth wrangler.[53] The final placements on the 1835 Tripos were thus somewhat of a shuffling of the junior soph standings, but Sylvester ultimately did not compete with his classmates on that final examination. After keeping the Michaelmas term of 1833—the fall term of what would have been his third year—Sylvester was not in residence at St. John's again until the Lent term of 1836. He missed two full calendar years.[54]

A Two-Year Hiatus

Ostensibly because of illness, Sylvester's irregular university career most likely stemmed from a combination of causes.[55] First, Sylvester's terms of non-residence began just after St. John's College had taken its strong stand against John Haviland's initiative to exempt medical students from subscription to the Thirty-nine Articles, and it coincided with the period of heated debate—both at Cambridge and nationally—on the question of allowing Jews to take degrees. St. John's College under the leadership of its conservative master, James Wood, and described by Bristed as a "hot-bed of bigotry"[56] thus provided a less than congenial atmosphere for a sensitive Sylvester likely stung by the fifth place junior soph finish. Moreover, Sylvester's father died in Liverpool in 1834, and although the circumstances surrounding his death are unknown, family members including Sylvester may have attended him in his final days and months. Finally, in 1835, Sylvester, writing from "Cambridge, January 1835" as "A Member of the University," brought out a *Collection of Examples on the Integral Calculus.*[57]

This 132-page study guide, published by the Cambridge house of J. and J. J. Deighton, consists of a series of 135 examples worked out in complete detail.[58] These range from the elementary example one—$du/dx = ax^4$; $\therefore u = ax^5/5 + c$—to examples comparable in difficulty to number 98, namely, $\int_x 1/[(1-x)^2] \log x$. After recognizing the latter problem as one of the form $\int_x p(dq/dx) = pq - \int_x q(dp/dx)$, where $p = \log x$, and $dq/dx = 1/[(1-x)^2]$; $\therefore dp/dx = 1/x$, and $q = 1/(1-x)$, Sylvester provided the detailed solution[59]:

$$\therefore \int_x \frac{1}{(1-x)^2} \log x = \frac{\log x}{1-x} - \int_x \frac{1}{x} \cdot \frac{1}{1-x},$$

$$= \frac{\log x}{1-x} - \int_x \frac{1}{x} - \int_x \frac{1}{1-x},$$

$$= \frac{\log x}{1-x} - \log x + \log(1-x),$$

$$= \log x\left(\frac{1}{1-x} - 1\right) + \log(1-x),$$

$$= \frac{x \log x}{1-x} + \log(1-x) + C.$$

He explained what motivated him to compile these examples in his brief preface:

> The great desideratum in the different works on this subject, appears to be a collection of the best examples thoroughly worked out, a want which tends to make expertness in one of the most powerful instruments of analytical investigation, a matter of long and difficult acquirement . . .
>
> The writer knows well from experience, the want of such an assistant, and feels confident that a careful reading of the present publication will save much time, that it will tend to impress on his memory the most important integrations, and confirm him that quickness of eye, which, in examinations is found so mainly useful. (5)

If Sylvester was following the standard practice of dating his preface from the location at which it was written, then he must have been well enough to have traveled to Cambridge to arrange for his booklet's printing. He was also apparently in adequate health to conceive of, concentrate on, and compose the series of examples. Although this does not represent conclusive proof that reasons other than health contributed to his irregular attendance at St. John's, it does cast some doubt on that accepted explanation. The facts that Sylvester had the work printed privately and that he reentered St. John's during the Lent term of 1836 as a pensioner[60] may also reflect the settling on him of part of his father's estate. Sylvester's health may have precipitated his hiatus from Cambridge; his father's death may also have contributed to it; but the Cambridge atmosphere, perceived as inhospitable to one who felt he had so much to prove, may have prolonged it for two years.

Back at St. John's

Despite the efforts from 1833 through 1835 to remove the disabilities for Jews at Cambridge, the university to which Sylvester returned in January 1836 had not changed its position on this issue. Sylvester would be allowed to sit for the Tripos examination one year hence, but he would not be able to take a degree because he could not in all good conscience swear "on the true faith of a Christian" to his adherence to the tenets of the Church of England as the degree ceremony required. With this knowledge, he nevertheless proceeded to prepare himself both for the ordeal ahead and, perhaps, for a livelihood afterward.

On his return, he was reassigned to his old team, but the personnel had changed.[61] His former tutor, Richard Gwatkin, had left for a vicarship in Barrow-on-Soar in 1833 and had been replaced by Henry Hughes, while Bushby still attended to training in the classics, and John Hymers now supervised the mathematical lecturing. Hymers then took over from Hughes when the latter accepted a living in Laytham in 1836.[62] The students, too, had changed. Most of Sylvester's former classmates had taken their degrees in 1835, and although some friends, like Colenso, were still at St. John's, Sylvester found himself in the midst of a new and largely unfamiliar group of people.

During his first term back, Sylvester attended the lectures of the Professor of Chemistry, James Cumming. These university lectures were completely outside the examination structure of the colleges, and their contents did not form part of the material tested on the Tripos. For these reasons, most reading men did not avail themselves of the opportunity to extend their educational horizons by such attendance unless they had a particularly strong interest in the area. That Sylvester enrolled in Cumming's course not only in the Lent term of 1836 but also during the Lent term of 1837—the same term in which he took the Tripos—suggests more than fleeting curiosity.[63] Cumming, an accomplished experimenter and interpreter of electromagnetic and thermoelectric phenomena, had been inspired by the work in the early 1820s of Hans Christian Oersted. His interests had led him to the invention of the galvanometer independently of both J. S. C. Schweigger and J. C. Poggendorff, and he reported on his new invention in 1822 in the first volume of the *Transactions of the Cambridge Philosophical Society*.[64] Moreover, he had earned a reputation as a teacher of remarkable clarity, who used experiments to excellent effect in his classroom.[65] Thus, in Cumming, Sylvester had a well-versed and talented guide to then-current advances in chemistry. The young scholar was beginning to draw connections between mathematical ideas of space, chemical and physical views of the constituency of matter, and philosophical conceptions of mind and matter, but he channeled these thoughts toward the principal goal. The Tripos was just a year away, and there was much to prepare.

The third-year course of college study tended to focus on topics like the integral calculus, astronomy, dynamics, and the latter parts of Newton's *Principia*, including optics and hydrostatics.[66] Moreover, early in the Lent term, disputations were set among those deemed likely to take honors in the Tripos the following year. Started in the seventeenth century as a way of determining the order of merit among the students, these public debates pitted a "respondent" arguing in syllogistic fashion and in Latin for a specified position against "opponents" on three questions set in advance by the respondent. In the 1830s, the first of these was usually from Newton's *Principia*, the second from mathematics or natural philosophy, and the third from ethics, with each student appearing twice as a respondent and six times as an opponent.[67] By the 1830s, though, this formality was more an entertainment than a serious academic exercise. In 1831, for example, one Trinity man totally confounded his opponent, "a Johnian and good mathematician but ignorant of the classics," with humorous Latin word play. The moderator even "entered into the spirit of fun and himself carried on the discussion."[68] Sylvester participated in these exercises five years later in 1836, but they were discontinued in 1839 because they had become little more than "an elaborate farce" and "a mere public performance of what had been already rehearsed" (183). In 1836, he had much weightier matters on his mind than participating in these disputations.

Sylvester had already burrowed into the integral calculus, as his printed study guide of 1835 evidenced. By 1836, he was thinking hard about Newton's ideas, and while he may have been studying his hydrostatics on the side, he was preoccupied with the earlier, more foundational aspects of the *Principia* and with synthesizing many of the ideas he was absorbing from his mathematical studies, from Cumming's lectures, and from his philosophical readings.

As is well known, Newton opened Book I of the *Principia* by laying out the various mathematical lemmas needed in the analysis of the motion of point masses. Of the eleven lemmas he gave, only his fifth was stated without any further justification: "all homologous sides of similar figures, whether curvilinear or rectilinear, are proportional; and the areas are as the squares of the homologous sides."[69] What a tease for the serious student steeped in the logical rigors of Euclid's *Elements*! If the lemma is not an axiom, then it requires a rigorous proof. Newton's oversight was thus Sylvester's challenge. What better way to begin establishing one's mathematical reputation than by filling in a gap in the work of the great Newton? In his second publication, another privately printed pamphlet, Sylvester sought to do just that.

Entitled "A Supplement to Newton's First Section, Containing a Rigid Demonstration of the Fifth Lemma, and the General Theory of the Equality and Proportion of Linear Magnitudes," the essay opened in rigorous fashion with a catalog and analysis of the "various definitions proposed for the right [straight] line, as the properties of the *right* line are to form the basis of our subsequent reasoning."[70] After dissecting Euclid's definition of a straight line, namely, "that which lies evenly between its parts,"[71] Sylvester concluded (rightly) that Euclid had tacitly committed a logical mistake by essentially defining the thing in terms of itself. He next analyzed Adrien-Marie Legendre's definition—"a right line is the shortest line that can be drawn between the extreme points"—arguing that

> in the first instance, this definition must be *preceded* by an admission that there is one line so short that none as short or shorter can be drawn; and, observe, the *converse* is not true, that "a line can be drawn so long that no line as long or longer can be drawn between the same points." But, waiving this *preliminary* objection, and looking to the terms actually employed, we observe that the right line is to be *determined* from its being the *shortest* line of *all the lines* that can be drawn between the extreme points.
>
> Now, "all the lines" are made up of *two* classes, "right lines" and "lines not right."
>
> The first class cannot avail the definition, consisting as they do of the thing itself to be defined, nor will the other class help it out (vi; his emphasis).

And the argument proceeded with this excruciating logical precision to the conclusion that "the definition (as such) must be rejected, as utterly worthless and of *no meaning*" (vii; his emphasis).

The self-confident, young mathematician, having thus shown both Euclid and Legendre to be defective in their definitions of straight line, gave faint praise to another of his textbook authors: "[John] Playfair's definition, although an improvement upon [Euclid's] in *other* respects, is still liable to *all* the objections named in the text" (vii; his emphasis). With the logical shortcomings of all of the standard definitions now in brilliant relief, only one thing remained. Sylvester had to put forth the "correct" definition. " 'A line is said to be right when no two of *its parts* can be made to enclose space,' or rather, 'coincide in two points without coinciding in all,' " he stated. "This is clear, logical and satisfactory, and will answer every purpose, with the aid of one very easily recognized principle, *viz.* 'That any two right lines may be made to lie in one and the same right line.' "[72]

Not realizing that his definition hinged on that of "part" (which he had not given) and so ultimately also fell short of the sort of strict logical standards he sought to enforce, Sylvester proceeded to base his sixteen-page demonstration of Newton's fifth lemma on what he viewed as the firm and rigorous logical foundation of a well-defined concept of straight (and, by negation, curved) line. As he concluded with more than a hint of mathematical superiority:

> This Proposition was originally enunciated by Newton without proof, and his example has been followed by the more judicious commentators, and with good reason,

seeing that the geometrical definitions in ordinary use are wholly inadequate to attach any *distinct* meaning, and much less therefore to lead to an *accurate proof* of the proposition announced in the text.

To compare lines of different curvature, (as a right line and a parabolic arc), with each other, by direct supraposition is impossible. Nor will it avail to conceive the lines divided into their elements, unless we are prepared to admit (contrary to all received notions) that a parabolic arc is made up of right lines, or a right line of small parabolic arcs. Thus we only multiply our difficulties by this procedure, just as if we should dash a mirror into fragments in the vain hope of annihilating the images in the glass.[73]

Through this youthful "Supplement to Newton's First Section," Sylvester provided a glimpse both of the mathematics he studied at Cambridge in the 1830s and of the mind of a serious, reflective, cocky personality at work. Even he seemed to sense the pedantry of his written exercise when he wrote rather defensively in its preface that

The embarrassment and inconsistencies to be found in the pretended demonstrations and explanations, given by *received* authors, of the Fifth Lemma, (the most important in the whole Section, of which it forms the very keystone) will amply justify the *intention* of this little work. If there be any who censure as trifling the manner of its *execution*, or sneer at the precision of its details, from such the author makes his appeal to the lovers of sound Logic, and to all who are *capable* of appreciating the *spirit* in which Newton composed his First Section, or Euclid his Fifth Book. (iv; his emphasis)

Still, he clearly felt that the authority of the "received authors" was not absolute, that he had something to offer that was superior to what they had provided, and that it was therefore his duty to make his insights known. In so doing, he was also announcing the arrival on the mathematical scene of a new author, a challenger to those unquestioningly "received."

Sylvester had apparently published a first version of this pamphlet during the Easter term of 1836. Like his 1835 study guide on integration, this initial edition came out under the authorship of "A Member of the University."[74] A burst blood vessel had reportedly forced him to sit out that Easter term,[75] although his mother's death on 28 May 1836 also called him back to Liverpool at that time. He could thus have certainly signed his name to the pamphlet, but he could not have legitimately identified his college. Under the circumstances, the designation "a member of the University" served, at least somewhat, to establish his credentials at the same time that the anonymity it provided protected his gentlemanly standing and effectively shielded him from outside criticism.[76] By the summer of 1836, with his latest family tragedy behind him and with his official reinstatement at St. John's for the Michaelmas term, he had a new run of the work printed with a new title page carrying its author's name and college affiliation for all to see.

In August 1836, Sylvester heavily annotated one copy of this newly printed version, evidencing not only his evolving thoughts on the ideas therein but also the restlessness of a mind constantly thinking, revising, perfecting, and connecting. In particular, on the back of the title page and spilling onto the bottom of the facing page iii, he recorded a newfound awareness of the shortcoming of his own definition of a straight line in one of numerous closely penned thoughts. "No definition of a right line," he admitted, "will hold unless it's [sic] points are taken up."[77]

More significantly, however, Sylvester began to formulate ideas intended for a sequel to the work, "speculations, which the author believes are new and important, upon the *origin* of Geometrical terms and the 'conceptions embodied in them:' subjects intimately connected with our notions of Matter and Space" (iii; his emphasis). Starting from the back of his pamphlet and filling some six-and-a-half blank pages from back to front, Sylvester scribbled what amounted to aphorisms encapsulating some of his widely ranging and developing ideas. Thoughts on mathematics, matter, space, even mind and human self-awareness poured onto these pages in a fast-flowing stream of consciousness.

Connecting his lectures on chemistry and geometry, Sylvester averred that "the question of the divisibility of matter is distinct from an inquiry into the divisibility of space"[78] and proceeded to conclude that

> The questions of Divisibility & Impenetrability of Matter are to be solved upon the same principles
>   — depending partly upon our definitions of matter and unity of matter which may refer either to unity of effect or unity of place
>   — partly on axioms drawn from universal experience of what we know—& extended by Induction to all cases—as for Example "Matter within other matter cannot come without except there be a solution of parts in that which is without"— Hence if that which is without cannot be divided that which is within can never come out. Thus Matter within a perfect sphere of matter can never come without it.
>   But we don't know whether this axiom extends to matter *coincident* in space with other matter—which forms a distinct case to be resolved by the *axioms* that Matter cannot be divided into an ∞^{te} no. of portions of matter similar & equal in all respects to itself—& so on for every case.
>   I merely give these as hints to those who can understand.[79]

These mathematicophysical musings quickly led to more theological reflections. "Mind or a man's personality is *said* to be where that is in which motion must take place in order for our sensations to be produced in the mind," Sylvester stated with indebtedness to John Locke. "∴ To say God knows all things is to say he is everywhere. Thus the idea of the omniscience of Diety comprehends the idea of his ubiquity."[80] Finally, his thoughts turned to more philosophical and existential matters. He wondered

> May (or is) not what is called coming to a consciousness of our own existence (recorded by others & experienced by the author to the best of his belief) be the coming to a consciousness of the separate existence of other persons no longer as images merely of the brain but as possessing wills akin to our own? a consciousness obtained from the production of effects in ourselves (similar to effects) known to be produced by our own will & generally not *except* by our will) without the action of our will as where Jesus saith "who has touched me, for I know that the Virtue is gone forth from me."[81]

The Cambridge-trained student, quoting from the Gospel according to St. Mark (Mark 5:30), further revealed that the required reading of the New Testament had made a certain impression upon a young man raised in the Jewish faith.

This 1836 work on Newton's fifth postulate—and the hand-written draft of ideas for a subsequent work—reflects an exuberant, twenty-one-year-old Sylvester bursting with ideas and captivated by the intellectual endeavor. Its reprinting with its author's name fully in evidence suggests a pragmatic young scholar fully aware of the fact that in five month's time, he would be allowed to sit for the Tripos but would then be ineligible for a degree and, more significantly, for a college fellowship. An obvious source of livelihood open to others would thus be closed to him. Publication presented one sure way to establish his mathematical credentials; a high score on the Tripos was another. Had he not returned to Cambridge, he could not have hoped to claim both of these distinctions. With his publication *and* a high wranglership, he might have an edge in establishing himself, perhaps, as a mathematical coach either in Cambridge or elsewhere. Many before him had done so on the basis of a high wranglership alone,[82] but that part of Sylvester's equation was still unknown.

The 1837 Tripos

Sylvester was back on the rolls at St. John's—again as a pensioner—in the fall of 1836 for the final push toward the Tripos. There was one term left before the ordeal and much to do. Come January, he would have to perform with pinpoint accuracy at top speed for five-and-a-half hours a day for five days if he was to make his mark.[83]

The kind of drilling that brought success under these conditions, in general, came not from college lecturers, oddly enough, but from private tutors. A reading man might engage in private tuition from his first year on, and he would certainly pursue it in earnest during his final two years. For this additional instruction, he would pay handsomely. At midcentury, students paid £14 per term for one hour of private tuition daily (except Sundays) and half that amount per term for three hour-long sessions a week, although noblemen and fellow-commoners paid more but sizars paid roughly half.[84]

It is unclear exactly how much private tuition Sylvester had during his Cambridge career, but he did have some, reportedly from Philip Kelland and John Colenso,[85] possibly from James Bowstead, and definitely from William Hopkins. Kelland, the senior wrangler in 1834 and until 1838 a fellow and tutor of Queen's College, attested to his "long acquaintance" with Sylvester and to his "ample opportunity of judging of his scientific and mathematical acquirements" in a testimonial written in October 1837 on the occasion of Sylvester's candidacy for the professorship of natural philosophy at University College London (see chapter 3).[86] Colenso, Sylvester's former classmate and the second wrangler and first Smith's Prizeman in 1836, served as a lecturer in mathematics at St. John's during the 1836–1837 academic year. Later in his life, Sylvester wrote that Colenso had "instructed him, 'long before the far-famed Zulu was heard of,' "[87] although because Sylvester and Colenso were friends and contemporaries, they probably did not have the usual coaching relationship. Given Colenso's showing on the Tripos just the year before, however, how better to get a sense of the test firsthand? James Bowstead, fellow of Corpus Christi and the second wrangler in 1824, may also have given Sylvester private instruction. In a testimonial also for the vacant chair at University College and also written in October 1837, Bowstead said that he had "been well acquainted with [Sylvester] for the last three years" and knew firsthand of "his great talents, and his extensive mathematical and philosophical attainments."[88] As an older member of another college, Bowstead would have been unlikely to have known Sylvester except as a private tutor. Moreover, his statement dates his acquaintanceship with Sylvester to 1834 and provides further evidence that Sylvester may not have been absent from Cambridge for all of the years 1834 and 1835. Finally, William Hopkins, seventh wrangler in 1827 and an established private tutor, remarked in yet another testimonial that Sylvester had been "for a short time a pupil of mine" (8). Hopkins, widely recognized as a "senior wrangler maker," would have been a very logical choice as private tutor for a student, like Sylvester, of demonstrated abilities. Moreover, Hopkins tended to watch carefully the results of the first year and to take on those students whom he deemed potential high wranglers, so Sylvester would have satisfied Hopkins's selection criterion as well.[89]

Sessions with private tutors were rigorous. The renowned coaches, like Hopkins, had earned their reputations for the successes of their students as well as for the comprehensiveness of both their teaching and their knowledge of past Tripos examinations. They "drove their teams" by giving them long series of exercises, by lecturing them on all the necessary topics, by providing them with numerous manuscript study tracts of carefully digested information, and by setting timed practice examinations literally to get them up to speed. During the tutorials, the students took down the lectures word for word to produce comprehensive notebooks for ready reference in the final stages of preparation before their Tripos. They also copied out by hand their own versions of the supplementary pamphlets as hedges against the trickier questions that might later come their way.[90] Finally, the best coaches, again like Hopkins, did more than merely "cram" their students; they provided them with broader frameworks within which to understand the mathematics and physics that they were working so hard to master. As Hopkins expressed it in 1841, "it is only when the student approaches the great theories, as Physical Astronomy and Physical Optics, that he can fully appreciate the real importance and value of pure mathematical science, as the only instrument of investigation by which man could possibly have attained to a knowledge of so much of what is perfect and beautiful in the structure of the material universe, and in the laws that govern it."[91]

Regardless of whom Sylvester may have studied with and when, he surely spent the Michaelmas term of 1836 hard at work on his final preparations for the Tripos. The subjects covered were well known: algebra, geometry, plane and spherical trigonometry, differential and integral calculus, Newtonian mechanics, gravitational and planetary theory, dynamics, hydrostatics, geometrical and physical optics, and, at least marginally, the theories of heat, light, and electricity and magnetism.[92] Likewise, students knew what texts to use in preparing these subjects, among them: William Whewell's *Elementary Treatise on Mechanics* (1819) and *An Introduction to Dynamics* (1832), the 1831 second edition of George Biddell Airy's *Mathematical Tracts*, Henry Parr Hamilton's *The Principles of Analytic Geometry* (1826) and *An Analytical System of Conic Sections* (1828), George Peacock's *Treatise on Algebra* (1830), John Hymers's *A Treatise on the Integral Calculus* (1831), William H. Miller's *An Elementary Treatise on the Differential and Integral Calculus* (1833), and Thomas Grainger Hall's text of the same name of 1834. Thanks to the efforts of Whewell, Peacock, Charles Babbage, John Herschel, and others in the 1810s and 1820s, the material Sylvester studied for his Tripos no longer focused as exclusively on Euclidean mathematics and on Newtonian mathematics and physics.[93] Continental developments—like those at the hands of Pierre Simon Laplace in celestial mechanics and Augustin Fresnel on the theory of light—had become part of the Tripos curriculum, even though other Continental research—such as the work of Hans Christian Oersted and André-Marie Ampère in electomagnetism and electrodynamics—was largely overlooked.[94] It was with knowledge of this sort that Sylvester faced the Tripos in 1837.

The year's moderators—the two men who set the lists of those qualified to present themselves for the examination and who then helped administer the test and mark the papers—were James W. L. Heaviside, fellow and tutor of Sidney Sussex, and Edwin Steventon, fellow of Corpus Christi. The two examiners—the men who proctored as well as graded—were Henry Philpott, fellow and tutor of St. Catharine's Hall, and Samuel Earnshaw, a recent M.A. of St. John's. Each of these men had been a high wrangler, and all had been Smith's Prizemen; they had come to perpetuate a system within which each had been successful.

Sylvester's ordeal began on Wednesday, 11 January 1837. He, along with the other Cambridge students sitting for the examination, left their colleges that morning for the imposing and notoriously cold and damp Senate House. Whewell gave a chillingly evocative description of the venue in a letter to his sister. "The examination," he explained, "is conducted in a building which happens to be a very beautiful one, with a marble floor and a highly ornamented ceiling; and as it is on the model of a Grecian temple, and as temples had no chimneys, and as a stove or a fire of any kind might disfigure the building, we are obliged to take the weather as it happens to be, and when it is cold we have the full benefit of it."[95] The Tripos began in this austere setting at 9:00 A.M. The first set of questions treated the more elementary aspects of pure mathematics, lasted for two-and-a-half hours, and constituted one of the so-called "bookwork" phases of the examination during which students were essentially expected to do little more than faithfully regurgitate statements and propositions from their various texts and lectures. A three-hour bookwork session on natural philosophy followed in the afternoon.[96] All students were present on this first day, restricted as it was to questions that did not require the calculus. The classical scholars there ardently tried to score well enough to secure at least the junior optime finish on the Mathematical Tripos that would qualify them to sit for the Classical Tripos, whereas the mathematical scholars strove to maximize their ultimate scores by not missing a single question.[97]

The timetable for the remaining days was the same as the first, but the subject matter shifted on Thursday first to bookwork in natural philosophy and then to problems, and back on Friday to bookwork in mathematics and more problems. The afternoon problem phases presented the students with questions that demanded more creativity than the morning bookwork but that still drew faithfully from the basic material. Except for the classical men trying for a double first, that is, a senior wranglership on both Triposes, only the mathematical men would generally remain beyond the third day.

For the diehards who persevered, the fourth and fifth days presented the greatest challenges, with questions in pure mathematics and natural philosophy of generally greater difficulty and sophistication. These were the days when the final order of merit was decided. At the end of the day on Saturday, the papers were collected and graded. By the start on Monday morning, the classes for the final round had been posted. Those in the first class went into the last day well placed for a wranglership, although some invariably fell back into the second class after the end of the morning session, while a few in the second possibly worked their way into the first. The last afternoon's session settled the matter, with those remaining in the first class competing against each other for their ultimate ranking.

Shooting as he was for a high wranglership, Sylvester was there to the end. He spent five days writing furiously, trying to answer as many of the 151 questions as fully and with as much panache as possible. It was an exhausting experience, mentally, physically, and psychologically. Bristed described the dynamics of the 1845 affair in terms that fairly crackle with the electricity of the moment:

> But now [the] best man [from St. John's] suddenly came up with a rush like a dark horse, and having been spoken of before the Examination only as likely to be among the first six, now appeared a candidate for the highest honors. E—[Robert Ellis, one of the examiners in 1845 and a wrangler in Sylvester's year of 1837] was one of the first that had a suspicion of this, from noticing on the second day that he wrote with the regularity and velocity of a machine, and seemed to clear everything before him. And on examining the work, he could scarcely believe the man *could* have covered so much paper with ink in the time (to say nothing of the accuracy of the performance), even though he had seen it written out under his own eyes. By-and-by it was reported that the Johnian had done an inordinate amount of problems, and then his fellow-collegians began to bet odds on him for Senior Wrangler. But the general wish as well as belief was for the Peterhouse man . . . . His backers were not disposed to give him up. "One problem of his will be worth half a dozen of the other man's," said they; and there were grounds for this assertion, some of the problems being more difficult, and therefore marked higher than others, so that four on a paper may pay more than ten.[98]

This description could almost have been about Sylvester's year of 1837. Sylvester, a Johnian, was most likely the cause of a certain amount of gossip and speculation. After all, he may well have been the first professed Jew ever to sit the Tripos[99]; he was taking the examination two years late, in some sense, on account of his irregular career at Cambridge; he had tied for fifth place the last time he had sat for a college examination; and he was undoubtedly known as a mathematical powerhouse. He surely had his backers, friends like Colenso, and he probably had his detractors, men from other colleges rooting for their own, men from St. John's rooting for other of the college's contenders, and perhaps men who preferred not to see a Jew as a high wrangler. Regardless, the 1837 Tripos like others before and after was caught up in a gaming spirit, not necessarily the most congenial of atmospheres in which to compete and not necessarily the best of environments for one, like Sylvester, who tended to feel deeply even unintended slights.[100]

When the final results were posted, the order of merit ran: senior wrangler, William Griffin of St. John's; second wrangler, Sylvester; third wrangler, Edward

Brumell, also a Johnian; George Green of Caius, fourth; Robert Ellis of Trinity, fifth; and Duncan Gregory also of Trinity, sixth. St. John's had swept the top places. Sylvester had earned a high wranglership, but he had not come in first.[101]

It had been a distinguished year. Griffin stayed on at Cambridge, immediately securing a fellowship at St. John's and setting up the very successful private tutoring practice he would run until he accepted a vicarage in Kent in 1848. Brumell, too, became a fellow of St. John's as well as a Johnian tutor. Green, as is well known, had done much of his original work—on the so-called Green's function and on Green's theorem—long before entering Caius in 1833 at the age of forty and died some seven-and-a-half years later in 1841.[102] Ellis and Gregory both became fellows of Trinity and founded the *Cambridge Mathematical Journal* in 1838. This would become the *Cambridge and Dublin Mathematical Journal* in 1845 and finally the *Quarterly Journal of Pure and Applied Mathematics* a decade later with Sylvester as coeditor. As this list suggests, Sylvester, like the other top six wranglers, would almost certainly have gone on to a fellowship and guaranteed livelihood at his college had his religious convictions not prevented it. Here was a case—precisely as the *Quarterly Review* had posited in 1834—of "a youth . . . who has satisfied all your own tests of merit to the uttermost—his talents are rare, his industry unwearied, his acquirements vast, his character spotless; but because . . . he has the misfortune to have a conscience and will not strain it by subscription, you . . . bid him seek his fortune as he can" (see epigraph note). Whereas the other high wranglers had a natural, almost predetermined course to follow, Sylvester had no clear-cut path before him. A price of dissent.

CHAPTER 3

The Hollow Walls of Academe

I want only position, a local habitation, and a name.
—J. J. Sylvester to Benjamin Peirce [22 May 1843]

Although Sylvester did not leave Cambridge with an official degree in hand in 1837, he *had* earned the distinction of a high wranglership, and he had two mathematical publications, which, although certainly juvenalia, established the earnestness of his commitment to mathematics. Sylvester left Cambridge with something more, however: personal connections and a set of scientific values and standards shared with others of what historian Susan Faye Cannon classically termed "the Cambridge Network."[1] "A loose convergence of scientists, historians, dons, and other scholars, with a common acceptance of accuracy, intelligence, and novelty," this network played a key role in establishing science as well as the individual sciences as fields of professional endeavor during the course of the mid- through late-Victorian period (30). Men like astronomer and rising senior scientific statesman, John Herschel; soon-to-be Master of Trinity, physical scientist, and philosopher of science, William Whewell; geologist, Adam Sedgwick; astronomer, George Biddell Airy; and mathematicians, Augustus De Morgan, Charles Babbage, and George Peacock; among others, defined the scientific—in contradistinction to the literary—node of the network largely through their curricular reform-mindedness directed toward Cambridge and, in the case of De Morgan, University College London. They fostered, from the 1830s on, a conception of the value of the applicability of mathematics to the various physical sciences, a view reinforced in examinations like the Mathematical Tripos. Moreover, they sought not only to broaden the curriculum by incorporating key aspects of Continental mathematics and the mathematical sciences into it but also to establish high disciplinary standards through their own research.

Their efforts have also been situated in the broader context of political and theological currents in British society. In seeking to introduce

Continental—and specifically French—mathematics into the Cambridge curriculum in the 1810s, certain more radical members of the Cambridge network
like Babbage and Herschel have been seen as openly challenging the traditional
linkage between the Anglican Church and the state through their calls for the
introduction of the abstract, analytical form of reasoning that had grown in the
republican hotbed of pre- and post-revolutionary France and that represented a
mindset at odds with that of an unquestioning Anglican.[2] These often religious
nonconformists—De Morgan among them—have been interpreted as calling
for the serious questioning of established authority, for the institutionalization of a meritocracy open to all regardless of social class, for the adoption of
reforming Whig as opposed to staid Tory positions in matters political, and for
the fostering and enhancing of a newly industrializing Britain.[3]

Although Sylvester's religion had made him an outsider at Cambridge, he
wanted nothing more than to fit in and to succeed in the kind of academic
world that the members of Cannon's Cambridge network envisioned and
that, at the same time, many of the radical political and scientific reformers
among them championed for more subversive reasons. By 1837, the year of
Victoria's ascension to the throne, there was as yet no scientific "profession"
in the sense that there were neither recognized avenues for training specialists
nor well-defined job opportunities for them.[4] Cambridge and Oxford, despite
numerous efforts at reform, still aimed mainly at instilling Anglican, Christian
values in their students. They functioned primarily as training grounds for
the Anglican clergy and for the nation's social elite, a situation that would not
begin to change until the second half of the century.[5] Moreover, even if the
Oxbridge tutorships and professorships provided (often marginal) financial
security for some, these posts were ultimately limited in number and restricted
to subscribers to the Thirty-nine Articles. As a Jew, Sylvester could not hold one
of them.

Still, science *was* "becoming more of a profession in England; that is, a
person who spent most of his energy on this activity might initiate, maintain,
or improve his middle-class status; might devote his life to this activity without
needing the justification of being a doctor or lawyer or clergyman in addition;
and might feel that his success was determined by the reputation he gained
among his peers, not by monetary returns."[6] Sylvester certainly wanted to be a
"professional" in this sense. Ideally, he would have liked to have spent his time
fully engrossed in what he would later term his "mathematical reveries."[7] Ideally,
he would have hoped that his worth as a mathematician would be decided by
his peers on the basis of his theorems. Sylvester, however, was a pragmatist. How
could he do this without "monetary returns"? Unless of independent means,
the aspiring mathematician of 1837 had to earn a living somehow. Sylvester's
next challenge was to figure out how he would make his way in the world and,
at the same time, ensure that he made his mark on mathematics.

An Unexpected Opportunity

In the first half of the nineteenth century, students tended to follow their Cambridge studies in one of three ways. Almost two-thirds of them entered the Anglican clergy, a career path that might or might not make them wealthy but that would secure for them that much sought "gentlemanly" status and acceptance by "society" in their environs.[8] At the same time, while almost a third of them came from the landed gentry, only 14%—eldest sons whose fathers made way for them—returned to the family property to take over the responsibilities of maintaining the estate and the family's social position. Finally, just over 20% left Cambridge to enter the newly emerging "professional class."

An outgrowth and extension of the traditional, eighteenth-century professions—the Established Church, the law, and medicine—the professional class was both firmly rooted in the evolving British middle class and definitive of a new concept of "society" within British culture. It consisted of the clergy, barristers, and physicians, but, as the century progressed, it also came to include teachers, especially professors, members of the civil service, army and naval officers, architects, engineers, artists and sculptors, and, by the end of the century, actuaries and others.[9] What it did not include were businessmen. Although they, too, were middle class, they were separated from the professional men socially, educationally, and philosophically.

Professional men tended to come from better social backgrounds, shared a certain standard of education defined by the Oxbridge ideal, and were perceived to work not so much for profit as out of a higher sense of duty. They "began to see themselves as in a special sense the descendants of the landed aristocracy and the new gentlemen of English society."[10] For someone like Sylvester, who was not of the landed gentry and for whom the Anglican clergy was not an option, a career in one of the professions would not only provide a secure and respectable income but also confer this new gentlemanly status. The problem was that the most prestigious of the nonclerical professions were expensive to enter. It could cost up to £1000 to prepare oneself for the law or medicine.[11] Army commissions had to be purchased and ranged from £450 for the infantry, to £800 or more for the cavalry, to more than £1200 for the Household regiments. Moreover, a candidate had to be nominated by a superior officer, at which point his name went on a list for the first available slot. Nominations tended to go to those deemed most companionable (thus not, perhaps, to Jewish mathematicians); slots could be long in coming (75). Neither the civil service nor teaching involved this type of expense and uncertainty, but the civil service turned on political patronage, involved drudgerous copying and summing, and offered little opportunity for financial advancement (until reforms in the 1850s instituted competitive examinations for entrance), and school teaching was poorly paid and intellectually unchallenging (81–84, 106–107). At the university level, positions were limited to Anglicans at Oxbridge (as well as at the University of Durham founded in 1837), and the thirty-year-old Augustus De Morgan filled the only chair of mathematics at nonsectarian University College London when the order of merit for the 1837 Tripos listed Sylvester's name at second. As Sylvester weighed his options in the summer and fall of 1837, however, an unexpected—perhaps undreamed of—career opportunity presented itself. Just before the start of classes, on 15 September 1837, the Professor of Natural Philosophy at University College, William Ritchie, died in his forty-seventh year of a fever caught a fortnight earlier.[12] The vacant post—a professorship of natural philosophy, astronomy, and civil engineering—was quickly advertised, and although it may not have been in mathematics per se, it was certainly a mathematically oriented position that a Cambridge high wrangler would have been well suited to fill.

By 23 October, Sylvester had learned of the opening and was in London to ready testimonials in support of his case. From the Inns of Court and the newly opened law office of his friend from Liverpool, William Drinkwater, Sylvester wrote to the Secretary of University College, Charles Atkinson, to "announce through you to the Council of the University College of London that it is my intention to offer myself as a candidate for the vacant Professorship."[13] At the same time, he called attention to the fact that "on its first opening I was a student at the London University but for the last six years have been connected with that of Cambridge where my conduct and character lie open to inquiry" (4). These bold pen strokes almost dared college officials to hold against him his short-lived, youthful association with the school. He was now a mature Cambridge man.

Several of the twenty-two testimonials that Sylvester submitted eight days later seemed crafted to address precisely this point. In particular, his former tutor, John Hymers, wrote jointly with the Rev. Yate, Dean of St. John's College, "to certify that Mr. Sylvester, during the time of his residence amongst us . . . has behaved himself soberly and regularly; and that in moral conduct we think him a fit person to have the superintendence of the instruction of youth."[14] Sylvester's classics tutor, Edward Bushby, echoed this sentiment, proclaiming his former student's "moral conduct . . . without reproach."[15]

If Sylvester was worried that the sins of his youth would come back to haunt him at this crucial juncture, he had nothing to fear relative to his scientific credentials. The other twenty testimonials all focused on his scholarly attainments and gave unqualified support to his case. Moreover, Sylvester had seemingly left no stone unturned. He had secured letters from the two moderators and the two examiners of the 1837 Tripos. He had four statements from fellows and tutors of St. John's; four more from outside fellows, tutors, and coaches; and another two from Cambridge men who knew of him through the Cambridge Philosophical Society. Three chaired professors—George Peacock, the Lowndean Professor of Astronomy and Geometry, James Cumming, the University Professor of Chemistry, and James Challis, the Plumian Professor of Astronomy—likewise wrote on his behalf. Finally, Sylvester obtained letters from two worthies outside of Cambridge—Olinthus Gregory, the same Professor of Mathematics at the Royal Military Academy in Woolwich who had examined the eleven-year-old Sylvester's mathematical attainments, and John Lubbock, a fellow of the Royal Society and soon-to-be Vice Chancellor of the University of London—as if to document that his reputation had already spread beyond Cambridge's walls. It was an impressive and extensive list of supporters, and everyone who wrote did so in glowing terms.

The moderators and examiners naturally addressed Sylvester's performance on the Tripos. James Heaviside, the Senior Moderator, "was *particularly struck*" by Sylvester's "talent and knowledge" and judged him "a person of very extra-ordinary philosophical abilities, and likely hereafter to contribute in extending the sphere of science."[16] Samuel Earnshaw, the Senior Examiner, shared this view and remarked curiously that "in that examination [Sylvester] gained the place of Second Wrangler, and satisfied the examiners that he was not sur-passed by any of his competitors in acquirement of philosophical learning and originality of talent."[17] Sylvester's abysmal handwriting may have masked his true performance and ultimately prevented him from earning more points than the senior wrangler, William Griffin, but Earnshaw, at least, thought that the better man had *not* won on the day. Among the Johnians and other Cambridge men who wrote, Hymers's remark was typical and in total accord with those of the Tripos officials: "I have no hesitation in saying, that if profound Math-ematical knowledge, and the possession of a mind singularly adapted for the investigation of scientific truths, be a passport to the situation for which he is now a candidate, Mr. Sylvester may entertain very confident pretensions to it."[18] Gregory held this view as well, but as the writer who had known Sylvester the longest, he had perhaps a unique vantage point from which to assess the candidate's merits. "I have known Mr. Sylvester from his boyhood," he wrote,

> and though from various circumstances our intercourse has only been of an interrupted character, yet I have traced the development of his intellectual and scien-tific faculties with peculiar interest. I regard Mr. Sylvester as a gentleman of great and original genius in reference to the abstruse sciences; and of a rich and ready inven-tion with regard to mathematical theories, to practical expedients and philosophical application. I believe his reading has been very extensive, and that he has formed his scientific taste upon the best models, while he retains that honest independent exercise of his own powers of investigation and research, which is best calculated to make him eminently useful as a public teacher in every department of pure and applied science to which he devotes himself.[19]

In amassing his testimonials, Sylvester had effectively enlisted the Cambridge network and beyond in his cause, and they had uniformly judged his scientific attainments in the highest possible terms.

The committee that received, read, and evaluated these references for the College Council was chaired by Sylvester's former teacher, the Professor of Mathematics, Augustus De Morgan. In all, six people applied. One provided no testimonials; another was actually seeking a headmastership at a grammar school! These were eliminated immediately. Sylvester thus had only three serious competitors, one from St. Andrews, one from St. John's College, Cambridge, and one from Trinity College, Cambridge. Neither the Scot nor the other Johnian particularly impressed the committee, but the Trinity man, Thomas Webster, seemed adequate to the task. The fourteenth wrangler in 1832, Webster had taken the M.A. three years later and had produced solid enough testimonials from Peacock and Challis in addition to William Whewell and others.[20] Still, De Morgan and the committee found "no evidence of great originality," despite "evidence of good reading and power of exhibiting results in simple and useful forms," and their report dispensed with him in one paragraph.[21] Overwhelmingly, they threw their support to Sylvester, but there *was* some explaining to do. As they put it, they were "convinced it would be their duty to recommend [Sylvester] in the strongest terms and in the most unqualified manner, if some other circumstances did not require consideration." These circumstances were the "most impetuous and irritable disposition" Sylvester had exhibited while a student at the College in 1828 and the fact that Sylvester had experience neither in lecturing nor in the preparation of laboratory demonstrations.

Sylvester had already anticipated the first issue, and, in the report, De Morgan pointed precisely to the character references from St. John's in support of the position that "since the period above alluded to, Mr. Sylvester has acquired and exercised self-control." As for the second matter, the committee argued that "the requisite degree of dexterity can be very soon attained by a man of talent" and, moreover, that "a properly trained mathematician possessing fluency of delivery (which Mr. Sylvester is known to several of the committee to possess) will speedily become fully competent to all the duties of the chair." They closed with their most compelling argument. In keeping both with the new spirit and goals of the Cambridge network and with the more radical, anti-Oxbridge agenda embraced by many at the "godless institution on Gower Street," they could not "forbear to remind the Senate of the desirableness, *ceteris paribus*, of securing the services of those who are likely to advance as well as to diffuse, the knowledge of their subject . . . and it could easily be shown that there has been more than one person who has left his name inseparably connected with the history of discovery, who had not, at the same age, given such decided proof of power as appears to have been exhibited by Mr. Sylvester." De Morgan and his committee shared a new conception (for England, at least) of the academic scientist as both a teacher *and* a researcher.[22] Sylvester, although untried, seemed an ideal candidate to fulfill this dual mission. The Senate apparently agreed, and by 21 November 1837 Sylvester had accepted the chair and made haste to settle his affairs in both Cambridge and Liverpool before taking up his new duties in January 1838.[23]

Professor of Natural Philosophy at University College London

Sylvester was the third person to hold University College's chair of natural philosophy. He succeeded William Ritchie, a Scot who had trained for the ministry before deciding to pursue a career in teaching. Ritchie had honed his skills in natural philosophy during a Parisian study tour that found him in the lecture halls of such early-nineteenth-century notables as Joseph Louis Gay-Lussac and Jean-Baptiste Biot. By all accounts, Ritchie learned his lessons well, returning to Britain to make somewhat of a name for himself as an experimenter. He published several papers in the *Philosophical Transactions of the Royal Society* on devices such as the photometer and the thermometer and on experiments aimed at understanding the phenomena of galvanism, electromagnetism, electricity, and heat. This research—and, presumably, his connection with Sir John Herschel through whom he had communicated his early work to the Royal Society—won him first a professorship of natural philosophy at London's Royal Institution in 1829 and then University College's chair in that same subject three years later.

Ritchie had had rather large shoes to fill at University College. The first incumbent in the chair, Dionysius Lardner, was renowned both as a popular lecturer and as a science writer. The public lectures and demonstrations he gave in association with his post regularly attracted large and appreciative crowds, whereas publication ventures like his 133-volume *Cabinet Cyclopædia* (1829–1849) united some of the best writers of the day in the common cause of presenting their special subjects to a wide, literate public.[24] Although not as flamboyant, Ritchie proved able in the classroom and presented natural philosophy to his students more from an experimental than from a theoretical point of view. Sylvester, the theoretician practiced as neither a lecturer nor an experimenter, thus assumed a chair that had strong traditions in both.

The University College Sylvester returned to as a professor was different, moreover, from the one he had briefly attended as a student almost 10 years earlier. The expansive, two-story, neo-Grecian building situated on eight acres of land now had its focal point complete: a massive, central, Corinthian portico of twelve columns supporting a pediment, approached by a grand flight of stairs rising from the ground floor to the first floor, and flanked by a well-proportioned dome. Inside, the ground-floor room where he had studied mathematics with De Morgan no longer existed, having burned two years earlier, but the semicircular natural philosophy theater still dominated the first floor of the building's right wing.[25]

The neighborhood had also changed. The City of London had further encroached from the southeast with the growing middle class occupying newly constructed row houses; the recently opened Euston Station just to the north of the university obscured the last glimpses of open fields and brought clamor and congestion. Moreover, Bloomsbury, so renowned for its writers in the later Victorian and Edwardian periods, had already begun to attract the likes of Charles Dickens and William Thackeray. In fact, by 1841, Sylvester would live at 22 Doughty Street; Dickens had lived at number 48 from 1837 to 1839 and had written *Oliver Twist* as well as *Nicholas Nickleby* there.[26]

Most importantly, however, what went on at University College had also evolved. The university had been controversial from its founding. The men behind it were social reformers. They abhorred the stranglehold that the Anglican Church had on English society and sought, through their Whig and often radical political agendas, to enable middle-class, predominantly dissenting students to craft a new professional élite, what Adrian Desmond has styled a "new middle management."[27] The university thus seethed with social and political change. It was dedicated to questioning the established political, religious, and intellectual authorities; it looked, unlike Oxbridge, beyond English shores and especially to France for its intellectual enlightenment; it actively sought to expose those secular and avant-garde ideas to the young men it trained to challenge established English orthodoxies and practices.[28] The University College Sylvester rejoined was most certainly *not* the Cambridge he had left. Its outward-looking and progressive ideals were exactly those that might allow him, as a Jew, to make a career for himself in academe. University College's new Professor of Natural Philosophy thus found himself in a politically radicalized, intellectually charged, increasingly urban, ever-modernizing environment that simultaneously supported the broader social agendas of its founders and the intellectual endeavors of the university, the nearby British Museum (founded in 1753), and the local literati.

Sylvester worked hard to fit into this new situation and to meet the expectations associated with it. He did extracurricular work, serving as an examiner for the first Flaherty Scholarship in mathematics and physics in October 1838 and taking drawing lessons—with his former Liverpool classmate and lawyer friend, Drinkwater—from the College's drawing master to improve his skills at the board.[29] In the classroom, he set up and performed experiments, gave his regular lectures, taught a special matriculation course for those working toward entry into the college, and participated in the program of evening lectures begun in 1839 for the benefit of schoolmasters in and around London.[30] As a lecturer, he did his best to treat the experimental side of his subject, but he tended nevertheless to emphasize the theoretical in keeping with both his natural predilections and his Cambridge training. His courses covered the standard topics from Newtonian mechanics—gravitational and planetary theory, optics and dynamics, statics and hydrostatics—with, like the Tripos he had taken in 1837, little emphasis on the theories of heat, light, electricity, or magnetism. That his three-year course of lectures was theoretically oriented is clear from the prerequisites he set. Entering students needed familiarity with algebraic notation, proportions, and trigonometry; those in the second year were expected to be fluent in conic sections, quadratic equations, and spherical trigonometry; third-year students required analytical geometry as well as the differential and integral calculus.[31] Despite the conscientiousness with which Sylvester approached his new duties, he never warmed to experiments, and only a modest number of students ever took his courses.

During the three-and-a-half years that he held the chair, University College suffered from low enrollments and insufficient revenues; the early 1830s had been particularly bad.[32] Since the professors depended on student fees for their livelihood, small enrollments necessarily meant a relatively small annual income. Sylvester, who probably made less than £150 a year while in London,[33] seemed particularly sensitive to the financial side of his position. In the fall of 1839, for example, he apparently dug in his heels over a student who had attended his course and then wanted to withdraw. Sylvester's refusal to relinquish the fees prompted the student's father to write to the college authorities explaining that his son was hard of hearing and could not follow the professor's lectures. As the father put it, "I regret very much that so much difficulty seems to be felt in a matter of comparatively very small importance—It cannot I think be either for the Interest of the Coll: at large, or Individual Profs: that such a narrow adherence to rights . . . should be maintained."[34] That prickly personality and almost painful sense of justice that De Morgan had assured the College was under control in the more mature Sylvester was nevertheless as much in evidence here as a concern over finances.

The fees and the experiments were not the only aspects of the post that Sylvester found uncongenial. As early as November 1838, he had written to Atkinson rightly complaining—on the basis of direct measurement and observation in the laboratory—that the temperature in his lecture theater was rarely as high as 52 degrees and sometimes as low as 46. This, coupled with water dripping in through the skylights, made the room "incommodious to the Lecturer, but absolutely unendurable to the students."[35] Five months later, he again wrote to Atkinson, but this time he seemed resolved eventually to leave his professorship altogether. In almost legalistic style, he asserted that, according to his understanding of the college regulations, he had the right as a professor to attend any of the college's courses free of charge. He further stated his intention "to obtain after the usual manner Certificates of . . . attendance with a view to qualifying [him]self for a Medical Diploma or Degree."[36] Although he did not earn a degree in medicine, the fact that he considered it provides strong indication of his dissatisfaction as Professor of Natural Philosophy as well as of his resolve to join, one way or another, the ranks of the "professional men" that University College was so intent on producing.

If Sylvester almost immediately came to dislike the teaching side of the University College post, the research side *did* prove rewarding for him. His earliest work, not surprisingly perhaps, was in mathematical physics. In fact, he had been thinking along these lines even before he knew about the vacant chair in that field at University College. While still in Cambridge following the 1837 Tripos, he had attended meetings of the Cambridge Philosophical Society. There, Philip Kelland, fellow and tutor at Queen's College, had presented his mathematical findings "On the Transmission of Light in Crystallized Media,"[37] and Sylvester had received his introduction to a research question that was "hot" both in Britain and on the Continent in the 1830s, namely, what is the nature of light? Is it a wave phenomenon? Or is it actually a stream of particles?[38]

This question had been debated since at least the seventeenth century, with Newton favoring the particulate or corpuscular explanation and Huygens providing wave interpretations. Newton's ideas held sway in Britain as well as in France throughout the eighteenth century, but in 1803 the issue once again came to the fore. In that year, the English natural philosopher Thomas Young discovered that light rays could set up interference patterns like water or sound waves. His findings in favor of the wave theory prompted many new experimental and theoretical investigations centered particularly on the phenomenon of double refraction. As is well known, a single light ray is generally refracted as a single ray when it passes from one nonopaque medium to another. There are, however, certain types of crystals that produce a double refraction, that is, a single incident ray results in two rays refracted at different angles. Moreover, these crystals have the property that if the incident ray enters the crystal from one particular direction—called the optic axis—then there is only a single refracted ray. Although Huygens had given a wave-theoretic explanation of double refraction in 1677, by the early nineteenth century, it had been shown that the corpuscular theory could account for it equally well. The natural philosophers were at a theoretical and experimental stalemate until 1813 when Edinburgh's David Brewster discovered that topaz, among other crystals, actually had two optic axes. Neither the wave nor the corpuscular theory of the day could account for this observation. A new explanation was needed.

That came eight years later in 1821. Working in France, Augustin Fresnel developed two explanations of the phenomenon grounded on a wave model—one experimental, one theoretical—that were so effective they proved difficult for his contemporaries to counter (210). The importance of Fresnel's results, as well as their acknowledged complexity, thus represented a mathematical challenge: simplify and streamline the mathematics; reduce the physical theory to its mathematical essence.

This problem attracted quite a bit of attention in Britain. In Ireland, the Professor of Mathematics at Trinity College Dublin, James MacCullagh, came up with a derivation in 1830, while William Rowan Hamilton deduced in 1832 that a continuum of rays forming a cone—not one or two refracted rays—resulted when incident rays came from particular directions (210–211). Hamilton's prediction based on the mathematics of the situation was confirmed experimentally by the Dublin natural philosopher, Humphrey Lloyd, later in 1832. The entire topic received extensive treatment in the *Philosophical Magazine* between 1833 and 1840 and at meetings of the British Association for the Advancement of Science in Cambridge (1833), Edinburgh (1834), and Dublin (1835). As a result of the Cambridge meeting, in particular, John Herschel and George Airy sent articles to the *Philosophical Magazine* in support of the wave theory.[39] It was in this intense research atmosphere that Sylvester broached Fresnel's theory and derived his own mathematical simplifications of it in his first published journal article in 1837 and 1838.[40]

Sylvester's approach in his paper, "Analytical Development of Fresnel's Optical Theory of Crystals," was algebraic with a geometric underpinning. In his view, "hitherto little has been done beyond finding and investigating the properties of the wave surface . . . Mr. Kelland . . . has incidentally obtained the difference of the squares of the velocities of a plane front in terms of angles made by it with the optic axes. I have obtained each of the velocities *separately*, and in a form precisely the same for biaxal as for uniaxal crystals."[41] This was not an earth-shattering finding, but it was nonetheless a *bona fide* simplification of the mathematical theory of double refraction: an original mathematical result. Its exposition, moreover, exemplified at least two traits that would come unmistakably to characterize Sylvester's mathematical writing and scholarship: a certain breathless haste and an impatience with bibliographic research. At the end of the paper's second installment, he confessed that "for all errors and imperfections in what precedes my excuse must be press of time and a total want of the materials to be derived from consulting works of reference" (19).

Three more applied papers followed in quick succession in 1838 and 1839.[42] In them, Sylvester utilized to good advantage what at the time were the novel concepts of surface and line integrals. Unfortunately, however, his work on these topics and the techniques he used were totally ignored by his contemporaries.[43]

Sylvester's brief tenure at University College also found him engaged in mathematics in a pure vein. Between 1839 and 1841, he threw himself into the theory of elimination.[44] Specifically, he considered two polynomial equations in one variable of degrees m and n, respectively, and asked when do these equations have a common root? Like the question about the ultimate nature of light, this question, too, had a long history. Newton had addressed it as early as 1707 in his *Arithmetica universalis*, and it had continued to occupy mathematicians—like Leonhard Euler and Étienne Bezout—throughout the eighteenth century. Bezout, in particular, had included an algorithm for deciding the issue in his popular *Cours de mathématiques* of 1764–1769. Moreover, in a very splashy result announced in 1829 but only published in 1835, Charles-François Sturm had provided an algorithm to settle the related problem of finding the real roots of an algebraic equation $f(x) = 0$ of degree n.[45] Specifically, Sturm applied the Euclidean algorithm to the function $f(x)$ and its derivative $f'(x) := f_1(x)$ to get the following sequence of equations:

$$f(x) = q_1(x)f_1(x) - f_2(x)$$
$$f_1(x) = q_2(x)f_2(x) - f_3(x)$$
$$f_2(x) = q_3(x)f_3(x) - f_4(x)$$
$$\vdots$$
$$f_{n-2}(x) = q_{n-1}(x)f_{n-1}(x) - f_n(x).$$

He then argued that [in the case of $f_n(x) \neq 0$] the number of real roots of the equation $f(x) = 0$ between the real values $x = a$ and $x = b$ is given by the difference of the number of sign changes in the sequence of so-called Sturm functions $f(x), f_1(x), f_2(x), \ldots, f_{n-1}(x), f_n(x)$, when $x = a$ and when $x = b$.[46] In his early works in pure mathematics, Sylvester sought not only more satisfactory methods for settling questions like the one Sturm addressed but also a deeper, algebraic understanding of the processes involved. As historian of mathematics, Hourya Sinaceur, has put it: "At a time when the Paris school was dominated, a few exceptions aside, by the analytic spirit, Sylvester wanted to promote the algebraic spirit to the point of subjecting all of analysis to it!"[47]

Relative to the question of when two polynomial equations have a common root, this predisposition led Sylvester to study both what he termed his "dialytic method" for arriving at its answer and the properties of what would later be called determinants. In particular, if $f(x) = a_m x^m + a_{m-1} x^{m-1} + \cdots + a_1 x + a_0 = 0$ and $g(x) = b_n x^n + b_{n-1} x^{n-1} + \cdots + b_1 x + b_0 = 0$ are two polynomial equations in x of degrees m and n ($m \neq n$, in general), respectively, then Sylvester showed that given the $(m + n) \times (m + n)$ determinant

$$
M = \begin{pmatrix}
a_m & a_{m-1} & \cdots & a_1 & a_0 & 0 & 0 & \cdots & 0 \\
0 & a_m & \cdots & a_2 & a_1 & a_0 & 0 & \cdots & 0 \\
\vdots & \vdots & \ddots & \ddots & \ddots & \ddots & \ddots & \vdots & \vdots \\
b_n & b_{n-1} & \cdots & b_1 & b_0 & 0 & 0 & \cdots & 0 \\
0 & b_n & \cdots & b_2 & b_1 & b_0 & 0 & \cdots & 0 \\
\vdots & \vdots & \ddots & \vdots & \vdots & \vdots & \vdots & \ddots & \vdots
\end{pmatrix},
$$

$f(x) = 0$ and $g(x) = 0$ have a common root, if det $M = 0$.[48] With respect to Sturm's Theorem, moreover, he provided an insightful algebraic analysis—in terms of the n complex roots of $f(x) = 0$—of the Sturm functions arising in the algorithm, claiming to have first considered this line of research during his two-year hiatus from St. John's and, thus, very shortly after Sturm's treatment of it had appeared in print.[49] This work, like his first study on Fresnel's theory, shows a mathematical mind of a fundamentally algebraic—as opposed to analytic—bent. This, too, would characterize Sylvester's research from the outset to the end of his long career.

The elimination-theoretic papers also reveal just how early two more traits of Sylvester's mathematical persona manifested themselves: his fascination— almost obsession—with mathematical language and his complexly nuanced sense of the importance of participating in the mathematical world that lay beyond the English Channel.[50] In the second installment of his paper on elimination theory, the twenty-five-year-old mathematician acknowledged his indebtedness to his colleague, De Morgan, for suggesting the now-standard term "order" and to both De Morgan and University College's new Professor of Jurisprudence, the able mathematician John T. Graves, for approving the (much less fortunate) designations "zeta-ic" and "zeta-ically." (Sylvester used the latter in the context of "zeta-ic multiplication," now understood as the formation and calculation of the determinant associated with a system of linear equations in n unknowns.) Sylvester closed his paper with what his later audiences would recognize as a signature footnote:

> I take this opportunity of retracting the symbol *SPD* used in my past paper, the letter *S* having no meaning except for English readers. I substitute for it *QDP*, where *Q* represents the Latin word Quadratus. On some future occasion I shall enlarge upon a new method of notation, whereby the language of analysis may be rendered much more expressive, depending essentially upon the use of similar figures inserted within one another, and containing numbers or letters, according as quantities or operations are to be denoted. This system to be carried out would require special but very simple printing types to be founded for the purpose.[51]

Revelatory of an ever-active mind that gloried in fine distinctions and shades of meaning, this passage reflects a Sylvester thoroughly swept up in his mathematics. Here was a Sylvester who wanted his readers to share in every twist and turn of his thought processes, who was certain they would be interested, and who tacitly sought their approbation for his special insights.

This passage likewise reflects a deep-seated conviction that mathematics was not a purely British preserve. Even at the age of twenty-five, Sylvester sensed at once the field's universality, the need to communicate beyond the confines of his own insular nation, and the nationalism associated with the international endeavor. In his paper's first part, he noted that "the reflections which Sturm's memorable theorem had originally excited, were revived by happening to be present at a sitting of the French Institute, where a letter was read from the Minister of Public Instruction, requesting an opinion upon the expediency of forming tables of elimination between two equations as high as the 5th or 6th degree containing one repeating term."[52] He then asked rhetorically, "Would it not redound to the honour of British science that some painstaking ingenious person should gird himself to the task? and would not this be a proper object to meet with encouragement from the Scientific Association of Great Britain" (1:44). Early in 1839, most likely between the fall and winter terms, Sylvester was in Paris, making himself known to French scientific society, learning what engaged French mathematicians, desiring his own country to make a good showing in the international mathematical arena, sensing at the same time that it could and should be doing better on that score. Sylvester shared in the outward-looking, professionalizing agenda of many of his colleagues at University College, and he was ambitious. He recognized that his own reputation could be built by feeding that ambition with the fruits of his mathematical researches and by disseminating them both at home and abroad.[53] But it was not merely a matter of ambition. Mathematical research, interacting with mathematicians and other men of science, participating in broader scientific society, these were all things in which Sylvester truly gloried. He was a social creature, who thrived on interactions with others.

During his time in the chair of natural philosophy at University College, Sylvester, in fact, pursued a variety of social contacts. He not only maintained his ties with long-time friends like William Drinkwater and Richard Wilson—calling at their offices and homes and undoubtedly sharing the occasional diversion—but also established new relationships with his colleagues and with men of science in the metropolis. Given his interest in chemistry, it is perhaps not surprising, for example, that he formed collegial ties with Thomas Graham, University College's Professor of Chemistry. Nine years Sylvester's senior, Graham, too, had joined the faculty in 1837. The two freshmen faculty members must have shared some of the same experiences as they prepared their respective classroom demonstrations and dealt with the college's fiscal limitations in trying to equip their laboratories. The fact that Graham described Sylvester as "a warm-hearted and most excellent man" suggests that they spent some time in each other's company, probably comparing notes on the college and discussing matters scientific.[54] Sylvester also sought out his counterpart at King's College London, the Professor of Experimental Philosophy Charles Wheatstone, as well as those scions of the Cambridge network, John Herschel and Charles Babbage.[55] These men and others formed the intersecting social and scientific circles within which Sylvester moved between 1837 and 1841.

That Sylvester, the newcomer, was welcomed into and esteemed by this society became clear almost immediately. On 25 April 1839, twelve of its members succeeded in winning the young mathematician's election to a fellowship in the Royal Society of London on the basis of the papers he had published between 1837 and early 1839. Graham, Wheatstone, and Herschel were among those who "of [their] own personal knowledge recommend[ed] him as deserving of that honor" and who felt him "likely to become a valuable & useful member."[56] They were joined by two mathematicians: Thomas S. Davies of the Royal Military Academy in Woolwich, and Baden Powell, Oxford's Savilian Professor of Geometry—one astronomer-actuary: Thomas Galloway of the Amicable Assurance Office of London—three London physicians: John Bostock, James Copland, and Herbert Mayo—and three additional worthies: Robert Grant, University College's Professor of Comparative Anatomy and Zoology; John Lee, a wealthy collector and benefactor of science; and Isaac Goldsmid, the Jewish financier who had so strongly backed both University College and Jewish emancipation.[57] Goldsmid's support was particularly interesting and significant. It undoubtedly reflected not only his regard for Sylvester's early scientific attainments and promise but also Goldsmid's own activism—mirrored, for example, in his efforts in founding University College—in the causes of Anglo-Jewry, and in social and educational reform.[58] Goldsmid's objectives were perfectly in tune with those both of Cannon's Cambridge network and of the political and social reformers among them: recognition within the evolving scientific community should be based on scientific attainments alone; they should be totally independent of religious persuasion or other extrascientific factors.

Despite Sylvester's success in the quest both to earn a living that allowed him to pursue his scientific interests and to make a name for himself in mathematics, he came to feel increasingly at odds in his position at University College. As he explained in a letter to Charles Babbage in May 1841,

> The nature of the materials with which I have had to deal and . . . the necessity of the case for attending what is called a course of Natural Philosophy in University College have hitherto cramped my exertions and . . . the free play of my zeal.
>
> My success as a lecturer in this department (whatever it may have been) is not to be taken as a measure of what I could effect in a situation more congenial to the early bent of my mind, the bias of my college studies, my sense of peculiar fitness and the spontaneous course of my meditations.[59]

The University College professorship had strengthened his resolve for an academic career, but one without the distractions of equipment and experiments and laboratories, one in which he could pursue mathematics. In England, however, no such avenues were open to him. The situation was different in the United States.

Looking beyond England

On 31 October 1840, the Professor of Mathematics at the University of Virginia, Charles Bonnycastle, died at the early age of forty-three. Bonnycastle, an Englishman, had been the first Professor of Natural Philosophy at Thomas Jefferson's university when it opened in 1825 but had moved into the chair of mathematics two years later when its first incumbent, Thomas Key, vacated it. Key, also an Englishman, had returned home to assume the professorship of Latin at the then newly formed London University and from there continued to take an interest in the educational experiment Jefferson had set up in Charlottesville.

Whether Sylvester first learned of the opening in Virginia through his colleague, Key, or whether he enlisted Key's support on seeing the position advertised, Sylvester's candidacy was heartily endorsed by the former Virginia professor in the spring of 1841, and officials in Charlottesville began making inquiries at University College about Sylvester and his association there.[60] In particular, the University of Virginia's governing Board of Visitors wanted to know two things: first, whether Sylvester's willingness to leave University College reflected any sort of problems between him and the school and, second, whether Sylvester was qualified for a university-level mathematics post. De Morgan was charged with transmitting the College Senate's resolution on the first point and with giving his assessment of Sylvester's mathematical qualifications.

In a letter to the Visitors dated 22 May 1841, De Morgan quoted the following Senate resolution:

> That the Visitors of the University of Virginia be informed . . . that Professor Sylvester's intention to retire from this College, in the event of his obtaining the Professorship of Mathematics in that University, is not the consequence of any disagreement either with the Council, his colleagues, or his pupils, nor arising out of any circumstance in any way affecting his character as a gentleman or a man of science, both which they believe to be unimpeachable; that he will carry with him, in the event of his obtaining what he seeks, their warmest good-will and best wishes for his success and prosperity.[61]

He then went on to offer his unequivocally positive views of Sylvester, the mathematician. "No person of his years in this country," he stated, "has more reputation than Mr. Sylvester as an original Mathematician, or bids fairer to extend the exact sciences by his labours. From my own knowledge of what he has done, I can most safely say that he is a Mathematician of great power, well acquainted with the most modern forms of the science, and very zealous in the prosecution of his inquiries. By these qualifications he will certainly make his name well known as an original cultivator of Mathematics, to the credit of any institution with which he may be connected" (24). If the Virginia officials had worried that Sylvester was in some way being forced to leave University College or that he was more a natural philosopher and less a mathematician, this letter surely assuaged those fears.

For his part, Sylvester approached the application process in 1841 as shrewdly as he had in 1837. He amassed another impressive array of testimonials that addressed his fitness as a mathematical researcher, as a colleague, and as a teacher, at the same time that they underscored the mathematical reputation he was earning. Sylvester's research was held in uniformly high regard. Herschel pronounced his work on elimination theory "of much interest"; Babbage saw in his publications to date "a capacity for discovery, and an originality of invention, which would do credit to any University in either of our respective countries"; James Ivory, noted for masterful applications of the infinitesimal calculus to physical problems and from whom Sylvester had drawn in his work in 1838 and 1839, wrote of the candidate's mathematical "ingenuity and originality"; John Graves echoed them all in finding his elimination-theoretic results "to exhibit unusual ingenuity and originality."[62]

As a colleague, Sylvester was likewise esteemed. Robert Latham, the Professor of English at University College and a friend from Sylvester's Cambridge days, affirmed in an unsolicited testimonial that "in London the character of Mr. Sylvester has conciliated the respect and friendship of all his colleagues, and the confidence of his pupils."[63] Three of those pupils also wrote. John Coombe, a fellow of St. John's College, Cambridge in 1841, had entered the College in 1836 and had studied mathematics privately with Sylvester there. Coombe declared Sylvester "eminently capable of conducting any mathematical class or examination, . . . particularly happy in methods of illustration which he resorts to, and very successful in rendering himself understood by his pupils in any Mathematical investigation."[64] Robert Cullen and John Robson, two assistant masters at University College's preparatory school, attended Sylvester's lectures in natural philosophy there and similarly "[bore] grateful testimony to the zeal, assiduity, gentlemanly deportment, and distinguished talents" of their professor.[65] As when he had applied for the University College professorship, Sylvester did his best to secure testimonials that would not only make his case convincingly but also anticipate any questions that might arise during the selection process.

As if these letters did not provide testimony enough to his fitness for the University of Virginia position, Sylvester also included two more that reflected both his resourcefulness and his growing reputation beyond the London–Cambridge axis. Sylvester had made himself and his work known to the mathematicians at Trinity College Dublin, in particular, the Professor of Mathematics, James MacCullagh, and fellow and Assistant Professor, Charles Graves.

From a scientific point of view, this contact was quite natural. Sylvester's first paper on the mathematical theory of double refraction was in precisely the area that MacCullagh had been cultivating so successfully since 1830, whereas Graves was a mathematician with purer leanings, who also happened to be the brother of Sylvester's University College colleague, John Graves.[66] From a more pragmatic point of view, though, Trinity College Dublin held out the possibility for something that had been denied Sylvester in England and that could prove essential for the further advancement of his career, especially in the United States: a degree.

Trinity College Dublin, unlike Oxford and Cambridge, had not required religious tests of its candidates for degrees since the antitransubstantiation oath aimed specifically at Catholics was abolished in 1794.[67] Moreover, the College had what amounted to a reciprocity agreement with the two English institutions and especially with Oriel College, Oxford, and St. John's College, Cambridge; any graduate in good standing of the English schools enjoyed the rights and privileges of the Irish school, and vice versa.[68] Sylvester, a Cambridge graduate in all but name, could thus benefit from this arrangement, if he had supporters in place to set the necessary wheels in motion. Apparently he did, for at the Summer Commencements on 6 July 1841, he was present to receive not only his B.A. (*ad eund Cantab*), that is, on the basis of his Cambridge work, but also an M.A.[69] In general, the latter was awarded, if applied for, no sooner than three years after the B.A., provided the student had been morally upstanding and had brought no disgrace upon the institution. Because Sylvester had for all intents and purposes earned his Cambridge B.A. in 1837, he effectively met the M.A. qualifications as well.[70]

Although the extent of Graves and MacCullagh's involvement in Sylvester's Trinity College degrees is unclear, it is certain that they both knew his work and regarded it highly. Graves wrote to Sylvester that "I have read several of your papers, and enjoyed the pleasure of conversing with you on scientific subjects: and with this knowledge of your high attainments, I feel quite sure that your zeal and perseverance in the pursuit of science, will raise you to a proud place amongst the Mathematicians of the present country."[71] MacCullagh likewise acknowledged Sylvester's "high talents" but modestly concluded that he did not "feel [himself] entitled to pass judgment on a brother Professor, especially on one so well known to the scientific world."[72] That Sylvester had indeed established a strong reputation in the short time since he had left Cambridge was not lost on the Board of Visitors at the University of Virginia. Based on their own reconnaissance and on the strength and extent of Sylvester's support, they appointed him to their vacant chair of mathematics at a meeting on 3 July 1841.[73]

News of the appointment reached Sylvester after his return from Dublin and just before he made his first appearance on the podium of a meeting of the British Association for the Advancement of Science to publicize more fully his earlier results on the theory of elimination and Sturm's functions.[74] He promptly—perhaps even precipitously—accepted, for he realized only later that the offer was, in fact, not unconditional. It was for an initial, one-year, probationary term. Sylvester was stung. How could the University expect him to make such a major move with no more than one year's job security? Did it not understand the true weight of the testimonials he had provided?[75] Only the intercession of Andrew Stevenson, U.S. Ambassador to the Court of St. James, prevented the mathematician from withdrawing his acceptance and remaining in England. The two men met several times during the fall of 1841 with Stevenson assuring Sylvester that the terms of his appointment in no way implied a lack of confidence in him and that they were in no way unique to his appointment. The diplomat finally won, but not without putting his tact and argumentative skills fully to the test. Writing to Joseph Carrington Cabell, a member of the Board of Visitors involved in Sylvester's appointment, Stevenson confessed that "it was with the utmost difficulty I could at last succeed in getting him to change his determination."[76]

Stevenson did not misrepresent the case to Sylvester. The University had made the same probationary proviso in offers earlier—in 1827 to Gessner Harrison for the professorship of classical languages, in 1830 to John A. G. Davis for the chair in law, in 1833 to Alfred T. Magill for the position in medicine. All these men were, like Sylvester, young at the time of their appointments, but they were, unlike him, also largely untried in the classroom.[77] At the same meeting in which the Board approved Sylvester's appointment, moreover, it also appointed a new Professor of Modern Languages, Charles Kraitsir, for one year initially.[78] Thus, although the authorities did not regularly insist on the one-year probationary term, their action in Sylvester's case was neither unprecedented nor unusual. There was, however, more to the story than Stevenson perhaps knew.

The University of Virginia had opened early in March 1825 and had been plagued from the beginning by student unruliness and unrest. On the nights of 22 June, 5 August, and 19 September 1825, students had staged rowdy demonstrations on the University grounds. These had culminated at the end of September when a mass of masked students assembled, chanted "Down with the European professors," and began hurling bricks and invective at those attempting to break them up. The elderly Jefferson decried these events as "vicious irregularities," and the faculty threatened to resign *en masse* unless the Board of Visitors provided for adequate policing of the University.[79] Although this situation was finally diffused, the students' complaints were not silenced. Not only did many students object—like the newspapers in New England, Philadelphia, and elsewhere—to the importation of professors from abroad,[80] but they also found that many of the rules and regulations imposed on them were unduly onerous. They were forced to arise each morning at 5:30 A.M.; the food was poor; they were provided neither with adequate linen nor with dependable cleaning services for their rooms; they had to wear a uniform at examinations, public days, and whenever they left the immediate university grounds; proctors patrolled their living quarters and reported all manner of infraction to the faculty. None of these rules was popular; all fostered an atmosphere of distrust and confrontation.[81]

Through the early 1840s and until the honor system was adopted in 1842, students regularly rebelled against the authority of the University. The 1825 disturbances were followed by major riots in May 1831 over the uniform law, in the fall of 1836 over the faculty's intervention in the formation of a student military company, and in 1838 over the faculty's refusal to allow a ball ostensibly in celebration of Jefferson's birthday. The 1836 riots, in particular, became so volatile that by November musket fire on the university grounds had become a common occurrence, faculty houses had been stoned, and faculty members had begun arming themselves, fearful for their lives and for those of their families. Although this, like all the other incidents, was finally quelled, it became a kind of *cause célèbre* among the students; in succeeding Novembers, they remembered the events of 1836 with raucous displays. These November demonstrations reached a tragic climax on 12 November 1840 when the Professor of Law, John Davis, tried to calm and disperse a noisy group of masked students gathered in front of his home. He was shot by one of the rioters and died of his wounds several days later. This was just one year before Sylvester arrived in Charlottesville to take up the duties of the chair of mathematics (2:301–310). In making one-year offers both to the Englishman, Sylvester, and to the Hungarian, Kraitsir, the Board of Visitors was exercising caution. It needed to see not only how the new professors would cope with its university's student body but also how that student body would receive and react to the professors. The last thing it wanted was to risk more student violence directed against the faculty.

In making these two offers, however, the Board did not perhaps anticipate backlash of a different sort—from the Richmond-based, Presbyterian journal, *Watchman of the South*. On 5 August 1841, the journal's editor argued at length against the two appointments on the grounds that, as a State institution, the University of Virginia should reflect its constituency, and that constituency was by definition not foreign and was "by professions Christians and not heathen, nor musselmen, nor Jews, nor Atheists, nor Infidels. They are also Protestants, and not Papists."[82] At base, the *Watchman* objected to what it viewed as the University's godlessness, referring to Jefferson and his collaborators as an "Infidel junto that once held fearful sway over the public mind and the public purse in Virginia."[83] Jefferson believed firmly in the separation of Church and State, and his adherence to this principle in founding the University of Virginia had been controversial from the start.[84] The *Watchman*'s editor clearly felt that in 1841, fifteen years after Jefferson's death, it was time for the University to look closer to home for its faculty, to abandon its nonsectarian principles, and to embrace the Protestant—in fact, Presbyterian—principles so dominant in the Commonwealth. Sylvester, the English Jew, and Kraitsir, the Hungarian Catholic, did not at all fit this vision of a professoriate of native-son Protestants.[85]

Had Sylvester known about this article and the prejudices it betrayed, he might have been even more uneasy about taking up the new position in Charlottesville. It was, however, the ongoing discussions between him and the Board via Stevenson over the probationary appointment that delayed his departure from England until well after the 1 September start of the university's academic year. Sylvester did finally set sail for the United States on 19 October 1841 aboard the Boston Steamer.[86] From Boston, he most likely sailed on to Baltimore continuing by steam train to Louisa, Virginia, the end of the line, and then by coach to Charlottesville.[87] The journey took the better part of a month, but by 22 November, he had assumed his duties and was in attendance at his first meeting of the faculty.[88]

Professor of Mathematics at the University of Virginia

The University of Virginia represented a new model not only educationally but also architecturally. Designed by Jefferson, it was a quadrangle open on the south to views in the distance of the Blue Ridge Mountains and crowned on the north by the remarkably proportioned, Palladio-inspired Rotunda. The ten so-called Pavilions, spaced five each down the western and eastern sides of a central "Lawn," housed the various "Schools" (or what today would be called departments) on their first floors and the professor of each School on their second. Two-storied, columned structures with capitals in the Doric, Ionic, and Corinthian styles and with varying architectural features, the Pavilions were intended to serve as life-size, three-dimensional lessons in architectural history. A series of single-storied rooms ran between and linked the Pavilions and served as student housing, while two lines of construction parallel to the eastern and western sides of the quadrangle and separated from it by kitchen gardens and pasturage—the so-called "Ranges"—each comprised more single-story student rooms and three large "Hotels" where the students took their meals. Jefferson's "academical village," an enclave in which faculty and students ideally lived and worked together, was situated about two miles west of Charlottesville, a town so small that even Jefferson deemed it "a place of so little resource" that one could not even procure "a bed, table, and chairs" there.[89]

The Jeffersonian ideal and the reality of his university were, however, two different things as the tumultous 1820s and 1830s attested. The neoclassical architecture served as a backdrop for a day-to-day scene that was far from idyllic. In the early years, the buildings defined not so much a lawn as a dirty, dusty, and often muddy promenade for boisterous young men, the slaves who served them, and all manner of livestock. It was into this environment—far from the urban life of London and with cultural practices foreign and even abhorrent to many outside the South—that Sylvester was greeted by the University community in November 1841.

The students staged an illumination in his honor, affixing lighted candles to the more than 100 pillars that line both sides of the quadrangle, burning on the lawn in between tar barrels pilfered from the university's stores, and, in general, raising an unholy racket.[90] The faculty attended Sylvester's inaugural address and found the new professor engaging, if nervous, but perhaps not exactly the sort of man it had been expecting.[91] One student, Robert Dabney, wrote to his mother on 15 December 1841 that

> They thought that a man all the way from London, recommended by great men, and titled lords and bishops, must be a wonder in every respect. They were looking for a splendid fellow, who was to take his place in the top notch of public estimation, at once and by storm, and to surpass everything that was ever seen in his mode of instruction, when lo! a little, bluff, beef-fed English cockney, perfectly insignificant in his appearance, and raw and awkward in his manners, only twenty-six [sic] years old, deficient in the faculty of giving instruction, and far below what we have been accustomed to in a lecturer. You may guess it required all efforts of the intended admirers to keep their countenances from falling. The students, indeed, made no secret of their disappointment; but the Faculty make much of him, and profess that they are highly delighted. He is, I should think, from what I have seen of him, which is but little, a fellow of excellent sense and good acquirements for his age . . . . He has a hearty, open countenance, and seems to be an industrious, lively fellow.[92]

Sylvester may not have fit the students' image of an Englishman worthy of the esteem of "titled lords," but he had come across as decent enough, and the faculty understood the import of his testimonials. It seemed a reasonably auspicious beginning.

Forty-seven students were enrolled in Sylvester's classes at essentially three mathematical levels: first junior, second junior, and senior.[93] Because the professorship of mathematics had been in limbo after Bonnycastle's death the year before and because Sylvester did not arrive until well after the start of classes, enrollments in the School of Mathematics were down from previous years. This meant that Sylvester's projected salary—derived from student fees of roughly $25.00 per faculty member per student—for his first year was also down from that earned by Bonnycastle.[94] Bonnycastle's average annual salary from 1835 through the 1839–1840 academic year was $2,917; Sylvester's first-year salary would have been about $1,175 with his home provided for, although this would presumably have increased once it became known that the School of Mathematics was again up and running.[95]

The texts and curriculum Sylvester exposed his students to had been set prior to his arrival and were fairly typical of the United States of the 1840s.[96] The first junior class studied arithmetic and algebra from translations of Adrien-Marie Legendre's standard texts in addition to the fundamentals of Euclidean geometry enhanced by models. The second juniors proceeded to translated texts by Silvestre Lacroix on algebra and by Legendre on geometry; the latter especially concentrated on synthetic geometry in addition to plane and spherical trigonometry with applications to navigation, that is, to the theory of projection. Finally, the senior classes mastered the differential and integral calculus from Bonnycastle's *Inductive Geometry* (1834) as well as from John Young's *The Elements of the Differential Calculus* (1833) and *The Elements of the Integral Calculus* (1839). For those advanced students of a more applied mathematical bent, there was also a course drawn from both the mechanics of Siméon-Denis Poisson and the first volume of Pierre Simon Laplace's treatise on celestial mechanics.[97] Sylvester finally had the opportunity to teach the kind of mathematics he had been trained in at Cambridge, undistracted by the vicissitudes of the laboratory.

There were, however, other distractions. At its meeting on 15 December 1841, the faculty was confronted with the report of a "very disorderly party of students" at a local tavern.[98] The situation had resulted in the tavern owner calling the constable, who encountered some seven loud, drunken, and verbally abusive students. "About an hour after [the constable's] arrival at the tavern, a very large reinforcement from the University came down—from 70 to 100—and a great deal of noise and disorder ensued with some damage to private property—some doors were broken by throwing of stones." Two of the seven primary offenders were thought to have incited the subsequent melee. One by one, the seven students were called in before the faculty and asked for their versions of the evening's events. As a consequence, the student judged the ring leader was dismissed from the university; judgment on his main collaborator, one Rees D. Gayle from Alabama, was put off until the next day after much discussion as to the appropriate punishment; and two more students were officially reprimanded with letters sent out to their parents or guardians. When the faculty reconvened the next day, it was in a more lenient mood. Instead of a one-month suspension, Gayle had to pledge that he would abstain from "intoxicating liquors and in case of any violation of this pledge on his part report it" to the chair of the faculty. Another student involved, William B. Wade of Mississippi, was also let off as a first offender. Both Gayle and Wade were in Sylvester's School of Mathematics, and by 1 January 1842, Sylvester was reporting both of them for excessive absences. Here were two students whose minds seemed not to be on their studies.

Sylvester was again forced to report both of these students for undue absence from his classes in January at the faculty's meeting on 1 February. Eight others also received negative mention in his report: one for being "inattentive, listless, and talkative," one for being "not generally well prepared" and "given to levity," and one, William H. Ballard of New Orleans, for being "disrespectful in Lecture room." The other professors filed similar reports; Kraitsir, the Professor of Modern Languages, cited more than twice as many students as Sylvester, and some of Sylvester's delinquents were delinquent in other Schools as well. The faculty took action in almost all of these cases, from notifying parents and guardians "that unless [the students] make speedy amendment in the prosecution of their studies, they will be dismissed from the University," to merely informing the parents and guardians of a problem, to conducting further investigations into the causes of the professors' complaints. Two problematic cases, however, those of Wade and Ballard, were postponed until the faculty's next meeting three days later on 4 February. If Sylvester was experiencing business as usual in the classroom in January, the same was, unfortunately, equally true outside of class.

Also at the 1 February meeting, Sylvester was charged $2.50 for five window panes broken in January in Pavilion 6, the building that housed his lecture rooms and private quarters. Sylvester objected to this charge, and the matter was "referred to the Proctor for investigation."[99] This was not a good sign, for it meant that Sylvester's house had most likely been the target of a malicious nighttime raid by a stone-throwing student. This kind of assault had been common at the university from its beginnings and always signaled student discontent.[100] Robert Dabney had, in fact, presaged such behavior in his letter to his mother on 15 December 1841. After describing the illumination to her, he added that "if [Sylvester] knew how much of this to set down to the students' love of frolic, and how much to their good-will to him, his gratulations to himself might be somewhat diminished. [The students] will probably give him a little insight into this matter, by stoning his house the first time he crosses their sovereign will, which will be very soon if he does his duty."[101] Sylvester *had* been doing his duty, reporting on the students in his School; the students had been doing what they viewed as theirs, confronting authority, especially a new, foreign authority barely older than they were.

All was not well between Sylvester and the university, either. When Sylvester arrived in November, he had taken up residence in his Pavilion amidst the Bonnycastles' books and furniture. Bonnycastle's widow and three children had continued to live in the Pavilion by courtesy until Sylvester's arrival, and arrangements for moving their belongings had not yet been made by January. This suited Sylvester well enough at first because his own furnishings had not yet been delivered to Charlottesville. When they came in early January, however, he rightly insisted that the Bonnycastles' things be removed so that he could finally settle into his new life. This caused something of a stir. The university's Proctor, Willis Woodley, was forced personally to move what he could outside onto the quadrangle. The books, though, were too heavy for him to deal with. He wrote to Lucian Minor, who, together with his brother, John, handled the university's legal affairs, to explain that most of Mrs. Bonnycastle's belongings were outside on the Lawn and to ask that arrangements be made to remove them to some local warehouse for storage. This sparked Minor's hastily scrawled note on the back of the letter: "Gone to Univ[ersit]y about the books, was engaged Thursday about the subject of this letter. The d[evi]l take the baby-ish selfishness which dictated the necessity."[102] Sylvester may have been right, but he had not been terribly diplomatic.

These incidents both inside and outside the classroom make clear that Sylvester's American honeymoon had been short. By February, it was definitely over. The realities of his life as a young, new, foreign professor at the University of Virginia in 1842 were daily being driven home to him. Discipline was a problem with a significant proportion of the student body; faculty members had to deal with this not only within their respective Schools but also within the university as a whole; Sylvester's sense of "the right thing to do" was not necessarily shared by all in his new circle. The faculty meeting on 4 February proved the validity of all three of these realizations.

The cases of Wade and Ballard that had been held over from the meeting three days earlier came before the faculty for discussion and action.[103] Wade, a student in the Schools of Mathematics and Modern Languages, that is, in the Schools of the two new, foreign professors, Sylvester and Kraitsir, had been not only absent repeatedly from his classes but also in trouble outside of class.[104] It was moved that he be dismissed from the university, but an amendment was made to the effect that the father be contacted and the son be dismissed only if he failed to make good progress thereafter. Sylvester voted for leniency and favored the amendment; Kraitsir opposed it. Since the vote was evenly split, the amendment failed. The vote on the original motion passed, with Kraitsir and Sylvester again on the same opposing sides of the issue, and Wade was dismissed. Ballard's case was different. Since he was accused of disrespect in Sylvester's classroom, he was called in to speak to the charge before the assembled faculty. Following his assurance that he had meant no disrespect, the faculty resolved "that Mr. Ballard be admonished by the Chairman, and, that further proceedings be dispensed with, in consequence of his statement to the Chairman."[105]

Meetings followed on 12 and 19 February in silence on the Ballard matter— the calm before the storm. Sylvester had clearly been seething since the faculty's acceptance of Ballard's statement at its meeting on the fourth. On 23 February, he presented the faculty with his own detailed account of what had happened in his classroom on 31 January 1842, and this was entered into the minutes verbatim. He described a locking of wills that had escalated out of control. Ballard was blatantly not paying attention to Sylvester's oral examination of the class, reading a book under his desktop. When it came Ballard's turn to answer, he did so "in a very unbecoming way; contradicting and interrupting as if he were disputing with" Sylvester. Sylvester called the class to attention, and Ballard looked up from his book momentarily before pointedly returning to it for the remainder of the class period. Sylvester recounted the scene that followed:

> On being called up after the lecture and privately recommended to pay more attention in the future—he answered (still with increased violence of demeanor and tone) that he understood the subject could follow the lecture without looking at me and had not his spectacles—On this as he was leaving it occurred to me to recommend him to bring them with him in future.—Hereupon he answered in a very violent tone, manner and language, but the exact words I do not remember— It is proper to observe that Mr. Ballard's mode of addressing me has been almost uniformly marked with insolence and defiance.
>
> I then felt it right to state, that such conduct must not be persisted in and that if Mr. Ballard could not alter his conduct he must cease to attend my lecture room.
>
> On this he answered with increased insolence and violence—but I cannot recall the terms employed—I answered that I had been in different parts of the world, but had never witnessed similar conduct in persons brought up amongst gentlemen.
>
> To this he replied that I was not to prate to him, to hold my jaw, that I might go to hell—and other abusive terms which I cannot recall.
>
> I declined altercation with him and ordered him to leave the room, which he declared he would not do.
>
> I here turned to such of the class as remained made some remarks on the disgracefulness and discreditable character of such conduct and language, and left the room.—Before my leaving Mr. Ballard was guilty of additional abusive language.

Not surprisingly, things were no better the next day. Ballard, according to Sylvester, "answered to his name on being called in such a way as to excite laughter in the whole class, and drew down from me the reprimand for persisting to adopt this peremptory tone in addressing me—'he forgot what was due to his own position and not mine.' The manner was such as to make me apprehend the probability of personal violence." Given the events particularly of November 1840, this last statement must have given Sylvester's colleagues pause.

Ballard was called in to present his side of the story. It was substantially the same, except for one key, interpretive difference. When Ballard refused to leave the lecture room as instructed, he understood Sylvester to say that "he was no gentleman or something equivalent—the words he does not recall— He considered that he was imposed upon, and spoken to in an authoritative manner—as an overseer speaks to a negro slave." In Ballard's view, Sylvester had, through his degrading tone, publicly offended his personal honor. The Southern gentleman's code of honor thus dictated that Ballard—not Sylvester— was actually the aggrieved party.[106]

The faculty dealt with this matter in meetings on the twenty-third and then again on the twenty-fourth after Sylvester strongly denied the accuracy of some of the student testimony. The students called in, all of whom adhered to the same code of honor, corroborated and sympathized with Ballard's account. The faculty ultimately judged that Sylvester's reprimand regarding Ballard's classroom behavior sufficed; no further action was taken against Ballard on this point. It did, however, separate this issue from the events that occurred after the lecture had finished. The faculty's understanding and appreciation of the students' shared sense of honor came through explicitly in the language of its second resolution on the Ballard–Sylvester incident: "In view of the transactions posterior to the termination of the lectures, and the reprimand of the Professor; while the Faculty cannot but reprehend the violent language indulged in by Mr. Ballard towards Professor Sylvester, being unable to determine in how far the remarks of the Professor *as understood by Mr. Ballard might extenuate his conduct in reply*—Resolved, that it is expedient that this much of the transaction in question be submitted to the Board of Visitors of the University at their next meeting for their decision."[107] This issue, in fact, never had to come before the Board.

Sylvester absented himself from the next two faculty meetings on 2 and 19 March. That he was angered by the faculty's failure to support his interpretation of the encounter with Ballard was clear; he communicated to his colleagues in writing at the meeting on the nineteenth, protesting again the faculty's handling of the charges. This was duly entered into the minutes, but the faculty did not budge on its earlier decision.[108] Three days later, the Board of Visitors received and accepted Sylvester's resignation from the faculty, making it effective 29 March 1842 "or at any earlier period that he may elect."[109] In so doing, it informed him "that . . . the Board has not deemed it necessary to investigate the merits of the matter in difference between himself and the Student Ballard, and does not mean to impute to Mr. Sylvester any blame in that matter" (446–447).

Sylvester's departure from Charlottesville was precipitous.[110] Although the evidence is inconclusive, it strongly suggests that the fears of bodily harm he expressed to the faculty on 23 February were not unfounded. Very likely between 19 and 22 March, Sylvester was accosted by a pair of brothers in one of his classes.[111] The same Robert Dabney who chronicled Sylvester's arrival in a letter to his mother, reportedly recounted the story this way. Following an oral examination in which Sylvester pointed out the errors of the younger brother, the elder confronted the mathematician in the name of family honor and demanded an apology on the threat of retribution. Sylvester, like the faculty after the riots in 1836, armed himself for self-protection by purchasing a sword cane. The two brothers, one with bludgeon in hand, approached Sylvester, again demanded an apology, knocked off his hat, and hit him squarely on the head. Sylvester drew the sword cane and struck a glancing blow off a rib with no damage done.[112]

If true, this encounter would certainly have been enough to prompt Sylvester's decision to leave the University immediately!

Many factors contributed to Sylvester's brief and ill-fated association with the University of Virginia.[113] If not outright xenophobia, then at least a pervasive sense of localism had worked against foreign faculty members—making them obvious targets of hostility—from the University's opening in 1825. The original Professor of Natural Philosophy, the Englishman Thomas Key, had been the first to resign in March 1827 with his countryman, the chair of ancient languages, George Long, following a year later. Both of these men had, in fact, initially tendered their resignations in October 1825 in the aftermath of the earliest student riots but had been persuaded to withdraw them.[114] Although some foreigners, like Sylvester's predecessor in the School of Mathematics, the Englishman Charles Bonnycastle, were ultimately accepted, others like Kraitsir's predecessor in the chair of modern languages, the German-born George Blaetterman, were universally reviled. Robert Dabney once again provided insight into these attitudes. He explained to his mother that "the best professors in the institution are native Virginians, and almost all the foreigners have had some difficulty with the students arising out of their own impudence, so that I hope the Visitors will learn after awhile to be satisfied with our own men."[115] The sense not only of local pride but also of local customs and mores comes through clearly here. The foreign professors were slow to learn the ways of the Southern gentleman and of the slave-holding South, and this naturally pitted them against adolescent students acutely aware of the differences. Religious preference further separated the predominantly Presbyterian Virginians from non-Protestant others. All these social and cultural factors made Sylvester even more of an outsider at the University of Virginia than he had been at Cambridge. Combine them with the fact that his professorship put a highly sensitive, twenty-seven-year-old mathematician in a position of authority over youths barely younger than he, and the result was an environment toxic to Sylvester. Having fled it, where was he to go?

Struggling against an "Adverse Tide of Affairs"

Sylvester had scarcely been in the United States four months when he left Charlottesville and his only source of income. He headed northward to New York City, where his elder brothers, Nathaniel and Sylvester, brokered lottery tickets. From there, Sylvester not only tried to secure another academic position but also made contact with scientifically minded men in the Northeast. By September, he had made the acquaintance of perhaps the best mathematician then active in the United States, Harvard's newly appointed Perkins Professor of Astronomy and Mathematics, Benjamin Peirce.

Only five years Sylvester's senior, Peirce had been educated at Harvard College but had received his real instruction in mathematics and natural philosophy proofreading Nathaniel Bowditch's English translation with mathematical commentary of the first volume of Laplace's *Mécanique céleste*.[116] In 1829, the book came out, and Peirce graduated, but this did not end the two men's association. As Peirce taught first at a grammar school and then as a tutor and finally professor at Harvard, he continued to work with Bowditch, who painstakingly translated and filled in the many gaps in the rest of Laplace's masterpiece. In all, Bowditch brought out the first three (of five intended) volumes with Peirce providing corrections and improvements. When Bowditch died in 1838, the fourth volume incomplete, Peirce finished the manuscript and saw it through the press the next year.[117] Here was a man whose interests and talents meshed well with those of the Tripos-trained, former Professor of Natural Philosophy, Sylvester.

The two men met late in the summer of 1842. Peirce invited Sylvester to come up to Cambridge for a visit and opened his home to the Englishman. That the trip had been a joy for a Sylvester who had left Charlottesville feeling less than welcome came through clearly in a letter to Peirce on 5 September. "I never remember more happy days than those I passed lately, under your roof, in the enjoyment of the pure pleasures which spring from the feeling of mutual good will and mental adaptation," Sylvester wrote. "It was the fault of your own hospitality and kind bearing, and of the good nature and charitable indulgence of others of your household, if I appeared to feel too much at my ease with you, for so short an acquaintance and appropriated myself more than my fair share of your company and conversation."[118] Sylvester and Peirce had spent time discussing fluid mechanics from a theoretical, mathematical point of view, a subject of Sylvester's earliest papers, and Sylvester ventured that "we could together make out a beautiful system of Mechanics much more satisfactory and more agreeable to common sense than any yet attempted at least so far as regards the deduction of laws from *first principles*" (116; his emphasis). There was just one problem; Peirce was at Harvard, and Sylvester was unemployed. Sylvester's desperation as well as his sense of what might be were palpable when he wrote that

> I would make almost any sacrifice consistent with independence to have the advantage of a long course of study with you.
>
> Without envy or jealousy by joining our forces in a constant and faithful cooperation in reading and reflection we might achieve wonders: Could any situation be made for me at your University? My expectations as to pay would be very limited as I know that your University cannot not [sic] afford large salaries. My first wish and desire is intellectual progress and if my services would be worth obtaining I should not set an exorbitant price on them (117).

Consistent with the ethos of the Cambridge network back in England, Sylvester just wanted to earn a living in a way that would allow him to pursue his mathematics. Despite Peirce's best efforts, however, there proved to be no opening for Sylvester at Harvard.[119]

By February 1843, other options appeared to open up. Sylvester traveled to Washington where he hoped to learn more about the national university that had been and was being debated in connection with the Smithson bequest (118). In 1836, the United States had received an unprecedented gift from the estate of the wealthy English gentleman of science, Sir James Smithson: $500,000 "to found at Washington, under the name of the Smithsonian Institution, an Establishment for the increase and diffusion of knowledge among men."[120] Because Smithson had provided no further indication of his intentions for the money and the federal government had never before dealt with a bequest, the matter had become endlessly tied up in debate. Some then in power—like George Washington, Alexander Hamilton, James Madison, John Quincy Adams, and others before them—had advocated founding a national university. Some favored establishing a natural history museum; some a series of popular lectures; some a national observatory; some a major library. In 1843, as Sylvester learned, the decision was far from being made.[121] Washington, like Cambridge, held no real possibilities for him.

It was, in fact, in New York City that real prospects seemed to have arisen on both the personal and the professional fronts. Sylvester, thanks to Peirce's introductions, had become part of the scientific and literary circle of Wall Street banker and *bon vivant*, Samuel Ward, and this had provided him with the kind of sympathetic social outlet that had been so lacking in Charlottesville.[122] Even more important, however, as early as 28 February 1843, Sylvester had alluded in a letter to Peirce of a new and serious romantic involvement going awry. In accepting Peirce's invitation to come back to Cambridge for a visit, Sylvester acknowledged that "the pleasure I have in your social and scientific sympathy will reconcile me to a *temporary* deadening of other feelings."[123] By 19 May, that "temporary deadening" seemed permanent. A dejected Sylvester had planned to set sail for England the next day but had finally changed his mind. He still hoped that things might work out. He confided to Peirce that

> I have as yet caught no glimpse of *the* one. Missed spending some hours last night where we should have met by a narrow chance. I am in an ill run of luck. *Before*, we were constantly chancing to meet. *Now*, as constantly missing.
>
> *Believe* in *runs*, i.e. that there is a providence and destiny directing hazard; believe I mean that there is a law of good and ill fortune and one chafes less at, and becomes much more reconciled to, a *series* of disappointments. They are then felt not as an aggregate of misfortunes but as an aggregate result of a single ill turn, parallel with some acquiesced in mental or bodily infirmity. This is not *theory* but a philosophy of consolation which flashed upon me as a reality only yesternight, and already in my own case reduced to practice.[124]

"The one" was a Miss Marston of New York City, and Sylvester was smitten, yet she seemed to have grown cold. Poetry may have given him solace in his love-sick anguish, but everything seemed to be working against him.[125] "Her friends," he wrote, "are dead set against me and it is to their opposition I must attribute all that has appeared cold and ambiguous in her."[126] He felt he still had a chance, if only he could stay in the United States. "If I leave," he said, "she is lost to me forever."[127]

A fortnight later, Sylvester's chances of remaining seemed to improve dramatically. On 5 June, he learned that Columbia's Professor of Mathematics, Analytical Mechanics, and Physical Astronomy, Henry Anderson, had tendered his resignation owing to his wife's illness. Sylvester immediately wrote to another new American acquaintance, Princeton physicist Joseph Henry, to ask not only for his endorsement but also for his help in manipulating the college's trustees behind the scenes.[128]

Sylvester had been introduced to Henry by letter through their mutual friend and Sylvester's former colleague at University College, Thomas Graham, and Sylvester had taken the opportunity to visit Henry in Princeton.[129] If Peirce was the foremost mathematician in the United States in 1843, then Henry was certainly the premier physicist. His work in electromagnetism, and especially his discovery (independent of Michael Faraday) of electromagnetic induction, had firmly established his reputation internationally. Sylvester, ever the savvy applicant, fully recognized the potential force of Henry's involvement in his case.

Henry willingly obliged and wrote to three people, one of whom was a Columbia trustee, on Sylvester's behalf.[130] Still, Sylvester was worried. He had heard that rumors were being spread to the effect that he was "unable to keep order in [his] class and on that account was *compelled* to leave the University of Virginia" and that this was being used as the basis for opposition to his candidacy for the Columbia post.[131] In effect, Sylvester was tacitly urging Henry to help him set the record straight through his own private influence. For his part, Sylvester contacted his former colleagues in Charlottesville asking them to write a letter clarifying the situation. They complied on 20 June 1843 in a way that Sylvester deemed "most handsome."[132] Wanting to ensure that Sylvester's chances for the Columbia chair not "be injuriously affected by erroneous impressions as to the circumstances of his separation from" the University of Virginia, the faculty "state[d] that his separation . . . was entirely his own voluntary act and occasioned by his dissatisfaction at the course which his colleagues thought it proper to adopt towards a student whom Mr. Sylvester had reprimanded for inattention in the lecture room and whom in their view of the circumstances they were unwilling to punish to the extent Mr. Sylvester required."[133]

In the end, it was not rumors about Sylvester in his former post but his religious heritage that cost him the Columbia position. One of the trustees on the selection committee told him point blank that the election of a Jew "would be repugnant to the feelings of every member of the board" and that, moreover, "they went not at all on the grounds of [Sylvester] being a foreigner but would have acted the same, had [he] been born of Jewish parentage in" the United States.[134] In reporting this outcome to Henry on 13 July, Sylvester could not contain his emotion. "My life is now pretty well a blank," he wrote, "and my only effort is to sustain existence."[135]

Sylvester remained in New York City another four months. Rumors of a possible opening at the University of South Carolina that had circulated in May had proved unfounded. He seemed to be out of options. Sylvester had left University College and the uncongenial professorship of natural philosophy to accept the Virginia offer, hopeful for a career in academe that would allow him actively to engage in mathematical research. It might have been but was not. The University of Virginia of the early 1840s was not ready for a young, English Jew; the Sylvester of the early 1840s was not prepared to cope with many facets of the culture of the Southern gentleman. When he set sail on 20 November 1843, his early hopes were a sad, distant memory. He was struggling "against an adverse tide of affairs" with no clear prospects for the future.[136] He still sought a "position, a local habitation, and a name." (See the epigraph note.)

CHAPTER 4

Actuary by Day . . . Mathematician by Night

*I have in a singular manner and with unexpe[c]ted good fortune
recovered my footing in the world's slippery path and have by a
succession of well directed efforts and happy opportunities obtained a
position which puts me quite at my ease in respect to this world's
goods and may serve as a secure landing place whereon to breathe
and calmly to survey and determine upon my future course.*
—J. J. Sylvester to Joseph Henry, 12 April 1846

*The theorem . . . was in part suggested in the course of a
conversation with Mr Cayley (to whom I am indebted for my
restoration to the enjoyment of mathematical life).*
—J. J. Sylvester, 1851

The twelve months of 1844 were among the longest of Sylvester's life. He had gambled on the prospect of life as a mathematician in the United States, and he had lost. Back in London, he had to start over at the age of twenty-nine. He was a mathematician; he felt that bred in the bone. Yet, he had done no real mathematics for two years, and he had no prospects for a professorship. Even if he could regain his former research momentum—and that was a big if—how would he earn his way in the world? Nothing was clear. Taking rooms at 53 Lower Brook Street, two blocks south of commercially bustling Oxford Street, he set to work rebuilding his life.

Mathematics yielded first. By April, he had published his first paper since 1842, reintroducing himself to the British scientific world through the pages of the *Philosophical Magazine*. On some "Elementary Researches in the Analysis of Combinatorial Aggregation," the work dealt primarily with the question, given *n* people, for *n* even, in how many distinct ways can the people be paired.[1] In what was becoming typical Sylvesterian style, however, this question was cast in terms of a whole new vocabulary of moduli, duads, synthemes, and totals to read: "*A system to any even modulus being given, to arrange the whole of the duads in the form of synthemes*; or in other words, *to evolve a Total of duad synthemes to any given even modulus.*"[2] Sylvester's analysis largely hinged on illustrating general principles of arrangement by systematically writing down the solutions of particular cases in tabular form. It was hardly an earth-shattering result, but it served as a sort of calling card in reestablishing his mathematical contacts.

On 11 April 1844, for example, Sylvester wrote to Charles Babbage, whose acquaintance he had cultivated during his short tenure in the chair of natural philosophy at University College London and whom he had tapped as a reference when applying for the position at Virginia. "I do myself the honour of enclosing for your acceptance a copy of the first number of a series of papers on Combinatorial Construction," he began, "and avail myself of this opportunity to transcribe my Theorem relative to Sturm's Auxiliary Functions which Mr. Graves informed me some time since you were desirous to see."[3] Sylvester's recent publication had given him a graceful way both to reenter Babbage's sphere and to recall for him some of Sylvester's best work to date, namely, the research from 1839 and 1840 in elimination theory that had been prompted by Sylvester's exploration of Sturm's Theorem.

Sylvester also reestablished contact with his former teacher and colleague, Augustus De Morgan. By early August, they had been exchanging both books and mathematical ideas. In particular, Sylvester had proposed a new line of investigation about which De Morgan "ha[d] doubts" but which he nevertheless felt was "worth thinking about."[4] Ever the professor, he was doing his best to encourage Sylvester in his work, but he would not be anything but honest when asked his opinion on a mathematical question.

A month later in September, Sylvester was in attendance at the annual meeting of the British Association for the Advancement of Science in York. Presenting a brief abstract "On the Double Square Representation of Prime and Composite Numbers," he used the opportunity to reconnect with Britain's assembled scientific community, to hear of the latest in mathematical research, and, with any luck, to get ideas for his own future work.[5] Whether or not he was inspired by the meeting or simply by his perusal of the literature, Sylvester published another brief note in the *Philosophical Magazine* before the year was out, in which, improving on a result in Legendre's *Théorie des nombres*, he gave necessary and sufficient algebraic conditions for a cubic polynomial equation to have all rational roots.[6] Year's end thus brought the promise of renewed mathematical vigor. Unexpectedly and thankfully, it also brought with it the solid offer of a job.

After engaging in negotiations during the fall, Sylvester began his duties in December 1844 as actuary for the newly formed Equity and Law Life Assurance Society in London at a decent annual salary of £250.[7] By midcentury, actuaries were well on their way to joining that increasingly desirable category of "professional men" who earned both a respectable income and a respected place in middle-class British society. If the emergent profession of the professor was effectively closed to Sylvester in 1844, that of the actuary would at least allow him to make his way in society *and* to do it using his mathematical abilities. It was not a career option he had previously considered, but it *was* an option. Beggars could not be choosers.

The Actuary: An Emergent "Professional Man"

The business of insuring lives according to sound mathematical prinicples had its origins in England in the mid-eighteenth century. In 1756, James Dodson, Augustus De Morgan's great grandfather, laid out in his unpublished *First Lectures on Insurances* how to calculate premiums to ensure the financial solvency of a mutual life insurance company. Six years later in 1762, the Society for the Equitable Assurance on Lives and Survivorships, known simply as the Equitable, opened as the first British company informed by Dodson's scientific ideas.[8] By 1830, 51 insurance companies were operating successfully in Great Britain, just "enough . . . to generate a demand for competitive premiums, and hence for actuaries to calculate margins of error, but not so many that competition burst the informal social bonds" that had come to connect those engaged in actuarial calculations.[9]

In fact, the period from 1820 to 1845 had witnessed the first steps in the actuaries' admission into the ranks of the "professional men," with companies actively seeking candidates for their actuarial positions from among those with demonstrated mathematical talent and achievement. All the better if the candidate was a member of the Royal Society to boot—like Benjamin Gompertz, a Jew hired in 1824 by the new Alliance Assurance Company, of which banking magnate, Nathan Rothschild, was a President.[10] This type of man not only reflected well on a company's reputation but also fell easily into a natural brotherhood of like-minded, well-educated, highly cultivated individuals. As Timothy Alborn has put it, "these actuaries established a kind of Royal Society colony in their new commercial surroundings" and, in so doing, helped to define theirs as another of the new professions.[11]

The 1840s, however, brought trouble. The years from 1830 to 1869 saw 321 offices open and 262 close for a mortality rate greater than two in three, with competition particularly fierce between 1840 and 1850.[12] Upstart companies wanted to capitalize on the public's increasing wealth, but established companies like the Equitable challenged their practices and their ethics. The situation looked so potentially volatile that the government intervened in 1841 and then again in 1843 in the form of Select Committees on Joint-Stock Companies to study the broader problems and to propose possible reforms. The result was the

Joint-Stock Companies Act of 1844, which mandated that all new companies—whether in insurance or in any other type of joint-stock venture—deposit a list of their directors as well as their balance sheets with the Registrar's Office. Rather than serving to regulate the insurance industry better, however, this act seemingly legitimized—through the government's imprimatur—any company that complied by filing the necessary paperwork.[13] From 1845 to 1848, in fact, 48 new companies opened for business under the act's provisions, but 45 of those subsequently closed, with an average life expectancy that proved to be less than 10 years and with many collapsing much sooner.[14] The Equity and Law Life Assurance Society's first full year of business was 1845, although it was officially founded late in 1844. It had entered the fray at its most fevered pitch.

The decade and a half from 1845 to 1860 found the emerging profession in turmoil. The informal social ties that had been cultivated by the actuaries associated with the older, more established companies were challenged by the relative flood of new practitioners attached to the upstart companies, many of whom did not have the same, first-rate credentials. These newcomers also challenged the more conservative business practices of the established firms, bringing with them new ideas for how to generate business and increase profit margins. Although some of these ideas were sound, many were not.[15]

Rash speculation by mathematically uninformed company officers fueled many of the closures in the late 1840s and sparked intense criticism in the press. The *Post Magazine* and its publisher, William Pateman, in particular, waged a campaign in 1852 aimed at informing the life-insurance-buying public of the potential dangers associated with certain types of companies. As Pateman saw it, "the facility with which new assurance offices are set on foot, the tempting promises with which each new scheme is baited, . . . have . . . converted a scheme of infinite utility and philanthropy into one of the greatest speculative projects of the present day."[16] And just so the gravity of the situation was driven home to his readers, he described it in terms they could not fail to understand. "Should we be considered in possession of our senses," Pateman asked rhetorically, "were we to risk our necks by travelling on a railroad, the engineer of which . . . promised to put the engine to its extremest speed, with the comfortable assurance that, however great the pressure in the boiler might become, it should be preserved just short of an actual explosion?"[17] The stridency of the voices of Pateman and others ultimately prompted the formation in 1853 of another Select Committee, this time specifically targeting insurance companies and their practices.[18]

The actuaries, too, recognized the problems plaguing their industry, and some of them mobilized in 1848 to form the Institute of Actuaries, a body that would serve not only to unite those in the emergent profession but also to set standards of professional competence through a series of three examinations. Those who successfully passed the battery of tests would earn the title "Fellow of the Institute of Actuaries" and would be free to append "F.I.A." to their name as a testament to their qualifications. By thus requiring a level of mathematical and actuarial knowledge for entrance into the fellowship, the founders of the Institute of Actuaries aimed to police their profession and thereby to close out in the future men like those who were jeopardizing its reputation in the late 1840s.[19] This was the unsettled professional environment Sylvester entered.

J. J. Sylvester, Esq., M.A., F.R.S., Actuary and Secretary

The Equity and Law Life Assurance Society opened its doors early in December 1844 at 26 Lincoln's Inn Fields, overlooking the park of the same name and just to the west of the labyrinthine complexes of buildings that make up London's Inns of Court.[20] This location perfectly suited the new concern in light of the founders' sense of the firm's niche in London's life insurance community. As they put it in an early prospectus, "life assurance being now generally adopted in marriage settlements, charges upon life estates, and as auxiliaries to other securities, a very large proportion of them are necessarily effected through the legal profession; and experience shows that the members of that profession place the greatest confidence in offices of which they constitute exclusively the proprietary and administrative bodies. Such is the constitution of this Society."[21] Thus, the directors and shareholders of the Equity and Law all had to be members of the legal profession, whether solicitors, barristers, or judges.[22] The idea was that these directors and stockholders would—out of self-interest—draw on their law practices and legal contacts to find suitable, reliable candidates for the new firm to insure. The Society's first resident director, barrister Francis Ewart, had already been associated with the short-lived London and Westminster Mutual before promoting the Equity and Law beginning in 1843. His tenure as director, however, proved brief.[23]

In April 1845 just months after the offices opened, Ewart was forced to resign because of intemperance.[24] Although a thoroughly unhappy situation for the barrister, this had totally unforeseen and positive consequences for the newly appointed actuary. As Sylvester put it a year later in a letter to his friend Joseph Henry back in the United States, "I began with being little more than a sort of Scientific Counsellor to the Society, but events have occurred to throw the whole responsibility of the Management into my hands so that I am now by Vocation a Man of Business . . . ."[25] In effect, Sylvester was the firm's chief executive officer in addition to being its actuary, but because he was not a lawyer, he could not, according to the firm's statutes, actually serve as a director. A new title was required, and Sylvester became "Actuary and Secretary" of the Equity and Law Life Assurance Society, responsible for "conduct[ing] the whole correspondence of the Office, superintend[ing] the Books, draw[ing] up the Minutes."[26] "In a word," he told Henry, he had been "transformed into a new character and ha[d] to perform a part which I should not twelve months ago have dreamt of undertaking" (16). He had "recovered [his] footing in the world's slippery path," but he was in charge of a risky business at a particularly delicate point in time. (See first epigraph note.)

From December 1844, Sylvester went about his duties at 26 Lincoln's Inn Fields, where he also lived in back rooms provided by the firm.[27] Earlier that year, as a kind of private subcontractor to the men behind the new venture, Sylvester had actually already performed the first task necessary in setting up such a business, namely, the calculation of figures based on available mortality data for the basic premium rates the company would charge for men and women of various ages.[28] This step was critical for the firm's solvency. Set the premiums too high, and potential clients would go elsewhere, but set them too low, and there might ultimately be insufficient funds to pay future claims. The firm's directors had apparently been impressed by Sylvester's actuarial piecework for them, because they had named him their actuary by October.[29]

The actual day-to-day routine at the Equity and Law was, however, more clerical and less mathematical. As each client was lined up, Sylvester tailored a premium rate based on the client's age, health, and general conditions of life. Although this required some mathematical calculations, the rest of the process involved drawing up the actual policy, collecting the premium, and turning over all the numbers to the company's bookkeeper.[30] As the chief executive officer, moreover, Sylvester worried about investing the company's assets in consultation with its directors.

In 1845, the first year of operation, the Equity and Law issued 117 policies valued at £106,910 that yielded just over £3,145 in premiums.[31] The business was going well, and Sylvester was clearly busy, but not busy enough to neglect his broader interests in the British scientific community and his place in it. In June, he again attended the annual meeting of the British Association for the Advancement of Science, this time held in Cambridge, and although he did not present a paper, he apparently did make at least one significant new contact.

In the summer of 1845, the recent Cambridge graduate and 1845's second wrangler, William Thomson (later Lord Kelvin), was just in the process of taking over from Robert Ellis as editor of the *Cambridge Mathematical Journal* and was transforming it, on the counsel of several influential advisers, into the *Cambridge and Dublin Mathematical Journal* in an effort to widen its purview.[32] By November, the thirty-one-year-old Sylvester was writing to the twenty-one-year-old Thomson to "hail with much pleasure the conjunction of Cambridge and Dublin Mathematics under [Thomson's] able guidance"[33] and to let him know that he had some material that Thomson might find suitable for publication. It involved ideas on the theory of combinatorial aggregation related to those he had published in the *Philosophical Magazine* in 1844, and Sylvester asked Thomson if this topic seemed suitable for the journal. "If so I think I can supply you with a series of short papers upon it for many months to come," Sylvester exuberantly announced. "The ideas are entirely new and will I believe be found at some future day to admit of important theoretical & also practical applications." Sylvester, himself an interesting avatar of "the conjunction of Cambridge and Dublin Mathematics," saw in the new journal the potential not only for encouraging mathematical research in Great Britain but also for his own establishment as a British mathematical force-to-be-reckoned-with.

Thomson greeted Sylvester's overtures positively. He liked the idea of "a continuous series" of papers "on a subject of original investigation" and shared Sylvester's sense that "the principal object of any scientific journal *should* be the publication of original investigations and discoveries."[34] He hoped, moreover, "that both *commercially* and mathematically, the Cambridge and Dublin Journal may be sufficiently prosperous to allow such a course to be followed." Like the British actuarial profession, the British mathematical research community was in a nascent state. Commercial viability was a concern, but so were standards. It was unclear in 1845 whether Thomson would be able to strike the kind of balance needed to make the journal both a financial and an intellectual success, but he and contributors like Sylvester were ready to take on the challenge.

Sylvester must have underestimated the amount of time he would have for the pursuit of his mathematical researches, however, for no papers on combinatorial or on any other mathematical matters were immediately forthcoming from him in the *Cambridge and Dublin* or in any other mathematical journal. The Equity and Law, however, continued to do a respectable business in 1846, issuing 145 policies valued at £136,685 and generating over £3,984 in premiums, so Sylvester had to keep on top of a lot of paperwork.[35] As he confided on 12 April 1846 in response to a letter recently received from Joseph Henry, "although I cannot help feeling that my mind is being frittered away on inferior objects   on the whole I think the habits of business will prove very advantageous in giving order & method—and I find much to admire in the system of arrangement, traditionally preserved in the transactions of all commercial matters and the mode of keeping books of an office."[36] "In fact," he added, "I find a gratification of an intellectual sort, in applying my mind to improving the details of our system. Where is there *not* science—i.e. truth & beauty if we will not only recognize the spirit through the forms?"

Sylvester was putting on an optimistic face for his old American friend, in describing what he nevertheless termed "my present (I trust temporary) profession" (16–17). The job was not giving him time for his mathematical research, but he did have "a pupil preparing for Cambridge living with [him]," was "quite independent in circumstances and in the possession of a very respectable not to say ample income," and was "able to indulge in the luxury of being liberal when inclination prompts or circumstances call for it" (17). Henry, though, had also asked Sylvester how he had been faring in the aftermath of his failed relationship with Miss Marston, and his reply to that was more wistful than optimistic. "As regards the certain person you name," Sylvester wrote, "I never allow it now to disturb my thoughts. Happy shall I be—at least I think so—when it shall please Heaven to provide me with a suitable partner—at least I know I am unhappy enough for want of one—unless when so fully occupied that the whole energies of the soul are turned outwards" (17).

At least Sylvester was "fully occupied" in 1846. In addition to running the firm, on 29 July, he entered the Inner Temple to prepare himself for the Bar.[37] One of the four Inns of Court, the Inner Temple was located on King's Bench Walk, a brisk ten-minute stroll from the Equity and Law's offices on Lincoln's Inn Fields. As a Cambridge high wrangler and the holder of a B.A. and an M.A. from Trinity College Dublin (TCD), Sylvester had little to do to qualify for entry. Applicants had to produce three testimonials: a certification of classical attainments in Latin, Greek, history, and general literature based on a highly informal examination given by a barrister of the Inner Temple who had been selected by the benchers for that task; a testimonial signed by the candidate and a bencher or by two barristers attesting that the candidate was "a gentleman of character and respectability, and a fit and proper person to be admitted a member of the Inner Temple"; and a statement of intent to enter that included the applicant's name, age, address, and "condition in life."[38] Because Sylvester held degrees from TCD, he was officially exempt from the assessment of classical attainment, which was, at any rate, perfunctory at best. The only problem was that the exemption hinged on the presentation of a certificate attesting to the fact that the degree had indeed been taken. During the course of his transatlantic moves, Sylvester had apparently lost the documentation he had received on earning his Dublin degrees, and this had necessitated a letter to his friend, Charles Graves, late in 1845 to try to secure another as he readied himself for his legal studies. In his reply of 26 November, Graves reported apologetically that the only way Sylvester could obtain another certificate was to incur "the heavy expense of £11.10.0." "I unfortunately," he added, "though perfectly cognizant of the fact [of Sylvester's TCD degrees] am forbidden by College rules to give a certificate of it."[39] Whether Graves's letter sufficed or whether Sylvester actually paid for the new certificate from TCD, the Inner Temple acknowledged Sylvester's prior academic work. At least the other two testimonials were less trouble. Gilbert Henderson, a barrister of Lincoln's Inn, and Edward Rushton, eldest son of a Liverpool bookseller and a barrister of the Inner Temple, attested formally to Sylvester's character, and the other piece of paperwork was nothing more than a brief form letter in which Sylvester filled out his contact information.[40]

The remaining requirements were financial. A candidate like Sylvester who had taken degrees at TCD only had to be of three and not five years' standing to become eligible for his call to the Bar; did not have to pay the deposit of

£100, which would have, at any rate, been returned without interest when he was called; but did have to pay just over £29 for admission and £7 per year for entrance into commons. The latter was especially important, for, to be called to the Bar, a candidate had to dine in hall a specified number of times each term for twelve terms or a full three years.[41] For the Inner Temple that meant "two days in each of two separate full weeks of the term," where "a bottle of port is allowed to each mess of four, and a comfortable and substantial dinner is provided." As this dining requirement makes clear, the mid-nineteenth-century English Bar continued to place much stock in sociability and in the kind of conversational—as well as rhetorical—expertise that could be further honed over a dinner table of adepts. It was also important for future barristers to interact with each other and with the members of the society, since, in the final analysis, it would be precisely those members who would pass judgment regarding the actual call to the Bar and with whom one would be working afterward. Sylvester would have had no trouble finishing up his paperwork at the close of the Equity and Law's business day at 4:00, donning the requisite black, sleeveless and hoodless gown, and making the last call to dinner at the Inner Temple by 5:30 four evenings a term.

Interestingly, the dining arrangements were much more highly regulated than the actual studies for the Bar. The lack of educational standards within the Inns of Court had, in fact, prompted the formation of a Select Committee on Legal Education in 1846. According to that Committee's report, "the Inns of Court provided no legal education 'worthy of the name,' and barristers were given no test of legal competency."[42] In 1854, a Queen's Counsel and member of the Irish Parliament quipped that to qualify for the Bar in England all that was required was "that a man should eat and drink and be able to write his name."[43] Normally, a student would "enter the chambers of special pleaders, equity draughtsmen, or conveyancers, where they [would] acquire experience in the actual practice of the law,"[44] but not even that was mandatory.[45] In Sylvester's case, with the responsibility of running the Equity and Law's offices, there would have been little or no time during the day for participating actively in this sort of apprenticeship, anyway, although, had he had the inclination to study, he would have had full access to the library of the Inner Temple to read up on the law in his off hours.[46] Moreover, in response to the Select Committee's recommendations, by 1848, students at the Inner Temple had to attend a required course of lectures on the principles of common law, although they were not tested on the material it contained.

In studying for the Bar, Sylvester was thus engaged in a balancing act of a different sort from the one that kept his company's books in order on a daily basis, but he was motivated to pull it off. Without a legal credential, he would never be able to assume a position as a stockholding director of the Equity and Law, and since that was precisely where his fortunes seemed to lay in 1846, he needed to do whatever it took to better his position.[47]

Still, he also wanted to leave his options open relative to returning to the life—as he conceived it should be—of a research mathematician. If a congenial teaching position should open up, he wanted to be ready, and, for him, that meant maintaining his mathematical contacts and, even more importantly, pursuing his mathematical researches. Both objectives were difficult, given the demands on his time. For example, after failing completely to deliver on his boast to Thomson in the fall of 1845 to shower the *Cambridge and Dublin* with combinatorial material, Sylvester had been reduced by the summer of 1846 to keeping in touch with him by soliciting his advice on a tutor for the son of a new acquaintance. Thomson had written Sylvester, perhaps inquiring about the promised series of papers but also telling him that he had been in town, and Sylvester had responded with the "hope that it may not be long before you pass thro' town again & that you will in that event favor me with a call."[48] He was silent on the matter of forthcoming submissions.

In 1847, when Sylvester was finally able to get back to a bit of research, however, he sent three short papers not to the *Cambridge and Dublin* but to the *Philosophical Magazine*.[49] Containing statements of results—as opposed to *proofs* of them—on conditions under which equations of the form $Ax^3 + By^3 + Cz^3 = Dxyz$ have no rational solutions, and concerning pure as opposed to the applied mathematics Thomson increasingly seemed to foster, these papers may have struck Sylvester as somewhat other than what Thomson wanted for his journal. He certainly did not want to risk having them rejected.[50] In fact, even he had to admit in the first paper that "pressing avocations" had kept him "from entering into further developments or simplifications at this present time" and to justify why he was nevertheless "putting forward these discoveries in so imperfect a shape."[51]

As Sylvester explained, he had managed to carve out enough time from his duties at the Equity and Law and from his legal studies for "a rapid tour on the continent" that had taken him to Paris. There, in conversation with his "illustrious friend M. Sturm," the new ideas had struck him, he had explained them to Sturm, and Sturm had immediately "ma[d]e them known . . . to the Institute" on Sylvester's behalf. The only problem was that, in his haste, Sylvester had made a mistake and so was "very anxious to seize the earliest opportunity of correcting" his error in print, hence the publication in the *Philosophical Magazine*. Yet, more seemed to be at stake than merely correcting an error. "I venture to flatter myself," he closed the first of the three papers, "that as opening out a new field in connexion with Fermat's renowned Last Theorem, and as breaking ground in the solution of equations of the third degree, these results will be generally allowed to constitute an important and substantial accession to our knowledge of the Theory of Numbers."[52] What better way to secure one's mathematical reputation—and internationally at that—than to make headway on what was then a more than two-hundred-year-old unsolved problem? What better way to position oneself to make a move out of business and back into academe—should the occasion arise—than to have solid results in a vein like this to one's credit?

Number theory continued to hold Sylvester's mathematical attention throughout 1847. Late November, in fact, found him once again trolling for ideas in Legendre's massive *Théorie des nombres* as he carried on business at the Equity and Law and fulfilled his obligations at the Inner Temple. It also found him sharing his mathematical interests with another lawyer-in-training, Arthur Cayley.

The twenty-six-year-old Cayley, almost seven years Sylvester's junior, had been senior wrangler on the 1842 Cambridge Tripos and had taken a fellowship at his college, Trinity, before entering Lincoln's Inn in 1846 to study law. Located on the east side of Lincoln's Inn Fields, this Inn of Court was literally yards from Sylvester's home and office at the Equity and Law. Just how or when Sylvester and Cayley met remains unclear,[53] but on 24 November 1847, 1837's touchy and outgoing second wrangler was writing to 1842's reserved and retiring senior wrangler to tell him of an interesting theory he had just been reading about in the second volume of the *Théorie des nombres*. From his desk at 26 Lincoln's Inn Fields, Sylvester described two of Legendre's examples that, he thought, might be "very congenial to [Cayley's] present line of thought," and he made Cayley "a cadeau of the subject not doubting that it will turn to good account in your able hands."[54] The two budding barristers—temperamentally so different—had already been discussing their mathematical interests, and although Cayley was still "My dear Sir," Sylvester knew that he had found a kindred spirit with whom he wanted to be friends.

For the time being, though, mathematics had to take a back seat to business. It had been another profitable year for the Equity and Law, with 1847 bringing in another 145 new policies valued at £129,988 and generating just over £3,752 in premiums.[55] The firm was still small, but it seemed well on its way to success, and this was due in no small measure to Sylvester's actuarial recommendations as well as to his business efforts. As the directors put it in their report at the first general meeting of the Society held in February 1848, "feeling that the true interest of the Society would be best promoted by a cautious selection of the risks undertaken," they "have constantly declined all which appeared to them doubtful or speculative, preferring, by the exercise of caution, to give the Society, in its infancy, a solid foundation, rather than to meet the shareholders with a large amount of business which might in a short time entail heavy losses" (3:26). Theirs was a fiscally conservative company based on their own good judgment and Sylvester's sound actuarial principles. They were one of the "good" new life insurance companies, not one of the rash and overly speculative fly-by-night firms that were increasingly cause for alarm. Perhaps the time had come for actuaries at the "good" companies to make common cause in an effort to regulate and police their growing industry.

The Institute of Actuaries

On 15 April 1848, some 28 men, both actuaries and managers of life insurance companies, gathered in London at the offices of the Standard Life Assurance to discuss the merits of forming some sort of association among them. Opinion was sharply divided. Representatives of some of the oldest firms were vocal in their desire to maintain merely the kind of informal social ties that had developed particularly in the 1820s and 1830s, whereas representatives of some of the solid, newer offices urged the formation of an institute that would set professional standards through meetings, teaching, and a system of examinations. With no consensus reached, the meeting adjourned, but not before a committee of 10 was chosen to consider the matter further and to report back at some later date.[56]

Although it is not clear whether Sylvester was present at this first meeting, he was one of the 66 actuaries in attendance at the follow-up meeting held at the offices of the Guardian two months later on Saturday, 10 June. The same Thomas Galloway who had supported Sylvester's candidacy for a fellowship in the Royal Society in 1839 was in the chair and had been one of the committee of 10 whose report was under consideration. It was, by one contemporary account, a "disorderly" meeting, with much heated discussion of the document before them (12). As they put it in their report, the committee "after mature deliberation" found "that it is not expedient at the present time to lay down rules for the formation of a Society of any description, but . . . they are of the opinion that occasional meetings . . . may be held with advantage, if they should from experience be found to establish uniformity in dealing with certain points which are of constant occurrence in Assurance Offices but in respect of which there is much diversity of practice" (11). Essentially, the committee's report called for no real action—but merely for the maintenance of the *status quo* of informal social meetings—and those assembled flatly rejected it. It was decided to form a new committee; a ballot was drawn up; a vote was taken. Sylvester was one of the 15 newly elected.

The new committee began deliberations the following Saturday and promptly and unanimously decided to form a subcommittee of five "to prepare a draft scheme with rules and regulations and the reasons for the necessity of the same" (16). Sylvester, together with Peter Hardy of the Mutual, Charles Jellicoe of the Eagle and Protector, Jenkin Jones of the National Mercantile, and Francis Neison of the Medical and Invalid, constituted this subcommittee, which immediately set about its work. Ten days and four subcommittee meetings later, the full committee met to consider the subcommittee's written, thirty-five clause recommendation for the formation of an Institute of Actuaries and, after some minor tweaking, resolved to have the report printed and distributed for discussion and a vote at a general meeting to be held on 8 July.

More than 50 people attended that next meeting. After some preliminary verbal tussling, it was moved to consider the report clause by clause. Only minor changes resulted, and the report was passed, making 8 July 1848 the official birth date of the Institute of Actuaries and Sylvester one of its midwives.

The Institute, as conceived by the subcommittee of five, had the most self-conscious of professional goals. It was to serve as the focal point for the improvement and diffusion of mathematical theories and data collection techniques connected with the practice of life insurance. It was to work for a uniformity of practice across the field and in all the country's offices. It was to confer status on members of the profession not only by recognizing their achievements through prizes and awards but also by administering a regular battery of examinations to certify competence in the field. It was to maintain a library and rooms for reading and conversation, in addition to providing a venue for the discussion and settlement of disagreements "whether theoretical, official, or professional."[57] In short, for Sylvester and his fellow committee members, the Institute of Actuaries would provide the industry with a means to keep its own house in order.

To get the new institute off the ground, the same committee of 15 that had proposed the successful resolution was elected to constitute the Enrolment Committee. By statute, anyone who was an "Actuary, Assistant Actuary, or person performing the office of Actuary or Assistant Actuary to the Government or to any Life Assurance, Annuity, or Reversionary Interest Society in Great Britain or Ireland, in existence or completely registered on or before the 8th day of July, 1848" had a right to petition the Institute for a fellowship, provided they did so before 1 September 1848 (20). Their petition would then be reviewed by the Enrolment Committee; a vote of four-fifths or more of the committee would result in the petition's acceptance and the candidate's admission to the fellowship. The same rules were in place for anyone who was a "Manager, Managing or Resident Director, Secretary, or other chief officer of any Life Assurance Society," except their title would be "associate" rather than "fellow" of the Institute (20). Prior to the 1 September deadline for charter fellowship and associateship, the Enrolment Committee met six times to decide on applications and to plot strategy for getting the word out more effectively to the nation's actuaries.[58] Once the deadline had come and gone, the committee continued to meet regularly to consider new applications and to collect the annual fees for membership: £3.3 for fellows and £2.2 for associates. When the newly formed Institute of Actuaries held its first general meeting on Saturday, 14 October 1848, the committee was able to report a membership of 94 fellows—Sylvester among them—and 34 associates with dues collected to that point of over £287.

Also at that first meeting, John Finlaison, the Government Actuary, was named President by acclamation, and ballots were drawn up for the election of the sixteen-person council to consist of four vice presidents, a treasurer, two secretaries, a registrar, and nine additional members. When the balloting was completed, Hardy, Jellicoe, and Sylvester from the subcommittee of five and David Jones of the Universal had all been elected vice presidents (28, 41). Clearly, Sylvester's judgment and his efforts on behalf of the fledgling professional society had been appreciated and respected by his peers.

Sylvester worked hard for the new Institute in its early months, serving diligently on the Council as it dealt with recruitment, the drafting of bylaws, the matter of securing rooms and setting a schedule of meetings, the creation of a library and a journal, and the establishment of the educational program as well as the system of examinations, among other issues (41). Relative to the latter task, in April 1849, Sylvester and three other Institute members were appointed examiners and charged with "prepar[ing] a Syllabus for the approval of the Council" for the competency examinations that the Institute proposed to initiate (216). In July, their recommendations and the procedures they had laid out for giving the examinations had been approved. There would be three examinations: the first, a mathematical examination in "Arithmetic and Algebra, the elementary doctrines of Probability, Simple and Compound Interest, and in the theories of Life Assurance and Annuities"; the second on vital statistics, that is, on the elementary principles of the construction and arrangement of tables of mortality; and the third on bookkeeping and on "such general questions on Official routine and detail as may from time to time appear necessary to the Examiners" (217). The first examinations were to be held in the late spring of 1850. The Institute of Actuaries was well on its way to achieving the professional goals in which Sylvester and his fellow organizers so firmly believed.

Divided Loyalties: Mathematician or Actuary?

The year and a half from the first organizational initiatives aimed at an Institute of Actuaries to the end of 1849 had found Sylvester with his loyalties seriously divided. He was heavily involved in the professionalization efforts of London's actuaries. He was keeping commons at the Inner Temple and otherwise working to qualify himself for the Bar. He was doing his job at an Equity and Law that had expanded to open a branch in Manchester.[59] He was trying to sustain some sort of mathematical research agenda. There simply were not enough hours in the day. Something was going to have to give.

Sylvester seemed resolved that it was *not* going to be mathematics. Although he published no papers in 1848 or 1849, he continued to talk mathematics with Cayley, whom he now addressed with the familiar "Dear Cayley." Their friendship had grown as they had shared mathematical ideas and discussed the mathematical literature. In April 1849, Sylvester was clearly torn as he tackled the project of drawing up the syllabi for the examinations of the Institute of Actuaries, while trying to keep up with his own mathematical interests. Writing to Cayley about some of his new ideas on differential operators on 18 April, he also had to admit that "I have not been able to do more than dip into Plücker but I shall be exceedingly obliged if you can leave the books with me for some time longer without inconvenience to yourself."[60] He would find the time to see what gems lay in the work of the German geometer; he just did not know quite when.

With the syllabi drawn up and the procedures for the Institute's examinations ratified, however, the Equity and Law demanded his attention. Not only did he have the usual day-to-day paperwork to take care of, but he also had to do some painstaking actuarial work in preparation for the quinquennial report that was due at the end of 1849 and that he was responsible for writing. Would there be profits to distribute to the stock- and policyholders? If so, how much, and who would get what? With a total of 542 policies in force valued at £515,254, it was going to be a big job for one person, and it would have to be done carefully. When the report was prepared, however, the news was good. There was a £7,272 surplus for distribution, but Sylvester had to handle that process, too.[61]

Still, as he crunched the numbers and ground out the report, he was thinking deeply about mathematics. On 28 November 1849, he followed up on a letter he had written to Cayley the day before on some new ideas on determinants as applied to geometry. "I think that I can put the substance of the latter part of my yesterday's communication in a much clearer point of view than I then succeeded in doing," he announced.[62] Page after closely written page followed, in which he laid out his thoughts to Cayley as to the conclusions he could draw from "our method." "We now see . . . ," he noted. "Now to come back to the 6×6 determinant obtained . . . ," he continued. "The above remarks . . . do not perhaps give altogether a correct view," he admitted, "but the Result is I believe substantially correct." After more and more statements, partial statements, and calculations, he concluded—apparently exhausted, judging from his virtually illegible handwriting—by asking Cayley to "excuse my troubling you with these details arising from my [illeg] to have the aid of your judgement and criticism of this [illeg] [illeg] and the advantage of your [illeg] detection of the weak or fallacious points of the argument." Late in November 1849, Sylvester was already using Cayley as a mathematical sounding board, a role that he would subsequently play so often in their long mathematical association.

Sylvester was a natural extrovert as a mathematician; he thrived on mathematical interchange. In isolation, as he had been from 1841 when he left for the United States until 1849 when his friendship with Cayley had blossomed, he languished, unable to focus, unable to sustain any kind of research momentum. With an appreciative audience, however, he rose to the occasion with seemingly innate ease. As his surviving letters from the end of 1849 reveal, he had found that audience in the talented, taciturn, and introverted Arthur Cayley.

If Cayley was becoming Sylvester's key mathematical inspiration by the end of the 1840s, another correspondent from this period would ultimately play a critical role in Sylvester's broader professional life. Sylvester had met Lord Henry Brougham at least as early as 1837 when, present in Liverpool for the opening of the Liverpool Mechanics' Institute, Sylvester had given him a copy of his 1836 pamphlet, "A Supplement to Newton's First Section."[63] Brougham, one of the founders of University College London, an advocate of the movement for workingmen's education, and an active Whig politician, had long-standing interests in mathematics and physics as well as in the removal of all remaining disabilities for dissenters. Sylvester, a Jew, the former Professor of Natural Philosophy at University College, and increasingly a mathematician of repute, was thus a correspondent who embodied Brougham's intellectual as well as his political concerns. On 29 December 1849, Sylvester responded to an earlier letter from Brougham, in which the physical theories of the eighteenth-century polymath, Roger Boscovich, had apparently been under discussion.[64] Sylvester had been flattered and honored to have been approached by Brougham on such topics and thanked him for "the rare privilege of giving the expression of my convictions to one so well qualified to comprehend general views and to discern between the true and false." In the years to come, this acquaintanceship, borne of common interests in science, proved crucial in behind-the-scenes politicking aimed at overcoming the barriers that Sylvester's Jewish faith placed in his path (see "Establishing a Mathematical Routine" and chapters 5–7).

At the beginning of 1850, however, these concerns were still far from Sylvester's mind. He was catching fire mathematically with Cayley fanning the flames. He may have been an actuary by day, but was he inexorably becoming a mathematician by night? The tension between the pragmatic demands of business and the siren call of mathematics were strongly in evidence.

As Sylvester put the finishing touches on the quinquennial report of the Equity and Law and prepared for the meeting of the directors and shareholders at which it would be presented in March 1850, he tried to snatch some time for himself to keep up with the mathematical literature. On 8 January, he reluctantly returned to Cayley some borrowed numbers of Crelle's *Journal für die reine und angewandte Mathematik*. He had not had the chance to do more than glance through them, but he did see that one carried a recent paper by his friend. "I shall be very glad," he told Cayley, to borrow that number again "in order to study your paper which I have not yet had an opportunity of doing."[65] The paper, most likely Cayley's treatment "Sur le problème des contacts," cast a geometrical problem about intersecting conics in determinant-theoretic terms.[66] By early February, Sylvester was also hard at work on determinant-theoretic ideas that were growing from his algebraic geometric discussions with Cayley.[67]

As he pushed through his mathematical ideas that spring, Sylvester distanced himself from all professional activities not directly linked to his duties at the Equity and Law. In particular, he withdrew from his leadership role in the

Institute of Actuaries, attending only the March meeting of its Council and opting finally not to participate in June as an examiner in the first round of qualifying examinations for aspirants to the credential of F.I.A. that he had been so instrumental in shaping.[68] These commitments simply took too much of his spare time, spare time he now chose to devote as exclusively as possible to mathematics.

Sylvester's decision essentially to leave the further development of the Institute of Actuaries to others within the profession soon began to have important mathematical consequences for him. For one thing, it gave him the opportunity to reengage with the broader mathematical community as defined by the Royal Society. Although he had been a fellow since 1839, his sojourn in the United States followed by the demands of his position at the Equity and Law had prevented him from participating with any regularity in the Society's affairs. In May 1850, he reengaged in its social side, attending the Society's *soirée* on the evening of Saturday, 18 May, and hobnobbing with the numerous attendees who included Prince Albert, Lord Brougham, Charles Babbage, and Charles Wheatstone.[69] By July, however, Sylvester's engagement was more purely professional when he lent his expert opinion as a referee to the editors of the *Philosophical Transactions.*

The Royal Society, founded in 1662, had only instituted the policy of formally and blindly refereeing papers in 1832 in the wake of a series of reforms aimed at improving British scientific standards, and, 18 years later, its conception of what was to be deemed publishable was still in flux.[70] Sylvester's assessment of "On the Solution of Linear Difference Equations" by the Rev. Brice Bronwin reflected this unsettled state.

Bronwin's paper concerned what had developed into a kind of British, mathematical cottage industry since the 1840s, namely, the calculus of operations. In the wake of mathematical reform efforts by Cambridge mathematicians like Charles Babbage, George Peacock, John Herschel, and others in the 1810s, the Newtonian notation and approach to the calculus had largely been superceded in the Cambridge curriculum by Continental techniques that had drawn from the Leibnizian tradition. Young students of mathematics exposed to this, for them, new way of treating the calculus began to explore—from an algebraic point of view—the behavior of the notation. They observed, for example, the similarities between the law of exponents and the behavior of the composition of functions. Thus, on the one hand, $x^m \cdot x^n = x^{m+n}$, and on the other,

$$\frac{d^m}{dx^m}\left(\frac{d^n u}{dx^n}\right) = \frac{d^{m+n}}{dx^{m+n}}u$$

as well as $f^m(f^n(x)) = f^{m+n}(x)$. How then, they asked, can symbols of operation be manipulated and to what extent can they be manipulated independently of the symbols on which they operate? The answers to these and related questions formed the algebraic calculus of operations and engaged such aspiring mathematicians as the Trinity College, Cambridge students, Duncan Gregory and Robert Ellis, who founded in 1837 the *Cambridge Mathematical Journal* for the encouragement of young researchers like themselves. In fact, the *Cambridge Mathematical Journal* quickly became the principal publication outlet for research in this new area.[71]

Bronwin, too, had published some of his earlier work in this field there, but in 1851, he had decided to seek acceptance for his research from his country's oldest scientific journal, the *Philosophical Transactions*.[72] In Sylvester, he found a referee lukewarm at best. Sylvester stated that "whatever merit the paper possesses must be sought for in" the formulas it lays out since "the examples to which they are applied possess no intrinsic interest and are devoid of any pretension to simplicity or probable direct utility."[73] "On the whole," he concluded, "I am of the opinion that the Memoir in question without possessing a high degree of merit might be inserted in the Philosophical Transactions without discredit to the character of the Society." It was hardly a rousing endorsement, and it did not reflect the high standards Sylvester had seemed to embrace in writing to Thomson about the *Cambridge and Dublin* in 1845. Apparently, even this tepid assessment was enough for the Royal Society, though. Bronwin's paper appeared in the *Philosophical Transactions* in 1851.[74]

Sylvester's critique of Bronwin's mathematical efforts came on the heels of his own reentry into print. By early in the summer of 1850, he had produced a substantial piece of work, his first paper in three years, "On the Intersections, Contacts, and Other Correlations of Two Conics Expressed by Indeterminate Coordinates," and he offered it to Thomson for publication in the *Cambridge and Dublin*.[75] After having boasted five years earlier about possessing an abundance of quality material, Sylvester had at long last produced something he felt worthy of the journal and its editor's ideals.

In the paper as well as in his correspondence with Cayley through the fall of 1850, Sylvester considered two conics U, V in three variables x, y, z namely, $U = ax^2 + by^2 + cz^2 + 2a'yz + 2b'xz + 2c'xy = 0$ and $V = \alpha x^2 + \beta y^2 + \gamma z^2 + 2\alpha'yz + 2\beta'xz + 2\gamma'xy = 0$ and asked under what algebraic conditions do these two conics intersect? His solution involved an analysis of the roots of the equation he denoted $\square(U + \lambda V) = 0$, where $\square(U + \lambda V)$ is the determinant

$$
\begin{vmatrix}
a + \alpha\lambda & c' + \gamma'\lambda & b' + \beta'\lambda \\
c' + \gamma'\lambda & b + \beta\lambda & a' + \alpha'\lambda \\
b' + \beta'\lambda & a' + \alpha'\lambda & c + \gamma\lambda
\end{vmatrix}
$$

that he would later dub the discriminant.[76]

In effecting his results, Sylvester drew from a number of key sources. Not surprisingly, given their growing mathematical and personal friendship, he drew from Cayley, corresponding with him regularly and in detail as his ideas solidified.[77] He also responded to Cayley's printed pronouncements, "adopting," for example, "Mr Cayley's excellent designation" of "quadrangle" to refer to "the four points of intersection of the two conics" under consideration.[78] Cayley had, in fact, introduced this term in the very paper Sylvester had noted—but had been unable to read and study—when he returned Cayley's copies of Crelle's *Journal* early in January. With freedom for mathematical pursuits in the spring, Sylvester had apparently made up for lost time and finally given Cayley's paper its due.

Another study with which Sylvester had familiarized himself in pursuing his researches was George Boole's 1841 "Exposition of a General Theory of Linear Transformations."[79] Boole, one year Sylvester's junior, was largely self-taught in mathematics and had only the year before "introduced himself" to the Cambridge mathematical world through a paper published in the *Cambridge Mathematical Journal*.[80] In his "Exposition" of 1841, Boole had analyzed, among other things, certain expressions in the coefficients of a homogeneous polynomial in two variables that remain unchanged under the linear transformation of the polynomial. To take the simplest example, he noted that for a binary quadratic form $Q = ax^2 + 2bxy + cy^2$ and a linear transformation T of determinant $r \neq 0$ that takes Q to $R = Ax^2 + 2Bxy + Cz^2$, the expression $b^2 - ac$ in the coefficients of Q—that he obtained through a process involving partial differentiation—remains unchanged up to a power of r under the action of T on Q. To put it symbolically, he showed that $B^2 - AC = r^2(b^2 - ac)$. Although Sylvester utilized more general aspects of Boole's work on linear transformations in his 1850 paper, this phenomenon of immutability—that Boole had uncovered in several settings in his paper and that Sylvester would only later term invariance—would soon come to dominate Sylvester's mathematical thought.[81]

Sylvester had, indeed, made a solid showing on the pages of the *Cambridge and Dublin*, but even he had to admit that "in this research I have only partially succeeded."[82] Still, a partial success is a success, and, for someone like Sylvester trying to get back into mathematical research, it is a good kind of success because it immediately suggests further lines of inquiry. Sylvester spent the rest of 1850 exploring the ramifications of this work and writing up his results. In all, he published seven papers that year, his most prolific to date,[83] with Cayley providing encouragement and critique through both their correspondence and their frequent evenings of "mathematical conversation."[84]

The year also ended with another of his distracting occupations behind him. On the evening of 22 November 1850, Sylvester was called to the Bar at the Inner Temple. His near contemporary, William Ballantine, could have been writing for Sylvester when he described the end of his own legal studies. "At last," he exclaimed, "the labors which led to the bar . . . were over. I had eaten the requisite number of dinners." In the hall that night, "the batch to be 'turned off' were summoned to the bench table. We were each presented with a glass of wine, and a speech was made to us by the treasurer, giving us good advice and wishing us prosperity in our forthcoming career; and so we were launched upon the sea."[85] Of course, it was a sea that Sylvester had no real intention of sailing, but if he really was bound to stay an actuary, then at least this new credential qualified him to become a director of the Equity and Law—should such an opportunity arise—and thereby secure even further his station in society.

The law was thus far from Sylvester's mind in the months following his call to the Bar. December 1850 found him finishing up the Equity and Law's annual financial report and pushing the algebraic geometrical research he had begun in the spring as he worked to gain some sense of the similarly spirited work being done on the Continent.[86] Apparently, he and Cayley had been discussing the findings of the Königsberger mathematician, Otto Hesse, on the problem of transforming a homogeneous cubic f in three variables x_1, x_2, x_3 into the canonical form $a^3 x_1^3 + b^3 x_2^3 + c^3 x_3^3 + 6abcd x_1 x_2 x_3$, for a, b, c, $d \in \mathbb{R}$. On 26 December, Sylvester wrote to Cayley to recount his own method for effecting this transformation, which used the discriminant-theoretic techniques Boole had developed in his "Exposition" of 1841, and to inquire "whether it is identical with Hesse's solution."[87]

In an 1844 paper, Hesse had considered the more general case of a homogeneous function f of degree m in n variables $x_1, x_2, \ldots, x_n$ and had introduced the $n \times n$ determinant $|(\partial^2 f)/(\partial x_i \partial x_j)|$, which Sylvester would later dub the Hessian.[88] In the course of his investigation, Hesse had shown that if T is a linear transformation such that $\det T = r$, if ϕ is the Hessian of f, and if ϕ' is the Hessian of $T(f)$, then $\phi' = r^2 \phi$.[89] This was another example of the phenomenon that Sylvester later termed invariance, except that the Hessian was an expression in the coefficients *and* the variables of f that remained unchanged under linear transformation. By 1850, and under the watchful eye of a Cayley to whom he was "indebted for [his] restoration to the enjoyment of mathematical life," Sylvester was coming into contact with many of the ideas that would determine his future mathematical course (see second epigraph note).

By year's end, he had also seemingly answered for himself the question that had implicitly been guiding his actions. Mathematics was his intellectual priority. That was where his true interests lay. He would have to continue—at least for the present—at the Equity and Law to have a roof over his head and food on his table, but it was as a mathematician and not as an actuary that he would make his name in the world. Neither the law nor the Institute of Actuaries need occupy any more of his time. He had research to do, even if he also had policies to write and rates to determine. Two out of four.

Establishing a Mathematical Routine

As the year 1851 opened, it was even clearer that Sylvester had actually resolved to try to extricate himself from the business world altogether and support himself in some way that would allow even more time for his mathematics. In this quest, he sought the intervention of his "Brother Geometrician," Lord Brougham, asking for a letter of recommendation for some sort of governmental position.[90] Although Brougham declined on this occasion to write, Sylvester responded to him gratefully, if somewhat obsequiously, on 6 January 1851. "I felt much indebted by your kind and prompt attention to my request although unable to comply with it, which I am now very glad that your Lordship was not at the trouble of doing as in consequence of the great reduction which has been made in the value of the appointment I have withdrawn my application."[91] He would have to continue to keep his eyes and ears open for new and sufficiently lucrative job opportunities that would give him the time he wanted for his mathematics. If, as he ultimately hoped, a viable *academic* post materialized, however, he would need to be ready to make his move, and that meant, to him at least, building a mathematical dossier of unquestionable strength and quality.

The winter and spring of 1851 found him as fully absorbed in this quest as a full-time actuary could be. A string of papers closely related to his work at the end of 1850 ensued, in which he pushed his algebraic—in this case determinant-theoretic—interpretation of the points of contact of lines and surfaces of the second order. He gave a fairly clear articulation of what was becoming his overarching research agenda in a paper on this subject that appeared in the *Philosophical Magazine* in February. "Geometry, to be properly understood," he asserted, "must be studied under a universal point of view; every (even the most elementary) proposition must be regarded as a fact, and but as a single specimen of an infinite series of homologous facts." "In this way only," he continued, "(discarding as but the transient outward form of a limited portion of an infinite system of ideas, all notion of extension as essential to the conception of geometry, however useful as a suggestive element) we may hope to see accomplished an organic and vital development of the science."[92] Like Descartes before him, Sylvester saw algebra as the key for unlocking the secrets of geometry.[93]

Sylvester made explicit the connection between determinants and the geometrical questions of contact in the very title of his follow-up paper "On the Relationship between the Minor Determinants of Linearly Equivalent Quadratic Functions" (see second epigraph note). There, although the fundamental geometric problem of determining the contacts of conics had motivated the paper, the research was purely and abstractly determinant-theoretic. Sylvester first developed a compressed notation for the usual square array representation of the determinant and then showed how to manipulate his new notation to form determinants of determinants (what he called compound determinants) as well as to compute minors of determinants and other constructs. In the context of these remarks, he also reintroduced a term he had mentioned in print in passing in November 1850, namely, matrix.[94] As he defined it, a matrix is "a rectangular array of terms, out of which different systems of determinants may be engendered, as from the womb of a common parent."[95] In the matrix as the underlying structure and in the determinant as a key construct based on it, Sylvester saw unlimited possibilities. "What is the theory of determinants?," he asked rhetorically. "It is an algebra upon algebra; a calculus which enables us to combine and foretell the results of algebraical operations, in the same way as algebra itself enables us to dispense with the performance of the special operations of arithmetic. All analysis must ultimately clothe itself under this form" (246–247).

If Sylvester was a mathematician by night engaged in what he would later describe as his "mathematical reveries,"[96] he was still an actuary by day concerned with the events that were then shaping the business and financial profiles of the nation. Long a dedicated reader of the periodical press, Sylvester, on 8 April 1851 read—amidst the financial and other business news reported in the *Times*—an account of the stunning experimental proof of the rotation of the earth recently given by Léon Foucault. This article piqued the broader scientific interests of the Sylvester who had been trained in the Cambridge applied mathematical tradition and who had also served as University College's Professor of Natural Philosophy.

Foucault, who had first tested his pendulum in the basement of his home in Paris early in January, had presented his findings to the Paris Academy of Sciences in February after repeating the experiment with a larger apparatus in the meridian hall of the Paris Observatory. Not long after that presentation, an even larger setup was installed in the Pantheon and opened to the public. As the *Times* reported on 8 April 1851, "the experiment now being exhibited in Paris, by which the diurnal rotation of the earth is rendered palpable to the senses, is certainly one of the most remarkable of the modern verifications of the theory."[97] After describing the installation—a pendulum composed of a fine wire attached to a sphere four or five inches in diameter suspended from the Pantheon's dome just over the center of a table some twenty feet in diameter and marked with degrees, minutes, and seconds—the *Times* writer went on to explain how the experiment worked. "Since, then, the table thus revolves, and the pendulum which vibrates over it does not revolve, the consequence is, that a line traced upon the table by a point projecting from the bottom of the ball will change its direction relatively to the table from minute to minute and from hour to hour, so that if such a point were a pencil and that paper were spread upon the table, the course formed by this pencil during 24 hours would form a system of lines radiating from the center of the table, and the two lines formed after the interval of one hour would always form an angle with each other of

15°, being the 24th part of the circumference." "Now, this," the writer continued, "is rendered actually visible to the crowds which daily flock to the Pantheon to witness this remarkable experiment." Unfortunately, that was not, in fact, what the Parisian crowds saw, and Sylvester, the former physics professor, could not resist pointing out the error.

Writing to the *Times* from his office in Lincoln's Inn Fields the following day, he set the matter straight in no uncertain terms. "Not only is the explanation" that was given in yesterday's edition "completely erroneous," Sylvester wrote, "but the facts themselves are misstated. It is not true that the plane in which the pendulum vibrates will travel at the rate of 15 degrees per hour—this would be only the case at the pole."[98] He went on to explain that "the rate of motion of the plane of vibration at any place intermediate between the pole and the equator," like Paris, "may be shown, upon geometrical principles not easily made intelligible to the common apprehension, to vary [inversely] as the sine of the latitude." Thus, contrary to what was stated on the pages of the *Times*, "the motion would be at the rate, not . . . of 15, but in round numbers of only about 11½ degrees per hour." Because Foucault's experiment was being conducted in Paris just as the architecturally spectacular Crystal Palace in London was being filled with the marvels of modern science and technology before its official opening on 1 May, Sylvester suggested further that "the Crystal Palace, when cleared of its contents, would afford an excellent site for making the experiment on a great scale, in the presence of the assembled *savans* of Europe." He then closed his letter in a self-satisfied and superior tone, confessing that he was "induced to hope for the insertion of this correction of a statement in which the threads of truth and error are intermingled in a manner calculated to baffle the best wits of an ordinary reader to unravel because I feel that under cover of the vast circulation and great authority of *The Times* any errors finding their way into its columns are likely, if unnoticed, to pass into general acceptance and belief." Sylvester was certainly not one of those "ordinary readers." *His* was not a wit to be baffled by the laws of physics!

Not so, however, for one B.A.C. whose letter appeared in the *Times* on 24 April. He had read both the accounts of the experiment in the press and Sylvester's letter on the subject and still had to "confess that [he] remain[ed] unconvinced by the reality of the phenomenon."[99] He began his somewhat lengthy argument against the effectiveness of Foucault's experiment by noting that, in his view, "except at the pole where the point of suspension is immovable, no result can be obtained." "In other cases," he continued, "the shifting of the direction of passage through the lowest point that takes place during an excursion of the pendulum, from that point in one direction and its return to it again, will be exactly compensated by the corresponding shifting in the contrary direction during the pendulum's excursion on the opposite side."

It was one of several sophomoric mistakes B.A.C. made in his letter, and Sylvester felt bound to point those out, too. Writing on the twenty-fifth, he declared first that "there is no force in the objections of your correspondent 'B.A.C.' to the conclusions of M. Foucault" and then proceeded to explain the principles once again and in much greater detail "for the satisfaction of that numerous and respectable body of thinkers who form an intermediate class between those who are incapable of any proof except what appears directly to the senses"—like B.A.C.—"and the elevated few"—like himself—"who can comprehend the force of an analytical investigation."[100] "Before any mathematical tribunal in the world," Sylvester sniped, an argument like the one B.A.C. made "would be laughed out of court, and the plaintiff nonsuited, without the defendant being put to the trouble of making any case in disproof of the allegations offered in its support." Sylvester, buoyed perhaps by his new mathematical routine and the recent successes it had brought, was feeling feisty to the detriment of a hapless B.A.C.

This touchy self-assuredness manifested itself again both in print and in private less than a month later. In May 1851, Sylvester's latest paper, a "Sketch of a Memoir on Elimination, Transformation, and Canonical Forms," appeared in Thomson's *Cambridge and Dublin*. In it, Sylvester sensed he was on to something.

For the past year, his research had been shaped largely by his efforts to understand and interpret algebraically the geometry of the intersection of conics. As noted, his quest had led him to Boole's 1841 "Exposition," and he had seen there Boole's procedure for generating expressions in the coefficients of a homogeneous polynomial in two variables that remain unchanged under the linear transformation of the polynomial. He had also talked with Cayley and learned that in 1845 his friend had published a memoir "On the Theory of Linear Transformations" in which he, too, had discovered a class of functions, so-called hyperdeterminants, which, when evaluated at specific values, were "gifted," like Boole's expressions, "with a similar property of immutability."[101] Sylvester himself had followed the lead Cayley had taken some six years earlier and had devised yet another technique—he called it compound permutation—for generating Cayley's hyperdeterminants. Finally, Sylvester had recognized an analogous "property of immutability" in Hesse's 1844 work. Was it possible, Sylvester wondered, to find "a common point of view" to link Hesse's functional determinant and "the common constant determinant of a function," that is, the discriminant?[102] "It was not long," Sylvester wrote, "before I perceived that they formed the two ends of a chain of which Hesse's end exists for all homogeneous functions, but the other only when such functions are algebraical."

Sylvester found this connection more than interesting; he foresaw in it, as in the theory of determinants, almost limitless possibilities. The "Sketch," however, merely alluded to these. Sylvester did not yet have more than a glimmer of what might emerge. He focused instead on the problem of transforming homogeneous polynomials in two variables x and y of degrees three, four, and five into their respective canonical forms, namely, $x^3 + y^3$, $x^4 + y^4 + Kx^2y^2$, and $x^5 + y^5$, where K is some constant.[103] His main tool was the Hessian, and he felt that he had some claim on its development. In particular, he saw in Hesse's research what he viewed as a double appropriation of his own earlier work, and he felt that he should be given credit in connection with Hesse's find. "I take this opportunity," he wrote in his "Sketch," "of entering my simple protest against the appropriation of my method of finding the resultant . . . by Dr Hesse, so far as regards quadratic functions, without acknowledgment, four years after the publication of my memoir in the *Philosophical Magazine*: the fundamental idea of Dr Hesse's partial method is identical with that of my general one."[104] "Still more unjustifiable," he went on, "is the subsequent use of the *dialytic* principle, by the same author, equally without acknowledgment, and in cases where there is no peculiarity of form of procedure to give even a plausible ground for evading such acknowledgment. It is capable of moral proof that what I had written on the matter was sufficiently known in Berlin and at Königsberg, at each epoch of Dr Hesse's use of the method."[105] At this stage, as he tried to build up a portfolio of first-rate mathematical results and as he sensed he was gaining real momentum, Sylvester found it particularly important to claim credit where he felt it was due. This was equally true in his evolving working relationship with Cayley.

As Sylvester stated in his "Sketch," he had succeeded in the spring of 1851 in developing a technique—compound permutation—for producing functions with the immutability property. In the "Sketch," however, he gave no indication of what the technique involved or of how to effect it. He had not yet had the chance to explore further the ramifications of the new process, but he did feel the need to stake his claim to it in print because he sensed here, too, that something important was afoot.

In what had become their almost daily mathematical conversations, Sylvester had naturally shared with Cayley his evolving thoughts. When Cayley inquired whether it might not be appropriate for him to write a paper on the relationship between Sylvester's new technique and his own earlier method of hyperdeterminants, Sylvester grew prickly. "As you appealed to me on the subject," he stated, "I must say that I do not think that you are justified in publishing your views on the Method (as applied to Hyperdeterminants), of *Compound Permutation of Umbral elements* founded on or suggested by my communications to you on the subject which were meant as confidential, until I have first published my own account of the matter."[106] Sylvester's conception of his mathematical relationship with Cayley at this point was one of a sounding board all right, but a sounding board in a sound-proof room. Every idea—or at least every idea that Sylvester deemed sufficiently important—was to be considered confidential until Sylvester had had his say in print. "It will then be right," he continued, "for you to point out whatever part of the idea you may think is included in your former printed papers and to suggest any generalizations."

Sylvester's sense of propriety, his need at this point to claim priority, and perhaps his frustration at not having more time to devote to his researches resulted in a complicated working relationship with Cayley. In no way was it to be construed that they were collaborating. They each had their own separate ideas, and it was incumbent upon each of them to develop and publish those ideas separately. Yet how to disentangle an ongoing mathematical conversation? Who got credit for being the "first"? Must credit be assigned to each new thought? Apparently Sylvester thought so. He laid things out for Cayley this way. "I believe you acknowledge your inspirations upon that point [an aspect of compound permutation] arose out of accidental observations on my part," Sylvester asserted (34). Score one for Sylvester. "But I owe to you," he continued, "the first simplified *statement* of its application in a particular case which however I repeat, it is quite certain from the direction my researches had taken, I must . . . have necessarily arrived at" (34–35; his emphasis). Score one for Cayley? Not exactly. In Sylvester's view, he really ought to be justified in laying claim to ideas, to which, with 20-20 hindsight, he would have eventually been led without Cayley's premature intervention. It was, indeed, a complicated working relationship, but Cayley, more settled and secure and certainly more reserved than his friend, acquiesced to let Sylvester largely define the rules. Sylvester had been lucky to find a friend and a mathematical equal like this, especially as he set out on what was looking like a real mathematical adventure.

CHAPTER 5

Into the Invariant-Theoretic Unknown

*I discovered and developed the whole theory of binary canonical
forms for odd degrees, and, as far as yet made out, for even degrees
too, at one evening sitting, with a decanter of port wine to sustain
nature's flagging energies, in a back office in Lincoln's Inn Fields. The
work was done, and well done, but at the usual cost of racking
thought—a brain on fire, and feet feeling, or feelingless, as if plunged
in an ice-pail.* That night we slept no more.
—J. J. Sylvester, 1870

The first six months of 1851 had finally found Sylvester settling in to a
sustainable mathematical routine, and they had been productive. He had
begun to glimpse new and unexpected mathematical vistas in his intriguing
ideas on immutability, and he increasingly sensed that he was on to something
big, something truly reputation making. He wanted nothing more at this juncture than to be a mathematician, to realize his conception of that category,
to devote himself fully to his mathematical research. That, however, did not
seem to be in the cards. The *professional* mathematician in that sense did not
exist in England—not even at Oxbridge—but that need not stop him from seeing himself—from defining himself—as a mathematician first and foremost,
an actuary incidentally. As he pursued the range of ideas that would come to
constitute the new field of invariant theory by the mid-1850s, Sylvester more
exclusively became Sylvester mathematician. He also increasingly realized that
that persona could no longer coexist happily with Sylvester actuary. It was a
time of transition as it was a time of stunning mathematical creation.

The Emergence of a Theory of Invariants

After dropping hints of an actual *theory* of "immutability" in his "Sketch" in
the May 1851 number of the *Cambridge and Dublin*, Sylvester embarked on a

single-minded mathematical pursuit. As in 1850, he refused to allow the Institute of Actuaries to distract him with its request that he serve as an examiner for its fellowship competition in June 1851.[1] Instead, he spent the summer hard at work, and in regular communication with Cayley, on the whole constellation of ideas and techniques that a theory of invariants might involve: the theory of elimination, the properties and consequences of linear transformation, the manipulation and theory of determinants. By the end of August, this research had developed to the point that a new and stable vocabulary was needed both to express the evolving body of mathematical knowledge and to lay claim to it effectively. On the twenty-fifth, Sylvester thus "submitted" to Cayley in writing a list of terms "for approval & ratification."[2] These included what Sylvester called "primitives," like "invariant," "determinant," and "discriminant" as well as "other terms," like "transformation," "coefficients" of transformation, and "modulus" of transformation. None was given with a definition. No definitions were needed. The terminology had evolved over the course of the summer of 1851 as the two friends talked, often late into the night. They had been creating and speaking their own private mathematical language.

Their intention, however, was not for that language to stay private for long. For his part, Sylvester began to introduce it and the ideas it embodied in the September number of the *Cambridge and Dublin*. In "On the General Theory of Associated Algebraical Forms," he considered a homogeneous polynomial and a linear transformation of its variables of determinant one. In particular, he defined—for the first time in print in his works—the notions of an invariant and a covariant associated with the given form. For Sylvester, an invariant was "a function of the coefficients of the form, as remains absolutely unaltered when instead of the given form any linear equivalent thereto is substituted," whereas a covariant was a similarly immutable expression except one in the coefficients *and* the variables of the form.[3] He continued by noting that "of course if the determinant of the coefficients of the transformations correspondent to the respective equivalents be not taken unity as supposed in this definition, the effect will be merely to introduce as a multiplier some power of the determinant formed by the coefficients of transformation." This was a more properly mathematical—if fundamentally narrative—statement of the immutability property he had been observing in so many distinct settings. In fact, he had found it yet again in recent work of the French mathematician, Charles Hermite, on so-called "formes-adjointes" and had mentioned this discovery in his paper as well.

Sylvester's reference to Hermite's work highlighted another aspect of the actuary's parallel life as a research mathematician. He had begun to strengthen his ties with mathematicians in France in an effort both to master the latest French algebraical work and to make known outside the confines of the British Isles his own mathematical results. For example, in another product of his summer research harvest—this one published in the August number of the *Philosophical Magazine*—Sylvester pushed the theory of determinants so essential in the emergent theory of invariants. There, he hailed the "New Algebra (for such, truly, it is the office of the theory of determinants to establish)" and noted that "in a recent letter to me, M. Hermite well alludes to the theory of determinants as 'That vast theory, transcendental in point of difficulty, elementary in regard to its being the basis of researches in the higher arithmetic and in analytical geometry.' "[4]

Hermite, eight years Sylvester's junior, had graduated from the École polytechnique with a *baccalauréat* and *licence* in 1847 and had taken a position there the following year as a *répétiteur* and admissions examiner. His work in the 1840s on elliptic and hyperelliptic functions had attracted the favorable attention of Jacobi in Germany, and by the 1850s, he was establishing himself as one of the rising stars in French mathematics for his work, especially on the theory of quadratic forms. Given that his then-current research interests overlapped with those of Sylvester, Hermite was a natural contact for a Sylvester seeking to make his work more broadly known.[5]

Another member of the French mathematical community with whom Sylvester established contact at least as early as 1851 was Olry Terquem, the editor of the *Nouvelles annales de mathématiques*, a journal founded in 1842. Writing "On Extensions of the Dialytic Method of Elimination" in the September number of the *Philosophical Magazine*, Sylvester acknowledged that he had "been induced to review" his ideas on the method "in consequence of the flattering interest recently expressed in the subject by my friend M. Terquem, and some other continental mathematicians, and because of the importance of the geometrical and other applications of which it admits."[6] Terquem and others on the Continent were finally taking notice of Sylvester's decade-old elimination techniques just as their originator was coming to recognize the key role of these techniques in the establishment of a theory of invariants.

Although the allusions to this theory that he gave in his papers in August and September of 1851 were brief and more suggestive than detailed, Sylvester explicitly promised his readers more to come. In September's paper "On the General Theory of Associated Algebraical Forms," he "propose[d] to enter much more largely into the subject generally" in the very next—the January 1852—number of the *Cambridge and Dublin*.[7] "More particularly," he continued, "I shall describe the new method of Permutants, including the theory of Intermutants and Commutants (which latter are a species of the former, but embrace Determinants as a particular case), and their application to the theory of Invariants." What the readership of Thomson's journal might have made out of all of these new and undefined terms is anyone's guess. What is clear is that Sylvester's use of them here was a tease. The interested reader would have to stay tuned, but the show would be well worth the wait. Sylvester was not merely going to give his readers isolated results; he planned to lay out for them the beginnings of an entire theory of invariants.

In November, as he labored on this major work, he once again baited his audience—although this time the audience not of the *Cambridge and Dublin* but of the *Philosophical Magazine*—with more hints of things to come. He wrote sensationalistically "On a Remarkable Discovery in the Theory of Canonical Forms and of Hyperdeterminants" and "prepare[d] the way for the more remarkable investigations which form the proper object of this paper, by giving a new and more simple solution of" the linear transformation that takes a homogeneous polynomial of degree $2n + 1$ in x and y to the canonical form $\sum_{i=1}^{n+1} (p_i x + q_i y)^{2n+1}$, for p_i, $q_i \in \mathbb{R}$.[8] Sylvester had announced this result and had given a more cumbersome proof of it in a tract published privately earlier in the year.[9] By November, he had "discovered and developed the whole theory of binary canonical forms for odd degrees, and, as far as yet made out, for even degrees too, at one evening sitting, with a decanter of port wine to sustain nature's flagging energies, in a back office in Lincoln's Inn Fields." (See epigraph note.) It was an elegant simplification in determinant-theoretic terms of his earlier work, and it led him to uncover invariants in unexpected elimination- and determinant-theoretic settings.

For example, in his efforts to analyze the situation for a homogeneous polynomial of even—as opposed to odd—degree in two variables, he took the small special case of the biquadratic or degree four polynomial

$$ax^4 + 4bx^3y + 6cx^2y^2 + 4dxy^3 + ey^4, \tag{1}$$

for a, b, c, d, $e \in \mathbb{R}$ and asked under what circumstances this could be reduced to the canonical form

$$(fx + gy)^4 + (hx + ky)^4 + 6m(fx + gy)^2(hx + ky)^2, \tag{2}$$

for f, g, h, k, $m \in \mathbb{R}$. By elimination—that is, in this case, by a clever algebraic manipulation of the equations gotten by setting corresponding coefficients in equations (1) and (2) equal—Sylvester ultimately obtained a monic cubic equation with no squared term. Surprisingly and interestingly, the linear and constant coefficients of that cubic were invariants of equation (1), namely, $ae - 4bd + 3c^2$, which is homogeneous of degree two in the coefficients of equation (1) and $ace - 2bcd - ad^2 - eb^2 - c^3$, which is homogeneous of degree three in them, respectively. Solving the cubic then gave him all the information he needed to effect the reduction of equation (1) to (2).[10] This result, and his successful extension of it to the case of a homogeneous polynomial of degree eight in two unknowns, led him "to generalize this remarkable law, and to demonstrate the existence and mode of finding $2n$ consecutively-degreed independent invariants of any homogeneous function of the degree $4n, \ldots$ a result, whether we look to the fact of such invariants existing, or to the simplicity of the formula for obtaining them, [that is] equally unexpected and important, tending to clear up some of the most obscure, and at the same time interesting points in the great theory of algebraical transformations" (273).

Sylvester presented his "remarkable" result and its various ramifications carefully, taking the opportunity to establish further some of the basic concepts he would need in the exposition of the fuller theory he was then currently preparing. He opened, for example, with a more precise definition than he had given in September of "invariant." Given a homogeneous polynomial in two variables $f(x, y)$, Sylvester now explained, "if the coefficients of the function $f(x, y)$ be called a, b, $c \ldots l$, and if when for x we put $lx + my$, and for y, $nx + py$, where $lp - mn = 1$, the coefficients of the corresponding terms become a', $b' \ldots l'$; and if $I(a, b \ldots l) = I(a', b' \ldots l')$, then I is defined to be an invariant of f."[11] With this as his point of departure, he proceeded to detail both his proof—one involving elimination techniques in determinant garb—and several spin-off ideas from it.

In the context of one of the spin-offs, he lectured his readers once again on what he viewed as an equally important aspect of the evolving theory—its vocabulary—and sketched what might be called his philosophy of language. After justifying, for example, the choice of the term "discriminant" for the special determinant used in analyzing, among other things, the intersections of two conics, Sylvester averred that "progress in these researches is impossible without the aid of clear expression; and the first condition of a good nomenclature is that different things shall be called by different names. The innovations in mathematical language here and elsewhere (not without high sanction) introduced by the author, have been never adopted except under actual experience of the embarrassment arising from the want of them, and will require no vindication to those who have reached that point where the necessity of some such additions becomes felt."[12] Those who give birth to new lives have the privilege of naming their offspring and, in so doing, of establishing their parentage. Sylvester, who many years later would refer to himself unabashedly as the "mathematical Adam,"[13] viewed naming as a clear way of establishing that ever-important matter of priority, yet the "high sanction" that his proposed terms had received was from Cayley. The two of them were in this together.[14]

The matter of "credit where credit was due" that had been so evident in Sylvester's letter to Cayley six months earlier in May 1851 and that was so fundamentally shaping their working relationship was also clearly in evidence in Sylvester's account of his "remarkable discovery." The results were Sylvester's. The vocabulary was Sylvester's. But, the paper was also peppered with references like "Mr Cayley has pointed out to me a very elegant mode of identifying . . . " and "Mr Cayley has made the valuable observation that . . . ."[15] If Sylvester was going to insist that others give him the credit he felt he deserved, then he was going to have to play by the same rules. After all, he and Cayley were engaged in a constant conversation about this whole panoply of ideas, and both were contributing to it. Moreover, at least in these critical, early stages of what Sylvester ultimately hoped would be his career as a research mathematician, he was focused on reading and trying to absorb the relevant literature, and that kept the matter of who had done what very much in the foreground.[16]

November and December of 1851 found Sylvester pouring himself into his mathematical work, "but at the usual cost of racking thought—a brain on fire" (see epigraph note). Mathematics, not his actuarial work, consumed him, yet he *was* still an actuary. While that may have meant that he had to perform the increasingly drudgerous task of preparing the year-end report, it did not mean that he had to keep up what had become the façade of participating in the broader professional agenda of a field he hoped fervently to leave.[17] In December, he made his final break, attending his last meeting of the Council of the Institute of Actuaries, despite the fact that his colleagues elected him once again to a vice presidency for 1852.[18]

Sylvester had more important things to think about. On 5 December, he had received a letter from Cayley that amounted to the birth certificate of their theory of invariants. His friend's message was short and sweet. "Every Invariant," Cayley stated, "satisfies the partial diff[erentia]l equations

$$
\left(a\frac{d}{db} + 2b\frac{d}{dc} + 3c\frac{d}{dd} \cdots + nj\frac{d}{dk} \right) U = 0
$$

$$
\left(b\frac{d}{db} + 2c\frac{d}{dc} + 3d\frac{d}{dd} \cdots + nk\frac{d}{dk} \right) U = \frac{1}{2}nsU
$$

(3)

(*s* the degree of the Invariant) & of course the two equations formed by taking the coeff[icien]ts in a reverse order. This will constitute the foundation of a new theory of Invariants."[19]

Crafting the "New Algebra"

When Sylvester received this letter, he had already been absorbed in writing up his first attempt at a systematic exposition of some of the main features of that new theory for the January number of the *Cambridge and Dublin*. As he explained to the readers of his paper "On the Principles of the Calculus of Forms," "the primary object of the Calculus of Forms is the determination of the properties of Rational Integral Homogeneous Functions or systems of functions: this is effected by means of transformation; but to effect such transformation experience has shown that forms or form-systems must be contemplated not merely as they are in themselves, but with reference to the ensemble of forms capable of being derived from them, and which constitute as it were an unseen atmosphere around them."[20] The constituents of this derivable "ensemble of forms" were precisely those expressions—like invariants and covariants—with the immutability property.

In his paper, Sylvester explored a variety of computational techniques for finding invariants, at the same time that he introduced a befuddling array of new terms to distinguish among them. Unlike in his vocabulary-establishing letter to Cayley in August 1851 where definitions were unnecessary, Sylvester opened this published work with explicit definitions of some of the basic concepts—linear transformation, determinant, discriminant, and resultant—admitting that "it may be well at the outset to give notice to my readers of the exact meaning to be attached to" these terms.[21] He then proceeded on a neologistic romp, employing terms like originant, concomitant, contravariant, plexus, commutant, and others. In particular, he recast—in a completely general setting—one of the results he had obtained in his November 1851 paper, "On a Remarkable Discovery in the Theory of Canonical Forms and of Hyperdeterminants," on homogeneous functions of degree $2n$ in two variables.[22]

As noted previously, Sylvester had been interested in that paper in the conditions under which equation (1) could be reduced to the canonical form (2), but he had also asked the question, when can equation (1) be transformed more simply into the sum of two fourth powers of linear expressions in x and y? The answer to the latter question reduced, he showed, to calculating when the determinant

$$\begin{vmatrix} a & b & c \\ b & c & d \\ c & d & e \end{vmatrix} = ace - 2bcd - ad^2 - eb^2 - c^3 \tag{4}$$

vanishes. He found this such an interesting construct that he gave it a name, "hereafter refer[ring] to a determinant formed in this manner from the coefficients of [a homogeneous binary form] as its catalecticant" even though, he added pedantically, "meicatalecticant would more completely express the meaning."[23] (His French friend, Terquem, would later caution him that "terminology . . . is your *strength* as well as your *weakness*. You have too much of a propensity to create new words. It would be well for you to forget about Greek."[24]) As Sylvester noted, Boole had actually been the first to produce equation (4) and effectively to show that it was an invariant of the biquadratic form (1), whereas Cayley, in the course of his conversations with Sylvester, had realized that all catalecticants are invariants.[25] New results, new invariantive phenomena, and new theoretical settings were beginning rapidly to coalesce.

The publication of "On the Principles of the Calculus of Forms" seemed to ignite an invariant-theoretic fire under both Sylvester and Cayley, the flames of which spread quickly across the Irish Sea. At Trinity College Dublin, Irishman

George Salmon, five years Sylvester's junior and two years Cayley's senior, had taken up his *alma mater*'s Donegal lectureship in mathematics in 1848—the same year he had first met Cayley—in addition to the fellowship, tutorship, and lectureship in divinity he had held since 1841. Realizing instantly the inadequacy of the textbooks available to his elementary mathematics students, Salmon wrote *A Treatise on Conic Sections* and quickly made a name for himself as a clear expositor with sound pedagogical techniques.[26] In his efforts to get back into mathematical research in 1850, Sylvester had begun to encounter and make use of Salmon's geometrical researches, and by February 1852, the two mathematicians had entered into an active correspondence.[27]

Early in 1852, Salmon was working on a new book, *A Treatise on Higher Plane Curves*, that would serve as a more advanced sequel to his text on conic sections. Because several of the papers Sylvester had published in 1850 and 1851 had dealt precisely with that theory from an elimination- and determinant-theoretic point of view, it is little wonder that the two men recognized in each other kindred mathematical spirits and entered into a mathematical dialog. February 1852 thus found Sylvester writing a follow-up letter to Salmon in which he offered a couple of mathematical observations and computations that he thought might be of use to Salmon in his work.[28] By the end of month, though, Sylvester had begun to draw Salmon into his invariant-theoretic researches.

In a letter dated the twenty-fourth, Salmon acknowledged receipt of a second copy of a reprint from Sylvester, most likely of his paper "On the Principles of the Calculus of Forms," and confessed that he had not acknowledged receipt of the first copy because he had not yet succeeded in fully understanding the paper's contents. "It seemed to me much less difficult than some of your preceding papers," he noted, but "I found that I could not read it satisfactorily without having your other papers to refer to."[29] "It tantalizes me," he continued, "to see so much that I am anxious to get at, enclosed in so hard a shell. I think all that is wanting to make it clear is a little illustration of your general statements, by particular examples." The Irishman may have been intrigued, but the Englishman's published work with its arcane vocabulary and results cast in such broad and sweeping terms proved baffling to the uninitiated.

Salmon and Sylvester continued in regular communication by post. For his part, Salmon sought Sylvester's help in working out a number of the sticky points he encountered as he continued to draft his book manuscript on higher plane curves, while Sylvester sent his replies—often in invariant-theoretic terms—and drew Salmon into that evolving theory. On 23 March 1852, for example, Sylvester wrote to Salmon, admittedly "at some risk of fatiguing your attention & interest," to relate an "improved & generalized statement" of an invariant-theoretic discovery that he had also just communicated to Cayley.[30] What had been two points determining a straight line—Cayley and Sylvester—was now becoming three points—Cayley, Sylvester, and Salmon—defining an invariant-theoretic triangle. As Sylvester and Cayley became ever more deeply involved in their theory-building exchange, they increasingly piqued Salmon's interest in it through their ongoing, three-way correspondence.

February had found the two Englishmen involved in pushing the basic invariant-theoretic concepts in new directions. On the fifth, Sylvester had laid out for Cayley his notion of so-called orthogonal invariants—that is, expressions in the coefficients of a homogeneous form of degree n in x and y that remain invariant under the action of a linear transformation T (implicitly over the field of real numbers $\mathbb{R}$) such that $T(x) = ax + by$, $T(y) = cx + dy$, and $T(x)^2 + T(y)^2 = x^2 + y^2$—and he sketched some of their properties. In particular, he had noted that "the partial diff[erential] equations," the equations in (3), that Cayley had laid out two months earlier "show immediately that the general Invariants are functions of the orthogonal Invariants; Hence may clearly be derived a process for forming the general ones when the orthogonal ones are given."[31] After working through the implications of this somewhat further, Sylvester closed with the question "Is every general Invariant a Rational Integral function of the Orthogonal Invariants?" As this letter makes clear, by the winter of 1852, Sylvester and Cayley had entered into an analysis of the structure of invariants. What were the invariants of a given binary form? Could they be expressed in terms of invariants that were, in some sense, more fundamental? Were orthogonal invariants those more fundamental ones? If so, how do they go together to form the full set or fundamental scale of invariants for a given binary form? Eight days later on the thirteenth, Sylvester thought he was closing in on an answer. "I believe that I am within sight," he wrote, "of the long-expected solution of the Law of construction for the fundamental general scale."[32]

Although this envisioned solution was not immediately forthcoming—this would, in fact, prove to be a problem that would drive the British approach to invariant theory for decades to come—Sylvester continued to work at a frenetic pace throughout the winter and spring of 1852, communicating daily with Cayley, working through new examples, developing new parts of the theory. By the end of March, he was not only ready to begin writing up his next installment of "On the Principles of the Calculus of Forms" for the *Cambridge and Dublin*, but he was also trying to make another move that would, he hoped, leave him more time for his mathematical work.

Patent reform had briefly engaged the Parliament during its 1851 session, and talk of the institution of a new system of patent examiners had caught Sylvester's attention. While some provisions actually passed in the House of Commons in that year, they were delivered to the House of Lords too late for consideration, and Sylvester's hopes—expressed to his powerful friend Lord Brougham, who had actually recommended him for one of the examinerships—had to be put on hold.[33] The matter returned to the floor in the 1852 session and eventually resulted in the passage of an act that changed how patents for inventions were granted.[34] Sylvester, still on the lookout for opportunites that might release him from what he increasingly viewed as the drudgeries of the Equity and Law and provide him more time for his mathematics, wrote to Brougham once again on 23 March 1852 to suggest himself as a candidate for an examinership should the act then currently under debate in Parliament finally become law. "If you would kindly exercise your influence," he asked, "by mentioning me as a person whom you think well suited to be an Examiner under the intended new Patent Act and whom you would wish to see in that position, I can hardly doubt that the success of my application would be insured."[35] The act passed only later in the session, and although Sylvester again petitioned Brougham to remind those in power of his earlier recommendation, Sylvester ultimately remained in his position at the Equity and Law.[36]

He was, in fact, becoming increasingly successful in balancing his own "bifarious occupation" of actuary and mathematical researcher.[37] The same day he wrote Brougham about the examinership, mathematical letters went out to both Salmon and Cayley. More mathematical letters to Cayley followed on the twenty-fifth and twenty-sixth, in which Sylvester was not only hammering out invariant-theoretic ideas destined for publication in the next—the May 1852—installment of his paper "On the Principles of the Calculus of Forms" in the *Cambridge and Dublin* but also generally engaging in his daily theory-building give-and-take with Cayley.

On the evening of 24 March, for example, he had received a letter from Cayley extending some of the theory that they were then discussing. The next day, Sylvester wrote back to inform him that the extension actually followed immediately from some of his own work and proceeded to show how. Ever mindful of the matter of priority and keen not to be outdone in the development of the new theory, he closed by assuring Cayley "that I was perfectly familiar with the spirit of this proof if not with the actual example itself before seeing your Note of yesterday evening, although not immediately recognizing the method under a foreign dress."[38] The next day, Sylvester was investigating yet another aspect of the evolving theory, and after penning a letter to Cayley that contained a detailed calculation to justify a conjectured theorem, he wrote yet again in a separate "P. S." to declare that, in fact, "there is no doubt as to the truth of the theorem before sent" and adding that "I think this a good step gained in the way of suggesting extensions of the Invariantive theory & it is a pretty application of Elimination Principles."[39]

The theory was, indeed, growing rapidly, and Sylvester was devoting almost every spare minute to its development. Not even an extra evening with Cayley could be squeezed in at this point. On Monday, 29 March 1852, Sylvester realized that he had double-scheduled himself for dinner that evening, having invited Cayley after having already accepted an invitation in the country. "I beg you to pardon the oversight," he wrote on Monday morning, "and shall be very glad if you will come on Friday next at the same time instead of this evening. I name Friday because of having just begun my paper for Thomson's Journal which I must devote the next one or two evenings to, in order to get it done in a reasonable time."[40]

That paper and the further ideas it spawned occupied Sylvester well into what proved to be a mathematically charged month of April. He and Cayley were in constant communication, with letters passing between them virtually every day—and sometimes more than once a day—as they explored questions like, given a homogeneous binary form of degree six—what they termed a binary sextic or binary 6-ic—what are its invariants?[41] He and Salmon were also engaged in an almost daily exchange as Salmon queried him relentlessly on various invariant-theoretic points. On the seventh, for example, Salmon, who was doing his best to understand a theory in flux, posed several prescient questions while freely admitting his naïveté. "You scarcely know yet," he told Sylvester, "how great the amount of my ignorance is in all of this theory. I have not yet studied your papers systematically but have only got a few ideas here & there. I want to know how you can fix a limit to the number of invariants."[42]

Things *were* happening fast, almost too fast. On 11 April, Sylvester wrote to Cayley to "correct two mistakes of my own, the other of yours." Then on the fifteenth, he had to confess first to having "omitted in my note of yesterday to have added a necessary restriction to" a theorem they had been discussing and then later in a "P. S." to "revoke my former note which was too hastily written."[43]

In between these mistakes, Salmon had come to London to meet Sylvester for the first time and to talk with both Sylvester and Cayley in person. His visit, like his correspondence, was a spark. On the evening of Tuesday, 13 April, the three men had talked well into the night about the theory's latest developments. The next day, as he was making his way back to Ireland, Salmon wrote Sylvester from the train as it traversed Staffordshire with further thoughts on their discussion. "You will doubtless have discovered since that you understated your theorem last night," he opened.[44] After articulating and justifying the fuller theorem— a connection between one of the many evolving invariant-theoretic concepts and the geometrical problem of locating the double points of a homogeneous ternary form of degree n (a so-called ternary n-ic)—Salmon enthused that "I think this theorem so pretty that it was almost worth coming from Dublin to be told it."

It had been quite a high mathematical time, which for Salmon had even been capped by a postprandial adventure. Cayley, the inveterate walker, had volunteered to accompany his Irish friend back to his hotel after their mathematical evening, but as Salmon recounted to Sylvester, "we progressed Westward" from the Inns of Court "in a satisfactory manner till we got to the church" of St. Clements Danes "just beyond Temple Bar. There we made a complete circuit of the church. [Cayley] had got on some interesting topic so that I never observed it & he left me with my face turned toward the East & I did not discover my mistake till I found myself in St Paul's Church Yard." For this, Salmon asked Sylvester to "give Cayley a good scolding for me."

The correspondence with both Cayley and Salmon continued unabated through the month of May as the theory grew. Sylvester wanted more than this British audience, however; he was seeking even more actively an audience for his work on the Continent. With the reprints of the first installment of his paper "On the Principles of the Calculus of Forms" fresh in his hands at the end of February, he had fired off a copy and a letter to his German friend, Carl Borchardt, on the twentieth. Borchardt, three years Sylvester's junior and a *Privatdozent* at the University of Berlin, had earned his doctoral degree at Königsberg under Jacobi in 1843 and was an up-and-coming figure on the German mathematical scene. When Sylvester's letter and paper arrived, though, he had been so busy with his teaching duties and other commitments that he had left it unacknowledged, and this had prompted another letter from Sylvester on 1 April 1852 inquiring whether the earlier communication had gone astray. Sylvester was clearly anxious for feedback on—and praise of—his work. Embarrassed and perhaps somewhat harrassed on 6 April after receipt of the second letter, Borchardt explained his delayed response and promised Sylvester that he would get to the paper as soon as he could, although that would still not be for several weeks. He wanted, he said, "to study [the work] with the attention it required . . . and it is precisely because I wanted to give you only an entire and complete reply that I have put off doing it."[45] Borchardt did, however, have time then to return to a touchy topic that he and Sylvester had already broached in correspondence, namely, Otto Hesse's alleged appropriation of Sylvester's work without attribution.

Recall that less than a year earlier in his 1851 paper "Sketch of a Memoir on Elimination, Transformation, and Canonical Forms," Sylvester had effectively accused Hesse of having plagiarized several of his techniques of elimination. Hesse, a slightly older contemporary of Borchardt's at Königsberg, had taken a lectureship on the heels of earning his doctoral degree there in 1840 and had moved up to a professorship in 1845. Borchardt thus knew him well and took the opportunity both to assure Sylvester once again of Hesse's unimpeachable character and indirectly to caution him against rushing to judgment. "If Mr. Hesse had known your paper on the elimination between three equations of degree n when he published his paper on the elimination between three equations of the second degree," Borchardt argued, "he would have committed plagiarism, and I know him too well to think him capable of it. That changes nothing with respect to your priority, but much relative to the moral judgment on Mr. Hesse. In mathematics, it happens all too often that, owing to an insufficient knowledge of the literature, results are published as new which have already been obtained by others. Such oversights must certainly be corrected, but if in each case of this kind one charged plagiarism, one would be unjust." If Sylvester wanted his work better recognized in Germany, sending it to Borchardt was certainly one step toward achieving that goal. Making harsh and accusatory statements in print about a German fellow worker was not.

Sylvester took a much more professional and collaborative tone toward his German counterparts in the second installment of "On the Principles of the Calculus of Forms" in May 1852. Already in the paper's first installment in January, he had noted in passing results with invariant-theoretic interpretations in the work of both Gotthold Eisenstein and Siegfried Aronhold.[46] By the time the paper's continuation came out in May, Sylvester was becoming even more aware—thanks largely to Cayley's command of the literature—that there was actually a parallel development of a theory of invariants under way in Germany.

As noted, as early as 1801 in his *Disquisitiones arithmeticæ*, Gauss had shown that the discriminant of the binary quadratic satisfied the invariantive property. In the late 1830s and early 1840s as Sylvester moved from his professorship in London to the uncertainties of life in the United States and as Cayley pursued his degree at Cambridge, Gotthold Eisenstein, two-and-a-half years Cayley's junior, had taken it upon himself first to master Gauss's text and then to extend to binary cubic forms results Gauss had obtained for binary quadratic ones. In December 1843, Eisenstein finished his first papers in this vein and, in so doing, not only isolated as a key theoretical element what Sylvester would only later call the Hessian of the binary cubic form (recall the definition given in the preceding chapter) but also demonstrated that it satisfied the invariantive property.[47]

Approaching similar questions at just the same time, but from the more geometrical, less number-theoretic point of view of the behavior of third-order plane curves, Hesse also came upon his Hessian, although he cast it in the general context, as we have seen, of a homogeneous function of degree m in n unknowns.[48] Five years later, Hesse's student at Königsberg, Siegfried Aronhold, picked up on his mentor's earlier work and explicitly cast it in an algebraic light. As he put it in the introduction to his 1849 paper,

> In the twenty-eighth volume of this journal [Crelle's *Journal*], Hesse exposed a series of problems which were as important for the theory of third-degree homogeneous functions in three variables as they were interesting for algebra, in that they furnish the first example of a new kind of higher equation [that is, an invariant]. . . .
>
> I have undertaken researches on third-degree homogeneous functions in three variables so as to ascertain the most important algebraic relations of their coefficients and the functions which are coordinate and subordinate to them. These researches have led to new characteristics of these functions which I shall take the liberty to share in a detailed discussion.[49]

Although he restricted himself to the case of the ternary cubic form, Aronhold, in 1849, was asserting the existence of the same kind of "common point of view" that Sylvester had alluded to in his "Sketch of a Memoir on Elimination, Transformation, and Canonical Forms" in May 1851 and that he had been singlemindedly fleshing out ever since. Both had recognized—Sylvester independently but later than Aronhold—the potential of crafting a *theory* of invariants.

By May 1852, Sylvester had come to realize that he and Cayley had been anticipated by Aronhold in some of their findings. In particular, in the continuation of "On the Principles of the Calculus of Forms," Sylvester noted that, relative to the partial differential equations in (3) that every invariant must satisfy, "M. Aronhold, as I collect from private information, was the first to think of the application of this method to the subject; but it was Mr Cayley who communicated to me the equations which define the invariants of functions of two variables. [And t]he method by which I obtain these equations and prove their sufficiency is my own."[50] A complex priority statement followed in which he sorted out results from his own work as well as from the work of Cayley and Eisenstein.

A problem other than priority was also beginning to rear its ugly head, however. Sylvester found it difficult to understand Aronhold's work—developed as it was in terms of a notation very different from what he and Cayley were employing—and he made a direct appeal to Aronhold on this score. "It is extremely desirable to know whether M. Aronhold's equations are the same in form as those here subjoined [that is, effectively the equations in (3)]. . . . Should these pages meet the eye of that distinguished mathematician he will confer a great obligation on the author and be rendering a service to the theory by communicating with him on the subject." For that matter, Sylvester continued, "I shall feel grateful for the communication of any ideas or suggestions relating to this new Calculus from any quarter and in any of the ordinary mediums of language—French, Italian, Latin, or German, provided that it be in the Latin character."[51] This was an international call to invariant-theoretic arms that nevertheless resulted in little international cooperation. For the most part, the British—and especially Sylvester—continued to develop their techniques in isolation from the German approach, and vice versa.[52] Ultimately, this resulted in an active competition—again, especially on Sylvester's part—in which nothing less than national honor was at stake (see chapters 9, 10, and 12.)

If Sylvester harbored somewhat ambivalent feelings about the Germans, he felt a much more natural bond with the French.[53] After all, he had been going to France at least since the late 1830s; he had begun to cultivate a number of mathematical relationships there, notably with Sturm and Hermite; and he had been elected earlier in 1852 to the Société philomatique de Paris, a general scientific and literary society in which many French notables participated. In June 1852, after a rewarding but mentally taxing spring that at one point had prompted him to tell Salmon that he was going to have to break off his investigations owing to exhaustion,[54] Sylvester wrote to Irénée-Jules Bienaymé in Paris asking once again for his help in distributing recent reprints of his work to key members of the French mathematical community. Bienaymé, a mathematician and an active participant in the Société philomatique, had already served this purpose in February, distributing copies both of Sylvester's 1851 paper "On the General Theory of Associated Forms" and the first installment of "On the Principles of the Calculus of Forms" to Hermite, Michel Chasles, Eugène Catalan, Joseph Serret, Joseph Bertrand, Augustin-Louis Cauchy, and the Institut de France.[55] In June, the list had shrunken, but new recipients were included. This time, Sylvester asked that Bienaymé distribute copies of the second installment of "Principles" to Hermite but also to the Société philomatique and to Joseph Liouville, editor of the *Journal des mathématiques pures et appliquées*. As he explained, "I wish M. Liouville to have a copy because I am told that M. Eisenstein of Berlin has sent to Liouville's Journal the same kind of matter as is in my Section VI."[56] That section contained precisely those differential operators in equation (3) crucial in identifying invariants as well as Sylvester's elaborate priority statement regarding them. Sylvester apparently wanted Liouville to understand the issue both of independent discovery with respect to Aronhold and of prior discovery with respect to Eisenstein's forthcoming paper. Sylvester recognized that crafting the new theory of invariants on the international—as opposed to merely the national—stage was going to be complicated, but his latest ideas were important. Potentially reputation making, they merited the effort.

Sylvester actually wrote his first paper to be published in a foreign journal around this critical time as well. Entitled "Sur une propriété nouvelle de l'équation qui sert à déterminer les inégalités séculaires des planètes," it appeared in Terquem's *Nouvelles annales de mathématiques* in the fall of 1852 and contained a very nice determinant-theoretic result that arose in the problem of determining the secular inequalities of the planetary orbits.[57] Sylvester proved that if A is an $n \times n$ (real) symmetric matrix, then, in terms that would only be developed later, the roots of the characteristic equation of A^p, for p a positive integer, are the roots of the characteristic equation of A each raised to the pth power. Yet, although his proof was crisp and clean, Sylvester had to acknowledge that the result itself "is a particular case of a more general theorem proven by Mr. Borchardt, for arbitrary determinants, and which becomes the theorem proven above when the determinant is symmetric."[58] It was just another example of how small the mathematical world could be with respect to independent discovery and of how important it was to be aware of the broader literature.

The intense invariant-theoretic work of the spring of 1852 had in some sense played itself out by mid-May with tantalizing, but hard-to-generalize, results on the invariants of specific examples of ternary and quaternary forms—that is, homogeneous polynomials in three and four variables—of low degrees.[59] Although it was too early for them to realize this, Sylvester and Cayley had developed and were developing methods for finding and analyzing invariants that tended to work primarily in the context of *binary* forms. Little wonder, then, that the mentally spent Sylvester took a temporary breather from his invariant-theoretic researches.

The month of June found him distracted not only by the movement in Parliament toward the actual passage of the new Patent Act but also by a number of social visits and by word from Hermite of his extension of Sturm's Theorem from one to several functions. On Sunday the sixth, for example, Norman Ferrers of Gonville and Caius College, Cambridge, had breakfast with him to discuss the editorial transition of the *Cambridge and Dublin*. Thomson, who had, for a variety of reasons, become increasingly disillusioned with editing the journal in the years following his move to Glasgow in the fall of 1846 (recall the discussion in the previous chapter), had finally secured a replacement in Ferrers, and Ferrers was undoubtedly keen to get advice on his new undertaking from various of the journal's strong, regular contributors.[60] Two weeks later, Thomson himself was in town and once again breakfast was on the agenda.[61]

Sylvester was also still seeing a lot of Cayley. He extended invitations to him to both of the breakfasts and could hardly contain his excitement on Saturday, 19 June when he heard of Hermite's new result. "I have great & glorious news from Hermite," he wrote, "which I shall be glad to make you participate in if you can find time to look in upon me at 5 o'clock this afternoon" at the actuarial office at 26 Lincoln's Inn Fields.[62] Although Sylvester spent some time thinking about this new work of his French friend, as he explained in a letter to Salmon on 16 July, "my attention has of late from various causes . . . been so called off from the Calculus of Forms . . . that . . . I feel almost a stranger to my own most recent speculations."[63] He had managed to clean up a few odds and ends of the theory in a paper in the *Philosophical Magazine*, and he had published a couple of additional papers,[64] but he largely spent the rest of 1852 going through the motions at the Equity and Law and puzzling over a phenomenon in the theory of invariants that he had termed the problem of syzygies.

The Problem of Syzygies

As early as 1845 in his groundbreaking paper "On the Theory of Linear Transformations," Cayley had noticed a peculiar fact about the invariants of the binary quartic form (1). There were three nontrivial invariants then known— the discriminant K that Boole had calculated in his 1841 "Exposition," the catalecticant (4) denoted J, and $I = ae + 4bd + 3c^2$—and Boole had noticed that these invariants actually satisfy the relation $K = I^3 - 27J^2$.[65] Contemplating this whole constellation of results in 1845, Cayley had realized that, in fact, I and J are even more remarkable; they form what would only later be called a minimum generating set of invariants for the binary quartic.[66] He was also quick to understand that, relative to the problem, as he stated it, of "find[ing] all the derivatives [that is, invariants] of any number of functions, which have the property of preserving their form unaltered after any linear transformation of the variables," "there remains a question to be resolved, which appears to present very great difficulties, that of determining the *independent* derivatives, and the relation between these and the remaining ones."[67] Although Cayley most likely did not realize this in 1846, the matter of showing the "relation" between the "independent"—what would today be called the "irreducible"—invariants in a minimum generating set and the construction of other invariants associated with a given binary form would also hinge on determining what, if any, *algebraic* dependence relations—what Sylvester later called syzygies—might exist among the irreducibles. In fact, it was the latter problem that ultimately presented the very greatest difficulties.

Although Cayley did a little work in this direction in 1846, as we have seen, he soon moved on to other things. He was drawn back into invariant theory only in the early 1850s by the infectious enthusiasm of his friend, Sylvester. By the summer of 1852, with a theory of invariants burgeoning with techniques, phenomena, and open questions unimagined in the mid-1840s, Sylvester had begun to tackle the syzygy problem.

On Friday, 20 August, Sylvester still could not get out of his mind the ideas that he and Cayley had been discussing before Cayley's departure on a hiking excursion in Wales. In a closely penned, eight-page letter sent to Cayley "at the top of Snowdon or Plinlimmon, Wales," he wrote that "the theorem which we were discussing just before your departure still continues to haunt me."[68] They had been mulling over how effectively to decompose invariants—the tip of the iceberg of the problem of finding a minimum generating set of invariants and its related problem of syzygies between the members of such a set—but the computations were hard, and the techniques that worked in the case of the binary quartic form seemed difficult to generalize. After six pages of relentless calculation and a passing conviction that "the process is universal," Sylvester had to admit defeat and note that "the method will require further modifying." Still, he remained undaunted. "I freely allow the imperfection of the mode herewith submitted," he closed. "But one must creep before one can walk: these are to be regarded as merely tentative essays, preliminary flutters toward getting a familiarity with the nature of the ground to be gone over before rising freely on the wing." The ideas simply needed to gestate. They were, as yet, far from fully formed.

Sylvester stewed over this problem through the fall of 1852 as he occupied himself with, what were comparatively, a variety of invariant-theoretic trifles.[69] There were also a few social diversions. In town, Ferrers had tea with Sylvester on Tuesday, 16 November, and Cayley was invited to join in.[70] Sylvester was also in touch with his former teacher and colleague, De Morgan, recommending that he read Joseph Serret's *Cours d'algèbre supérieure*.[71]

By the time the first number of the 1853 *Cambridge and Dublin* came out in February, Sylvester had to admit, though, that he had little to report. His "Note on the Calculus of Forms" opened with a disclaimer. "Accidental causes," he explained, "have prevented me from composing the additional sections on the Calculus of Forms, which I had destined for the present Number of the *Journal*. In the meanwhile," he assured his readers, "the subject has not remained stationary."[72] In October, for example, he had finally figured out a procedure for determining the answer to the following problem: given three homogeneous quadratic equations in four unknowns together with one homogeneous linear equation in four unknowns, when does this system of four equations have a nontrivial common solution? He alerted his readers to this finding on so-called resultants and to three other new findings in February, but provided neither proofs nor details.[73] He was too distracted. He had already begun the systematic assault on the problem of syzygies that would result by June not only in his formidable paper, "On a Theory of the Syzygetic Relations of Two Rational Integral Functions," but also in a number of spin-off papers on not unrelated matters in the calculus of forms.[74]

Sylvester was distracted as well by the political happenings of the day. On 24 February 1853, Lord John Russell, speaking before the House of Commons, re-sounded a call that had been heard at least since the 1830s for the removal of civil and religious disabilities for Jews. By this time, such disabilities had already been removed for Protestant Dissenters and Catholics, so, in essence, only Jews remained at this political disadvantage. One particularly eggregious case—and the one that at least partially motivated Russell's efforts—was that of Jewish banking magnate Lionel Rothschild, who had been elected Member of Parliament for the City of London in 1847 and again thereafter, but who was debarred from taking his seat by the requirement that he swear an oath "on the true faith of a Christian." In reelecting Rothschild, London's electorate consciously risked depriving itself—and *did* deprive itself—of Parliamentary representation to make the point that it had chosen the man it deemed best for the post.[75] It was thus a hot-button issue that Russell broached, and not surprisingly, it once again generated heated debate in Parliament throughout the months of March and April.[76]

Another matter of concern during this session of Parliament was national finance. On 8 April, William Gladstone, Chancellor of the Exchequer, introduced into Parliament various resolutions aimed at dealing with a national debt that had burgeoned owing primarily to a prolonged agricultural depression. His lengthy presentation met with an enthusiastic response, and on the eighteenth, he rose again to present his plan in detail. Five hours later, he had finished a speech punctuated repeatedly by cheers from the floor. In it, he called for a number of measures, among them, a progressive reduction of the income tax, an inheritance tax on real property, and, of particular interest to Sylvester, a number of measures involving government stocks and annuities (46–60). Ever hopeful of a new circumstance that would allow him more time for his mathematical researches, Sylvester saw in Gladstone's ideas a possible avenue into the government employ, especially if, this time, the Parliament publicly and officially recognized the fitness of Jews by passing Russell's bill. It could be a propitious moment.

On 23 April 1853, Sylvester wrote to Lord Brougham, one of the proponents of Russell's initiative and ever Sylvester's most influential contact, to offer his services. "I have thought," he wrote, "at this conjuncture when the Government are proposing to remodel the finances of the Country & to consider under some form to introduce the grant of life assurances as part of their scheme for dealing with the National Debt, I might from my attainments as a Mathematician & knowledge of business acquired during eight years painful experience as an Actuary be possibly able to render good service to the government & to the Country could I be brought into relation with the former as their confidential adviser or at least consultee in matters of calculation connected with the proposed changes."[77] Sylvester was certain that, if only Brougham would intervene on his behalf with an introduction and recommendation to Gladstone or to some other person of influence, he could help the country save or gain "many tens of thousands a year" given his "sound knowledge of principles, considerable experience in the details of computation and . . . inventive power for appreciating the effect of or devising new & useful combinations." And, just in case those qualifications were deemed insufficient, Sylvester let Brougham know that he had been informed that his name was about to appear "(unsolicited) on the list of those one of whom is to be selected a successor" to "my friend M. Chasles the eminent geometer" as a coresponding member to the Section of Geometry of the Paris Académie des Sciences.[78]

It was not, however, just the possibility of helping the nation that was on Sylvester's mind. He also had more self-centered motivations. "In a position more favorable to tranquility & satisfaction of mind," he confessed to Brougham, "I feel a strong persuasion of being capable of producing works that should prove me more worthy of the high distinction thus placed within my reach" by the Paris Academy. A government post, which would be a greater possibility if Russell's bill were successful, could, he thought, provide that peace of mind. It may have been wishful thinking to suppose that a place in the government would have been more congenial to mathematical research than life at 26 Lincoln's Inn Fields, but in the spring of 1853, Sylvester was stretched thin trying to balance the practical necessity of running the Equity and Law and his own personal need to pursue his mathematics, especially at a time when his mathematical researches—motivated by the problem of syzygies—were generating such a formidable yield. Six days after Sylvester wrote so hopefully to Brougham, however, the fate of Russell's bill was decided. It went down in defeat and with it Sylvester's hopes that the government's discrimination against Jews might officially come to an end.

Throughout this politically charged time, Sylvester had at least managed to maintain his mathematical focus. On Thursday, 16 June 1853, he presented before the Royal Society the enormous memoir "On a Theory of the Syzygetic Relations of Two Rational Integral Functions, Comprising an Application to the Theory of Sturm's Functions, and That of the Greatest Algebraical Common Measure" that he had composed during the spring. A month later, Cayley submitted his anonymous referee report on the paper to the Society for its *Philosophical Transactions*. The report was, perhaps not surprisingly, strong. Sylvester's paper, Cayley wrote, "appears to me to contain investigations of very great interest and importance in the branch of Mathematics to which they relate and as well on account of the value of the results actually obtained as from the progress which a paper so suggestive of further investigations is likely to lead to."[79] It was a gargantuan piece of work that ultimately ran to 141 printed pages, and, interestingly, it was the first that Sylvester had ever submitted to the Royal Society, despite the fact that he had been a fellow since 1839.[80]

In addition to its length, this paper was distinguished in Sylvester's mathematical output to date by its degree of organization. His multipartite 1852 paper "On the Principles of the Calculus of Forms," for example, had been open-ended. Sylvester started it knowing some of what it would contain, but he added on sections as new material developed. It was unashamedly a work-in-progress. His 1853 effort, on the other hand, was more of a well-thought-out and reasoned whole, despite its almost overwhelming magnitude. This was reflected in the lengthy introduction, in which Sylvester carefully outlined the contents of each of the paper's five sections. Characteristic of his ever active mind, however, new ideas did arise after the memoir's actual presentation before the Society, and this necessitated the addition of some supplementary material to the third and fourth sections before the paper went to press. Yet, even those additions were directly to the points of the sections and in keeping with the paper's overall structure. Only in the short "General and Concluding Supplement" did he permit himself to tack on some additional thoughts out of sequence. The paper represented, for Sylvester, a remarkably disciplined presentation, a fact even he seemed to appreciate. "Some of the theorems given by me in this paper," he wrote, "have been enunciated by me many years ago, but the demonstrations have not been published, nor have they ever before been put together and embodied in that compact and organic order in which they are arranged in this memoir,—the fruit of much thought and patient toil, which I have now the honour of presenting to the Royal Society."[81] Sylvester had, indeed, revisited some of his earlier work, in particular, on Sturm's Theorem, in an effort to gain insights into what the invariant-theoretic problem of syzygies might involve.

Consider the case of a binary form and the problem of determining a minimum generating set of invariants for it. As noted above, the elements in such a set could satisfy algebraic dependence relations that might prove difficult to detect. For example, suppose A, B, C, D are four invariants determining a minimum generating set for some binary form and suppose they also satisfy the algebraic dependence relation $16D^2 = AB^4 + 8B^3C - 2A^2B^2C - 72ABC^2 - 432C^3 + A^3C^2$.[82] At least two questions immediately arise: when do such dependence relations exist? and by what means can they be determined explicitly? In the cases of the binary quartic form and of the ternary cubic form, that is, a homogeneous polynomial of degree three in three variables, Salmon and Sylvester had already been discussing this issue as early as April 1852.[83] A year later, Sylvester tried to tackle the general problem by considering a radically simplified version of it.

Recall (from chapter 3) that Sturm's Theorem involved the calculation of a series of so-called Sturm functions derived from the Euclidean algorithm applied to an nth degree polynomial $f(x)$ and its first derivative $f'(x)$. Considering now "$f(x)$ and $\phi(x)$, two perfectly independent rational integral functions of x," Sylvester proposed in "On a Theory of Syzygetic Relations" to analyze each of the Sturm functions $f_i(x)$ at each stage of the Euclidean algorithm as "a syzygetic function of the two given functions" $f(x)$ and $\phi(x)$.

Consider, for example, the Euclidean algorithm applied to these two functions just to the second step:

$$f(x) = q_1(x)\phi(x) - f_2(x)$$
$$\phi(x) = q_2(x)f_2(x) - f_3(x)$$

These two equations can be rewritten as

$$f_2(x) = q_1(x)\phi(x) - f(x)$$
$$f_3(x) = q_2(x)f_2(x) - \phi(x),$$

or, substituting the first equation into the second, as

$$f_2(x) = q_1(x)\phi(x) - f(x)$$
$$f_3(x) = q_2(x)(q_1(x)\phi(x) - f(x)) - \phi(x)$$
$$= (q_2(x)q_1(x) - 1)\phi(x) - q_2(x)f(x),$$

where now both of the Sturm functions $f_2(x)$ and $f_3(x)$ have been expressed as a sum of products of the original functions $f(x)$ and $\phi(x)$, that is, expressed syzygetically in terms of them. Clearly, continuing this process, all the Sturm functions can be so realized.[84] Sylvester seemingly hoped that, by analyzing and exploring the behavior of syzygetic relations in this limited context of two homogeneous polynomials in one variable, light would be shed on the much more complicated problem of syzygies in the context of the invariant theory of binary—to say nothing of ternary and higher—forms. Although that hope went largely unfulfilled in 1853, many new results did ensue.

In line with his earlier work from 1839 and 1840 in which he had successfully expressed the Sturm functions arising in the case of a polynomial $f(x)$ and its derivative $f'(x)$ in terms of the n complex roots of $f(x) = 0$, Sylvester, in 1853, considered the more general setting of an mth degree monic polynomial $f(x)$, an nth degree monic polynomial $\phi(x)$, and a syzygetic relation between them denoted $\tau_i(x)f(x) - t_i(x)\phi(x) + D_i(x) = 0$, for $D_i(x)$ of degree i in x. The question Sylvester answered in the affirmative was, can $\tau_i(x)$, $t_i(x)$, and $D_i(x)$ be expressed explicitly in terms of the n complex roots of $f(x) = 0$ and the m complex roots of $\phi(x) = 0$ (458–483)? He next took these results and reinterpreted them in the particular case of Sturm's Theorem, that is, taking $\phi(x) = f'(x)$ (483–510).

With these algebraic preliminaries out of the way, Sylvester turned to a proof and novel interpretation of a theorem he had stated without proof in a paper in the *Philosophical Magazine* at the end of the summer of 1852, namely, his law of inertia.[85] For $Q = Q(x_1, \ldots, x_n)$ a real quadratic form of rank r, the law of inertia states that there exists a real nonsingular linear transformation that takes Q to $x_1^2 + \cdots + x_p^2 - x_{p+1}^2 - \cdots - x_r^2$, where p is uniquely determined. In "On a Theory of Syzygetic Relations," Sylvester not only provided the proof of this law but also showed how to use it—in conjunction with the extension of Sturm's Theorem from one to several functions that Hermite had effected in June 1852—to give "an *instantaneous* demonstration . . . of the applicability of my formulæ for M. Sturm's functions for discovering the number of real roots of $f(x)$, without any reference to the rule of common measure."[86] In other words, Sylvester managed to translate Sturm's *analytic* result into purely *algebraic* terms via an application of the law of inertia to a particular quadratic form that he termed the Bezoutiant and that he derived from a particular symmetric matrix. As he put it, "we are led to the following remarkable statement. '*An algebraical equation of any degree being given, an equation whose degree is one unit lower may be formed, all the roots of which shall be real, and of which the number of positive roots shall be one less than the total number of real roots of the given equation.*'"[87]

Sylvester then closed this algebraic *tour de force* with "a consideration of the properties and affinities of Bezoutiants . . . regarded from the point of view of the Calculus of Invariants," that is, "under a purely morphological point of view."[88] Returning to the evolving theory of invariants *per se*, he showed that the Bezoutiant, interpreted now for a system of two homogeneous *binary* forms f and ϕ of degree m, is a covariant of the system (549–551). In particular, it is one of a whole family of lineo-linear covariants—that is, covariants with the property that each summand is linear in the coefficients of both f and ϕ—of which Sylvester claimed to have the minimum generating set (554–559).

That Sylvester had high hopes for where the theory he had presented might go was clear in his final paragraph. "The foregoing theory," he told his readers, "took its origin . . . in meditations growing out of the celebrated theorem of M. Sturm. There appear to be several directions in which a development or extension of the subject matter of that theorem may be sought. Thus a theory may be constructed relative to a single function of one or more variables . . . Or, again, a theory may be formed in which the number of functions is always kept equal to that of the variables. . . . Finally, . . . we may construct a theory of a system of homogeneous functions equal in number to the variables in them" (571). Any and all of these cases had the potential to shed further light on the evolving calculus of forms and its associated theory of invariants. As Cayley put it in his referee's report, it was, indeed, a paper "suggestive of further investigations."[89]

For his part, Sylvester continued to think throughout the fall of 1853 not only about the implications of Sturm's Theorem *per se* in the determination of the real roots of a polynomial equation but also about the algebrization of the components arising in the Sturmian analysis.[90] On 6 September 1853, for example, he wrote to Salmon to tell him about the "very remarkable law" he had just found "concerning the Quotients which appear in Sturm's theorem."[91] September also found him working on yet another continuation of his paper "On the Calculus of Forms" for the *Cambridge and Dublin*, lecturing on Sturm quotients at the annual meeting of the British Association for the Advancement of Science (BAAS) in Hull and admitting that he had not had the time to prepare the report on the state of the theory of determinants that the BAAS had charged him with a year earlier at its meeting in Belfast.[92] He was clearly overcommitted, but he was intent on making his presence known in a British scientific community that had developed around journals like the *Cambridge and Dublin* as well as around both the Royal Society and the BAAS. His mathematical research had also never been stronger, but he was tired.[93] The mental exertions of 1852 and 1853—as he pushed the theory of invariants, conducted the business of the Equity and Law, and tried to secure a new position in life more amenable to the hard concentration deep mathematical work required—had exhausted him. It was time to regroup.

From Actuary to Academic?

Sylvester published eight short papers on as many separate topics in 1854.[94] He had lost his mathematical focus, and he had little time to devote to regaining it, given that he had taken on even more actuarial duties as consultant to the Law Reversionary Interest Society formed in 1853.[95] With new mathematical ideas now fewer and farther between, the demands of business took precedence over mathematical research.

Sylvester was, however, unwilling to let what he hoped would be a temporary redirection of his energies affect his visibility. In March 1854, when the Royal Society sent him a paper to evaluate, he thus continued to assert his mathematical authority as a referee. The paper was on the oscillatory theory of a simple pendulum—the very same topic that had so exercised him in 1851 that he had engaged in a polemic in the *Times*—and his judgment was once again brutal. "The reasoning . . . upon which all the conclusions of the author hang appears to me to be entirely gratuitous and the resulting formula which forms the basis of all which follows . . . is certainly erroneous," he informed the Royal Society.[96] Moreover, he was outraged at what he perceived to be the incompetence of the referee who had evaluated the paper before him and who had left pencilled marginalia on the review copy. "I ought to notice," he felt bound to state, "that the party who preceded me in the inspection of the paper . . . must have performed his task in a very perfunctory manner or it would have been otherwise impossible that he should have asked the question (enclosed in the cartouch on page 3)." Therefore, in Sylvester's view, "the Society are . . . bound to consider the opinion of the writer whoever he may be and whatever his opinion may have been (after so clear a betrayal of inattention) as entitled to little or no weight." The unspoken corollary was, of course, that Sylvester's judgment should categorically prevail!

Sylvester also took care in this mathematically slow period to continue to promote his work in quarters he deemed useful and worthy. On 14 April, for example, George Boole wrote from his position at Queen's College Cork in Ireland to thank him for sending a reprint of "On a Theory of Syzygetic Relations." Although Boole had to confess that he had not yet had a chance to go through the massive paper in detail, he nonetheless assured Sylvester that "I know sufficient of its general spirit & ideas to satisfy me of its importance."[97] Coming as it did from one who had in some sense initiated the British approach to invariant theory back in 1841, this was welcome praise. All the more so, because it arrived at a time when Sylvester was feeling vulnerable mathematically.

That vulnerability must only have been enhanced in early May 1854 when Cayley read his paper, "An Introductory Memoir upon Quantics," before the Royal Society.[98] The first paper he had presented to the Society since Sylvester had successfully proposed him for a fellowship two years earlier in June 1852, it was important.[99] Unlike Sylvester's rambling and open-ended series "On the Calculus of Forms," Cayley's "Introductory Memoir" systematically and succinctly laid a foundation for a theory of covariants (of which invariants are, recall, a special case) on pillars of differential operators analogous to those in (3) that he had discovered in 1851. Moreover, he introduced the combinatorial concepts of degree, order, and weight associated with a covariant that would become so crucial in the subsequent British attack on invariant-theoretic problems.[100] Sleek and elegant, Cayley's paper reflected a mathematical style very different from that of his friend, Sylvester.

By the end of May, Cayley had made even more progress. On the twenty-seventh, he reported to Sylvester on his efforts to determine the covariants of the binary quintic form. He had succeeded combinatorially in ruling out a number of odd orders for which there were no covariants and asserted that "all that remains to be found is therefore whether there are any covariants" of orders three, five, and seven.[101] He confidently added that "I shall accomplish this without much trouble." Although his confidence was ultimately unfounded— the binary quintic form would present difficulties unforeseen at this naïve stage in the development of the British approach to invariant theory (see chapter 9)— it must have weighed on a Sylvester who was making little mathematical progress of his own. He certainly did not wish his friend ill, but he *was* competitive, and his own work was paling in comparison.

Sylvester's thoughts had still not returned to his mathematical research by July, but they were very much focused nonetheless on mathematics and on the possibility, finally, of a career more reflective of both his self-image and his research ambitions. At the Royal Military Academy to the southeast of London along the Thames in Woolwich, the Professor of Mathematics, Samuel Christie, had retired and a search for his successor had begun. It would necessarily be a political appointment made through the War Office, but it was a teaching post in mathematics, and, unlike professorships at Oxbridge or seats in Parliament, it did not require the taking of an oath "on the true faith of a Christian." Maybe, Sylvester thought, he could actually have a shot at the job, especially if, once again, he could count on the political and behind-the-scenes support of Lord Brougham.

He could. By early in the month, Brougham had already set the wheels in motion. He had not only written a recommendation on Sylvester's behalf but also enlisted the support of the Secretary of State for War, Sidney Herbert, in Sylvester's cause. Herbert, in turn, had contacted Hew Ross, Colonel-Commandant in the Royal Horse Artillery and a key figure in the educational reform of the Royal Military Academy then under way. On 7 July 1854, Ross wrote to Herbert explaining the selection procedure. The Academy's Lieutenant Governor, Major-General Griffith Lewis, was ultimately responsible for hiring decisions, so Ross had forwarded to him both Sylvester's letter of intent and

Brougham's letter of recommendation. "I need hardly say," Ross added in closing his letter to Herbert, "that it will give me great pleasure if among the numerous candidates Mr. Sylvester should be found the most eligible."[102] An astute politician, Herbert read and reported on this to Brougham for exactly what it was. "It is a guarded answer," he said, "as such answers ought to be, but I hope that the attainments of Mr. Sylvester give him a good claim."[103] Brougham immediately forwarded both letters to Sylvester so that he, too, could see where things stood.

While these letters were passing through the corridors of military and political power, Sylvester was frantic to put together his case. At a brief personal meeting on Sunday, 9 July, he had forgotten to ask if Brougham might return the testimonial letters and other supporting documents Sylvester had presented to him in soliciting his support. Sylvester needed to forward them on to Ross, and time was of the essence. Among the materials were a complete collection of his published works to date, a copy of Francesco Brioschi's book-length treatise on determinants in which Sylvester's work was mentioned favorably, the two numbers of the *Nouvelles annales de mathématiques* in which Sylvester's recent papers had appeared, and a number of the *Comptes rendus* of the Paris Académie des Sciences, presumably the one recording the Academy's meeting on 13 March 1854 in which Sylvester's name was placed in the *second rang* for the distinguished position of foreign corresponding member to the Section of Geometry.[104] Reflective of the awkward position in which he felt himself, Sylvester added subserviently that "the documents in question of course do not include the copy of my paper bound in calf which I did myself the honor of presenting for your acceptance."[105]

In sending these materials first to Brougham and then on to the electors for the professorship, Sylvester clearly wanted to highlight the *international* character of his growing reputation. In Sylvester's view, candidates for academic jobs such as the one for which he was applying should be judged precisely on the basis of the quality of their published works and of the reputation thereby engendered. A case should thus be made even stronger if the candidate had succeeded in making a name for himself internationally. Sylvester, one of the new professionals and one who had already suffered the injustice at Cambridge of the adoption of criteria for advancement other than merit, had taken an attitude toward academe that was prevalent in Prussia and in France but that had not yet been embraced at Oxbridge. The mathematically better man—as proven through his published work and his documented reputation—should win. The question was, did the Woolwich electors share these values?

Sylvester continued to build his calculatedly international case throughout the month of July. In particular, he enlisted the help of his French friend, Charles Hermite, in securing a letter of support from one of the German mathematical greats, Peter Lejeune-Dirichlet in Berlin. Writing to Dirichlet on 25 July 1854, Hermite provided an interesting read on the job situation in England as compared with that on the Continent. "Mr. Sylvester, whose discoveries you know in elimination, on Sturm's Theorem and on the study of algebraic forms," he explained, "is actually a candidate for a professorship at the military Academy which is for England the analog of our École Polytechnique. But his works are less known among the high functionaries charged with choosing a professor than among the geometers of France and Germany. . . . You are aware, my dear Sir, how far England is from offering its *savants* as large a number of official positions as France or Germany. It is thus a unique opportunity that offers itself at this moment to Mr. Sylvester and the extreme importance that must be attached to such an affair will excuse, I hope, my approaching you."[106] Although Hermite was correct in his assessment of the uniqueness of the opportunity and of the relative lack of governmental patronage of the sciences in England, he was only technically correct in drawing an analogy between the Royal Military Academy and the École polytechnique. Whereas the French institution had, by midcentury, a decades-long commitment to appointing first-rate research mathematicians to its teaching staff, the same could not be said of its English counterpart. Research had not yet been adopted as a criterion in England for academic appointments.

That fact became all too apparent to Sylvester on 2 August, when he learned that the military electors to the Woolwich professorship had chosen a mathematically inferior candidate who had had a prior connection with the Academy. Sylvester's exact contemporary, the Reverend Matthew O'Brien had been educated at Trinity College Dublin and at Caius College, Cambridge, where he had been third wrangler on the Mathematical Tripos in 1838, the year after Sylvester's second place finish. Since 1844, O'Brien had served as Professor of Natural Philosophy at King's College in London and, since 1849, as lecturer in astronomy at Woolwich as well.[107] It had thus been, in some sense, an inside job, and Sylvester could barely contain his disappointment and his outrage in a letter to Brougham a week later. His case had been supported, he declared, "in the strongest language in which a recommendation could be clothed" by some of the most renowned mathematicians in Britain and abroad: Sir William Rowan Hamilton, Charles Graves, and George Salmon in Dublin, Philip Kelland in Edinburgh, and James Challis at Cambridge in the British Isles; Jean-Victor Poncelet, Michel Chasles, Jean-Marie-Constant Duhamel, Joseph Serret, Joseph Bertrand, and Charles Hermite in Paris; and with "letters . . . also written but too late to be sent in by the great Lejeune Dirichlet in Berlin, Professors Peters & Hesse, Königsberg, Professor Joachimsthal, Halle, [and] Professor Thomson, Glasgow and also from distinguished pupils testifying to my teaching powers."[108] The tacit question was, with support like this, how could his application have possibly failed? Deep-down, he knew the answer. His had been the wrong kind of case. Although he had not neglected to secure letters from former students, Sylvester had crafted his application around his research and international reputation. These factors, however, really mattered little to the authorities in 1854; it had been more important to be known to the electors, to have a connection with the Academy.[109] As Sidney Herbert put it in a letter to Brougham marked "Private," "the Ordnance people . . . have taken one of their own men . . . . I am sorry for it for I [would] like a little fresh blood in there."[110]

There would be no "fresh blood" in mathematics at the Academy in 1854. Sylvester had had such high hopes. The position would have been for him not only a way out of business but also a way into mathematical research. As he explained to Sir John Lubbock, the Vice Chancellor of University College during Sylvester's tenure there and an astronomer and mathematician of some note, the position would have "afford[ed] me a secure haven against the storms and distractions of the world and enable[d] me to devote as I desire to do all my powers for the remainder of my life to the service of Mathematical inquiry."[111] After years of trying to fulfill his actuarial duties at the same time that he pursued his mathematical research, after keeping his eyes open for possible new situations only to have his hopes repeatedly dashed, this had been his chance at the one kind of job an active research mathematician should have, in his view, and at an institution where, at least officially, he was not debarred on account of his religion . . . and he had not gotten the call.

The fall of 1854 was hard for Sylvester. On 3 September, he turned forty, still not having assumed what he felt was his appropriate place in the world. Cayley continued to make progress in the theory of invariants, writing to him on 12 October that he was "approaching the general solution of the problem" of determining the number of linearly independent covariants in terms of the number of irreducible covariants of the various degrees and then announcing, "Eureka," he had the general proof.[112] By contrast, Sylvester published a brief piece "On Multiplication by Aid of a Table of Single Entry" in the *Assurance Magazine,* reporting on an easy algorithm for using some of the standard sets of tables to multiply any two numbers by addition and subtraction alone. It was the kind of calculation that came up repeatedly in actuarial work and so merited wider diffusion, in Sylvester's view. As he put it, "any saving of labour of this kind, however slight in itself, when multiplied by the number of arithmeticians continuing their labours through countless ages, gives rise to an accumulation of savings which an ordinary feeling of benevolence must show the importance of not disregarding."[113]

The truth of this statement was then currently being driven home to him as he painstakingly prepared the second quinquennial report for the Equity and Law. Once again, the news was good. The company had continued to flourish, with the number of policies in force having risen from 542 at the time of the first quinquennial report in 1849 to 986 in 1854 and with the surplus distributed to shareholders up 350% over the five-year period.[114] The company was growing, nourished by Sylvester's actuarial practices, but all he could think about was leading a new and different life.

Late in November, he tried to make one key, personal change through a letter to Barbara Smith, the granddaughter of the liberal Unitarian and antislavery activist, William Smith, the illegitimate daughter of Benjamin Smith and his mistress, Anne Longden, and already an outspoken advocate of women's rights at the age of twenty-eight.[115] Sylvester had likely met Smith through her younger brother, Benjamin Leigh, perhaps the "pupil preparing for Cambridge" who was living with Sylvester at 26 Lincoln's Inn Fields in April of 1846 when Sylvester reported optimistically on his then-new career as an actuary to Joseph Henry back in the United States.[116] Smith *frère* had gone on to Jesus College, Cambridge in 1848, had come out thirty-first wrangler on the Mathematical Tripos in 1852, and had entered at the Inner Temple, only a stone's throw away from Sylvester, later the same year. That Sylvester had not only followed his progress but also sought to promote him was evident in the fact that he referred to Smith in print late in 1852 as "a promising young geometrician."[117] Just when Sylvester may have met Smith's sister, Barbara, is unclear, but on 21 November 1854, he wrote to her "conscious of an ever increasing charm in [her] society and conversation" and avowing that "I should regard myself as most favored by Providence were it possible for me to believe that you could accept the offer of my attachment and lifelong devotion."[118] At the time of this proposal, however, Smith had just refused another proposal and had already met John Chapman, the editor of the *Westminster Review*, the publisher in October of 1854 of Smith's anonymous essay, "A Brief Summary in Plain Language, of the Most Important Laws Concerning Women," and a notorious womanizer with whom Smith would have a tempestuous affair that would culminate in the summer of 1855 and precipitate her temporary breakdown.[119] Smith valued Sylvester as one of the friends in her intellectual, liberal, and dissenting circle, but she did not love him. "Let me hope," he closed, "that whatever other ill effect this note may produce it may not occasion me the loss of your friendship which I know how to prize at its full worth." At this key juncture in Sylvester's life, Smith chose not to marry, although she did remain his friend and correspondent. (See chapters 8 and 9.)

Sylvester's disappointment in love was followed almost immediately, however, by yet another hope of a new situation. Robert Edkins, the Professor of Geometry at Gresham College in London, had died on 11 November, creating an academic opening there. Founded in 1597 with an emphasis on scientific education for those in the capital city without the leisure to pursue studies at either of the ancient universities, Gresham College had had a long but checkered history. Its professors—of divinity, astronomy, geometry, music, law, medicine, and rhetoric at the time of its founding—were paid by the corporation to give free lectures to the public. It was thus not a college in the usual sense.[120]

Moreover, by the nineteenth century, it had been outpaced by newer institutions like University College London and seemed to have little sense of direction (46–51). Still, from Sylvester's point of view, a professorship there would have nevertheless meant liberation from the world of business and access to that of research.

Unfortunately, everything happened too fast. Sylvester was asked, with barely 48 hours notice, to give a probationary lecture before the selection committee on Monday, 4 December. Unable to secure an extension, he had to drop everything—the preparation of the quinquennial report included—to write a lecture that he "commenced and finished at a single sitting of a few hours' duration; the Author being so pressed for time that he had not even enough of it at his disposal to write out a fair copy of the manuscript."[121] Not surprisingly, the presentation did not go as well as Sylvester would have hoped, and he immediately had a version of the lecture printed so as not to "be wanting in proper deference to his judges, especially to such of them as were unable to attend in their places on the day of probation," by failing "to afford them an opportunity of considering it in print" (2–3). The ploy came to naught, despite the fact that the lecture in its printed form has been called "a powerful essay on geometrical science."[122] Morgan Cowie—Johnian, 1839's senior wrangler and second Smith's Prizeman, but no active researcher—got the call.

Except for the financial success of the Equity and Law, 1854 had been a year of successive disappointments for Sylvester. His frustration came out clearly in a letter penned hastily to Cayley in the heat of preparing the Gresham lecture. He had been trying to think about invariant theory, but, he told Cayley, "it is of course impossible for me not to require *time* for doing the task which is set before me."[123] "Add to this the impediments arising from the cares of life & business & not infrequent fits of Disgust & tedium causing long intervals of inactivity," he continued, and the picture was complete. He hated his job and had failed repeatedly to break out of it. He had again lost in love. He had also seemingly been abandoned by a fickle mathematical muse. He felt increasingly trapped by circumstances as 1854, at long last, drew to a close.

CHAPTER 6

A New Beginning

I have now all that I most desire—leisure, a position & definite
object in life, and a competence sufficient for my wants with
unlimited opportunities for self-improvement and
rendering myself useful to others.
—J. J. Sylvester to Lord Brougham, 21 September 1855

The new year opened with Sylvester and the country in a dark mood, albeit
for different reasons. England had been at war in the Crimea for nine
months, allied with the Turks, the French, and ultimately with the Sardinians
against long-standing threats of Russian aggression on the Ottoman Empire
that, if unchecked, could have resulted in naval commandeering of Mediterranean seaways so critical to British imperial concerns. In January 1855, news
from the front was not good. Winter had set in with British troops, under-
provisioned and ill-clothed, hunkered down between Balaklava and the key
Russian port of Sebastopol. Conditions were unspeakable and the death toll
high, despite the widely reported efforts of Florence Nightingale and others.
Countrymen back home read in horror day-by-day accounts transmitted tele-
graphically from the war zone and began seriously to question the leadership
in the field, in the War Office, and in the Parliament. To contemporaneous eyes,
the impossible was happening; the economically, technologically, and culturally
advanced British were being pummeled by the backward Russians.

From its opening day on 23 January 1855, Parliament was in turmoil, with
calls from the opposition for, among other things, close scrutiny of the Army
with an eye to its preparedness and a formal inquiry into how Lord Aberdeen's
coalition government had been administering the war. There was much drama.
One prominent member of that government, the same Lord John Russell who
had unsuccessfully introduced the bill to remove disabilities for Jews in 1853,
resigned as Foreign Secretary on the twenty-fifth. Addressing a packed House
of Commons the next day, he acknowledged the unthinkable nature of the

British situation in a speech repeatedly punctuated with raucous cries of "Hear, hear, hear!"[1] "No one can deny the melancholy condition of our army before Sebastopol," he admitted. "The accounts which arrive from that quarter are not only painful, but they are horrible and heartrending." Yet, he continued, "if you had been told as a reason against the expedition to the Crimea last year, that your troops would be seven miles from a secure port . . . and that at seven miles' distance they should be in want of food, clothes, and of shelter, to such a degree that they should perish at a rate of from 90 to 100 a day, I should have considered such a prediction as utterly preposterous, such an objection as fanciful and unjust." The situation was truly appalling, and by February 1855, Aberdeen's government had fallen to be replaced by a Liberal coalition under the bombastic and popular Lord Palmerston that would last, with the exception of a fifteen-month interlude in 1858–1859, until Palmerston's death in 1865.

Like the country, Sylvester fell into a state of introspection, reevaluation, and change in the winter, spring, and summer of 1855. What had gone wrong? What steps could be taken to put things right? How best could those steps be implemented? Both in the macrocosmic context of a nation at war and in the microcosm shaped by one man's efforts to be a mathematician, these questions were still open.

Changing Responsibilities

Already in the fall of 1854, Sylvester had taken one step that he hoped would cement his reputation within the emergent mathematical community in Great Britain and in Europe. In February 1854, Norman Ferrers, Thomson's successor as editor of the *Cambridge and Dublin Mathematical Journal*, was under increasing pressure from his publisher, Macmillan & Co., to cease operations; the journal had long been both difficult to produce and insufficiently profitable from the company's point of view. As he had on several occasions since assuming the editorship in 1852, Ferrers consulted not only Thomson but also Sylvester on how best to handle the problem. "Sylvester has often told me," he reported to Thomson, "that he is quite sure the publishers of the Philosophical Magazine [Taylor & Francis] would be glad to take it up and would advertise it extensively and energetically."[2] Clearly, Sylvester refused to entertain the notion of canceling the journal. For him, it was simply a matter of finding the right publisher. He was confident that the mathematical endeavor to which he was so firmly committed could and would be supported, and that confidence proved justified. The journal found a new publisher (although the London firm of John W. Parker and Son and not Taylor & Francis) and acquired a new name and editorial staff. By October 1854, plans were under way for the inaugural issue early in 1855 of the *Quarterly Journal of Pure and Applied Mathematics* with Sylvester and Ferrers as editors assisted by Stokes in Cambridge, Cayley in London, and Hermite in Paris.

The new name was significant. This would not be a journal linked to national geography as its two predecessors, the *Cambridge* and the *Cambridge and Dublin*, had been. It would take its place alongside the similarly named *Journal für die reine und angewandte Mathematik* in Germany and the *Journal de mathématiques pures et appliquées* in France as a vehicle, in its editors' words, for "communicat[ing] a general idea of *all* that is passing in mathematical circles, *both at home and abroad*, that can be of interest to Mathematicians as such."[3] In other words, like mathematics itself, it would transcend national boundaries and serve "Mathematical Science." There was, however, an overtly nationalistic agenda as well. "At a period when Mathematical Science is putting forth new powers and induing itself with a more perfect and vigorous form of organisation,—when it is daily extending its domain over the laws which govern the material universe, and at the same time opening out fresh and unthought-of paths of research in the regions of abstract speculation,—it has been felt," the editors admitted, "that it would be little creditable to English Mathematicians that they should stand aloof from the general movement, or else remain indebted to the courtesy of the editors of foreign Journals, for the means of taking a part in the rapid circulation and interchange of ideas by which the present era is characterised." Sylvester, first as one of those responsible for "opening out" the "fresh and unthought-of" path of invariant-theoretic research and now also as editor of the *Quarterly Journal*, saw himself as poised to wield power in the international mathematical arena.

Unfortunately, the timing could not have been worse for taking on such a new responsibility. The fall of 1854 had followed immediately on the failed Woolwich application, had continued with the intense labor at the Equity and Law preparing the quinquennial report, and had ended with both the refused proposal of marriage and the failure at Gresham College. Little wonder, then, that as the new year of 1855 opened, Hermite had heard nothing from Sylvester since early in October of 1854, despite the fact that the Frenchman had obliged his English friend by sending material for the new journal.[4]

With the quinquennial report behind him in the winter of 1855 and with the impending March deadline for the appearance of the journal's first number fast approaching, Sylvester was more mathematically engaged. In particular, he reworked for publication in the first number a paper "On the Change of Systems of Independent Variables" that he had presented earlier to the Royal Society for its *Proceedings*, although, as he confessed, "want of leisure prevented me then, and still prevents me, from producing the proof of the theorem, or the investigation by which I arrived at it."[5] He had managed to do better than polish up earlier material before the first issue went to press, however, sketching his discovery of a new calculational result in partition theory and working to stimulate new mathematical results by posing "A Question in the Geometry of Situation."[6]

Sylvester also sought and received help and advice for the journal from a number of members of his extended mathematical circle. First and foremost, his friend and editorial "assistant," Cayley, provided two articles, posed a problem, and shared some correspondence on cubic forms, all of which appeared in the first number.[7] Similarly, William Spottiswoode, an 1845 Oxford graduate and mathematician-cum-printer who shared Sylvester's research interest in the theory of determinants, published two short works on applicable mathematics,[8] and Sylvester's former teacher and colleague, De Morgan, submitted a paper but, perhaps more importantly, supplied a steady stream of suggestions on the sorts of issues the journal might address.[9] Writing on 14 February, for example, De Morgan offered that "a small space of the Journal might well be given to correcting errors of demonstration in fundamental theorems—and insufficiencies. There are more," he added, "than any one dreams of who does not come again & again into contact with the demonstrations in teaching."[10] William Thomson, too, supported the reincarnation of his former journal with an applied paper "On the Thermo-Elastic and Thermo-Magnetic Properties of Matter."[11]

Despite all this support, Sylvester was still uneasy. In a letter dated 13 March 1855 belatedly expressing his enthusiasm for a submission from Thomson, he conveyed his anxiety about getting out the first issue. "Many apologies for the delay," Sylvester wrote. "I had to see the printers & to speak to Ferrers. We shall be *most glad* to have your paper for the present number and if you send it by the 21st it will enable us to keep our engagement with the public to be out on the 31st Inst, which as we have a character to gain for punctuality (as successor to the Cam[bridge] Journal) is most desirable."[12] Reliability was certainly important but so was quality, and Sylvester did not hestitate to brag a bit to Thomson about the submissions slated for publication. "There will be some good papers by De Morgan, Cayley, Hermite, . . . and one of mine containing the solution (not the proof) of the general problem of the change of systems of any number of variables for any number of variables and connected together by any number of equations." Moreover, Sylvester immodestly noted, his own paper "completes the Diff[erential] Calculus as an Engine of Development." In Sylvester's less than impartial eyes at least, the journal would clearly be off to a most auspicious start.

Although it missed its target publication date of 31 March—it had indeed proved to be the case as De Morgan had consoled that "first numbers are always difficulties"—the *Quarterly Journal* was out before the end of April, and De Morgan, for one, was already providing material for the next number.[13] The journal's appearance brought with it more than the opportunity to convey new mathematical ideas, however; it revealed a new dimension of Sylvester's mathematical persona. Although he had long aligned himself with mathematicians in

France and had more recently made some overtures either directly in the case of Borchardt or indirectly as with Dirichlet to mathematicians in Germany, Sylvester had expanded his mathematical sphere to Italy as exemplified by the "Extract from [a] Letter of Signor Enrico Betti to Mr. Sylvester" that appeared in the journal's first number.[14] Sometime, perhaps in the late summer or early fall of 1855 after his unsuccessful candidacy at Woolwich, Sylvester had journeyed to Italy and had met Betti, among others, in Florence. Nine years Sylvester's junior, Betti, at the time a Florentine high school teacher, had made a name for himself as a mathematician through his work in Galois theory and in the related theory of substitutions. In fact, it was his latest result in the latter area about which he had written Sylvester in March of 1855 and which Sylvester had included in the "Correspondence" section of the journal.[15] This marked the beginning of a relationship between Sylvester and the mathematicians of Italy that continued to strengthen in the 1850s and 1860s.

With the immediate trials of publication over, Sylvester worked to clear his desk of a different mathematical matter, yet another paper to referee for the Royal Society, and he was more than a bit testy. It was a paper by Spottiswoode, who had just done him the good turn of supplying material for the *Quarterly Journal's* first number, but it was on Sylvester's special subject of invariant theory, and he did not find it up to snuff. "I feel placed in a most difficult position in being called upon to pronounce upon [the paper's] fitness for insertion in the Transactions," Sylvester wrote. While he had to admit that "the memoir shows great industry and has some points of interest; that it contributes materially to promote the existing state of knowledge of the subject in which it treats is open to question."[16] Cayley, the other referee, concurred with his friend's assessment, but Sylvester did more in his report than merely cast doubt upon Spottiswoode's paper. Showing the strain of producing the first number of his own journal, he "beg[ged] to add" to the Society's Committee of Papers "that the task of arbitrating upon papers of the class which have been from time to time submitted to me is so thankless in itself and the principle upon which the selection is made is in my humble opinion so objectionable as letting in the influence of fear of giving offense & favors . . . that I must beg most respectfully but most urgently to convey . . . my exceeding unwillingness to be again called upon to act in the capacity of referee upon papers submitted for insertion in the Transactions."

This outburst reflected more than the release of anxiety pent up over production matters during the winter and spring. Sylvester now had his own journal to run and to supply with quality material. There, he, together with Ferrers and the other editors, set and controlled standards unfettered by extramathematical concerns like ruffling the feathers of a Society member, even if they *were* fettered by those like regularly filling their pages. Unsuitable submissions could simply be returned; refereeing them was not required. His sense of the differences between his own independent journal and a society publication like the *Philosophical Transactions* came through clearly in his report. "If Mr. Spottiswoode on full consideration prefers that the paper should appear in the Transactions I do not think that the insertion can be refused to him," he told the Committee of Papers. Sylvester viewed that as Spottiswoode's right as a fellow, even if "he would act more wisely and be consulting best the interests of his own reputation" by delaying "until he has . . . produced a result which shall do more ample justice to his powers as a discoverer." The *Quarterly Journal*, unlike the *Philosophical Transactions*, reflected, at least in principle, a purer meritocracy where only the quality of the mathematical research mattered. It was that kind of meritocracy—blind to religion and social class—that Sylvester had been seeking since his undergraduate days at Cambridge.

He had also long been searching for a viable alternative to his position at the Equity and Law Life Assurance Society. On 22 May 1855, he finally made the break, resigning his full-time post and accepting the position of consulting actuary.[17] This meant the loss of a set income but, at the same time, the liberty to pursue his mathematical researches and to focus on his new editorial duties. It also meant that he would need to find a livelihood that would not sacrifice his new-found freedom. Fortunately, a prospect actually presented itself in mid-June.

Since coming to power in February 1855, Lord Palmerston's new government had been busy enacting reforms aimed at rectifying the egregious situation that the Crimean War had been and was daily proving to obtain in the British Army. One of those reforms involved the Royal Military Academy and the selection of its future officer corps. Effective July 1855, each candidate for a place at the Academy would have to pass a battery of rigorous subject examinations of which the one in mathematics was mandatory. Successful candidates would thus be determined on the basis of their combined scores. No longer would entry be based on a marginal "pass" examination and the personal nomination of the Master-General of the Ordnance. No longer, in other words, would political and social connections outweigh intellectual merit in securing the future engineers and artillerymen for Britain's Army.[18] This represented a major change, and, not surprisingly, it was not universally applauded. On 23 June 1855, for example, "A Parent" wrote to the *Times*, wondering accusatorily if it was "possible that Government will inaugurate their military reforms with a breach of faith?"[19] "Parents have, trusting to promises made by the Master-General," this parent continued, "educated their sons with a view to admission into the academy, after an examination conducted with a certain fixed standard; and are they now to be told that this standard is no longer to be fixed, but that their sons must compete with all the world?" The answer was quite simply "yes," and Sylvester resolved to try for a place as one of the examiners charged with evaluating the mathematical competence of those into whose hands Britain would ultimately entrust its military. Once again, however, he would need help behind the scenes.

Writing to Lord Brougham on 14 June 1855 Sylvester announced that he had submitted his application for one of the examinerships and solicited his support, asking if Brougham would write on his behalf to Lord Panmure, the Secretary of State for War ultimately in charge of making the selection.[20] As he had so many times in the past, Brougham willingly complied, and it was an exuberant Sylvester who wrote back to him on 10 July to impart what for once was good news. "Your interference [?] with the authorities has been successful & I have this day received the notification of my appointment as examiner at Woolwich from L[or]d Panmure," he wrote. "I am perfectly confident (from what has happened to me in other quarters and on other occasions) that nothing less than the most powerful influence such as your Lordship can command would have availed to obtain for me such an appointment in the public service and to have countervented the operation of the prejudice of race & religion."[21] There was, however, another point to be made. "The emolument is I believe inconsiderable," Sylvester conceded, "but not so (I venture to submit) the moral of the result as constituting a practical admission (the first I think in this country) that a Jew as such shall not be debarred from public situations for which he is competent." It was a moral as well as a personal victory for Sylvester, and it had come at a critical moment. His responsibilities were changing. He was moving out of business and into mathematics—as a researcher, as an editor, and now as an examiner. The year before, O'Brien had been named to the professorship at Woolwich largely because of his prior association with the school. Could this possibly be Sylvester's entering wedge into the system as well?

Despite, or perhaps because, of his triumph, Sylvester spent a hectic and emotionally draining July and August. His examination in mathematics had been set, administered, and graded by mid-August, although the final admissions results had not yet been announced. With scores of aspirants for the 60 places, the grading, in particular, had been a travail. Moreover, just as he was fully embroiled in the examination process, he had to move from his lodgings at 26 Lincoln's Inn Fields and find a new place to live. He was thus in an "unsettled state of mind" when he wrote on 13 August to report belatedly to a Cayley in transit to the BAAS meeting in Glasgow on his doings and his whereabouts.[22] "I truly regret that I have been able to avail myself of late so little of your society . . . [b]ut it has been as I have said, the necessary result of . . . the dislocation of my habits of 10 years standing." He had moved to "the private address of 68 Mortimer St. Regent St." midway between but north of the Inns of Court and Hyde Park. No longer would he and Cayley be able simply to cross through the Inns to carry on their mathematical discussions. At least with the examinations over, Sylvester could try to think about some mathematics to discuss, but even those efforts were "at an absolute standstill." "Happy you," he added wistfully, "who never flag, and never weary!"

A fortnight after penning this letter, Sylvester had intended to leave London for a time in order to regroup, but his plans most unexpectedly changed. Matthew O'Brien, who had been awarded the professorship of mathematics at Woolwich just the year before, had died on 22 August, and the post was once again open. As Sylvester told Cayley on 8 September, "I am *again* an applicant which keeps me in town."[23] Less than a week later on the sixteenth, what had for so long eluded Sylvester, an English professorship of mathematics, was finally in his grasp. Writing triumphantly to Lord Brougham, he announced "that I have this day been appointed Professor of Mathematics at Woolwich by Lord Panmure."[24] It was a dream come true, and he thanked his political benefactor for his many efforts on his behalf. Even his help in securing the examinership, so seemingly inconsequential in the grander scheme of things, had, Sylvester thought, played a key role in his success. "Fortunately," he explained, "during the brief period of my examinership opportunities occurred (which I was careful to improve) of evincing a disposition to make myself useful beyond the strict line of my duties and thereby securing the good opinion and the good will of Mr. Monsell who is the official adviser of L[or]d Panmure in all these matters; so that I had the advantage both by direct and reflected light of the influence which you put forth in my favor." Thus freed from the yoke of business and secure in academe, Sylvester could finally devote himself unencumbered to his research. As he put it rhapsodically to Brougham, "I hope henceforth (apart from all meaner objects of ambition) to consecrate my life and powers to the study & humble imitation of the spirit of the great masters of our science, to serve and if possible advance [it] which is in my mind one of the noblest objects of ambition which one can propose to oneself. Excuse me my Lord," he continued, "if in the first moments of satisfaction at a success almost despaired of, I indulge in a somewhat superfluous but I can assure you a perfectly sincere expression of my innermost feelings in relation to it." Eight days later on Monday, 24 September 1855, Sylvester embarked upon his new career.[25]

The Royal Military Academy, Woolwich

The Royal Military Academy had begun humbly in a single room in Woolwich in 1741 with two masters, each of whom lectured six hours a week to "the raw and inexperienced people belonging to the Military branch of [the Ordnance] office in the several parts of mathematics necessary to qualify them for the service of the Artillery and the business of the Engineers."[26] Almost immediately, the student body expanded from officers to non-commissioned officers to cadets, until by the late 1850s, the Academy had grown to include "an imposing pile of buildings" to house and educate some two hundred cadets aged seventeen to twenty and to consist "of 18 officers on the military staff, and some fifty professors and masters in the civil and educational corps" teaching theoretical subjects ranging from pure and applied mathematics to modern languages to fortification to landscape drawing to history and geography as well as the practical subjects of surveying, artillery, and fieldworks.[27] Cadets typically spent between two-and-a-half and three years at the Academy, taking subject examinations twice yearly and proceeding only if they scored a sufficiently high number of points. At the end of the full course of study, those students who ultimately achieved the most cumulative points on their final examinations received commissions in the engineers, with those in the lower tier proceeding to the artillery.

On paper, at least, it was a well-regulated system. The day-to-day reality of the situation was somewhat different. "The moral tone of this military college has never, we regret to say, been of a very high order," stated the author of an article on "Military Education" published in *Blackwood's Edinburgh Magazine* in November 1857. "Excellent men have been at the head of it, and the ability of the professors and teachers appointed to instruct admits of no question. Yet few right-minded officers look back upon the years spent in the cadet barracks except with disgust. It is not very difficult to account for the circumstance. Long after Continental nations had seen the absurdity of pressing upon boys the sort of training which belongs to men . . . we . . . persisted, both at Woolwich and elsewhere, in our endeavor to accomplish an impossibility."[28] Colonel Joseph Portlock, the Academy's Inspector of Studies since 1851, concurred. In his view, the first problem was that the officers in charge of the boys tended to be too young and inexperienced themselves, and although "such officers may indeed be able to superintend drill," they fell short when it came to "moral training." "We know how to deal with boys so long as we recognize their boyhood," the argument continued, "but we no sooner dress them up in uniform, and affect to treat them as soldiers, than we lose all moral control over them. They smoke, drink, swear, and fall into other vices, not because they are overcome by any irresistible temptation, but because they look upon such acts as tokens of manhood." The Woolwich of the 1850s—not unlike the Virginia of the 1840s—could thus be a rather raucous place where adolescents tested the limits of the rules imposed upon them.

Among those rules was the mandatory course of study in mathematics. As noted, after 1855 to gain admission to the Academy, cadet candidates were required to take and score sufficiently well on an examination in mathematics. This test covered geometry (the first six books of Euclid's *Elements* plus several additional topics), arithmetic, algebra (including quadratic equations, arithmetic and geometric series, and logarithms), plane trigonometry, mensuration, conic sections, and elementary mechanics (including the basic principles of statics, dynamics, and hydrostatics). It was thus assumed, wrongly as it turned out, that all entering cadets had an adequate mastery of these more elementary topics in pure and applied mathematics. Their Academy course in mathematics then moved on to the more advanced topics of spherical trigonometry, the differential and integral calculus, practical mechanics (including the calculation of strains and the pressure of earth on retaining walls as well as the theory of work), elementary optics, and descriptive geometry.[29] No cadet could receive his commission until he did sufficiently well in these subjects to proceed to and pass the final examination. For those cadets for whom mathematics was not that strong a suit, this requirement proved onerous. Suffice it to say that the mathematics classroom was perhaps not their favorite place!

The Woolwich of the 1850s also found itself under a Parliamentary microscope. After the Crimean War finally came to an end on 30 March 1856, the Parliament repeatedly addressed the issue of military reform in the wake of Britain's dismal, although ultimately triumphant, performance in the conflict. On 5 June, for example, Sidney Herbert, who had lobbied hard but ultimately unsuccessfully for reforms as early as 1854 as Secretary of State for War under Lord Aberdeen, returned to the floor of Parliament to re-sound his call, this time specifically targeting military education. "Recollect," he told his colleagues, "that by every day's delay you raise up fresh difficulties; the recollection of past evils will pass away; people will begin to acquiesce in the old routine; and above all, you create fresh barriers to the efficiency of the army by admitting more uninstructed and incompetent officers; you create vested interests, and postpone to an indefinite period the ultimate attainment of this desirable object."[30] His speech, greeted with loud cheers of approval at its conclusion, was consonant with actions that had already been taken by Lord Palmerston's government.

Early in 1856, Palmerston's Secretary of State for War, Lord Panmure, had formed a Commission on Training Officers for the Scientific Corps with the charge first to visit and study not only the various military-training institutions in Britain but also those in France, Prussia, Austria, and Sardinia and then to make recommendations on how best to improve the education of Britain's officers.[31] A year later, the so-called Yolland Commission issued a detailed report based on its observations and on extensive testimony given by professors and instructors, artillery officers and engineers, and other members of the British Army. Of the professors, the newly hired Sylvester testified candidly that "my experience in the Academy has left me under the painful impression that the Cadets are just beginning to take an interest in their studies and to comprehend their scope, when they are about to quit them, too often never to be resumed." Moreover, he added, "the attainment of the cadets on leaving the Academy . . . are insufficient, meagre, and unsatisfactory."[32] Two main changes resulted from this and other, similar testimony. First, the average age for admission into the Academy was increased to between seventeen and twenty; before the institution of the open entrance competition in 1855, the Master-General of the Ordnance had often admitted boys barely in their teens. Second, a Council of Military Education was formed in 1857 that eventually oversaw and regulated the syllabi, teaching, and examining at each of the institutions responsible for the Army's education.[33] Sylvester had thus joined the civilian teaching staff of the Royal Military Academy at a pivotal point in its history.

Professor of Mathematics Once More

Some ten miles as the crow flies from the center of London to the southeast, Woolwich was a pleasant journey from London that passed through Deptford and Greenwich along the southern shore of the winding Thames.[34] By train, it was barely more than a half-hour away from the hustle and bustle of the capital city. The Academy, stretched along the southeastern side of the broad Woolwich Common on which cadets regularly performed their drills, was an expansive but symmetrical and well-proportioned complex. In its center stood a four-story, battlemented castle with four lofty domed turrets. Single-story, crenellated arcades extended to the left and right, connecting the central building to two three-story compounds that carried on the military architectural themes. For its Professor of Mathematics, the Academy provided a "commodious" private house "with a good garden" at K Woolwich Common and an initial annual salary of £500 supplemented by medical coverage and the right of pasturage on the green.[35] The salary made it possible, furthermore, given the demands of maintaining an entire house, for Sylvester to employ a womanservant to keep things in order.[36] In taking the Woolwich job, Sylvester had thus come up in the world in more ways than one. As he put it to Lord Brougham, "I have now all that I most desire—leisure, a position & definite object in life, and a competence sufficient for all my wants with unlimited opportunities for self-improvement and rendering myself useful to others" (see epigraph note).

Sylvester's classes met Mondays, Tuesdays, and Wednesdays, leaving him in principle four consecutive days each week for his own mathematical research. In practice that first year, however, he was run somewhat ragged performing the duties of the professorships of both mathematics and natural philosophy. Combine that with the tasks of getting out the December 1855 and the March 1856 numbers of the *Quarterly Journal*, and he had his hands full.

Still, his new duties did not prevent him from thinking at least a bit about mathematics. His work in the natural philosophy classroom, for example, inspired a self-proclaimed "Trifle on Projectiles" as well as a not unrelated letter to the editors of the *Philosophical Magazine* on "Professor Galbraith's Construction for the Range of Projectiles," but, all in all, the return to a professorship of natural philosophy after a fifteen-year hiatus had not proved happy.[37] Liberated from his double load of teaching by the end of the spring term in 1856, Sylvester tried to regroup, attending a gala affair in June hosted at

Mansion-House by the Lord Major and Lady Mayoress of London in honor of "those whose names had been rendered illustrious by their labours in various departments of science" and spending part of the late summer in Paris.[38] On his return, he reported to Cayley on holiday in the Lake District that "I have made a start in bringing order into my paper on Cubic forms" and hastened to assure him that "you will find me restored to Mathematics on your return & I shall hope to see a good deal more of you than I have been able to do in the last twelvemonth as I am relieved from that portion of my work which pressed so intolerably upon my spirits and hindered me from entering into any mathematical speculation."[39] Indeed, in the fall of 1856, Sylvester succeeded in getting into an intellectual, social, and professional routine that involved a mixture of mathematizing, regular interaction with his friends, and teaching and examining the Academy's young cadets.

Anxious to begin establishing that routine in mid-September, he anticipated both seeing Cayley again and beginning the new academic year when he wrote to his friend on Tuesday, 16 September to chide him for not having come as promised to visit in Woolwich the preceding Sunday. "When shall I see you?," he asked. "Hermite, Terquem, as I have told you, Tortolini, all have charged me with messages for you."[40] Writing ten days later, he was still trying to get Cayley to come to Woolwich. He had invited Wheatstone, to dinner on Thursday, 2 October and hoped that Cayley would join them.[41] He also had some mathematics to tell Cayley about—"a new species of invariants" associated with the ternary cubic form that "have been in my possession 7 years."

Cayley did come, and he brought with him one of his own evolving ideas on transformations of cubic forms to discuss.[42] Although it is not clear how Wheatstone may have fit into a conversation at times dominated by the mathematical intricacies of the cubic, Sylvester was clearly energized by the interchange. Eight days later on the tenth, he announced to Cayley that "I have made a prodigious and most unexpected discovery in my theory of cubic forms." After sketching the result, he averred that "if I am not laboring under a delusion, a most wonderful step in advance has been achieved in this theory."[43]

Cubic forms continued to occupy him throughout the fall as he taught his cadets and edited material for the fifth number of the *Quarterly Journal* that would appear in January 1857. Sylvester was "delighted" when Cayley proposed to come back out to Woolwich on Saturday, 25 October to continue their discussions about the cubic and further coaxed him on the twenty-third with the possibility that "by that time I shall have a grand new development to discuss with you."[44] Sylvester thought that he might have found a way of making ternary cubic forms depend on quaternary cubic forms—that is, cubic forms in four variables—and, by so doing, have a procedure in hand for determining rational points of ternary cubics of the form $x^3 + y^3 + Az^3$. This was the same type of problem that Sylvester had made limited headway on nine years earlier in 1847.

There was still work to be done, and, for him, that meant talking through his evolving ideas with whomever would listen. Like Cayley, John Graves, Sylvester's longtime friend and former colleague at University College, had a sympathetic ear. On 24 October, Graves took a moment out from his legal work at the Poor Law Board to let Sylvester know that "it is very likely that I made several mistakes when I last saw you in speaking from memory" on cubic forms, "a subject wholly diverse from the current of my recent thoughts."[45] It was not, however, a subject "diverse" from the thoughts of Sylvester's most faithful sounding board. A week later on Friday, 31 October, Cayley wrote to tell Sylvester that "I like the theorems in your letters very much" and proposed that "if you are so inclined to walk Blackheath ways viâ the Shooters Hill road the whole way, i.e. from your side of the common on Sunday, I will walk that way to meet you."[46] The Cayley family home was in Blackheath to the west of Woolwich, and Cayley, the inveterate walker, was ever ready for a stroll.

Although it continued well into November, the intense mathematical conversation between Sylvester and Cayley ultimately led, on Sylvester's part, only to a brief note published without proofs in Barnaba Tortolini's *Annali di scienze matematiche e fisiche*, Sylvester's first paper in an Italian journal.[47] Matters at the Academy required his attention, not the least important of which was trying to fit in and to make new friends.

Sylvester soon got to know his neighbor at L Woolwich Common, Colonel F. M. Eardley Wilmot, the Superintendent of the Royal Gun Factory located at the Woolwich arsenal.[48] Late in October 1856, Sylvester wrote to Eaton Hodgkinson, Professor of Mechanical Principles of Engineering at University College London and a well-known expert on the strengths of materials, to see if he might be willing to give Wilmot a demonstration of his machine for testing the strength of metal beams. Hodgkinson complied, and Sylvester and Wilmot made the short trip into London to meet him at noon on Saturday, 1 November.[49]

Sylvester also had to establish a good working relationship with the mathematical masters who worked under him and who thereby saved him much of the aggravation of drilling the cadets in their mathematics. Of them, John Heather, was just one year Sylvester's junior and had been a fellow sizar with him at St. John's in 1832 and 1833 until he transferred to Peterhouse in 1834. Earning his B.A. in 1839, he joined the mathematical staff at Woolwich the following year and remained in his post until 1860.[50] Three of the other masters, George Boddy, Frederick Vinter, and William Racster, were also Cantabridgians, the first two Johnians and the third a Peterhouse man. Vinter, the sole wrangler among them (having come in third in 1847), only stayed at Woolwich from 1855 until 1858 when he became Sylvester's counterpart at the Royal Military College, Sandhurst. Sylvester nevertheless found himself at the head of a remarkably homogeneous mathematical team.

From late November through early December, that team labored over the examinations in the various classes. By 10 December, the ordeal was once again over, and Sylvester had a chance to address some of the correspondence that had piled up during the examination period. In particular, he penned a short note to Lord Brougham, who had written him a week earlier about a mathematical matter, to let him know of his travel plans and how to reach him. "I propose leaving for Italy (to spend my Christmas vacation there) on Saturday next," he explained, where "I shall probably remain until the last week in January between Naples and Rome returning to Woolwich to resume work on the 9th of February 1857 when our Vacation ends."[51] "Any communications with which you may be pleased to honor me," he added, "may be addressed to me Poste Restante Naples." With his affairs in order, Sylvester set off on his journey to warmer Mediterranean climes, where he hoped to cement further his relations with the mathematicians on the Italian peninsula.

His first stop was the port city of Naples, where he visited Giuseppe Battaglini, then a thirty-year-old private instructor of mathematics; in 1860, Battaglini would be named to the new chair of higher geometry at the University of Naples that would result as part of the reforms aimed at unifying the Italian states. During their conversation, Battaglini had occasion to ask "l'illustre Prof. Sylvester," as he termed him, what he thought about the work of his Florentine colleague Betti. In a letter to Betti, Battaglini quoted Sylvester verbatim: "Sig.r Betti has a very good mind but his language is not so clear: I spent two nights on his researches on groups without managing to understand them; Cayley, who has perhaps the most perspicuous mind in Europe, has never understood Betti."[52] Battaglini went on to urge Betti, as a friend, to make the effort to express his ideas in "the same clear and brilliant form so admired in the immortal works of Euler and Lagrange." While Betti's work unquestionably presented a challenge to the reader, one can only wonder what the Italian might have said especially about the Englishman's researches in invariant theory with all of their neologisms and arcane vocabulary!

From Naples, Sylvester continued on to Rome. There, Barnaba Tortolini, six years Sylvester's senior, held a number of professorships in the mathematical sciences at various Roman institutions of higher education. Known primarily for his work in analysis, Tortolini had founded the *Annali di scienze matematiche e fisiche* in 1850 in an effort both to spur mathematical research in Italy and to make Italian discoveries more widely known in Europe.[53] As editors, then, Tortolini and Sylvester had similarly nationalistic and internationalistic goals, which it was only natural for them to want to discuss.

By 22 January 1857, Sylvester had returned to Naples, having stayed on in Rome longer than he had originally intended.[54] Waiting for him at the Neapolitan Poste Restante was a letter from Lord Brougham, in which, in a rare turning of the tables, Brougham was asking Sylvester for a favor. His nephew had resolved to try for a place at the Royal Military Academy, and Brougham was asking Sylvester's advice on how best to proceed, given the reforms that had been and were still under way at the Academy. In particular, Brougham wanted Sylvester's thoughts on a suitable tutor to prepare his nephew for the entrance examination, and he tossed out a name. Sylvester responded that the tutor Brougham mentioned "belongs to the class of trainers under the *old system* of examination," while "the new system of competitive examinations" requires "a different and more extended course of preparation." Sylvester, in fact, recommended John Heather, the senior mathematical master working under him, as "a most *zealous* and efficient instructor and trainer." "In the event of your nephew going to Mr. Heather or indeed to any one else at Woolwich," Sylvester closed, "I shall be able to keep an eye upon him & report his progress from time to time."

Sylvester was back in England before the end of January. He had returned, however, not to Woolwich but to London, where the Athenæum Club served as his social hub. Established in 1824, this gentlemen's club located at 107 Pall Mall, just a stone's throw from the Royal Society, served as a meeting point for the male component of London's political, scientific, literary, and artistic élite. Inside the large and elegantly appointed Greek revival building, members gathered for coffee in the morning, lunch and tea in the afternoon, and dinner in the evening, spending the hours in between in quiet conversation, in writing, or in reading in what was widely considered the best and most comprehensive club library in London.[55] Sylvester had only recently been elected to membership in 1856 under the auspices of the club's so-called Rule II which provided for "the annual introduction of a certain number [limited to no more than nine] of persons of distinguished eminence in science, literature, or the arts, or for public services."[56] This circumvented "the normal procedure of waiting for election after being proposed, seconded and supported by Members" and provided Rule II members with a certain cachet (7). The historian and critic, Thomas Carlyle, for example, had been admitted under the rule in 1853, and the celebrated portrait painter, George Richmond, came in under it in Sylvester's year of 1856 (5, 28). The Athenæum was thus a place where Britain's male intelligensia could count on engaging in interesting and varied conversation with a congenial and diverse group of companions. It was also a haven for making and maintaining a variety of social ties. From the Athenæum, then, Sylvester wrote to his longtime Irish friend, Charles Graves, on 26 January 1857 about a missed opportunity to meet up in London.

Sylvester had sought Graves out, presumably at the home of his brother, John, only to learn that Charles was no longer in town; Sylvester was anxious to talk to his friend about the possibility of his nomination to the Athenæum. Now that Sylvester was a member, he could certainly propose others, and Graves, one of his oldest friends and a respected member of the faculty at Trinity

College Dublin, would have made an excellent addition to the club. There was more on Sylvester's mind than club membership, however. All was not right at the Academy. He confided to Graves that Henry Moseley—Johnian, mathematician, Anglican canon, and administrator for the War Office of the competitive admissions examinations instituted for the Academy in 1855—had "been appointed one of the members of the Court of Inquiry which is to meet on Tuesday next to go into the [illeg] between the Military at Woolwich and myself."[57] Although it is not clear exactly what this Court of Inquiry was investigating, it may well have had to do with Sylvester's setting of grades in conjunction with the examinations at the end of the previous term (see below).

Back at Woolwich for the start of classes in February, Sylvester spent a busy spring teaching, preparing for the volume of the *Quarterly Journal* that would appear in May, and succeeding little in his own mathematics. The summer and fall were also mathematically quiet. A bit more dabbling on cubic forms was followed by a brief foray into the theory of elliptic functions, a subject that, momentarily at least, "strongly and strangely enchanted" him.[58] Sylvester was clearly groping for mathematical direction, but he simply did not find it in 1857. No papers appeared in print that year under his name.

January of 1858 found him once again on vacation following the close of another term at the Academy. This time, he stayed closer to home, going down to the seaside, he told Cayley, "for the sake of my health & the sake of the sea air."[59] The mathematical muse had still not returned to him, but he had the journal to think about, and he was in regular correspondence with the self-taught mathematician and country prelate in Croft in Warrington, Thomas Kirkman.[60]

Sylvester and Kirkman had been in mathematical contact at least since the end of Sylvester's first summer at Woolwich in 1856. From that time, Kirkman had written him regularly with queries related to and updates on his (primarily) combinatorial researches, and although, as Sylvester confessed to Cayley in 1856, the information he sometimes solicited "quite transcends my mathematical lore," the two men had developed a certain mathematical camaraderie and personal friendship.[61] In a New Year's greeting written on 2 January 1858, for example, Kirkman chided Sylvester on a key aspect of his personal life. "I regret to learn from your letter," he wrote, "that one so formed as you are for connubial bliss is taking so little pains to attain it, & is still endeavouring to reconcile himself to the fate of a bachelor."[62] In addition to this revelation, Sylvester had apparently also told Kirkman about a teaching possibility at the Staff School that the Army had just created at Aldershot. The school targeted officers stationed there and aimed to provide them with further training in practical geometry as well as in surveying, military sketching, fortification, and photography.[63] Kirkman, who had lamented to Sylvester that his pastoral position brought in only £150 a year, was anxious both to improve his financial situation and to move into an actual teaching post. Yet, as he put it, "only a man has a Lord or two at his back, he has no chance of reaching the ear of power from a place like this." Sylvester, who had so frequently called on and received help from Lord Brougham, must surely have appreciated Kirkman's sentiment.

Letters continued to arrive from Kirkman during the winter of 1858, even though, as he realized, Sylvester "was very busy at present" with both his duties at the Academy and with the journal.[64] These responsibilities also interfered on more than one occasion with the standing engagement Sylvester had made to visit every Friday with Cayley to benefit from his "mathematical inspirations."[65] Nevertheless, the two did manage to meet at the Royal Society, at Cayley's law chambers at 2 Stone Buildings, and perhaps even at the Royal Society Club, a dining club formed in the eighteenth century "for the convenience of certain members who lived in various parts, that they might assemble and dine together on the days when the Society held its evening meetings" and to which Sylvester had been elected in the summer of 1857.[66]

Unfortunately, Sylvester ultimately profited little, even from the company of his faithful mathematical friend. In August, with examinations for yet another open competition for entrance into the Academy set and graded, another brief note without proofs appeared on cubic forms, but it contained results he had actually obtained and communicated to Cayley two years earlier.[67] He closed it in the "hope to have the tranquility of mind ere long to give to the world my memoir or a fragment of it, 'On an Arithmetical Theory of Homogeneous and the Cubic Forms,' the germ of which, now, alas! many weary years ago, first dawned upon my mind on the summit of the Righi, during a vacation ramble" (109).

By November 1858, however, a bit of new mathematical light did seem to be dawning, with the inspiration coming not so much from Cayley but, interestingly, from Kirkman. In two short notes published back-to-back in the November issue of the *Philosophical Magazine*, Sylvester sketched some new thoughts on the theory of equations of the form

$$ax + by + cz + \cdots + lw = n, \tag{1}$$

where $a, b, c, \ldots, l, n \in \mathbb{Z}^+$. In the first, he considered the solutions $(x, y, z, \ldots, w)$ of (1) and showed that they "may be made to depend on algebraical equations whose coefficients are known functions of $a, b, c, \ldots, l$ and n."[68] As he explained, "the fact is somewhat surprising, the proof easy, being an immediate consequence of" a theorem on the partition of numbers. Moreover, it represented "a considerable advance upon the conception . . . of the explicit representability of the mere number of the solving systems $x, y, z, \ldots$ by general algebraical formulæ. By this new theorem," he continued, "we pass, as it were, from the shadow to the substance." And how had Sylvester come to think about these matters in the first place? "It was my valued friend Mr Kirkman's Manchester memoir on partitions," he admitted, "which first drew and fixed my attention on the subject."[69] Kirkman had been thinking about partitions at least since 1854 and had been writing up his findings primarily for the publications of the Manchester Literary and Philosophical Society.[70] With their correspondence firmly established in 1856, Sylvester had thus become aware both of Kirkman's evolving thoughts and of the works he was publishing.

Of course, Sylvester was also well aware of the applications that Cayley had made of partitions in some of his further work in the theory of invariants. In particular, in an undated letter to Sylvester sent some time between the fall of 1854 and the winter of 1855, Cayley had sketched foundational results on determining the number of linearly independent covariants of any given degree and order associated with a given binary quantic, results that he was able to cast in partition-theoretic terms.[71] He then presented the fuller exposition of his findings to the Royal Society on 24 May 1855 in his paper, "A Second Memoir upon Quantics." It was thus only natural that partitions would also pique Sylvester's interest.

Sylvester's new partition-theoretic finding almost immediately got him thinking about a classical problem Leonhard Euler had termed "the problem of the virgins," that is, given two equations with integer coefficients in n variables as in equation (1), how many simultaneous integral solutions does the system have? Euler had recognized that the answer to this question could be recast combinatorially in terms of a particular generating function, and Sylvester realized that Euler's combinatorial equivalence had yet a further partition-theoretic interpretation that was perfectly generalizable to systems not just of two equations as in (1) but, in fact, of m such equations.[72] It was this latter result that Sylvester had discovered in the fall of 1858—"led," in his words, "as by a higher hand"—and published in his second note in the November number of the *Philosophical Magazine*.[73]

These results had finally struck a research chord in Sylvester, yet his end-of-the-term examining duties at the Academy forced him to put aside his own mathematical work for the moment and concentrate on his professorial responsibilities. As always, the grading was both an onerous and a thankless task. Many of the cadets simply had not learned their mathematical lessons sufficiently well, and so he could not, in all good conscience, reward them with large numbers of points. This presented a problem, however, with respect to the cumulative system of grading. Cadets only received their commissions if they earned a sufficient number of points in mathematics and their other subjects, first on the probationary and then on the passing-out examinations. Because one of the criticisms of the Royal Military Academy during the Crimean War had been its inability to provide sufficient numbers of officers sufficiently quickly for the war effort, there was a certain pressure to see the cadets through successfully.[74] As a man of principle, however, Sylvester refused to bend under such pressure.

This refusal resulted in December 1858 in another clash with the military authorities. Writing to Cayley on 22 December, Sylvester reported on the "most disagreeable affair" in which he was then currently engaged over his manner of grading.[75] The military authorities, he told Cayley, were trying "to dictate to me my mode of giving marks for our examinations," and the situation was dire. "The conflict," he continued, "is likely to result [?] in the loss of my post and throw me too in the *wide* world." He was clearly ready to lose his job rather than buckle under the pressure to assign grades that were undeserved.

One reality that Sylvester did not seem ready or able to face in this matter, however, was the issue of his own effectiveness as a teacher. From his point of view, the cadets simply were not applying themselves adequately, but there was another side of this particular coin. Alfred Drayson, one of Sylvester's colleagues at the Academy, was Professor of Surveying and Topographical Drawing. As such, he necessarily relied on the Professor of Mathematics to provide the cadets with the basic mathematical tools they would need in his more practical class, and Sylvester let him down. Writing in his memoirs of a career at Woolwich, Drayson recounted that Sylvester was "an individual who was, perhaps, one of the best mathematicians in Europe, but who had not the gift of imparting knowledge."[76] He simply could not "descend to the small matters to be attended to when teaching." As a result, the cadets came to Drayson "unable to work out a simple problem in plane or spherical trigonometry," and he "was compelled to teach these two subjects before [he] could proceed to give instruction in their application." By way of excuse for their shortcomings, the cadets told Drayson "that they could not learn from this mathematician, as he was too profound for them, and if they asked him for an explanation of any difficulty, his explanation, instead of making the case more simple, made it more complicated, and [they] left his presence more puzzled than before." It was clearly a frustrating situation for all concerned.

Fortunately for Sylvester at least, the storm about his grading practices soon blew over, and he was able to return to his partition-theoretic thoughts. By April 1859, with another term at the Academy almost behind him, he had announced that he would give a series of public lectures at King's College London beginning in June on the new insights he had been obtaining on partitions throughout the winter and spring. In response to Cayley's assurance that he would be in attendance, Sylvester acknowledged that "if I obtain an audience of 6 I shall be quite content."[77]

Sylvester gave the first two of what grew into seven lectures on 6 and 9 June, and he apparently need not have worried about attracting a sufficient number of auditors. According to the account that appeared in the *Times* on Saturday, 11 June, "nearly 60 attentive listeners" were in attendance to hear Sylvester's results on the problem of double partition, that is, "in how many ways can a given pair of numbers . . . be composed of a given set of couples" and, more generally, on that of so-called "compound partition."[78] The reporter had clearly been impressed by Sylvester's performance. "We are not aware of any other similar attempt being made in London at expounding in public lectures a very abstruse branch of mathematics," he concluded, "but we may fairly say that the lecturer's clear and forcible manner of exposition was perfectly successful, and, if we may judge from the satisfaction expressed by" the audience, "the verdict was unanimous."

Sylvester finished up the seventh and final lecture on 11 July. At each of his presentations, he had distributed printed outlines of the day's material that he had prepared with the help of Andrew Noble, an able mathematician and an officer in the Royal Artillery then at Woolwich performing ordnance tests as part of the controversy surrounding the Army's switch from smooth-bored guns to rifled artillery.[79] Sylvester, unfortunately, never managed to ready a full exposition of this partition-theoretic work for publication. Another year closed, in fact, without a single published paper from his pen.[80]

A Roving Mathematical Eye

If Sylvester had been less than prolific mathematically during his first four years at the Academy, his next two years there—1860 and 1861—brought a profusion of mathematical ideas in a variety of areas and more than two dozen publications. He still had his usual duties—his teaching, his setting and marking of examinations, his editing of the *Quarterly Journal*—but now he also managed to sustain a certain mathematical momentum. His research was not, however, focused.

The winter and spring of 1860 found him thinking fleetingly about number theory. In four short notes to the *Comptes rendus* of the Paris Academy of Sciences in February, March, and April, he considered the quantity $E(x)$, that is, the greatest integer less than x and gave a number of algebraic interpretations involving it, inspired by the work of Gotthold Eisenstein. In his third note, presented to the Academy at its meeting on 26 March, for example, he formed the functional relation

$$F(p, q, k, l) := E\left(\frac{p}{q}\right) + E\left(2\frac{p}{q}\right) + \cdots + E\left(l\frac{q-1}{k}\frac{p}{q}\right),$$

where $p, q, k, l \in \mathbb{Z}^+$ and where $q - 1$ is assumed to be divisible by k, and sketched some of the algebraic relations that exist between such functions F under various assumptions on $p,\ q,\ k$, and l.[81]

By the late summer of 1860, these ideas had been supplanted by thoughts on group theory sparked by Cayley's efforts to realize groups in terms of generators and relations and by Cayley and Sylvester's joint effort to read and understand one of Cauchy's papers on permutation groups.[82] Discussing his evolving ideas with Sylvester, Cayley had asked his friend for help in producing interesting examples. On 11 August, the vacationing Cayley wrote to thank him for a number of letters in which he had worked out the ramifications of a group of order 20 and to share with him observations about groups of specific small orders.[83] Five days later on the sixteenth, Cayley was still thinking about Sylvester's group. In modern terminology, he had managed to take Sylvester's original presentation and show that the group was, in fact, the semidirect product of a cyclic group of order 4 and one of order 5 or, in other words, a group of order 20 generated by elements e and f satisfying the relations $e^4 = 1, f^5 = 1$, and $fe = ef^3$.[84] He also continued their discussion of Cauchy's analysis of the group of substitutions on six letters—S_6 in modern notation—and its subgroups. In particular, he focused on a subgroup of order 120 living inside S_6 and proved—to his great surprise—that it is isomorphic to S_5. "I had no idea," he confessed to Sylvester, "but that the group of 120 for the six things was really distinct from the group of the substitutions of 5 things, whereas they turn out to be identical." Sylvester, however, knew this quite well, a fact of which he informed Cayley immediately. "I did not know that you knew that the group of 120 . . . was identical with the group of substitutions of five things," Cayley wrote back. "But the premises on which the conclusion rests are undoubtedly altogether yours & the conclusion is immediate."[85] Cayley would nip *this* potential priority flap in the bud!

While Sylvester was discussing group theory with Cayley, he was also thinking about mathematics in a more immediately applied vein not unrelated to his teaching at the Academy. In 1834, the French military engineer, mathematician, teacher, and ultimately General, Jean-Victor Poncelet, had published a paper in Crelle's *Journal* in which he gave a method for approximating square roots of homogeneous quadratic forms in two variables. As he fully recognized, this had immediate calculational applications to the problem of determining the resultant of two forces in the plane and hence to problems in mechanical and practical engineering.[86] Sylvester saw how to generalize Poncelet's result to homogeneous quadratic equations in three variables and thereby obtain an application to the case of three forces in space. He both announced this finding at the annual meeting of the BAAS in Oxford and published it in the September issue of the *Philosophical Magazine*.[87]

A month later, he was back on that journal's pages with more on his new result. In a move so typical of his mathematical mindset, Sylvester had stepped back from the immediate theorem and considered it from a purely algebraic point of view. As he put it in his follow-up paper, "hitherto Poncelet's theorem has been regarded as a method *sui generis* and complete in itself; but in truth it is but the first germ or rudiment of a vast and prolific algebraical theory;

and not only so, but the principle which it contains admits of applications of the utmost value in various dynamical and analytical questions, which it is surprising should have been allowed to lay so long dormant."[88] Ever on the lookout for algebra at the core of mathematics, Sylvester had found it even in this unexpected place. "Henceforward," he proclaimed, "Poncelet's theorem figures no longer as a detached method, a mere stroke of art in aid of the computer, but becomes integrally attached to the grand and progressive body of doctrine of the modern algebra."

Sylvester was still thoroughly captivated by these issues when he invited his new friend, Thomas Archer Hirst, to dinner and an evening of mathematical conversation at his home in Woolwich for Monday, 8 October.[89] The thirty-year-old Hirst had earned a Ph.D. in geometry at the University of Marburg in 1852 before returning to shaky job prospects in England, settling in London by 1859, and securing a post as mathematics instructor at the University College School. He had managed quickly to enter into the capital city's scientific and mathematical milieux thanks to his friend, the physicist and Professor of Natural Philosophy at the Royal Institution, John Tyndall. Not surprisingly, among the worthies to whom Tyndall introduced him were the mathematicians, Cayley and Sylvester, and he was quickly befriended by both.[90] The mathematical soirée that Sylvester was planning for the eighth, in fact, was to be a foursome comprising Hirst, Cayley, himself, and his "French mathematical friend," Camille Jordan. Jordan, eight years Hirst's junior, had recently completed his studies at the École polytechnique and had been sent to England by the École des Mines in the summer of 1860 to, as Sylvester put it in a letter of introduction to Charles Babbage, "study Mining and other practical applications of Mechanics."[91] Sylvester was thus doing his best to expose the young Jordan to the local mathematical scene and to show him a measure of English hospitality. He would also expose him to his own then-current mathematical ideas stemming from Poncelet's theorem, for, as he told Hirst, "I shall have something *Very striking* to tell you . . . when we meet."[92]

The evening must have been a success. When Sylvester wrote up his third and final paper on this subject, he remarked that "my friend, M. Jordan, of the École des mines (author of a remarkable thesis on *groups*), has developed some interesting geometrical consequences arising out of the study" of one of the equations involved in the analysis of Poncelet's theorem, "which I hope he may be induced to publish."[93] By thus singling Jordan out on the pages of the *Philosophical Magazine*, Sylvester was styling himself a senior mathematical statesman, calling to the attention of the mathematical publication community that the journal defined the work of a young Frenchman whom he viewed as promising.[94] By the end of 1860, however, it was no longer a matter of self-styling. Sylvester, together with the noted naturalist, William Carpenter, had been awarded the Royal Medal of the Royal Society of London in recognition of their eminence within the British scientific community. They had thus received one of the community's highest and most valued honors.[95]

The year also closed for Sylvester with one more work inspired by his teaching at the Academy. As noted, one of the topics that the would-be engineers studied in their advanced course in pure and applied mathematics was the pressure of earth on retaining walls. In the December issue of the *Philosophical Magazine*, Sylvester considered the idealized situation of "the pressure of *Mathematical earth*," meaning "earth treated . . . as a *continuous* mass separable by planes in all directions, but whose separating surfaces exert upon one another forces consisting of two parts, one of the nature of ordinary friction, the other of so-called cohesion."[96] In what has been described as "his most original contribution to applied mathematics," Sylvester extended the work in this area of William Rankine, since 1855 the Regius Professor of Civil Engineering and Mechanics at the University of Glasgow, by analyzing the problem from the point of view of lines of pressure and thrust within the body of earth contained by a particular configuration of revetment walls and calculating the maximal and minimal thrusts within the system.[97]

With this paper in print, with the end of another academic term, and in the afterglow of his recognition by the Royal Society, Sylvester spent January 1861 in Paris thinking mathematical thoughts of yet another kind. From the applied mathematics of the fall of 1860, he moved first in the winter of 1861 back to number theory in work stemming from his arithmetic analysis of the Bernoulli numbers, namely, those numbers arising as coefficients of the polynomial s in x defined by $s_k(x) := \sum_{i=1}^{x} i^k$. Personally presenting the first two of three short notes on his findings before the assembled *savants* of the Paris Academy of Sciences on 28 January and 4 February,[98] he also interacted individually with a number of his French mathematical friends, notably Liouville and Chasles. In particular, when both Frenchmen confided to Sylvester that they were anxious to promote Cayley's mathematical attainments within Parisian mathematical and scientific circles, Sylvester wrote promptly to tell his friend that "I think it would tend to your advantage in a matter of real importance, to appear in person [in Paris] and you could bring a good supply of your memoirs with you."[99] Ever conscious of the importance of the international as well as the national imprimatur, Sylvester did not want Cayley to miss out on a prime opportunity further to secure his reputation abroad.

Paris and such thoughts of scientific reputation making were soon a distant memory, however. Back in Woolwich for the winter term, Sylvester focused on his teaching and on yet another new research direction. During the course of his reading, Sylvester had happened upon August Ferdinand Möbius's *Lehrbuch der Statik*, where, among many other things, Möbius had proven that, given a body in three-space and four forces acting on it in equilibrium, the lines along which the forces act generate a hyperboloid.[100] Sylvester generalized from this to consider six lines in a solid such that forces acting along them are in equilibrium and showed how, given five of the lines, the sixth could be constructed so as to pass through any given point or to be situated in any given plane.[101]

These geometric ideas seemingly obsessed Sylvester throughout the spring of 1861 as he worked to elaborate a full theory. Although many days, especially at the end of March and the beginning of April, found Cayley's mail stuffed with multiple letters from Sylvester on the subject, little ultimately came of his efforts. His own frustration at having so little success was palpable in a letter penned to Cayley on 30 April. "I am at my *wit's end*," he exclaimed. "If you should take sufficient interest in the question to induce you to think of it and could throw any light upon the cause of the apparent contradiction you would *greatly oblige* me by doing so."[102] Two days later, Cayley replied, acknowledging that "you are certainly on the horns of a dilemma as to the 6 lines—either your theorem is false—which I do not believe—or else the relation of the 6 lines is not a descriptive one. Of the two I prefer the last."[103] Sylvester's frustration level had only increased the next week when he wrote to tell Cayley that "I have been so beset with Examination business and so unwell with a fresh attack of Influenza that I have not had energy yet to read through your valuable letters but hope to do so by tomorrow and to have a long talk with you thereon, whether previously read, or unriddled to me vivâ voce on the occasion by yourself."[104]

Matters only worsened as the month of May wore on. Following a conversation with Cayley, Sylvester wrote testily on the sixteenth "that you ought not to have doubted (when I said) my recollecting having had a familiarity with the form of the $=^n$ to the lines of curvature on a surface; although I was unable to dictate the expression I have [n]ever given anyone I hope the least reason to doubt my perfect reliability in the article of veracity in matters great or small."[105] Ten days later, it was no longer just his mathematical research that had run afoul. Beside himself on Monday, the twenty-sixth, he was forced to cancel plans not only to have Cayley to his house the next night for a mathematical evening but also to have Kirkman spend time with him that week at Woolwich. "I am in trouble and perplexity relative to mundane affairs," he told Cayley. "Please tell Kirkman when you see him . . . that . . . I shall be so much engrossed . . . with the unpleasant business I have on hand added to inimical examinations . . . that I am quite unable to fix for a meeting but shall hope to see him at the Club before he leaves town." He closed this letter ominously and mysteriously. "The news of my troubles only reached my knowledge yesterday evening after I had seen you."[106] Had what had seemed like a promising, new beginning at the Royal Military Academy in September of 1855 suddenly reached a cruel and bitter end in May of 1861? Was this the end of the career he had finally realized as a research mathematician?

CHAPTER 7

At War with the Military

I have borne with great patience and forebearance a long course of
systematic encroachment upon the rights and duties of my office
hoping that time would work a change and anxious for calm of mind
to pursue my private studies and spin my mathematical reveries.
—J. J. Sylvester to Lord Brougham, 18 June 1861

Sylvester's letter to Cayley of 26 May 1861 was certainly not the first indication of friction between the military authorities at Woolwich and their Professor of Mathematics. The incompatibility of the objectives of the two sides had become increasingly obvious since January 1857 when Sylvester reported trouble to Charles Graves. That had merely been the first skirmish. A second had occurred in December 1858. With 20–20 hindsight, the problem is clear. The military officials were charged with efficiently providing an educated officer corps, whereas the mathematician viewed his professorial duties as involving research and teaching, in that order. Moreover, grades on mathematical examinations were supposed to reflect, in Sylvester's view, real attainment in mathematics, but the Academy's military leaders merely required that they reflect a level of expertise sufficient for the demands of life as an engineer or artilleryman in the field. Sylvester, in their view, was not charged with training future mathematicians nor was this the criterion on which he should judge his students. He was simply supposed to attest that the future officers in England's Army would have enough mathematical expertise to do their jobs. Little wonder, then, that with standards so fundamentally at odds, Sylvester, the civilian professor, ultimately came into conflict with the career military men who controlled the Academy.

First Battle

Sylvester's "troubles" in late May 1861 *were* serious. They involved nothing less than what he viewed as the professor's fundamental right of academic freedom.

Earlier in the month, Sylvester had submitted what he considered a routine report to the Governor of the Academy, Colonel Edmund Wilford, in which he proposed a change in the textbooks to be used in conjunction with his courses.[1] Although he realized that his report would be discussed by the military staff in charge of the Academy, he had no reason to doubt that his wishes in the matter would be honored. He was wrong. His suggested changes met with opposition, most probably because the texts he proposed were deemed too sophisticated for the Academy's cadets.[2] This would have amounted to little more than a matter for negotiation had Wilford and his staff merely discussed their concerns with Sylvester. Instead, without his knowledge of their opposition, the military men had forwarded the mathematician's report directly to the Duke of Cambridge, the General Commanding-in-Chief of the British Army and the titular head of the Academy, for a decision.

As Sylvester told Cayley, he learned of this action on the evening of 25 May, and he was livid. First, he had had no idea that his report would be handled substantively anywhere but in house. Second, he had been given no chance to justify his recommendation further. The teaching of mathematics, in his view, was being taken out of the hands of the one person entrusted with it, namely, the Royal Military Academy's Professor of Mathematics.

By 1 June, Sylvester had drafted a letter of appeal to Wilford in which he justified his recommendation, requested that that justification immediately be forwarded to the Duke of Cambridge, and laid out a number of allegations of past abuses by the military authorities at Woolwich with respect to the program in mathematics. It was imperative, to Sylvester's way of thinking, that both sides of the issue be considered seriously before any decision was made at the highest level. Apparently, Sylvester's request was honored only perfunctorily; his appeal was forwarded on the fourteenth, *after* the Duke of Cambridge's decision to reject the mathematician's proposal had been sent off on the twelfth. Five days later, Sylvester had finally received the news.

On 17 June, he wrote to Major General W. F. Forster, the military secretary of the Horse Guards (that is, the seat of administration for the Army in London under which the Academy fell in the chain of command) and the man who had communicated the Duke of Cambridge's directive. In his letter, Sylvester laid out not only the above sequence of events but also his case for a reconsideration of the decision.[3] That case hinged principally on the fact that Sylvester's appeal had "been irregularly and illegally . . . kept back and not allowed to leave the Governor's hands until the decision of His Royal Highness on the subject matter of the appeal had been obtained." This decision, moreover, "is a matter of the deepest importance to the office I have the honor to hold," he declared, "as upon the ultimate decision of His Royal Highness will depend whether the Professor of Mathematics is to be maintained in the position he has always hitherto held or whether his status and functions are to be made identical with those of the

Masters under him." "Under these circumstances," he continued, "I respectfully request that you will submit to his Royal Highness my dutiful prayer that he may be pleased to order that all proceedings connected with the execution of his command may be suspended until an opportunity has been afforded me of making good the allegations contained in my statement of the first instant and of proving as therein shewn the long continued and systematic course of the encroachment which has been pursued with reference to the discharge of my duties as Professor and Examiner by the present Governor and Inspector of the Academy."

Having thrown down this gauntlet, Sylvester, perhaps not surprisingly, wrote the next day to Lord Brougham both to detail the situation and to solicit his advice. "My position for the last two or three years has been an exceptional and delicate one," he confided, "being left under the authority of the Military Governor whose displeasure I had incurred because I would not accept his command as a sufficient authority for wrongfully altering the Marks I had allotted to a Cadet at an examination."[4] "On that occasion," he explained, "my resistance was sanctioned by the Duke of Cambridge and Col. Wilford, the Governor, severely censured for his interference, but he remained Governor and I subject to his authority." It was not a good situation. Sylvester, the *civilian*, in pursuing to its end the matter of principle back in December 1858 about his autonomy in assigning grades, had jumped the *military* chain of command all the way to the top and the Duke of Cambridge, and this had brought a humiliating rebuke to Wilford from his ultimate superior. Who was in charge of the Royal Military Academy? The civilian professors or the military authorities? It was, at least with Sylvester as one of the counterweights, a delicate balance that had repeatedly been tipped. Yet, as Sylvester put it to Lord Brougham, "I can most truly and conscientiously affirm that my sole desire has been to live in peace with [Wilford] and everybody else . . . hoping that time would work a change and anxious for calm of mind to pursue my private studies and spin my mathematical reveries" (see epigraph note).

Brougham sympathized with Sylvester's situation and agreed to write on his behalf to the same Sidney Herbert who had argued so vociferously—and successfully—in Parliament in 1856 for the reform of military education and who had been at work to implement those reforms ever since. Going to Herbert, again the Secretary of State for War under Lord Palmerston, would thus be a blatant move to try to exert political pressure on the Academy. As Sylvester put it to Brougham, Herbert's support "would be of the *greatest possible advantage* in securing for me a fair trial of the points at issue."[5] Moreover, Sylvester the lawyer noted that "my engagement was with the War Office and nothing can be done against me by the Military Authorities without Lord Herbert's ultimate approval and sanction."

Following Brougham's advice, Sylvester put together his case for submission directly to Herbert, including what he called "*pièces justificatives* which Lord Herbert can consult if he should feel so inclined in support of the assertions or some of them made in the statement."[6] Once again, Sylvester was going outside the chain of command to defend his actions, and, not surprisingly, once again, he quickly incurred the wrath of the Woolwich authorities. Writing to Brougham on 16 July, he explained that the case he had assembled for Herbert's benefit had apparently reached not him but the Duke of Cambridge and his Council of Military Education. As noted, the latter had been created in 1857 as part of Herbert's reforms to oversee military education in Britain, and its composition had evolved to include the Duke of Cambridge as President *ex officio*, a Major-General as Vice President, and four additional members only one of whom was a civilian.[7] This body was thus almost exclusively a military one, and, as Sylvester reported to Brougham, they "are very angry with me for having made an independent application to the Secretary for War and are likely to make me feel in a very serious form the might of their displeasure. My application to Lord H.," he continued, "is construed as an attempt to 'exercise coercion' over the judgement of the Commander in Chief."[8] From Sylvester's point of view, of course, it had been nothing of the sort. Rather, he had "merely [sought] to secure a fair hearing of [his] case before civilians who could appreciate the claims of a civilian to just and considerate treatment" and thereby assure that his case "might not be left to the arbitrariness of an adverse and prejudicial tribunal." Sylvester's perspective on the matter had clearly shifted from the local view of the professor versus the administration to the global view of the civilian versus the military establishment. From either perspective, it was a battle that he felt compelled to fight.

Ten days later on the twenty-sixth, things seemed to be looking up. Sylvester was in town at the Athenæum and thinking once again about mathematics. This time, it was the mathematical theory of arrangement—or tactic as he called this branch of combinatorics—that had caught his imagination, and he was writing to an absent Cayley both to convey his latest ideas and to catch him up on the saga with the Academy. Relative to the latter, he reported that "my affair in connection with the Academy seems likely to turn out *entirely* to my satisfaction but still detains me in town and will probably absorb the whole of my vacation."[9]

Still, he wanted to leave nothing to chance. Herbert, in rapidly failing health, had stepped down from his post as Secretary of State for War and had been replaced by Sir George Lewis. It was thus now Lewis and not Herbert who needed to hear Sylvester's side of the story. On 1 August, Sylvester wrote to ask Brougham whether he knew Lewis and, if he did, whether he would "communicate to him the same favorable impression as you produced with regard to me upon Lord Herbert."[10] "Upon the view [Lewis] may take of the subject," Sylvester added not without some hyperbole, "the whole future of my life probably depends." All this politicking behind the scenes apparently paid off, at least in the short term. By mid-August, Sylvester reported to Cayley that "my work at the Academy . . . is again on."[11]

Sylvester's 1861 battle with the military authorities at Woolwich had begun innocuously enough over a textbook recommendation but had quickly escalated to a thorough airing of the mathematician's grievances against a military administration that he perceived as stifling the academic freedom of its civilian professoriate. At the heart of the controversy was the mathematician's deep conviction in academic freedom, a conviction that the military authorities in charge of the Academy did not share. Sylvester's victory thus helped to establish the principle at Woolwich and to set a precedent for at least a measure of professorial autonomy in the classroom and in the examination theater. All was not sweet in victory, however. By engaging in the tactic of jumping the chain of command and even of going outside the military and into the political arena, Sylvester had, from the point of view of the Woolwich authorities, not played by the rules. Although this ultimately served to pit him, from 1861 onward, against the military men in charge of the Academy, in August 1861, all he knew was that he had won and could get back to the "mathematical reveries" that his battle had so rudely interrupted.

Temporary Truce

In the spring of 1861 as he labored, largely unsuccessfully, on the geometrical problems associated with a body in equilibrium in three-space, Sylvester had also revisited some of the work that he and Cayley had discussed in the summer of 1860 on substitution groups. In particular, in May 1861, Sylvester had published a "Note on the Historical Origin of the Unsymmetrical Six-Valued Function of Six Letters" in the *Philosophical Magazine*. There, he both highlighted results in a permutation-theoretic vein that he had obtained as early as 1844 and claimed priority for having stated the not unrelated "fifteen school girls problem."[12] That counting or "tactical" problem, to use Sylvester's terminology, was cast this way: if fifteen school girls walk three abreast during each of the seven days of the week, then how can they be arranged daily so that no two walk abreast more than once? In a somewhat different guise, this had been a prize problem posed in the *Ladies' and Gentlemen's Diary* in 1844; it had been solved by Thomas Kirkman two years later, long before he and Sylvester were acquainted.[13] In his paper in the May issue of the *Philosophical Magazine*, Sylvester told his readers "that, in connexion with my researches in combinatorial aggregation [of 1844], . . . I had fallen upon the question . . . which has since become so well known, and fluttered so many a gentle bosom, under the title of the fifteen school-girls' problem."[14] He went on to assert that "it is not improbable that the question, under the existing form, may have originated through channels which can no longer be traced in the oral communications made by myself to my fellow-undergraduates at the University of Cambridge long years before its first appearance . . . in the *Ladies' Diary.*" It was a stretch, but Sylvester, ever concerned with priority, apparently could not resist the temptation to stake even this shaky claim.

The months of June, July, and August—as his battle with the Woolwich authorities raged—saw the publication of three more short papers in the *Philosophical Magazine* on "tactical" matters of counting and arrangement, but by mid-August, this work had led him back to group theory *per se.*[15] He had succeeded in expressing a particular group of order 54 in terms of generators and relations, and he was anxious to get Cayley's opinion of it. Writing to his vacationing friend on the seventeenth, Sylvester expressed his hope that "if this example is new to you it may serve to excite some fresh thought on Groups in your mind and that you will give me the benefit of your reflexions upon it."[16] For his part, Sylvester was trying to get back into group theory, and he sensed he could use a boost from Cayley. "I have almost entirely lost the recollection of my previous researches in Groups," he confessed, "but hope to come back upon the old track." If only he could regain his former momentum, he felt sure "that the true theory is all to make" and that he could "do something towards its improvement." At least with the battle now over with the Academy, he could focus once more on his mathematical life.

The journal was one concern. It was late in coming out, and Sylvester had written to Ferrers to inquire as to the problem. When Ferrers, who was busy correcting the proof sheets, finally had a chance to reply, he reported that it was just a routine delay and assured Sylvester that "we have plenty of matter."[17] As this letter makes clear, Sylvester and Ferrers had established what was ultimately an inequitable division of labor in running the journal, but one which played to their respective strengths. They both received and evaluated submissions. For Sylvester, in particular, dealing with the purely mathematical side of the operation was congenial enough. As far as the mundane details of actually running the journal were concerned, however, he had neither the time nor the patience. Ferrers alone handled the headaches, like reading the proof sheets, of actually producing the issues. In Ferrers, as in Cayley, Sylvester had luckily found a collaborator willing and able to accommodate himself to an idiosyncratic and strong-willed Sylvester.[18]

Another concern was the September meeting in Manchester of the British Association for the Advancement of Science. Before all the unpleasantness occurred at the Academy, Sylvester had planned to attend and had even made arrangements to stay with some friends in the industrial city. On 26 July, even though he seemed to be winning his battle against the authorities, he still feared that Academy matters would ultimately prevent him from going, but a month later, with things completely resolved in his favor, he was finally free to participate.[19] "I shall have the pleasure of meeting you at Manchester," he wrote to Cayley with an air of relief and satisfaction, "where I intend to edify the natives by a short paper 'on Skew Commutants and Hyper-Pfaffians.'" It was to have been a talk in which he presented new material, an extension of some earlier determinant-theoretic work of Cayley's, but this line of research ultimately came to naught. Sylvester did speak in Manchester but on the geometrical work from the spring that he had already published in the *Comptes rendus.*[20]

By October, the events of the summer were sufficiently behind him that he was once again able to make some progress mathematically. He had returned to combinatorial ideas not unrelated to work he had done in 1858 on equations of the form $ax + by + cz + \cdots + lw = n$ (for a, b, c, ..., l, $n \in \mathbb{Z}^+$) and had developed methods that allowed him to extend a result in this vein that Cauchy had proved in 1844.[21] He wrote up his findings in a letter to Joseph Serret that was presented to the Paris Academy at its meeting on 7 October. Two weeks later, Serret presented a follow-up piece, in which, using similar methods, Sylvester gave the solution of the combinatorial problem "to find the number of substitutions of n letters that can be represented in terms of a given number r of cyclic substitutions of odd order."[22]

If Sylvester had found the peace of mind by October 1861 to return to his own researches, that peace was seriously disturbed on the twenty-third when work at the Academy essentially came to a standstill. The morning started badly. At breakfast in their dining hall, the cadets, who habitually complained about the food, were confronted with "a particularly disgusting sample of egg."[23] One of the cadets, fed up with this standard of fare, revolted by tossing his plate on the floor. In what was virtually a chain reaction, other cadets followed suit with "the 'squashy' sound of eggs bursting on the floor boards." When the officer on duty, not realizing exactly what was going on, called for the cadets to leave the dining hall, they "threw [their] chairs and forms down, seized [their] caps, and trooped out of the hall, whooping and yelling like demons."

The day only went from bad to worse. During the class hours that immediately followed breakfast, the cadets were unusually restless, their professors trying in vain to keep them focused on their studies. Nor had the excitement of the morning's revolt dissipated by the time of the daily midday drills on the parade ground. The cadets assembled with their rifles in hand, but first one then another then another dropped his rifle "or a busby went rolling on to the parade" (99). It was pandemonium, with the officers in charge once again unable to bring the situation under control. Lunch followed at 1:00, but at 2:00 when the cadets were supposed to return to their studies, all but a few stalwarts refused. When, at last, they were forced into their classrooms, no serious work was done. At the end of the day, the cadets were confined to their barracks pending the findings of a Court of Inquiry, and the next morning, Wilford gave the entire corps a dressing down for its behavior.

This show of authority, however, did not stop the mischief. Wilford, who had assumed command of the Academy shortly after Sylvester's arrival there, had been unpopular with the cadets from the beginning because of his imposition of rules perceived as draconian. He had forbidden them to have furnishings such as carpets or armchairs that would have offered them a modicum of comfort; he had denied them the privilege of taking tea in their rooms by forcing them to march to the dining hall for this refreshment; he had required them first to stand at attention for roll call before lights-out and then to make their ways to bed in the dark (91–92). Wilford's lecture to the cadets following the "mutiny" of 23 October 1861 thus only fueled further the discontent of a particularly rowdy corps.[24]

Several nights passed quietly. The calm before the storm. A field gun was rolled to the parade ground, charged, rammed with a loaf of bread, aimed at Wilford's house, and fired. Although the doubly baked projectile fell short of its mark, the commotion more than roused the Academy. More pranks followed (100–101). The situation had gotten so out of hand that it received repeated mention in the *Times* from the first incident on 23 October through the end of the year. Letters to the editor appeared from parents, cadets, and others, with some sympathetic to the Woolwich authorities but many on the side of the students in their grievances if not in their insubordination.[25] In the end, the Court of Inquiry ruled, not surprisingly, against the cadets. On 22 November, the Duke of Cambridge's ruling was read before a full assembly. In particular, "he wished [the cadets] to understand that no amount of scientific knowledge would be of use, to make them efficient officers, unless they learnt to subject themselves to discipline; that their coming to the Royal Military Academy, was a voluntary act on their part, and they must consequently submit to the rules in force at the time of their admission."[26] It was certainly not one of the Academy's finest hours, nor did it reflect well on Wilford. In fact, just months later in 1862, he was replaced as Governor. The cadets—and to some extent Sylvester as well—got the last laugh.

In the immediate aftermath of both this ruckus and the end-of-the-term examinations, Sylvester went to Paris where four short notes from his pen were read at meetings of the Paris Academy on four consecutive Mondays beginning on 30 December 1861.[27] As was his habit while in the French capital, he renewed his contacts within Parisian scientific and mathematical circles and made still others. On this particular trip, in addition to seeing his old friends Chasles, Hermite, and Poncelet, he was invited to dinner at the home of Urbain Jean Joseph Leverrier, the Director of the Paris Observatory and the astronomer famous for his mathematical discovery (independent of John

Couch Adams) in 1846 of the planet Neptune. Reporting back to Cayley on the evening, Sylvester had to admit that "the dinner at Leverrier's was rather dull."[28] "Leverrier was of course not uncivil," he continued, "but was still far from being an agreeable host & besides he had had a vehement discussion at the Institute . . . which it may be presumed had not tended to sweeten his temper: . . . altogether I like him the least of all the savants I have met in Paris either as regards personal deportment or political principle (or rather I should say political unprincipledness if there is such a word)." Sylvester was certainly not alone in his dislike of Leverrier. The Frenchman was known for his political conservatism as well as for his authoritarianism and was often at odds with his colleagues.

From Paris, Sylvester continued on for an extended leave in Italy. Taking the night train, he had arrived in Milan "after a very agreeable journey in sleighs over the St. Gothard via Basle & Fluellen and Andermatt" (109). From Milan, he had moved on to Florence, which he used as his home base for the six weeks from early February through mid-March of his Italian stay, making side trips to Siena, Pisa, and Pavia. Thanks to Betti, through whom he made "numerous acquaintances not only with the Savants but with the Signori and Signore," his time in Florence was diverting and included tours of the city's artistic and architectural masterpieces as well as a "ball at the Prefecture (Palazzo Vecchio)."[29] He was on holiday after a trying year, but he was determined to make it a working holiday. As he confessed to Cayley, "I have done no mathematics thus far but intend D.V. to make them my chief occupation here treating everything else as subordinate and endeavoring to check the roving propensities which if I indulged would lead me on to Rome, Naples & Sicily."[30]

He was also in Italy to cement further his relationships with the mathematical scientists of the newly unified Italian nation.[31] In particular, he visited Betti, who had left his high school teaching job in Florence in 1857 for a university post in Pisa and who, in 1862, was deeply involved in national politics as a Member of Parliament for his native town of Pistoia. Similarly, Sylvester had sought out Francesco Brioschi in Pavia only to learn that he was then in Turin serving as General Secretary of the new Ministry of Education. These two of his Italian mathematical friends now had positions of great influence in shaping the educational structures and academic values of a united Italy. It was, Sylvester realized, a pivotal moment in Italian history, and he sensed its gravity. "Italy is making wonderful strides toward unity and regeneration," he told Cayley. "It will I believe become one of the greatest nations in the world and play a principal part in the future history of civilization. My warmest affection, sympathy and hopes accompany her in the fulfillment of the noble destiny she proposes to herself to accomplish.[32]

By mid-March, the time had come to embark on the return trip to England. In Pisa on the nineteenth, he journeyed first to Livorno, then to Genoa to meet with the mathematician, Placido Tardy, and then to Turin, finally reaching London via Paris on the thirtieth. It had been a wonderful trip. He had spent many pleasant hours with Betti as well as with the University of Pisa's Professor of Algebra, Giovanni Novi, who, as he reported proudly to Cayley, was "writing a large work on Algebra in which he will introduce the theory of Invariants and all the latest novelties of the subject."[33] The theory that he and Cayley had developed in the first half of the 1850s—and that was being developed largely independently on the Continent—would now, a decade later, reach Italy in a codified form and open itself up for further development in the international mathematical arena. It was a gratifying thought.

Despite Sylvester's hope to make a good deal of progress on his research while in Italy, he ultimately managed to get little done in the spring or summer of 1862 with matters at the Academy continuing to distract him. On Tuesday, 24 June, for example, he met personally with the Secretary of State for War, Sir George Lewis, to discuss a proposed redefinition of the terms of his professorship. The government wanted to see a significant increase in the number of contact hours Sylvester would be responsible for, and negotiations were under way between the mathematician and the War Office. Writing to Lewis in the third person later in the day, Sylvester had to correct a mistake he had made during the course of their conversation. "In admitting that 18 hours would be the maximum that could fairly and reasonably be demanded of him in consideration of an addition of £100 per annum to his present salary of £550 per annum," Sylvester had made a calculational error.[34] "The full amount of time for which he was liable under his original engagement," Sylvester noted, "was 13 hours in Winter and 14 hours in summer averaging 13½ hours in the week. The numbers of attendances for which under the suggested possible arrangements for the future he would become liable being 6, the amount of time corresponding thereto would be nearly as possible *16* hours per week and not *18* as he inadvertently was led to state." If his teaching duties were going to increase, then he certainly did not want to be compensated at a salary rate any *less* than he was then currently making!

The implications of these negotiations weighed on Sylvester throughout the summer and into the fall. In a letter to Cayley sent from the Athenæum on 10 July 1862, for example, he had to apologize for his lengthy silence, admitting ruefully that "I have been so entirely unemployed in Mathematics that I have felt ashamed to encounter you."[35] By October when the BAAS met in Cambridge, the situation had hardly improved. Sylvester spoke, but his result was little more than an advanced calculational exercise that he elaborated further in the *Philosophical Magazine*.[36] When the distractions of the examination period were finally behind him in December, moreover, he was still prevented from focusing as exclusively as he would have liked on his researches. The wheels that had been set in motion at least at the time of his meeting with Lewis in June had continued relentlessly to roll. Sylvester had once again to act, and decisively, if he hoped to preserve the position to which he had been hired at Woolwich.

Second Battle

This time, the issue was "job description." Sylvester understood his agreement with the Academy to involve teaching on the first three days only of every week and for a fixed and limited number of classroom hours. This schedule had allowed him the flexibility, in principle and oftentimes in practice, to pursue his own mathematical researches concurrently with the performance of his Academy duties. It had been, from the beginning, one of the attractions of the post for Sylvester, and he viewed it as his contractual right.

In 1862, however, the new Secretary of State for War—in concert with the Woolwich military authorities and the Council of Military Education—had determined that the cadets would benefit from more hours in the mathematics classroom. Thus, as part of the ongoing reform of military education, it had been proposed to change the job description of the Professor of Mathematics to provide for this increase in instructional time. In June, those changes were still under discussion, and Sylvester had been asked for and had given his opinion personally to Lewis. By December, the details of an actual proposal seemed close to being finalized, but Sylvester was both stunned and angered to learn how things had evolved in the intervening months. He could barely contain his rage in a letter to Hirst on the twenty-third. "In violation of the contract under which I was engaged and have acted for a long series of years," Sylvester stormed, the authorities "lay claim to the actual disposal of the whole of my time in lieu of the stipulated limited portion of it for which I am lawfully liable and as a *first instalment* require of me to give from 18 to 20 hours per week next term to the business of tuition."[37] "I need hardly say," he went on, "that I intend standing on my contract and shall refuse submission to this unjust demand throwing upon the Horse Guards the responsibility of dismissing me for disobedience to their unlawful demands if they are so disposed." In fact, when he posted this account to Hirst, Sylvester was in the Devonshire coastal resort of Torquay to meet with "an important M.P.," quite likely the Liberal member from Limerick, William Monsell, "to engage him" in the fight "against the Military Rulers of the Academy including the Duke of Cambridge." As he had demonstrated during the course of his first battle with the military authorities, Sylvester would hesitate neither to use any and all influence he could muster nor to take on even the most powerful in battling what he viewed as blatant injustice.

The new year found him trying to get on with his life and his mathematics as he once again awaited the outcome of politicking behind the scenes. On Saturday, 3 January 1863, for example, he wrote to Cayley to fix a day and time for their next meeting, telling him both of some new results he was working on relative to the catalecticant and of the "select party" he would attend on Tuesday, the sixth, in honor of his friend from undergraduate days, "the Heresiarch Bishop Colenso."[38] John Colenso, who had been consecrated the first Bishop of Natal in 1853 and who had made a scholarly name for himself through his translation into Zulu of the Bible, had drawn sharp criticism from the Church of England just months before when he publicly questioned the literal interpretation of the Bible in light of his familiarity with non-Western traditions.[39] (Sylvester's perhaps tongue-in-cheek use of the word "heresiarch" early in 1863 proved prophetic; Colenso was formally found guilty of heresy and excommunicated during the course of 1863 and 1864, although he refused to acknowledge the ruling and continued to preach before a supportive congregation.) The dinner in Colenso's honor thus undoubtedly represented as much a social gathering as a show of support from his friends, and Sylvester would not miss it, working to schedule his meeting with Cayley either before the party on the sixth or afterward on the evening of the seventh.

A month went by before Sylvester had anything further to report of his situation at the Academy. The news was both good and bad. Writing to Cayley on 8 February, he seemed to be between a rock and a hard place. "The Sec[retar]y of State for War *has decided* in favor of my interpretation of the terms of my engagement," he reported.[40] That was the good news. The bad news, however, was that they were proposing to increase his teaching load from five to eight attendances weekly. This was much more teaching than he and Lewis had discussed earlier in June. The increase from 13½ to 16 hours a week would have amounted to essentially one additional attendance; now they were proposing, in fact, three additional attendances. Although they asked him to propose a salary increase that he felt would be commensurate with the extra work, as he told Cayley, "my wish would be to say ∞; as no money could compensate for the possible effect of an arrangement annihilating my capabilities for independent work." To make matters worse, he explained, "the proposal is however made to me bonâ fide and in a very conciliatory spirit, and it is very difficult to know how to evade it." Yet, somehow to evade it was most definitely his desire. Maybe he had an idea.

Writing also to Hirst on the eighth, Sylvester asked "would any consideration be possible according to which you could take two evening attendances of 2 hours each off my hands?"[41] Sylvester assured him that there would never be any more than 15 students in any given class, and he even offered the use of his house on Woolwich Common where Hirst "would have a bedroom and sitting room appropriated to [his] exclusive use." "Excuse the liberty I take in asking you to consider the possibility," he closed, "and oblige me with an answer at your earliest convenience." If Sylvester could get the authorities to agree to hire a part-time professor, then he would at least have some chance of carrying out his research agenda. This could be the ace up his sleeve, but he would have to play the card carefully.

Sylvester gave no indication of this plan when he responded officially to the proposal of the Council of Military Education on 11 February 1863. His first move was to see what reaction his salary proposal would elicit. The proposition would not be cheap, and, perhaps, it would be deemed so expensive that the authorities would rethink the entire deal. It was at least worth a try.

Sylvester laid the case out in an effort to convey the high cost of the proposed new arrangements in personal terms. "I doubt if my strength will be found equal to the amount of work now called for and still more whether the wear and tear of an occupation so laborious and continuous supervening after I have passed the Meridian of my days will not prove fatal to the elasticity of mind necessary to the successful prosecution of those scientific pursuits which I set before all pecuniary considerations."[42] "Still," he continued more than somewhat disingenuously, "if the necessities of the Public service are considered to render this sacrifice on my part indispensible I will cheerfully lay all personal feelings aside and shall be ready upon an equitable basis of remuneration to give effect to the wishes of the Council on the subject." What would that "equitable basis" be according to the now forty-eight-year-old Sylvester who, for all intents and purposes, saw the Council's proposal as taking total control of whatever "free" time he had ever enjoyed? "I deem it but fair and reasonable," he explained, "that I should receive the same proportion of pay for the three additional attendances as I at present receive for the number of which I was originally engaged or at least that the additional remuneration should be based on the rate (which I had to accept) proposed by the Secretary of State for War for the one additional attendance." That meant that—because Sylvester was then being paid £110 per attendance—then either his salary should increase by £330 for the three additional attendances or at least by £300, given the earlier offer of £100 for one additional attendance. Either way, his salary would jump from its 1863 level of £550 to well over £800! The equity of this solution, he closed, "is confirmed by the opinion of all the scientific and other friends most competent to form an opinion whom I have had the opportunity of consulting on the subject." The War Office would clearly not be happy about this proposal, but what would be its next move?

While Sylvester waited to learn the answer to this question, he distracted himself both with a bit of mathematics and with politicking of a different kind. The mathematics, which he discussed with Cayley and with Hirst, involved determinants, and the politicking centered on the Royal Society. In 1862, Sylvester had been elected to serve a term on the Society's governing Council. This meant, among other things, that he was in more regular attendance at Society meetings, in general, played a more active role in the Society's affairs, and was involved in making a variety of behind-closed-doors decisions.[43]

In particular, the Council was in the process of choosing, in the winter of 1863, a distinguished foreign scientist for the honorific position of foreign member, and Hirst had a candidate. At least as early as December 1862, he had suggested that Jakob Steiner, the geometer and Professor of Mathematics at the University of Berlin, would make an excellent choice, and he had approached Steiner directly. As Sylvester put it to Hirst, "very glad am I for the sake of the R[oyal] S[ociety] that you have proposed to make us a present of so magnificent a fish from the other side of the Channel and that he is willing to be caught."[44] By late February 1863, Sylvester, in his capacity as a member of Council, had acted on Hirst's suggestion and had formally proposed the German geometer. The problem was that there were other candidates in both mathematics and in other fields of scientific study, so the case for Steiner would have to be made carefully and convincingly. "Cayley has suggested that you & he together might work out a résumé for my instruction and guidance as the advocate of Steiner's case," Sylvester informed Hirst.[45] "Would you confer with Cayley upon the subject?," he asked.

Sylvester, Hirst, and Cayley continued to plot strategy relative to Steiner's candidacy into March. On the eleventh, Sylvester wrote to tell Hirst that the election would take place on Thursday, the nineteenth, and he asked him to prepare, in particular, a clear statement on some of Steiner's geometrical papers.[46] Three days later, he wrote again, this time inviting Hirst to dine with him at his home in Woolwich so that they could "go into the question of Steiner's candidature at the Royal Society" immediately in advance of the vote.[47] Sylvester wanted to be prepared and who better to brief him than the geometer, Hirst. Unfortunately, when the votes were cast and counted, Steiner had lost to his Berlin colleague, Ernst Kummer. Hirst, as the initiator of Steiner's candidacy, was charged with informing him of the bad news. The German, however, died on 1 April just days after his sixty-seventh birthday, most likely unaware of the failed efforts of his English supporters.

When Sylvester wrote to Hirst two days after the Council's vote to ask "have you written to Steiner?," his mind was really less on Society business and more on matters much closer to home. He had heard back from the Secretary of State for War, and he was incredulous. He "has decreed that I *must* give the 8 attendances," Sylvester told Hirst.[48] "No mention is made of compensation for this proposed extinction of my scientific existence for to such the surrender of 20 hours per week to the work of grinding will practically amount." Sylvester simply would not accept this. The fight would have to continue. "Mr. Monsell will be willing to take my case into Parliament if he can find a strong man to cooperate with him," he explained, but more politicking would have to take place to secure that needed ally.

Sylvester had an idea. Sir James Stansfeld, the M.P. from Halifax since 1859, had been one of his students at University College and was both a Liberal and a dissenter. Sylvester would talk with him and try to solicit his support. In the meantime, he wondered if Hirst would, "*without delay*," exploit his friendship with Sir Stephen Cave, the Conservative M.P. from Shoreham, by asking Cave to "influence Stansfeld to lend his aid to check this act of high handed injustice." Perhaps if Stansfeld heard of Sylvester's plight from several corners, he would be even more likely to join forces with Monsell on Sylvester's behalf.[49]

Sylvester apparently need not have worried about the loyalty of his former student. On Sunday, 29 March 1863, Sylvester wrote to thank Hirst for his interventions and to tell him that Stansfeld "has given me his warmest and most valuable support in my recent transactions with the War Office authorities."[50] Moreover, two days earlier on Friday, Sylvester himself had "had an interview with Sir George Lewis" and could happily report that "there is now a very fair prospect of an equitable arrangement being come to with the Government on the basis that my attendances instead of being raised to 8 be *reduced* to 4, I surrendering a portion of my salary towards procuring the services of a substitute for the other 4." It was precisely that ace he had had up his sleeve since early February, and he had apparently played it to good effect with the Secretary of State for War. Still, "experience of the past warns me," he confessed, "not to be over sanguine, and the more so as the determined opposition of the Horse Guards has still to be overcome." Once again, his strategy had been to work the civilian politicians against the military in command of the Academy, but it was not over yet.

Nine days later on 7 April, Sylvester made his case before the Council of Military Education both for the reduction of his own teaching commitments and for the appointment of a substitute, whom he hoped would be Hirst. The meeting, however, went badly, and Sylvester left fully convinced that the support Lewis had seemed to give him had been nothing more than "a Mockery, a Delusion and a Snare."[51] "The truth is," he lamented to Hirst, that Lewis "has all along been playing a double hand, yielding to Military pressure on the one hand, and on the other fobbing off my friends and supporters with *professions* of his good intentions." The situation looked hopeless.

A month later, however, everything had changed. Lewis had died in the midst of the battle over Sylvester's claims, and a successor, Sir George Grey, had been appointed Secretary of State for War. Sylvester had, it turned out, been wrong about Lewis. In describing the last days of the battle in a letter to Lord Brougham on 8 May, he related that "one of the last acts of the late lamented Secretary of War [Lewis] (although at first under hostile influences he took a different view of my case) was to acknowledge the justice of my claim and his decision has been concurred in and carried into effect by his most estimable successor Lord De Grey."[52] It had been a long, hard war, but this, his second battle, had been decisive. "Only the sense of justice of my claims and the hardship of the treatment to which I had been subjected," he admitted, "could have supported me in so arduous and unequal a struggle protracted over a span of two years against the combined influence of the ruling Military powers." "I trust," he said in closing, "to make a good use of my recovered leisure for study to leave a lasting mark on 'The Algebra of the Future.'" That, of course, had been the goal that had for so long fueled his ambition.

Hard-Won Victories

Sylvester celebrated his victory by returning immediately to some mathematical ideas that had essentially been on hold since the end of November 1862. At that time, he had written jubilantly to Cayley, convinced that he had made a major research hit. "The grain of mustard seed is grown into a mighty tree," he announced. "I have struck the key to the extension of Gauss' quadrature method to Cubatures or even Biquadratures etc. . . . and the theory I am now launched upon gives promise of results of quite as general and important a character quâ the laying bare of a new trail of Algebraical gold diggings."[53] At issue was a method Gauss had devised for approximating the value of an integral of a function in one variable, and Sylvester believed he had found a way— involving purely algebraic means—to extend that method to double integrals of functions in two variables, triple integrals of functions in three variables, etc. As he explained in his first published account of his technique in the June 1863 number of the *Philosophical Magazine*, his extension of "Gauss's method of approximation from single to multiple integrals" hinges on "a method which invariably leads to the construction of a *canonizant* whose roots are all real."[54]

Given, for example,

$$f(x, y) = a_1 x^{2m-1} + (2m - 1)a_2 x^{2m-2} y$$
$$+ \left(\frac{2m - 1}{2}\right) a_3 x^{2m-3} y^2 + \cdots + a_{2m} y^{2m-1}, \tag{1}$$

a binary form of odd degree $2m - 1$, its canonizant is the determinant:

$$\begin{vmatrix} a_1 x + a_2 y & a_2 x + a_3 y & \cdots & a_m x + a_{m+1} y \\ a_2 x + a_3 y & a_3 x + a_4 y & \cdots & a_{m+1} x + a_{m+2} y \\ \vdots & \vdots & \ddots & \vdots \\ a_m x + a_{m+1} y & a_{m+1} x + a_{m+2} y & \cdots & a_{2m-1} x + a_{2m} y \end{vmatrix}$$

$$= \begin{vmatrix} y^m & -y^{m-1} x & y^{m-2} x^2 & \cdots & -x^m \\ a_1 & a_2 & a_3 & \cdots & a_{m+1} \\ a_2 & a_3 & a_4 & \cdots & a_{m+2} \\ \vdots & \vdots & \vdots & \ddots & \vdots \\ a_m & a_{m+1} & a_{m+2} & \cdots & a_{2m} \end{vmatrix},$$

where the latter is an expression of degree m in the variables x and y.[55] As Sylvester had shown as early as 1851 (compare the discussion in chapter 5), factoring the *canonizant* yields the reduction of f to its *canonical* form as a sum of m powers of degree $2m - 1$.[56]

In results communicated to Cayley early in 1863, however, Sylvester had succeeded in relating the canonizant to the catalecticant in an interesting way. The problem was that he had been so distracted by his troubles at Woolwich that he had had neither the time nor the peace of mind to publish his results. In that same letter to Cayley of 8 February in which he had related both the good and the bad news about the developments in his case, Sylvester had also given Cayley leave to "do as you please about my canonical theorem; it is not quite present to my mind having been displaced by subsequent ideas."[57] Cayley, who had set Sylvester's ideas in a slightly more general context, proceeded to publish them with due acknowledgment in the March 1863 number of the *Philosophical Magazine*.[58] Now, by June, with both another term and the Woolwich business behind him, Sylvester presented—with characteristic hyperbole—further results he had obtained along these lines. "Subsequent generalizations . . . have led me on, step by step," he proclaimed, "to the discovery of a vast general theory of double determinants, . . . constituting, I venture to predict, the dawn of a new epoch in the history of modern algebra and the science of pure tactic."[59]

In his paper, he considered, instead of a function $f(x, y)$ in two variables, a function $f(x, 1)$ in the single variable x and formed the canonizants associated with such functions of successive odd degrees, namely,

$$X_0 = 1, \quad X_1 = \begin{vmatrix} 1 & x \\ a & b \end{vmatrix}, \quad X_2 = \begin{vmatrix} 1 & x & x^2 \\ a & b & c \\ b & c & d \end{vmatrix}, \quad X_3 = \begin{vmatrix} 1 & x & x^2 & x^3 \\ a & b & c & d \\ b & c & d & e \\ c & d & e & f \end{vmatrix}, \ldots$$

He next formed the corresponding series of catalecticants

$$\lambda_1 = a, \quad \lambda_2 = \begin{vmatrix} a & b \\ b & c \end{vmatrix}, \quad \lambda_3 = \begin{vmatrix} a & b & c \\ b & c & d \\ c & d & e \end{vmatrix}, \ldots$$

and showed that "the resultant of X_i and X_{i-1} is an exact power of λ_i, which (as will at once be seen) is the coefficient of x^i in X_i."[60]

After laying out various ramifications of this result, Sylvester closed with a statement, which, even for him, was remarkable for its bravado. Speaking in the third person, he asked his readers to "let him be permitted also in all humility to add . . . , that in consequence of the large arrears of algebraical and arithmetical speculations waiting in his mind their turn to be called into outward existence, he is driven to the alternative of leaving the fruits of his meditations to perish . . . , or [of] venturing to produce from time to time such imperfect sketches as the present, calculated to evoke the mental cooperation of his readers, in whom the algebraical instinct has been to some extent developed, rather than to satisfy the strict demands of rigorously systematic exposition." The military authorities may have slowed him down mathematically over the past six months, but he was determined to get his ideas in print, even if that meant begging the indulgence of those who might comprise his readership. He most certainly asked a lot of both his audience and his editors!

If Sylvester's life at Woolwich seemed finally, in June of 1863, to be con-ducive to his sense of his dual mission as teacher and researcher, another key aspect of that life was about to change radically. For some fifteen years, a deep friendship and productive mathematical cooperation had developed between Sylvester and Cayley. It had begun when they were neighbors at the Inns of Courts, talking with and seeing each other almost daily. It had continued after Sylvester's move to nearby Woolwich, with the two men meeting up often in London or rendez-vousing on walks between Blackheath and Woolwich Com-mon. Letters had certainly supplemented these face-to-face conversations, but the immediacy of the one-on-one mathematical give-and-take with Cayley had always served as inspiration for the gregarious Sylvester. On 10 June 1863, how-ever, that immediacy was effectively lost. Cayley had won the appointment to Cambridge's new Sadleirian professorship of pure mathematics, a chair filled at a time when Cambridge University was making its first tentative moves toward the model of the modern, research-oriented university. Cayley would be in the vanguard of the new professor-researchers. Moreover, with the new position came the steady income and social respectability required of the middle-class Englishman who would be wed. Cayley wasted no time. On 3 September 1863, he married his intended Susan Moline in Blackheath, and the two moved to Cambridge to begin the 1863–1864 academic year.[61] Meetings with Sylvester would now be much fewer and farther between. Never again would they talk as two bachelor-mathematicians.

Although Sylvester was certainly happy for Cayley in both aspects of his changed situation (he had long known of Cayley's desire to leave the law for a post in academe and was surely aware of Cayley's affection for Susan Moline), he sensed the void that his friend's absence from London and its immediate environs would leave in his own life. Hirst was there, but the young man, almost sixteen years Sylvester's junior, was perhaps more of a friend and protégé than a friend and equal. Still, in the months after Cayley's departure and, in fact, for the next several years, it was to Hirst that Sylvester turned frequently for companionship and mathematical conversation.[62]

Early in October of 1863, for example, Sylvester had come back to ideas on integration. These were not unrelated to the work he had done in June and had then sketched at the end of the summer before the assembled scientists at the annual meeting of the British Association for the Advancement of Science at Newcastle-upon-Tyne where he had served in the honorific post of vice president.[63] It was to Hirst whom Sylvester wrote to share his new thoughts, closing his letter "Ever truly yours in the grace of God and the fellowship of Mathematick."[64] The sacred bonds of that fellowship were tested the next day when Sylvester thought out loud, as he had so often with Cayley, in a follow-up letter. "Many thanks for your note and suggested attention; but by good luck (for I am an awful blunderer) this time I am right," he announced.[65] "No I am wrong," he countered immediately. "I must state that my proof which at the moment I thought complete or perfect . . . is only so" in certain restricted cases. "I believe however that the theorem is *generally* true although wanting proof." November brought more of the same. "Some fatality appears to pursue me and involve me in continued mistakes," he confessed.[66] "I am quite wrong in all I wrote you . . . ; please destroy my letter which is an evidence of delusion and excuse the needless trouble I have given you." This was exactly the kind of dynamic that had characterized Sylvester's exchanges with Cayley; now it was Hirst to whom Sylvester unselfconsciously laid bare his mathematical soul.

As his correspondence with Hirst makes clear, Sylvester had perhaps more downs than ups in the fall of 1863 as he tried yet again to settle into a routine of teaching and research. His relative lack of success on the latter score might soon have left him despondent had his ego not received a boost early in December with word of a singular honor. A decade earlier in 1853, Sylvester had been mentioned as a possible candidate for the position of foreign corresponding member to the Section of Geometry of the Paris Academy of Sciences, but when the votes were finally cast in 1854, Jakob Steiner, and not Sylvester, had won the election (recall the discussion in chapter 5). With Steiner's death in the spring of 1863, however, the position had once again become vacant, and this time Sylvester had gotten the nod. It was with a great sense of gravity that he thanked the Academy on 11 December. "I am very aware," he wrote, "of the honor the Academy has done me by inviting me to communicate to it the results of my studies in the sciences which are within its purview and I hope to find the occasion to profit from this gracious permission."[67] As foreign corresponding member, Sylvester would no longer have to communicate his ideas to the Academy through a third party; he had been given the right to speak directly to his French scientific *confrères*.

As if energized by this new privilege, Sylvester soon prepared and submitted a note that was read for him at the Academy's meeting on Monday, 14 March 1864. Sometime after learning of his election to the Academy, he had abandoned the theory of integration for which he had made such grand claims in 1862— and yet in which he had made so little real progress—to return to a problem that had engaged him at least since the late 1830s and early 1840s, namely, that of analyzing the real roots of an equation of the form $f(x) = 0$, for $f(x) \in \mathbb{R}[x]$. When a new result on a particular special class of equations yielded almost immediately, he communicated it—without proof—to the Academy.[68] A month later, he had done even more. Writing to ask Joseph Bertrand if he would do him the honor of reading a new result before "your (dare I say our?) illustrious Academy," Sylvester informed him in passing that "I have proven not without some difficulty Newton's Rule up to the fifth degree inclusive."[69]

In his *Arithmetica universalis* published in a first edition in 1707 and then in a posthumous second edition in 1728, Isaac Newton had discussed how to isolate pairs of complex roots of algebraic equations but had given no formal justification of his techniques. Later in the eighteenth century, mathematicians such as Euler and Colin Maclaurin had picked up on and exploited "Newton's Rule" but had also provided no proof. The rule had become, in Sylvester's words, "the wonder and the opprobrium of Algebra," but by April 1864, he had succeeded in providing its elusive proof, at least for equations of degree less than or equal to five.[70]

Almost immediately, he wrote up and submitted a paper on his limited, but nevertheless important, result for consideration by the Royal Society for its *Philosophical Transactions*. The manuscript was massive, and it only grew larger over the course of the summer and fall of 1864 as Sylvester continued to think hard about the ramifications of his proof. Initially, the paper had two parts: the first constituted the proof of Newton's Rule for algebraic equations up to and including the fifth degree, and the second provided the proof of a special, but related, result that Sylvester had announced in the *Comptes rendus* in April, namely, a rule for providing the lower bound on the number of imaginary roots of equations of the form $\sum_{i=1}^{n} (a_i x + b_i)^m = 0$, for $m \in \mathbb{Z}^+$ and $a_i, b_i \in \mathbb{R}$. By the fall, however, Sylvester had added a third part in which he provided "absolute invariantive criteria for fixing unequivocally the character of the roots of an equation of the fifth degree, that is to say, for ascertaining the exact number of real and imaginary roots which it contains."[71] He also continued to add remarks and footnotes and figures to each of the paper's parts, frustrating thoroughly Royal Society Secretary, Gabriel Stokes, in his efforts to send a completed whole first to his referees and then to his printer.

This paper, however, represented a kind of watershed for Sylvester. It was of the utmost importance to him that it come out with all his evolving ideas incorporated. As he put it in a letter to Hirst enlisting his intervention with Stokes, "my former Syzygetic memoir which you have heard favorably spoken of grew up in the proofs in the same way as the present memoir—and were I to wait until assured that no more would arise to be added I should never publish at all . . . . It would be very painful and discouraging to me in my renewed activity to find that the pages of the *Transactions* were opened to me with reluctance or that objections of mere form were taken to the *modus operandi* forced upon me by the necessities of the case and it may be in part also by certain invincible idiosyncracies of mental constitution."[72] A man of large ego and one somewhat desperate in 1864 to demonstrate that he had not lost the ability to contribute significantly to mathematics, Sylvester expected nothing less than the indulgence of others in his foibles. One can only wonder to what extent those who submitted their work for *his* consideration for the *Quarterly Journal* were equally indulged.

Cayley at Cambridge and Henry Smith at Oxford refereed the paper's first two parts, and both found the results impressive. In Cayley's view, "the demonstration of [Newton's Rule] even in the case of the fifth order (the highest to which the paper relates) gives occasion for a most refined & interesting investigation."[73] Still, he "doubt[ed] whether a proof of the theorem in its generality must not be sought for in a different direction." Smith, too, thought that "the difficulty attending a complete demonstration appears to be very great," yet judged that, even given the restrictions on his proof, Sylvester had "made a most important addition to what has hitherto been known on the subject."[74] As for the paper's third part, Cayley refereed that as well, and although he found it "to contain much that is interesting & valuable" and so recommended its publication, he could not "help remarking that the manuscript is not such a one as should have been presented to the Society, the illegibility [being] such as not only to make the referee's task a very irksome one, but to throw undue difficulty in the way of his forming a correct judgement of the paper."[75] As was so often the case when his thoughts outran his pen, Sylvester's manuscript, composed in the heat of discovery, was so illegible and disorganized that even his long-time friend had found himself at his wit's end in trying to decipher it. It indeed reflected the "invincible idiosyncracies" of Sylvester's "mental constitution."

Sylvester recognized that all three parts of his paper contained important results, but, betrayed even by his handwriting, he was most excited about the discoveries contained in its third part. In his view, that part constituted "by far the most valuable portion of the memoir, containing as it does a complete solution of one of the most interesting and fruitful algebraical questions which has ever yet engaged the attention of mathematicians," namely, the problem of the invariant theory of the binary quintic form.[76] It had been almost a decade since Sylvester had actively engaged in any properly invariant-theoretic research, but, as he put it, "as all roads are said to lead to Rome, so I find in my own case at least, that all algebraical inquiries sooner or later end at that Capitol of Modern Algebra over whose shining portal is inscribed 'Theory of Invariants.'"

What Sylvester had succeeded in effecting was an analysis of the binary quintic form, namely, $f(x, y)$ as in (1) where $2m - 1 = 5$, that linked its roots and its invariants. His main tool was the canonizant, itself an irreducible covariant of the quintic of degree three in its coefficients $a_1, \ldots, a_6$ and of order three in its variables x and y and one that Cayley had explicitly enumerated in his paper, "A Second Memoir upon Quantics," in 1856.[77] There, Cayley had also noted that the binary quintic had one irreducible quadratic covariant, that is, there was one irreducible invariant expression of degree two in $a_1, \ldots, a_6$ and of order two in x and y. Drawing from this well-known theory in 1864, Sylvester showed, for example, that the canonizant is a perfect cube if and only if the quintic has three equal roots. He then reinterpreted this result in terms of the quadratic covariant, proving that when the quadratic covariant is a perfect square but does not vanish, the quintic has three equal roots.[78] He went on, in an algebraic *tour de force*, to give invariant-theoretic criteria for determining the number of real and imaginary roots of any given quintic (435–463).

The production of this paper continued through the 1864–1865 academic year as Sylvester simultaneously attended to his duties at the Royal Military Academy. Things were much improved there that year, on one front at least, as a result of Sylvester's victory in his second battle with the military authorities. As he had intimated to Hirst in March 1863, one of the proposals that had been on the table in his protracted negotiations with the War Office at that time had been the appointment of an instructor to take some of the teaching burden off of the professor. That instructor, the Dublin-born Morgan Crofton, was finally appointed for the 1864–1865 academic year, and his presence at the Academy provided Sylvester not only with much desired teaching relief but also with a certain measure of local mathematical companionship (see below).

That Sylvester's role as teacher-researcher could still be a trying one even with Crofton at the Academy was borne out in the fall term of 1864 just as Sylvester was in the thick, particularly, of the renewed invariant-theoretic researches that ultimately appeared in part three of his paper. The cadets had returned on 3 August, and those in their final year of study had begun their course in higher mathematics with Sylvester. Although foremost in their minds were, first, surviving Sylvester's class and, second, garnering a sufficient number of points on his final examination to assure the awarding of their commissions, the cadets apparently could not resist some fun at their professor's expense.

As noted, Sylvester was generally regarded as less than effective in his Woolwich classroom. According to one of the cadets in his class that fall, "order was usually badly kept in his Academy, and sundry measures of annoying him were indulged in with success by the cadets."[79] Among those was one that might be styled "the disappearing act." While Sylvester was deep in thought at the board, his back to his class, the cadets would drop down behind their desks, rendering themselves invisible from his vantage point at the front of the room. He would then, at some point, turn around only to "see the class-room half empty. This made him rush up and down, a movement which was prepared for by sprinkling the floor round his table with wax matches, which went off in succession as he stamped round, driving him quite wild." In another of their favorite antics, the cadets sneaked into his classroom before his arrival and filled his inkwell with chalk. This, of course, "clogged his pens and made him mad!"

These harmless pranks apparently got out of hand on 2 October, the very day Cayley wrote his referee's report on the third and final part of Sylvester's paper on Newton's Rule. The day before, over two hundred tons of gunpowder had blown-up in the Arsenal's magazine. The cadets "were just getting ready for parade when [they] heard the explosion, which shook the Academy like an earthquake, and then saw a great column of black smoke rising slowly and spreading out into a cloud in the sky."[80] This incident left the cadets rattled both literally and figuratively; mischief, in addition to smoke, was in the air. "The first symptom," according to a cadet present at the time, "was a disturbance in the class-room where Professor Sylvester presided. The corporal on duty failed to quell it, and the assistant inspector of studies had to be called in. Then followed a row . . . , which ended in the rustication of two cadets." By the end of the month, the situation had escalated when "one of the field-guns on the parade was fired towards the Governor's house, and all the swords which the cadets carried during punishment drill were thrown into the reservoir." As in the aftermath of the so-called "mutiny of 1861," a Court of Inquiry was convened and punishment exacted.[81] The term finally ended late in December with the Duke of Cambridge's customary visit to the Academy to view the cadets and to award both the academic prizes and the new commissions. On this occasion, however, he alluded also to the disturbances earlier in the term, cautioning the cadets that "they ought never to overlook the fact that to be able to command they should learn to obey. This was one of their duties . . . [and] [w]ithout it they could never be qualified to figure well in the roll of the valuable corps to which they might be appointed."[82] It had been another eventful year, but this time research had most definitely been overshadowed *neither* by teaching *nor* by any other happenings at the Academy.

Sylvester was not present to hear the Duke of Cambridge's admonishments to the wayward cadets. He had finished up his examination duties, taken care to try to smooth the way for a paper by Julius Plücker at the Royal Society, and left England to spend several weeks in Paris where he could enjoy personally his new status within the Académie des Sciences. At the time of his return to Woolwich in February 1865, however, the production of the paper on Newton's Rule and the ideas contained therein continued to absorb him. In particular, a certain fourth-order curve—he called it a bicorn because it "consist[ed] of four branches, coming together in pairs or two cusps, so as to form two distinct horns"—had arisen as a section of a particular surface associated with his algebraic and invariant-theoretic analysis of the binary quintic form, and both the surface and the curve had captured his imagination.[83] Always more of an algebraist than a geometer, Sylvester had a hard time visualizing especially the surface, but, as he told Hirst late in March, "I have thought a good bit upon this wonderful surface since last seeing you. . . . [Its] form . . . seems to be gradually growing up in my mind but it requires a prodigious effort beyond my present powers of conception to realize it in its totality."[84] Those efforts ultimately found expression in the paper on Newton's Rule. It indeed "grew up in the proofs" as it made its torturous way through the press in 1864–1865.

The development of further results for the *Philosophical Transactions* paper on Newton's Rule did not come ultimately to dominate the spring of 1865.[85] Sylvester also again engaged in some domestic politicking with Hirst at the Royal Society. This time, the issue was the reconsideration of Andrew Noble for a fellowship, and Sylvester felt a certain responsibility in the matter. In 1865, Noble was a civilian involved in the private ordnance company owned by captain of industry, Sir William Armstrong, but recall that, while an officer in the Army stationed at the Royal Arsenal in Woolwich, Noble had assisted Sylvester in preparing the outlines for his King's College lectures on partitions in the summer of 1859. At that time, as Sylvester explained to Hirst, he "solved one problem about which I was at fault in some researches connected with my theory of partitions [and] I did him an injustice in asking him to withhold it until my own memoir would be ready to appear. This memoir," however, "would make a volume—Heaven knows when it will appear."[86] Sylvester thus asked Hirst to "mention this in his favor if you think it will do him any good" at the time of the elections. Continuing to stew about the case and his own role in it, Sylvester wrote again to Hirst to confess that "I am very anxious about Noble. He is a capital fellow and w[oul]d certainly have got in last time had it not been for a mishap about his paper which was not delivered on time for hanging up owing to some unfortunate mistake for which no one can be found to accept the blame. I had nothing to do with it."[87] Despite Sylvester's concerns and Hirst's efforts, Noble did not make it in 1865, either. He would only be admitted to a fellowship in 1870, but he would go on to win a Royal Medal ten years later for his researches on explosives.[88]

The disappointment of Noble's failure at the Royal Society was soon replaced by the elation of new developments on the Newton's Rule front. On 11 June 1865, Sylvester wrote triumphantly to tell Hirst that "I have obtained a *simple and completely general proof* of Newton's Theorem. If you can arrange for my doing so," he continued, "I should have much pleasure in bringing it before the Mathematical Society of London about which you spoke to me some time back."[89]

The London Mathematical Society (LMS) had met officially for the first time at University College London on the evening of Monday, 16 January 1865.[90] At that inaugural meeting, Augustus De Morgan had been elected the society's president and Hirst its vice president, but the Society had the flavor more of a mathematical club for students associated with University College than of a professional, research-oriented organization. By June, however, mathematicians like Cayley, Sylvester, Spottiswoode, and Henry Smith, among others, had been recruited, and Sylvester had, in fact, further legitimated the new venue by presenting there his now fully general result. On Monday, 19 June, in a brilliant display of mathematical promotion, his note "Sur les limites du nombre des racines réelles des équations algébriques" was read for him before the Paris Academy of Sciences, and an English rendition was delivered by him personally to the members of the London Mathematical Society.[91] As he told Hirst, "the work of weeks of thought is condensed into half a dozen pages."[92]

It was a stunning accomplishment that even received notice on the pages of the *Times*. On 28 June 1865, the paper ran a piece under the title "A Mathematical Discovery" in which it announced "that Professor Sylvester, a mathematician whose reputation is as well established abroad as at home, has just made a great discovery in that science. This is no other than the proof of Sir Isaac Newton's rule for the discovery of imaginary roots of Equations."[93] "Certain it is," the writer went on, "that this rule has been a Gordian knot among algebraists for the last century and a half," being the only result in Newton's *Arithmetica universalis not* accompanied by a proof. Given this state of affairs, "authors became ashamed at length of advancing a proposition the evidence for which rested on no other foundation than the belief in Newton's sagacity."

Sylvester had provided the firm and mathematically incontrovertible foundation that had for so long been wanting, and on the evening of the twenty-eighth, he presented it to an even wider audience in a public lecture at King's College. He must have made an impression that night, for later in the year, readers of Mortimer Collins's novel *Who Is the Heir?* found one of the characters "looking over some notes he had made on Sylvester's proof of Newton's process for discovering the imaginary roots of an expression—'the beautiful child of Newton's youth,' as the poetic Professor styled it" during the course of the King's College lecture![94] Sylvester's notoriety had spread even to the gentle reading public.

With the successful conclusion of the King's College lecture, the frenzy of activity surrounding the proof of Newton's Rule finally subsided, and Sylvester moved on to different mathematical activities. One of those centered on a new journal that had been founded just the year before, the *Mathematical Questions . . . from the Educational Times*. The *Educational Times* had begun in 1847 as an initiative of London's College of Preceptors aimed at promoting mathematics among England's teachers and students. From its inception, it had carried questions and solutions submitted by its readers in an effort to stimulate mathematical thinking, but that department had ultimately proved so successful that the journal's editor, William Miller, decided to launch a separate and supplementary periodical, *Mathematical Questions . . . from the Educational Times*, in 1864. Sylvester supported Miller in his endeavor from the outset. In fact, not a year would pass from 1864 until 1899, two years *after* Sylvester's death, without the appearance of at least one of his mathematical questions.[95] In July 1865, however, Sylvester was prepared to have nothing more to do with the journal, not because of Miller but because of what Sylvester perceived as the publishers' abuse of their editor.

Vacationing in Wales on 16 July, Sylvester penned a blistering letter to C. E. Hodgson & Son denouncing their treatment of Miller and severing his ties with the *Educational Times*. "That you should not only avail yourself of Mr. Miller's invaluable services without giving him any compensation," he fumed, "but that you should also presume to express dissatisfaction with him for sending copies of the works he edits to his gratuitous contributors and endeavour to put pressure upon him to recruit for subscribers among the same appears to me so preposterous a proceeding that I must henceforth decline to be concerned in any shape or form with your journal."[96] For Sylvester, it was another matter of principle, but, fortunately, Miller was able to calm him down and keep him as a mainstay among his contributors.

Evidence of Miller's success followed almost immediately. In September, Sylvester attended the annual meeting of the British Association for the Advancement of Science in Birmingham in 1865, once again as one of the vice presidents and this time also as a member of the BAAS's governing council, a post he would hold for five years. He spoke as well, drawing his topic from his reading of the *Educational Times*. In 1863, the year before the *Mathematical Questions* began to appear under separate cover, Sylvester had found the following problem posed there—what is the probability that a triangle joining three points at random in a sphere is acute?—and had read the solutions to it provided by the poser, one Wesley S. B. Woolhouse, as well as by the man who became his associate at Woolwich in 1864, Morgan Crofton.[97] As Sylvester reported to those assembled in Section A in Birmingham, he had succeeded in generalizing these solutions to a whole class of questions in what he termed "form-probability," that is, in what would now be called "geometric probability." "The chance of three points within a circle or sphere being apices of an acute or obtuse-angled triangle, or of the quadrilateral formed by joining four points, taken arbitrarily within any assigned boundary, constituting a reentrant or convex quadrilateral, will serve as types of the class of questions in view," he explained. "The general problem is that of determining the chance that a system of points, each with its own specific range, shall satisfy any prescribed condition of form."[98] Using his new technique, Sylvester was then able to solve Woolhouse's problem in terms of "an easy single integration, the solution heretofore given . . . involving complicated triple integrals." Moreover, Sylvester was also able to apply his method to effect a solution when Crofton posed the question—"two points being taken at random within (1) a circle or (2) a sphere, what is the probability that their distance apart is less than a given line?"—shortly after the Birmingham meeting in the *Mathematical Questions*.[99]

These probabilistic considerations continued to occupy Sylvester through the month of October. On Saturday the fourteenth, for example, he wrote to tell Hirst that he intended to be at Monday's meeting of the London Mathematical Society to hear Cayley present a paper, but he also made a proposition for the next meeting to be held thereafter. "If you think the form suitable to the purpose of the Society," he ventured, "I should like to give notice that at the next meeting of the Society after Monday next, I will propose for *discussion* the following *question*: Does the problem of finding the probability of three points in an indefinite plane forming the angles of an acute angled triangle or of four points in an indefinite plane forming the angles of a reentrant quadrilateral admit of a *determinate* solution?"[100] Clearly, this idea of posing a topic for discussion—as opposed to presenting a paper—had not yet been tried, and Sylvester was unsure whether it fit with Hirst's conception of how the Society should function. Still, he found "the question . . . interesting under a logical and philosophical point of view," all the more so because "it has attracted a good deal of attention" with "many . . . having spoken to [him] about it." If not a discussion of it among those assembled at the London Mathematical Society, then where?

Geometric probability and debates with Hirst over the Society's evolving practices ultimately provided only temporary distractions from another concern that weighed heavily on Sylvester in October 1865. His sister Fanny, seven years his senior, wife of Lewin Mozley, and mother of their eight children, had come down with a case of bronchitis so severe that her life was in danger. One of his other sisters, Elizabeth, had cut short a trip to return to Liverpool to be with her, but things did not look good. "Fanny," she reported, "is too ill to know we are here or to see or speak to anyone" despite the fact that "everything is done that can be."[101] The doctor had been called in and had even stayed for two days around the clock, but he seemed to be fighting a losing battle. It was thus a supremely relieved Sylvester who, "thank God," was able to report to Hirst on 25 October that his "sister is for the time being quite nearly out of danger."[102]

With his mind at ease on that score, Sylvester was able to concentrate once again on matters of import before the Royal Society and on new mathematical ideas. The Royal Society's business was the most pressing. Hirst was backing his and Sylvester's long-time friend, the French geometer Michel Chasles, for the Copley Medal, and Sylvester was charged with writing up the case. "As a geometer," Sylvester proclaimed, "for the fecundity of his methods and clearness and originality of his conceptions he has never been surpassed."[103] Moreover, Chasles had made a major advance in 1864 and 1865 with his so-called theory of characteristics, a theory that allowed him to deal systematically with all questions involving conic sections and one which, according to Sylvester, "constitutes a pure geometrical Calculus as important for the new power and grasp which it gives in dealing with the most difficult questions of geometry as the theory of *fluxions* in Newton's time in conquering the thorny domain of Analysis or as the greatest discoveries of Davy or Faraday in enlarging and multiplying the methods and instruments of physical science." This was thus the time for the medal to go both to Chasles and to mathematics, and Sylvester urged Hirst to "leave no stone unturned. *Go early* to the Council; get hold of the ear of as many as you can in private and state the facts to them. I have observed that many are undecided until the last moment and go with the last hearty speaker."[104] The strategy worked. Chasles won the Copley Medal in 1865.

As the drama of this politicking unfolded, Sylvester was also in the thick of some new research in dynamics, another topic, like geometric probability, rather far from his usual interests and mathematical tastes.[105] At issue was a result that Johann Heinrich Lambert had published in Augsburg in 1761 and that provided a geometrical proof of the time required for a planet to sweep out an arc on an ellipse in terms of the associated radius vectors and the chord.[106] Lagrange had followed Lambert's proof with at least three distinct proofs of his own, one involving trigonometry, one depending on the behavior of integrals, and one more algebraic. "Notwithstanding this plethora of demonstration," Sylvester wrote in the announcement published by the Royal Astronomical Society in December 1865, "the following *direct* algebraical proof . . . may be deemed not wholly undeserving of notice."[107] That was ever the issue with Sylvester—finding natural and straightforward *algebraic* demonstrations where other kinds of proofs may have been given—and he had succeeded even in this uncharacteristic mathematical realm by using a simple argument cast in terms of determinants.

Sylvester continued to work out the ramifications of his new discovery through the months of November and December, discussing them regularly with Hirst as well as with his "ingenious coadjutor," Morgan Crofton, and presenting them more publicly before those assembled at the LMS for its final meeting of the year on 18 December 1865.[108] The very title of the paper that subsequently appeared in the January 1866 number of the *Philosophical Magazine*, "Astronomical Prolusions: Commencing with an Instantaneous Proof of Lambert's and Euler's Theorems, and Modulating through a Construction of the Orbit of a Heavenly Body from Two Heliocentric Distances, the Subtended Chord, and the Periodic Time, and the Focal Theory of Cartesian Ovals, into a Discussion of Motion in a Circle and Its Relation to Planetary Motion," reflected the mathematical stream of consciousness in which he was immersed. "Such is the fascination of the service of the Mathematical Muse," he confessed to Hirst, "the Muse whose service to my mind is the most worthy of the devotion of an immortal soul."[109]

Sylvester continued to serve that muse with works in mathematical physics throughout the year 1866. For example, on 19 March, just two months after his election along with Hirst and Cayley as a vice president, he spoke to the LMS "On an Addition to Poinsot's Ellipsoidal Mode of Representing the Motion of a Rigid Body Turning Freely Round a Fixed Point, Whereby the Time May Be Made to Register Itself Mechanically."[110] By June, he had presented a fuller written account for consideration by the Royal Society.[111]

The focus of this foray was a result published by Louis Poinsot in his 1834 treatise, *Théorie nouvelle de la rotation des corps*. There, he had determined, from a purely geometrical point of view, the axes of permanent rotation of a body and had given what René Taton later described as "a very elegant representation of rotary motion by the rolling of the ellipsoid of inertia of a body on a fixed plane."[112] The problem, in Sylvester's view, was that while Poinsot's formulation "exhibits the geometrical path of the body, it gives no image to the mind of the time in which any portion of such path is performed."[113] In 1866, Sylvester had seen how to recast Poinsot's analysis—in the case of a body on which no forces are exerted—so that "this imperfection could be remedied and the time put distinctly in evidence." In his referee's report on the version of this result submitted to the Royal Society, Oxford's Savilian Professor of Astronomy, William Donkin, had to admit that "I certainly did not expect to find that anything both new & important could be said on a subject so often handled. But on reading the paper I changed my opinion."[114] He had been particularly impressed by Sylvester's ability to provide "the geometrical representation of the time in the dynamical problem," a feature convenient especially in considerations involving the angular momentum.

The strength of Sylvester's research surge in the years from 1864 through 1866 was reflected not only in his publication record but also in the honors showered upon him. News of his election as foreign corresponding member of the Paris Académie des Sciences at the close of 1863 was followed in 1864 with word that the same status had been accorded him by both the Gesellschaft der Wissenschaften in Göttingen and the Royal Academy of Physical and Mathematical Sciences in Naples. The next year, 1865, his *alma mater*, the University of Dublin, awarded him an honorary LL.D. on the basis of his past achievements, and in July 1866, he was named a foreign corresponding member of the Berlin Academy of Sciences. The strategy that he had been using at least since the 1840s of actively working to make his work known abroad was clearly paying off. In his early fifties, he had categorically achieved the international reputation for his contributions to mathematics that he had so long desired.[115] It represented— like his appointments at Woolwich first as an examiner in 1854 and then as a professor in 1855—a personal victory over the societal mores toward Jews that had prevented him from taking his Cambridge degree and from entering into an academic career there in the late 1830s. He had beaten the odds. He had made a career and an international reputation for himself.

By July 1866, however, with his researches in mathematical physics having largely played themselves out, Sylvester felt somewhat alone and adrift. He was staying in Woolwich for the summer, and he longed for congenial companionship there. Cayley, of course, was now married and occupied with his wife, but perhaps Hirst would entertain the notion of coming for an extended stay. Writing with some hesitation on 14 July 1866, Sylvester told him that "I can truly say that it would give me the greatest pleasure if you would take up your abode with me for a few weeks in this agreeable part of the country (for so it truly is during the summer season)."[116] Moreover, he continued, "I could promise you perfect quiet, absolute independence and every facility for carrying out the execution of your paper, with ready sympathy and willingness to be made participant in your evolving theory. . . . You could have your own study room and bedroom and see as much or as little of me as you might desire." Sylvester missed the mathematical and personal comradeship he had shared with Cayley. He missed the frequent conversations and the dinners together. He was not good at being alone, and yet he faced an entire summer of relative solitude.

Not surprisingly, Hirst had his own plans. He intended to pay a visit to friends in Ireland, and he planned to attend the BAAS meeting, as did Sylvester, in Nottingham in August. Despite Sylvester's repeated invitations over the months of July, August, and September, Hirst ultimately demurred. As early as 1859, the very beginning of their acquaintance, Sylvester had been, according to Hirst writing in his diary, "excessively friendly, wished we lived together, asked me to go live with him at Woolwich and so forth. In short he was excentrically [sic] affectionate."[117] That Sylvester could, at times, be overbearing was clear. Hirst responded in 1866 by going about his own affairs, and Sylvester went about his.[118]

These included casting about for new mathematical ideas. Early October, for example, found him pouring over the new, third edition that had just come out of Joseph Serret's *Cours d'algèbre supérieure*, whereas in November, he was writing to Cayley about the properties of commuting matrices, and by December, he had dabbled in the calculus of operations as it applied to the theory of invariants.[119] He also continued his active involvement with the London Mathematical Society, presenting papers on 8 and 22 November and acquiescing to the will of its members that he succeed Augustus De Morgan and serve as its second president for a term to last into 1868.[120] He took his new duties seriously, participating in the Society's regular meetings and working to recruit new members like Sir John Herschel.[121]

Relative to his research, however, the new year of 1867 brought few fresh insights. A brief paper following up on his ideas on invariant theory and the calculus of operations appeared in the January 1867 issue of the *Philosophical Magazine*, and some thoughts on matrices came out in the journal's final issue of the year in December, but the intervening months had witnessed a research drought. In March, he had tried to think again about some of Plücker's geometrical work and to read a paper that Alfred Clebsch had published several years earlier, but to no avail.[122] By May, he was doing "no mathematics—ever intending and ever putting it off."[123] As he confessed to Cayley, "thus I have been too ashamed to call upon or to write to you but have been hoping all along to see or hear from you." "If I thought it would do any good," he continued, "I would ask you to pray for my rescue from this enslaving indolence and paralysis of the will for such it amounts to." He could not concentrate; he could isolate no new directions for his research. He was "(bodily well) but mentally or spiritually afflicted" and saw, despite all he had accomplished and all the rewards he had garnered, little hope for his future as a mathematician in 1867.

CHAPTER 8

The Uneasy Years

I am more than half inclined to go out to the Antipodes rather than
remain unemployed and living upon charity *in England.*
—J. J. Sylvester to Arthur Cayley, 15 February 1875

If elation and fanfare had accompanied Sylvester's work on Newton's rule for detecting imaginary roots in the mid-1860s, 1867 had found him increasingly despondent about what he perceived as his failing mathematical powers. The Sylvester who had so defined himself as a research mathematician—despite the fact that that category was only beginning to emerge in Victorian England—felt increasingly stripped of his mathematical identity. For so long, mathematics had consumed his mind and dictated how he spent his days and nights. Its retreat left a mental, and even physical void that he keenly felt the need to fill. But how? With what?

Sylvester poured his energies into a variety of concerns in the years from 1868 to 1876, and he did so with the vigor and enthusiasm that had characterized his engagement with mathematics. The only difference was that he could no longer count on one, virtually all-consuming—if fickle—passion to shape his life. He constantly faced the moment when his interest-of-the-moment would flag and he would have to find some other direction for his mind and his activities. These uneasy years opened with Sylvester focused on matters of general education at the Royal Military Academy and in England as a whole. In a sense, this initially unwanted, but ultimately welcome, turn to issues of education and educational reform mirrored broader political and intellectual trends, which had been evolving in England since at least the 1830s and which came dramatically to a head in the late 1860s and early 1870s.

Victorian England and Educational Reform

The late Georgian and early Victorian periods had witnessed much agitation both for and against the principle of national education. As early as 1832, the Whig government under Earl Grey had set aside allocations for public education—and, in particular, for the construction of new schools at state expense—that effectively began the state's involvement in national education. By 1839, a separate office for education, the Committee of Council on Education, had been established to oversee the disbursement of these funds and the operations of the new schools constructed. The next four decades—the first four of Victoria's reign—witnessed numerous initiatives aimed at ensuring an education for children of all classes, at monitoring the effectiveness of teaching and the efficiency of school administration, and, in general, at improving education at the elementary, secondary, and even university levels. The 1860s and 1870s, especially, saw both the formation of several important Royal Commissions to study a wide range of educational issues and the passage of legislation key to the implementation of the recommended reforms.

In 1861, the first in this spate of commissions, the Clarendon Commission, was convened to examine the operations of some of the country's most prominent endowed schools—the Etons, the Harrows, and the Rugbys. The commission's work and recommendations led, in 1864, to the passage of the Public Schools Act, which mandated various reforms of the administration of the targeted schools and their endowments. This was followed in 1864 by the creation of the Schools Inquiry Commission under Lord Taunton to look more comprehensively at England's schools.

The Taunton Commission united in common cause some of the country's most noted advocates for education and educational reform, among them, the celebrated poet, school inspector, and essayist, Matthew Arnold, and his brother-in-law, the Liberal Member of Parliament for Bradford, William E. Forster. These men, together with their fellow commissioners, focused on secondary education in England from both a pedagogical and an adminstrative point of view, but not ignoring other aspects of the English educational landscape. The findings of the Taunton Commission appeared in a massive, twenty-one volume report in 1868 and 1869. Among the problems isolated were the misappropriation to elementary education of funds originally endowed for the support of more advanced work and the concomitant inefficiency of endowed schools. In the commission's view, more efficiently run, state-supported, and state-monitored schools could profitably replace poorly run endowed schools in a large number of localities.[1]

After William Gladstone's Liberal Party came into power in December 1868, Forster successfully introduced an Endowed Schools Bill into the House of Commons to redress the Taunton Commission's concerns. It was enacted into law later that year, but not without much debate in the more conservative House of Lords. The act provided for the creation of a special commission charged with investigating each of the country's endowed schools and with tailoring plans to each individual school for the reorganization and administration of its various endowments. In particular, the Endowed Schools Act mandated that the endowments be considered nondenominational—contrary both to long-standing tradition and to the fact that the Church of England dominated the country's schools—unless the benefactor had made specific reference to a particular denomination in the initial bequest. This proved yet another important step in opening England's schools to children of dissenting families.[2]

With secondary schools largely in hand thanks to the Endowed Schools Act, Forster moved on in 1870 to elementary education. He and others in the House of Commons had been angling at least since 1867 for a nationwide system of mandatory tax support for elementary education. Sufficient groundwork had been laid so that, with the change of government in 1869, the time was ripe by 1870 for the passage of Forster's Elementary Education Bill. The act provided for public elementary schools nationwide that would be religiously neutral, supported by tax revenues, and subject to regular state inspection. Moreover, they would be overseen locally by school boards elected by the taxpayers. These school boards would have the power not only to determine where new public schools would be built but also to draw up by-laws for their districts that could mandate, among other things, compulsory school attendance and direct subsidy for children of the poor. The Elementary Education Act of 1870 thus represented a major step forward in both compulsory attendance and free education for children.[3]

Despite all of this concern for elementary and secondary education, higher education was not ignored. In 1854, the Oxford University Act was passed, effectively removing the religious disabilities of dissenters and opening Oxford to students of all creeds. Two years later, the Cambridge University Act effected the same result at England's other ancient university. After 1856, then, Jewish students, as Sylvester had been in the 1830s, were no longer debarred from entering or taking degrees at Oxbridge. The 1871 passage of the Universities Tests Act finally removed all remaining religious tests, namely, those for tutors, fellows, and professors. Even military education had come under a Royal Commission's scrutiny in 1868.

These governmental initiatives, as positive and far-reaching as they were, nevertheless affected only those in schools of various kinds. What of the adults who had had little or no education, formal or otherwise? What could be done to help them meet the challenges of an industrializing England with a growing, educated middle class? As early as the 1820s, Lord Brougham had been instrumental not only in the foundation in 1828 of London University for nonsectarian higher education of young men but also in the formation in 1823 in

London of the first so-called Mechanics' Institute for adult education of skilled working men.[4] As conceived by Brougham and his fellow radicals, men like Jeremy Bentham and James Mill, the Mechanics' Institute would provide scientific education at a sufficiently high level in the short run to improve working men's abilities to grapple with the increasingly industrialized work place. In the long run, however, the founders believed that this education would allow working men to make original contributions to science and thereby raise themselves into the higher social classes.[5]

Although this idealized vision was never realized, the London Mechanics' Institute served as a model for other mechanics' institutes throughout England, like the one in Liverpool that Brougham had been on hand to open in 1837. These provided lectures and courses at a low cost and in a wide range of subjects at a variety of levels, from the basics of reading, writing, and arithmetic, to the higher mathematics of algebra and geometry, to economics and politics, to literature and the basic sciences. By the 1850s and 1860s, moreover, these institutes had begun increasingly to recognize their potential as instruments of social and cultural—not just academic—development. For just a penny a person, a working family could attend a so-called Penny Reading and listen to dramatic renderings from Shakespeare or poetry recitations or even operatic or other high-brow musical selections to raise their level of cultural literacy.[6]

These sorts of purely paternalistic, top-down initiatives—organized by the upper and middle classes in an effort to benefit and raise up the lower classes—were soon supplemented by more truly cooperative ventures. Chartism, an organized working class movement of the late 1830s and 1840s that had aimed (unsuccessfully) to remove many of the political and economic disabilities of the lower classes, had nevertheless served to focus working men on their own plight and on ways of overcoming it. In London, these working men joined forces with middle-class reformers like Frederick Denison Maurice to found in 1854 the Working Men's College, an adult educational institution anchored initially in Maurice's twin principles of Christian brotherhood and traditional humanistic studies. According to Maurice's vision, the college would be more a place where men could enrich their minds (and, in so doing, enrich their lives) than a venue, like the Mechanics' Institutes, for higher vocational training. As it evolved over the course of its first two decades, however, the Working Men's College came to embrace a range of practical as well as humanistic subjects. Men could come to study history, literature, and politics, but they could also come to learn mathematics, English grammar, and drawing, the latter under such noted pre-Raphaelites as Dante Gabriel Rosetti and John Ruskin.[7] And, although the students paid small fees to cover the cost of books, equipment, and the college's building, the teachers, at least into the 1890s, taught their courses gratis.[8]

Sylvester was caught up, to varying degrees, in all of these educational reforms. One of them—the reform of military education—fundamentally and profoundly disrupted his life. Many of the others came to give it direction in those uneasy years when mathematics no longer faithfully served that purpose.

Reform and the Royal Military Academy, Woolwich

Early in 1868, a Royal Commission on Military Education was formed to follow up on the report Lord Panmure had made in 1857 in his capacity as Secretary of State for War in the immediate aftermath of the Crimean conflict. This new 1868 Commission, royal as opposed to departmental, was under Lord Dufferin, who from 1864 to 1866 had served as England's Under Secretary for India and in 1866 as Under Secretary for War. The latter political service as well as his respected diplomatic skills undoubtedly fitted him well to the task set for the commission of examining and evaluating the state of military education for those seeking commissions into the Army.

Military education in England took place in a variety of venues and comprised a multitude of subjects at various levels in 1868. In addition to the Royal Military Academy, Woolwich, which trained candidates for commissions in the artillery and engineers, the Royal Military College in Sandhurst provided the education needed for those going into the cavalry and the infantry, whereas the so-called Advanced Class at Woolwich and the Staff College served the ongoing educational needs of officers.[9] Even among these four institutions, there was clearly duplication, but did the system suffer from it? That was one thing that Lord Dufferin and his fellow commissioners set about to discover. They were also charged with determining whether miltary education was efficient and sufficient. Did the young men study the "right" subjects and at the "right" depth, or could some classes and even subjects be dispensed with and the time to commission shortened? And what about that majority of young men who bought their commissions, had only to pass an easy examination before entering their regiments, and essentially had no military education at all?

By the spring of 1868, the commission was well into its work, gathering the data needed to inform its final judgment by drafting and circulating series of targeted questions. Although the College in Sandhurst seemed to be the main target of its inquiries, the commission did not neglect the Academy in Woolwich.[10] In particular, it requested written responses of the Woolwich faculty to six specific queries. The commission wanted to know their opinions on admission standards, the optimal age for admission and time to completion of the course of study, the course of study itself and the method of instruction employed, the advisability of different courses for candidates for the artillery and those for the engineers, the organization of the Academy, and the desirabilty of combining the Academy in Woolwich and the College in Sandhurst. The commissioners also gave the respondents leave to offer any other suggestions they might have for the Academy's improvement.[11] As Professor of Mathematics, Sylvester had been queried by Panmure's Commission in 1856 just after he had taken up his position. By the late 1860s, after more than a decade of fighting with the Woolwich authorities and with a much stronger sense of the Academy's shortcomings, he seized the opportunity to air his more mature views before Dufferin's.

Basically, Sylvester had little problem with sixteen as the cadets' usual age at admission, or with their two-and-a-half year course of study, or with the common curriculum for the engineers and the artillery, or even with the admissions requirements, except in mathematics. There, he felt that "the course of instruction . . . would be susceptible of considerable improvement were more importance assigned to it as a condition for admission, and a higher scale of attainments, especially in pure mathematics, insisted upon at entrance."[12] If this were to become the case, "the time now bestowed in the earlier part of the course upon elementary mathematics might be greatly abridged, and an opportunity given for the introduction or more detailed study of subjects now either omitted altogether or only studied by a few as extra subjects, and the want of a knowledge of which is felt as a practical inconvenience in other departments of study at the Academy."[13] Clearly, Sylvester thought that too much time was currently being wasted in low-level mathematical instruction. He also deplored the fact that there were no institutional mandates for contact and cooperation between those faculty members whose subjects had mathematical components—the professors of practical geometry, artillery, fortification, and surveying and topographical drawing—and the professors of mathematics and mechanics. This was really a matter both of turf and control, for Sylvester explained that "at present [the latter] have no means of ascertaining with what degree of efficiency, completeness, or reference to *correct principles* such mathematical or mechanical applications are taught in the Academy."[14] All these comments find Sylvester arguing for a stronger place for mathematics in the curriculum and a concomitant increase in the influence of the Professor of Mathematics, but this was a two-way street. The professors in the more applied fields needed their students to be able adequately to *use* the mathematics they had been exposed to by the Professor of Mathematics. They needed the "means of ascertaining with what degree of efficiency" he was performing *that* task.[15]

Relative to the Academy's organization, Sylvester also had decided opinions. While he likened the suggestion of the amalgamation of Woolwich and Sandhurst to "linking the living with the dead,"[16] he similarly did not mince words about what he viewed as the Academy's inefficient and grossly overstaffed administration, an administration with which he had repeatedly locked horns. "The Academy groans under the weight of a triple government," he wrote, "one (itself heavily loaded) at Woolwich, a second at the Council of Military Education, and a third at the Horse Guards." And if the Commission could do little about the latter two prongs of the governmental structure, it could certainly recommend cutting out some of the administrative fat in Woolwich, where "eight officers [are] engaged to perform for 200, recently raised to about 220, cadets, . . . the same duty as formerly was discharged nominally by three, but virtually by two officers for 160 cadets." The Professor of Mathematics was simply asking the commissioners to do the arithmetic.

Related to organization was the issue of faculty autonomy; in Sylvester's view, there was none. He argued that the faculty—not just the administrators—should have the authority to eject troublesome students from their classrooms and that, moreover, there should be a joint faculty-military council to deal with matters like discipline, number of hours of instruction per subject, curriculum, textbooks, etc. As Sylvester explained, "under existing arrangements an order might come down from the Horse Guards, and no one on the spot would have the power of protesting against its being carried out."[17] Of course, this had been precisely the cause of so many of Sylvester's earlier run-ins at the Academy. When the local administration had balked at his request to change textbooks in 1861, he had had no defined line of recourse. When the War Office had wanted unilaterally to increase his teaching load in 1862, there had again been no clear procedure for protest. The existence of a council like the one he now proposed would at least provide a venue for discussion and a vehicle for oversight.[18]

Sylvester closed his written comments by raising three additional points for the commission's consideration: the low entering salaries of the mathematical masters under him as compared with the entering salaries of the junior military instructors; the encroachment into the evening hours of courses—like mathematics—in the core curriculum; and the lack of adequate time for cadets to engage in private, as opposed to course-directed, study. The commissioners must have read Sylvester's comments carefully, for when they deposed him on 27 April, they had a number of points that they wanted to take up with him and on which they wanted further elaboration. These centered specifically on the Academy's organization and the size of its staff.

One of the commissioners had picked up explicitly on and quoted Sylvester's written remark that the Academy "groans under the weight" of its administration. When asked to elaborate further, Sylvester heaped particular scorn on the Lieutenant-Governor, referring to him as "a little king, or god, who seldom or never comes into connexion with the ordinary mortals of the establishment."[19] Sylvester had been at war with Wilford and had gotten along no better with his successors. At least now he could speak freely on the matter, since six weeks prior to his deposition, the latest Lieutenant-Governor had left Woolwich and had not been replaced in the interim. The Professor of Mathematics argued that "the discipline and working state of the Academy have never been so good as during th[is] [six-week] period" and proffered the opinion that "no Lieutenant-Governor during my connexion with the Academy has ever exercised any moral influence" over the cadets.[20] In his view, the duties of both the Inspector of

Studies, who actually had contact with the students and the staff, and the Second Commandant, who served as an assistant to the Commandant, should be folded together into a redefined position of Lieutenant-Governor. In that way, the person in that post would meaningfully unite the interests of the faculty, the students, and the military.

Another commissioner, Lord Cecil, grilled Sylvester hard on the more general allegations of overstaffing. After rehearsing the issues with respect to the officers and servants, he homed in on the faculty, asking Sylvester point blank if he thought the professoriate was too large. His answer was curious. "I cannot state—we are not brought together—a public spirit is not developed. I can only speak of my own department, and I think that we are not at all over-officered; we have as much as we can do to get through with the work" (302). There would be follow-up to the comment about a lack of *esprit de corps*, but not before Cecil pressed with two hypotheticals. "Do you think that by a re-construction there might be a reduction?" "Yes," Sylvester answered. "Supposing that the classes and the classrooms were re-adjusted in your own department, could that be done?," Cecil asked. "I do not say that it would be impossible," Sylvester hedged, "but I do not think that any serious reduction could be made. When I first joined there were seven mathematical masters and myself as Professor. There are now only four mathematical masters and myself as Professor."[21] These responses seemingly piqued Cecil. "You say you have no knowledge of what might be done with regard to other studies; but you have been a very long time at the Academy?," he asked pointedly. "Yes," said Sylvester. "And I daresay that you have turned your attention to that subject, namely, as to the great number of professors as compared with cadets?" "Occasionally." "And possibly you have something which you can suggest to the Commission on that head?" Cecil's sarcasm was barely contained, and Sylvester's answer to the effect that he had really only considered his own department could not have pleased him. The tension was only broken when another commissioner returned to probe the issue of *esprit de corps* among the faculty. The deposition concluded shortly thereafter.

The Commission received written comments from, and formally deposed scores of individuals associated with the Academy in various capacities in the spring of 1868. It then went off to deliberate and to formulate its recommendations. These were expected to be made known some time the following year. Until then, everyone at the Academy would have to wait and hope that changes, if proposed, would not be too radical.

Sylvester, like his colleagues, returned to his normal routine following the excitement and activity occasioned by the commission, but in Sylvester's case, that routine largely lacked the stimulation of mathematical discovery. In August, he attended the annual meeting of the British Association for the Advancement of Science in Norwich and just managed to put together a very short note on analytic geometry for it.[22] His interest had focused on the so-called involute of the circle, that is, the curve gotten from a circle in the following way.

Consider a circle of given radius with center C and a fixed point A on the circle's circumference. Let P be another point on the circumference, and consider the tangent to the circle at P. Let Q be a point on that tangent such that the length of the line segment PQ equals the arclength PA. The involute of the circle is the locus of points Q so defined.[23] (See figs. A and B.) In Norwich, Sylvester proposed to generalize this well-known curve by determining and analyzing the properties of the family of successive involutes, where the second involute was the involute of the involute, and so on. His enthusiasm for this family of curves, what he was calling at Cayley's suggestion "cyclodes" by November 1868, continued through the fall of 1868, generating two companion pieces in the *Philosophical Magazine* for October and December.[24]

This work together with the uncertainties at Woolwich had taken their toll by December, however. When Sylvester finally returned a referee's report late in December that had been requested five months earlier by his friend, the Secretary of the Smithsonian Institution in Washington, Joseph Henry, he confessed that he had "little or no energy surviving in [him] for the performance of the most necessary acts."[25] He had gone to the seaside after the end of the academic term to try to recover his strength and to steel himself for the term to come, but his spirits were low.

Things looked better in the spring. His mathematical energies were somewhat renewed and focused again on cyclodes. At the BAAS meeting that convened in Exeter in August 1869, he reported briefly on his latest findings and even distributed copies to those present in the Mathematical and Physical Section (Section A) of what turned out to be his final word on the subject, a paper that had just appeared in the *Proceedings of the London Mathematical Society*.[26] Exeter, however, represented much more for Sylvester than the chance to pass out a reprint.

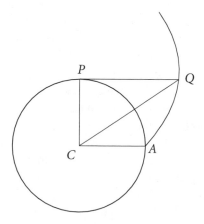

FIG. A.

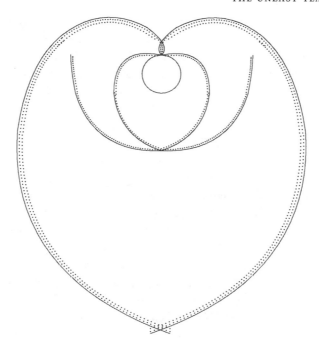

FIG. B. Here, the single-dashed curve denotes the involute of the circle, the double-dashed curve denotes an involute of an involute or a second involute. This figure was originally produced by William Spottiswoode and appeared in Sylvester's published paper, "On Successive Involutes to Circles-Second Note" (1868), in *Math. Papers JJS*, 2:642.

Mathematics Misrepresented and Misunderstood:
Taking on Huxley

The spring of 1869 had found Sylvester anticipating what he hoped would be the Dufferin Commission's positive and constructive recommendations for the reorganization of the Woolwich administration, the presentation of his researches on cyclodes before the London Mathematical Society, and engrossed in accounts of Forster's efforts to get the Endowed Schools Bill through Parliament. In its original form, the bill, like the Taunton Commission's report that inspired it, called for the creation of an Educational Council charged with examining and issuing certificates of attainment to students from the endowed schools. This group would thus have the power effectively to raise educational standards at endowed schools throughout England. By May 1869, Sylvester had decided that he wanted to serve on this proposed twelve-person examining body, six members of which were to be appointed by the universities of Cambridge, Oxford, and London and six by the government.[27] As usual, in his view, securing such an appointment would require an influential supporter, but Sylvester could no longer turn to Lord Brougham. His long-time advocate and supporter had died in 1868 at the age of eighty-nine after a long, if not always uncontroversial, career in public life. This time, Sylvester turned instead to another old friend, Charles Graves, now (since 1866) the Bishop of Limerick.

In a letter dated 15 May 1869, Sylvester explained to Graves his intentions and his predicament: "My desire is to be appointed one of the six members of the Educational Council who are to be nominated by the Government. Although educated at Cambridge and the second wrangler of my year, the religious disqualification under which I labor made my success negatory so far as regards my reaping any of the emoluments and honors of the Univerity so that access to the Council through that Channel is out of the question."[28] Sylvester then provided a brief account of his achievements to date (an indication that the two men had not regularly been in touch) and closed with a remark reflective of his uneasy state of mind. "I should value very highly the appointment which I solicit through your friendly intervention," he told Graves, "not only as a public recognition of good scientific work done in the past but also as affording a motive and a sphere for dedicating a portion of my spare time and energies to subserving the great cause of National education."[29] Sylvester was in transition. With his scientific work slowing down, he needed a new direction, something to occupy fruitfully the time he had previously devoted to mathematics, a cause in which he believed: national education.

Unfortunately, he would not find this new vocation on the Educational Council. After its second reading before the House of Commons, the bill went into committee and ultimately came out with many of its provisions deleted, among them, that for the proposed examining body.[30] Although Sylvester's past achievements would not be acknowledged with a post in the service of national education, they had already been recognized in another way. He had been chosen to serve as President of the BAAS's Section A at Exeter.[31]

High honors within the evolving British scientific community, section presidencies went to those recognized as Britain's leaders in the respective scientific fields. In the case of Section A, moreover, mathematicians *and* physicists vied for the title; in Norwich in 1868, the physicist, John Tyndall, had sat in the chair, whereas the last mathematician to do so had been William Spottiswoode in Birmingham in 1865. Sylvester's scientific peers in Britain had thus acknowledged his elevated place within their fellowship. The section presidency also brought with it the additional honor, but not the requirement, of addressing the section's assembled practitioners. Ironically, this recognition and this opportunity came at a time when Sylvester felt himself increasingly inadequate and powerless as a mathematician. So unsure and unsettled was he that he initially decided to forego his chance to speak. What could he say of interest, anyway? How could he compete in eloquence with predecessors like Tyndall? Still, if he opted not to speak, would it seem an admission of the widely perceived inability of mathematicians to communicate with a broader scientific audience, or, worse yet, might it reflect his own deepening sense of mathematical inadequacy? These thoughts plagued him in the weeks leading up to the meeting, and in the end, he resolved neither to let himself reinforce a stereotype nor to risk revealing his innermost feelings.[32] When Sylvester opened Section A on 18 August, he took on no less an adversary than the noted biologist, Thomas Huxley, and no less a topic than the nature of mathematical discovery and its implications for mathematical pedagogy. The nation's atmosphere of educational reform pervaded the BAAS's Section A when Sylvester rose to speak. Education, not new mathematical results, had focused his thoughts.

Thomas Huxley had become notorious in the years after the 1859 publication of Charles Darwin's *On the Origin of Species* for his tenacious and vociferous defense of the shy and retiring Darwin's ideas on evolution through natural selection. Styled "Darwin's bulldog," Huxley had gone on the defensive in print in virtually all mid-Victorian Britain's intellectual periodicals and had created a name for himself as both a biologist and a literary stylist. His talents also extended to the podium, where he was frequently called to edify and entertain audiences with his pithy and pointed commentary.[33] It was in an after-dinner speech delivered before the Liverpool Philomathic Society in the winter of 1868—and subsequently published in *Macmillan's Magazine*—as well as in an essay in the June 1869 issue of the *Fortnightly Review* that Huxley provided the ammunition for Sylvester's August attack.[34]

Huxley had chosen to address his Liverpool audience on a timely topic: scientific education. With so much discussion nationally about education in general, it was only natural that, as a scientist, he make a special plea for educational reform directed specifically toward science. As he saw it, scientific education should begin early with the goal of producing a scientifically literate public. "No boy nor girl should leave school," he argued, "without possessing a grasp of the general character of science, and without having been disciplined, more or less, in the methods of all sciences."[35] This level of scientific education would produce youngsters, who, "when turned into the world to make their own way, [would] be prepared to face scientific discussions and scientific problems . . . by being familiar with the general current of scientific thought, and being able to apply the methods of science in the proper way." In particular, early science education should focus first on geology, with an eye toward ensuring "a general knowledge of the earth, and what is on it, in it, and about it," then move to systematic botany and physics, and ideally proceed to chemistry and basic human physiology. Moreover, all this scientific knowledge should be acquired in an active, hands-on way. A student "should not merely be told a thing, but made to see by the use of his own intellect and ability that the thing is *so* and no[t] otherwise."[36] This educational process should take place, like the discovery process in science itself, through "induction; that is to say, in drawing conclusions from particular facts made known by immediate observation of nature."

Huxley contrasted this inductive quality of science with what he understood to be the deductive quality of mathematics. "Mathematical training is almost purely deductive," he asserted. "The mathematician starts with a few simple propositions, the proof of which is so obvious that they are called self-evident, and the rest of his work consists of subtle deductions from them." He pushed this position even more strongly in the *Fortnightly Review* essay, when he exclaimed that mathematics "is that which knows nothing of observation, nothing of experiment, nothing of induction, nothing of causation!"[37] These were the fighting words that drew Sylvester's fire in Exeter. Huxley may have been eminently qualified to hold forth on biology, but he had exceeded his expertise and overstepped his bounds in speaking out about mathematics. Huxley was completely wrong in his characterization of mathematics and its practice, and Sylvester had to set the record straight, for "the eminence of [Huxley's] position and the weight justly attaching to his name render it only the more imperative that any assertions proceeding from such a quarter, which may appear to me erroneous, or so expressed as to be conducive to error, should not remain unchallenged or be passed over in silence."[38]

Sylvester thus spoke out before those assembled in the lecture hall of Section A on what *he* knew mathematics to be. Mathematics, he explained, "unceasingly call[s] forth the faculties of observation and comparison"; "one of its principle weapons is induction"; "it has frequent recourse to experimental trial and verification"; and "it affords a boundless scope for the exercise of the highest efforts of imagination and invention" (654). Sylvester used the case of invariant theory to illustrate his point. How did invariant theory originate, he asked his listeners? "In the accidental observation by Eisenstein, some score or more years ago, of a single invariant . . . which he met with in the course of certain researches just as accidentally and unexpectedly as M. Du Chaillu might meet a Gorilla in the country of the Fantees, or any of us in London a White Polar Bear escaped from the Zoological Gardens."[39] "This single result of observation," he continued, "(as well entitled to be so called as the discovery of Globerigerinæ in chalk or of the Confoco-ellipsoidal structure of the shells of the Foraminifera) . . . has served to set in motion a train of thought and to propagate an impulse which have led to a complete revolution in the whole aspect of modern analysis, and whose consequences will continue to be felt until Mathematics are forgotten and British Associations meet no more."[40] Sylvester had chosen his analogy astutely and pointedly. Huxley himself had spoken to the working men of Norwich at the BAAS meeting just a year earlier in 1868 on the unexpected observation that chalk had resulted over vast periods of time from the deposition of the bodies of Globerigerinæ, and Sylvester had been present in the audience![41]

If Sylvester took Huxley to task for his views about mathematical practice, he agreed with the biologist wholeheartedly on the real issues at hand, namely, curricular reform and scientific education. "No one can desire more earnestly than myself to see natural and experimental science introduced into our schools as a primary and indispensable branch of education," he announced. He also publicly applauded Huxley's call for a hands-on pedagogical approach. In Sylvester's view, mathematics, in particular, had suffered from "the frozen formality of our academic institutions," where teaching styles are "traditional and mediæval" and where a "living interest" in mathematics "is so wanting."[42] He would "rejoice to see mathematics taught with that life and animation which the presence and example of her young and buoyant sister could not fail to impart, . . . Euclid honourably shelved or buried 'deeper than did ever plummet sound' out of the schoolboy's reach, morphology introduced into the elements of Algebra—projection, correlation, and motion accepted as aids to geometry—the mind of the student quickened and elevated and his faith awakened by early initiation into the ruling ideas of polarity, continuity, infinity, and familiarization with the doctrine of the imaginary and inconceivable."[43] Sylvester had to admit that his own early exposure to Euclid had made even him "a hater of Geometry," although he acknowledged that "there are some who rank Euclid as second in sacredness to the Bible alone, and as one of the advanced outposts of the British Constitution" (660).

In some sense, then, Sylvester could not hold Huxley accountable for his misperception and misrepresentation of mathematics. It was a product of a system of education in drastic need of overhaul, a system that had produced a popular perception of mathematics "after the 47th proposition of Euclid, as a sort of morbid secretion, to be compared only with the pearl said to be generated in the diseased oyster, or, as I have heard it described, 'une excroissance maladive de l'esprit humain.'"[44] Sylvester wanted the educational system transformed so that people could see mathematics as he did, could appreciate its power and beauty. From his perspective, "the world of ideas which [mathematics] discloses or illuminates, the contemplation of divine beauty and order which it induces, the harmonious connexion of its parts, the infinite hierarchy and absolute evidence of the truths with which it is concerned, these, and such like, are the surest grounds of the title of mathematics to human regard, and would remain unimpeached and unimpaired were the plan of the universe unrolled like a map at our feet, and the mind of man qualified to take in the whole scheme of creation at a glance" (659). Mathematics thus offered the means for uniting much (if not all) of human knowledge and, through its unassailable truth, for gaining insight into God's plan. It thus deserved a special place in the curriculum, and it needed to be taught right.[45]

His oratorial crescendo at a peak, Sylvester brought his remarks to a close and moved on to oversee the packed program of talks on astronomy, electricity, instrumentation, mathematics, meteorology, optics, and the theory of heat. His Section A was one of the most international of all the sections in Exeter, attracting at one time or another seven of the seventeen foreigners present at the meeting.[46] Moritz Hermann Jacobi, brother of the mathematician, Carl G. J. Jacobi, and a physicist at the Imperial Academy of Sciences of St. Petersburg, delighted Sylvester both with his presence at the opening address and by his anecdote of an episode his brother had related to him of how important the experimental method had been in some of his number-theoretic researches.[47] Gustav Magnus, Professor of Technology and Physics at the University of Berlin, got the properly scientific part of the conference off to an excellent start with a lecture on the "Absorption, Emission, and Reflection of Heat." The noted French physical astronomer and spectroscopist, Pierre Janssen, followed later, holding forth on how to obtain monochromatic images of luminous bodies. Yet another Frenchman, Auguste Morren of the University of Marseille, commented on Tyndall's discovery of the chemical action of light, and the German explorer, Georg Balthasar Neumayer, reported on his observations of a then-recent meteorite strike in Krähenburg. Finally, the Yale mathematician and astronomer, Hubert Anson Newton, together with his colleague and fellow astronomer, Chester Lyman, were in attendance throughout the week of lectures and actively took part in the frequent discussions.[48]

Sylvester was clearly in his element at the helm of this scientifically and geographically diverse group. He had delivered an address, although he had surely been preaching to the choir in Section A, that had been well received by a visiting scientific dignitary; he had presided over a meeting that had included lectures by some noted foreigners as well as by the likes of Peter Guthrie Tait, William Kingdon Clifford, Balfour Stewart, William Rankine, and the future Lord Rayleigh; and he had had the opportunity to show off a bit of his own research to these and other worthies.

The BAAS and Science Education

At Exeter, Sylvester was also at the peak of his involvement in the inner workings of the BAAS. Although he had attended his first BAAS meeting in Plymouth as early as 1841, just before leaving to take up his new post at the University of Virginia, his participation had been spotty until the Manchester meeting in 1861. From that point on, he was in attendance regularly throughout the 1860s, sitting as one of the vice presidents of Section A in the three consecutive years from 1863 through 1865 and serving as a member of the BAAS's governing council from 1865 to 1870.[49] At the Exeter meeting, in addition to his presidency of Section A, he also recorded, together with his fellow committee members, the latest in a series of reports on rainfall in Great Britain. Sylvester and his nine colleagues on the committee—among them, the mathematician,

James Glaisher, and the astronomer, John Couch Adams—oversaw a nation-wide network of data gatherers and then compiled and analyzed their findings annually for the BAAS. Sylvester had been appointed to this committee the year before in Norwich and ultimately served on it through September of 1873.[50] And if his service on the rainfall committee may have been a product more of his sense of broader service to the BAAS than of real interest in such metero-logical phenomena, the successful lobbying in which he took part in Exeter for the appointment of a committee charged with exploring ways to improve instruction in elementary geometry stemmed from the deep convictions he had so passionately articulated in his Section A address. The committee of twelve, which initially consisted of Sylvester, Cayley, Salmon, Hirst, Clifford, Henry Smith, and others, was appointed in 1869.[51] When it finally reported in 1873, it offered not a concrete solution to the problem but rather an endorsement of the work of the Association for the Improvement of Geometrical Teaching (AIGT). That group, composed almost exclusively of secondary school teachers, was established in 1871 with Hirst as its first president and sought to liberate English education from the stranglehold of Euclid's *Elements*.[52] Ultimately, the AIGT, not the BAAS committee, served as the primary catalyst for the curricular and pedagogical reform of geometry in late-nineteenth-century Britain, but both groups shared the view that change was desperately needed.[53]

Perhaps the most important decision making in which Sylvester took part in Exeter, however, centered on the council and its deliberations. In particular, two resolutions were brought before that governing body for debate and as potential action items. First, the council was charged with considering whether it should push strongly for the appointment of a Royal Commission to study the whole issue of support for scientific investigation. What institutions fostered it and how? What facilities were available for it? Were they adequate? What changes, if any, should be made to stimulate and facilitate scientific research? How should such changes, if necessary, be effected? Second, the council was mandated to take up the issue of science education. Had the government supported it ade-quately and fairly? If not (and that was clearly the assumption *a priori*), what steps should be taken to rectify the situation?[54] In passing these resolutions, the BAAS clearly realized that, as with the Clarendon Commission relative to privately endowed schools and with the Taunton Commission in the context of secondary education more generally, the formation of a Royal Commission proved critical in focusing national attention on a given set of educational issues. In this decade of educational reform, the time was ripe for scientific education and scientific research finally to take center stage, and the BAAS through its council was poised to act. It made its first move in February 1870.

Led by its president, the mathematical physicist G. G. Stokes, the full council formed a deputation that met on 4 February 1870 with Earl Grey and William E. Forster, the president and vice president, respectively, of the Committee of

Council on Education. The eighteen scientists, among them, Francis Galton, Joseph Hooker, Huxley, and Sylvester, called for the formation of a Royal Commission "to inquire into the relations of the State to scientific education and investigation."[55] In the scientists' view, "no such inquiry will be complete which does not include the action of the State in relation to science education, and the effects of that action upon independent educational institutions."

As president, Stokes made the formal presentation. In true statesmanlike fashion, he first gratefully acknowledged the government's support for science through such institutions as the Royal Observatory in Greenwich and the Royal College of Chemistry in South Kensington, but he quickly moved on to point out two glaring deficiencies: support for experimental as opposed to purely observational science and a rationalized program of science education. Relative to the first deficiency, Stokes made what might be considered the increasingly standard argument. First, experimental science was expensive. It was costly to outfit laboratories, so few private individuals could afford to carry out such work independently. Second, experimental science was time consuming. Individual scientists, almost all of whom had to earn their living in other pursuits in mid-Victorian Britain, could not hope to carve out of their "free time" the hours necessary to conduct complex experimental procedures. Many within the BAAS felt that a national institution for the support of laboratory science would go far to circumvent these problems, but only a Royal Commission could determine whether this was, in fact, a majority view and whether it was truly the most effective way to put state funds to use for scientific research. Thus, a Royal Commission "was what was regarded by many scientific men as a prominent want."

Relative to scientific education, again the BAAS held that only a Royal Commission would be well enough placed "to ascertain the amount of science taught in the various establishments throughout the country and whether the teaching of science requires to be supplemented by Government aid." The latter issue was especially touchy. Who would receive such aid? Would only certain sciences or institutions be targeted? What sort of oversight procedures would be implemented to ensure that state money was used effectively and efficiently? It was clear to the BAAS council members that these were precisely the sorts of questions that Royal Commissions addressed.

The scientists' case having thus been made, Earl Grey and Forster pressed them hard for their own personal opinions on the two main issues raised. The council, however, had agreed in advance not to speak individually, arguing that "the opinion of a commission would be far more authoritative" than their few voices. Still, Grey and Forster were not to be put off so easily. Grey wanted the council's opinion, for example, on how the recipients of government grants for experimental research should be selected should such grants be deemed desirable, whereas Forster wanted to know why they seemingly believed that funds from the Department of Science and Art were being improperly used. The Department of Science and Art had been formed in 1853 to oversee, among other things, school accreditation through a nationwide system of examinations and the allocation to accredited schools of state funds on a per capita basis. An 1856 reorganization had shifted the department into the Committee of Council on Education, so Forster may have been piqued by the suggestion that, as vice president of the committee, he was falling down on the job.[56]

These questions prompted little response from the BAAS deputation until Grey finally called their bluff. He confessed that "he felt no little difficulty as to the representation he should make to his colleagues, for he did not think that very great ground had yet been established for the appointment of a Royal Commission. If an inquiry of that sort was to be made it must be an important one, embracing a large subject. He should, therefore, wish to hear the views of other members of so great [a] body as the British Association." Three council members finally stepped forward to offer their views, each giving a different reason why a Royal Commission should be appointed. Huxley, the BAAS's president-elect, stated baldly that "the present organization of the action which the State takes to science was somewhat chaotic"; the chemist, Alexander Williamson, argued that a commission would underscore "the want of aid to utilize the means of higher scientific education"; and Sylvester gave examples of what he viewed as the misuse of government grants because of a lack of oversight.[57] Sylvester's comments, in particular, sparked a lively exchange in the *Times* between the mathematician and a correspondent styled "Armed Science."

Sylvester had apparently focused on an example of science in the military, namely, the so-called Advanced Class that had been designed to provide special training for those seeking "to qualify for the scientific appointments attached to the Artillery service."[58] Formed only seven years earlier in 1864, the Advanced Class involved instruction in applied mathematics and so required a qualified instructor. But Sylvester argued that that instructor, the Professor of Applied Mathematics, taught so few students a year that the expenditure marked true waste in government. The Advanced Class had, in fact, been a topic under discussion relative to the issue of streamlining and economizing within the Royal Commission on Military Education in 1868.

Some account of Sylvester's comments in the session with Earl Grey and Forster was apparently conveyed to the pseudonymous "Armed Science," perhaps by one of the two former military men, Alexander Strange and William Sykes, who served on the BAAS deputation.[59] In a letter that appeared in the *Times* on 10 February 1870, "Armed Science" stated that Sylvester "had brought forward, as an instance of the abuse of a Government grant for scientific education, the case of the advanced class of Artillery officers, which [Sylvester] stated consisted of seven officers, and cost the country £7000 per annum."[60] He declared Sylvester's facts to be egregiously in error in regard both to the cost (some £2157 and not £7000) and to the number of students (some 84 officers and 255 noncommissioned officers and not seven officers). "In view of the above facts," "Armed Science" concluded, "I think it will be allowed that there are few public departments which will bear comparison, as regards economy, with that under discussion."[61]

Sylvester countered the following day. "Armed Science"'s source had not been reliable. Sylvester had stated that the cost of each student officer was £700, not £7000 as reported. Moreover, he had since learned that the number of officers he had cited was, in fact, too high. "Now I am informed, and I invite denial of the fact if I am in error," he challenged, "that the actual number during the greater part of the past term was five, and has recently been brought down to four."[62] Thus, according to Sylvester's arithmetic (based, presumably, on rounding "Armed Science"'s budget figure of £2157 to £2200 and dividing by four), the government paid some £550 a year for a separate Professor of Applied Mathematics to teach just four or five student officers. Thus, "upon the most favourable supposition, each student officer has paid for him to one Professor alone £110 per head."[63] It would thus clearly be cheaper and more efficient for the government to send student officers to existing institutions like University or King's College for their advanced instruction rather than paying for an unnecessary duplication of effort. "It is high time that the guardians of the public purse should set their faces against such vicious reproduction of fresh *nuclei* of misapplied prodigality, surely destined to eventuate (if I may apply the words of an eloquent member of the deputation of which I formed a part) in 'endowed and decorated idleness.' "[64]

"Armed Science"'s rather limp reply appeared on the twelfth. The usual number of officers in the Advanced Class was eight, he conceded, but "this number may be reduced by casualties," and the next class was slated to consist of nine officers.[65] Sylvester had clearly won the argument, but he got in one final lick on the fourteenth: "For those four officers, . . . a director and assistant director of studies, with a staff of professors, lecturers, and subordinate officials are provided and paid for out of the public purse. Is not this the *reductio ad absurdum* of the principle of detached scientific establishments?"[66]

Sylvester's choice of example and the tenacity with which he argued the point reflected his own personal situation as much as his views on the abuses of government funding for science education. In August 1869, just when he was so triumphantly presiding over the BAAS's Section A in Exeter, Dufferin's Commission on Military Education had finally come in with its recommendations.[67] Its report called, among other things, for a reorganization and streamlining of the staff at the Royal Military Academy that (1) eliminated the professorship of mathematics Sylvester had held since 1855—creating instead a combined professorship of mathematics and mechanics—and that (2) required the mandatory retirement of civilian members of the teaching staff over the age of fifty-five, precisely the age Sylvester would turn on 3 September 1869.[68] Sylvester's worst nightmare had been realized; the opinions he had offered both in writing and in person during his deposition about the place of mathematics and the Professor of Mathematics at the Academy had been totally ignored. Thus, Sylvester's confrontation with "Armed Science" on the pages of the *Times*, coming as it did in February 1870, was just months before his fifth-sixth birthday when he would apparently find himself out of a job at Woolwich, unless those already on the staff would be grandfathered under the new provisions. The situation was unclear. At the same time, it was, however, crystal clear that the Professor of Applied Mathematics to the Advanced Class would continue to be paid handsomely to teach barely a handful of students a year. The injustice of it could have overwhelmed Sylvester completely had his involvement in the BAAS and in educational reform not kept his mind on other matters. The exchange with "Armed Science" showed that Sylvester's emotions were very close to the surface.

Poetry Soothes the Soul

Sylvester's work on the BAAS's educational initiatives filled only part of the void left by mathematics during the 1869–1870 academic year at the Royal Military Academy. On 22 December 1869, he made the short journey southwest from his home on Woolwich Common to the pastoral suburb of Eltham with its "ancient trees lining the roadside" and its "striking-looking houses" to be a featured speaker at a Penny Reading in the National School Rooms.[69] The working-class men, women, and children in attendance—several hundred in all—came not only to hear Sylvester recite his translation of Horace's ode, "Tyrrhena Regnum," but also to listen to songs, piano pieces, and readings from Shakespeare as well as from modern authors.[70] When it was his turn, Sylvester distributed the translation from Latin into English that he had had printed up on a single fly-leaf, provided background to the poem and its contents, and proceeded to intone:

> *Birth of Tyrrhenian regal line!*
> *In unstooped cask, a mellow brew*
> *Roses and myrrh those locks of thine*
> *Fresh pressed Maecenas! to bedew,*
> *Long wait thee here! Shake off delay,*
> *Nor spray-washed Tibur still gaze on*
> *Nor Aefule's slope, nor heights survey*
> *Of parricidal Telegon. (27)*

Fourteen more stanzas followed.

Just what his listeners may have made of his oratory is unclear, but Sylvester made much of it. For him, it had been a triumph: "the very National School children in the front rows, who poked their fun over Marc Anthony's Funeral Oration on Julius Caesar, were too much awed to laugh, and looked on and listened in rapt and solemn wonderment" (70–71). This Penny Reading sparked Sylvester's imagination, focused his creative energies, and plunged him headlong into poetry, something in which he had dabbled off and on since his American sojourn in the 1840s.

After his perceived poetical success in Eltham, Sylvester encountered the noted poet and educational theorist, Matthew Arnold, at their club, the Athenæum. It is not hard to imagine Arnold bent over a book in the club's opulently wood-paneled library only to be interrupted by a Sylvester brimming with enthusiasm over his evolving ideas on how best to translate poetry from foreign languages into English. The two men talked, and by 28 January 1870, Arnold had jotted down in his diary to "write Sylvester."[71] The poet must have given the mathematician some cause for optimism, for by July, Sylvester had dedicated his book, *The Laws of Verse or Principles of Versification Exemplified in Metrical Translations*, "to Matthew Arnold, Esq., D.C.L. somewhile Professor of Poetry at the University of Oxford, a consummate master of the art, in grateful recognition of much valuable criticism and generous encouragement received at his hands."[72]

Sylvester had spent the months from January to July absorbed in his poetry. Things had snowballed following the Penny Reading. First, he had appended explanatory notes to the Horatian ode. This had led him to new material and "into unintended analyses and discussions," until, "by successive stages and additions," he had "the simulacrum of a full-fledged book—to the surprise of no one so much as that of the originator of its being" (14). As Sylvester explained, the book dealt with "the technical or material part of versification (the art of rhythmical composition)," and this art, "like that of any of the other fine arts, is capable of being reduced to rules and referred to fixed principles."[73]

In particular, Sylvester found poetry to be dissectable into series of analytic triads. First, it consists of sound, thought, and words, or what he termed its "pneumatic," "linguistic," and "rhythmic" components, respectively. Although each of these had its own laws, Sylvester wanted to focus on the rhythmic component, which he trisected into the branches of "metric," "chromatic," and "synectic" (10). Of these, the synectic involved the "continuous" and was itself trisectable into "anastomosis," "phonetic syzygy," and "symptosis." Although Sylvester invented most of these terms and gave no rigorous definitions of them, he did characterize synectic and its "main branch," syzygy, as what gives "that coherence, compactness, and ring of true metal, without which no versification deserves the name of poetry" (13). Especially in translation, syzygy seemed to involve how best to choose the word that fit the meaning—at the same time that it preserved the tonal qualities—of the original. Syzygy involved a continuity or repetition of sound that somehow reflected the meaning of the line or stanza. As Sylvester put it, "the great law of Continuity (continuity of sound and continuity of mental impression) has been my guiding principle" (14–15).

If all of this seems, on the one hand, impossibly analytical and, on the other, difficult if not impossible to put into practice, Sylvester never promised "a systematic body of doctrine on the Art of Versification, but merely" an indication, "in the way of cursory comment, chiefly contained in notes to the text," of "the existence of such a doctrine, and the possibility of moulding it into a certain definite organic form" (9). The reader of *The Laws of Verse* was to intuit those laws, both from the original but anonymous poems which most probably were from Sylvester's own pen and from his running commentary of the thought processes that had led him to the "right" turns of phrase in the translations he provided from German, Italian, and Latin.

Consider, for example, his gloss on the opening stanza of the Horatian ode:

I propose as an alternative rendering what follows below:—

> *Tyrrhenian progeny of kings!*
> *In unstooped cask a mellow brew*
> *Roses and balm-drawn myrrh-droppings*
> *Thy hair Maecenas to bedew.*

I hestitated, and chopped and changed a long time, as my printers can too well attest, between the two readings, 'Birth of Tyrrhenian' and 'Tyrrhenian birth of'; and yet it is as certain as any proposition in Euclid can be that the former is the proper order of words. The latter, it is true, has in its favour a closer correspondence with the original, and the fact of the initial T being a crisper and grander opening sound than the B; but this cannot outweigh the double objection,—1st, of the *b* in 'birth' following the *n* in 'Tyrrhenian,' contrary to the laws of Anastomosis; and, 2nd, of the number and measure of 'Tyrrhe' being to the number and measure of 'Tyrrhenian birth of re,' as 1:3; whereas, in the contrary order, the corresponding ratio is 2:3,—which latter, by the principles of Symptosis (here applying to the clash or congruence of the open ē sounds) is preferable, especially at the opening of the piece, as being less suggestive of subdivision of measure (30–31).

The comment goes on, then again as long, distinguishing between various kinds of "clashes" of sound and debating the relative merits and demerits of "stirred" for "stooped"!

In all this, Sylvester was in earnest. In his egocentrism and as with his mathematical work, he truly thought that his readers would care about and be enlightened by his thought processes. Moreover, he was taking issue with and calling into doubt some of the standard translations of Horace's poetry such as those by John Conington in his posthumously published *The Satires, Epistles and Art of Poetry of Horace* of 1870.[74] Throughout the first half of 1870, Sylvester was just as serious about poetry as he had been about mathematics in the 1850s and 1860s. He needed to explain how he came to see the errors of others' ways; those errors needed to be corrected; things needed to be done right.

For Sylvester, then, *The Laws of Verse* represented a kind of intellectual high point, yet it was largely lost on his public. The reviewer in the *Times*, for example, commended the "little book" for its "spirit and accuracy of the poetical translations" but seemed unconvinced of Sylvester's "laws" of verse.[75] On the one hand, the failure to define the spate of new terms rendered the discussion largely incomprehensible, while on the other, the unsystematic presentation made the "laws" virtually unintelligible.[76]

As for Sylvester's enthusiasm for syzygy as a law at the heart of poetry, the reviewer was also doubtful: "This branch of Synectic would be more important if poetry, like music, appealed merely to the emotions produced through the sense of hearing. But poetry depends for its effect chiefly on the ideas it awakens; it deals with men and things, and with human actions and passions, and when a line falls on the ear the intellect is mainly employed in grasping the idea conveyed, the emotion responds almost wholly to this idea, and only a small part of the attention is left for time and sound."[77] There may be "laws" of verse, but they are more involved than Sylvester had seemingly allowed. He was too reductionistic, but, then again, "the mathematician especially has a firm faith that what seems most arbitrary or perplexed will be reduced by the requisite research to laws, the last expression of which will be a formula furnished by his own study." It was a curious little book, thought the reviewer, but perhaps the mathematician should stick to his mathematics.

The writer for the *Times* had confined his comments on *The Laws of Verse* to the part involving poetry, but that, in fact, was only two-thirds of the printed text. Next followed a reprinting of Sylvester's BAAS address, which, to his way of thinking, was not unrelated to his poetic exercises. Just as syzygy in poetry dealt with continuity of sound and meaning, so continuity was "the pole-star round which the mathematical firmament revolves, the central idea which pervades as a hidden spirit the whole corpus of mathematical doctrine."[78] Moreover, Sylvester believed in a universe governed by laws, and, just as laws rule mathematics or musical harmonies, so they should govern poetry, thereby "remov[ing] [it] from that indefinite region of taste which, like the so-called discretion of a judge, does not admit of being made the subject of rational discussion."[79] Continuity thus linked music and mathematics and poetry; all operated according to determinable laws. This was sure evidence of the underlying unity of all knowledge and human expression, a goal of the positivistic philosophy of science that Sylvester espoused.

Sylvester closed his slender volume with an exchange of letters he had had in the newly inaugurated scientific publication, *Nature*, in January 1870 with George Henry Lewes, Thomas Huxley, and others on the correct interpretation of Kant's doctrine of space and time.[80] Again, to Sylvester's way of thinking, the position he put forth in *Nature*, like his poetic translations, was meant to set the record straight. The "right" translations and the "right" philosophical interpretation were thus of a piece.

In his presidential address before Section A of the BAAS, Sylvester had remarked that "like his master Gauss, Riemann refuses to accept Kant's doctrine of space and time being forms of intuition, and regards them as possessed of physical and objective reality."[81] Sylvester later elaborated on this idea in *Nature* on 30 December 1869 in somewhat less than diplomatic terms, positing that "it is very common, not to say universal, with English writers, even such authorised ones as Whewell, Lewes, or Herbert Spencer, to refer to Kant's doctrine as affirming space 'to be a form of thought,' or 'of the understanding.' This is putting words into Kant's mouth . . . , words which he would have been the first to disclaim."[82] On 3 January, Lewes defended himself against this imputation "to stop *this* error from getting into circulation through the channel of *Nature*."[83] Although he acknowledged that Kant would indeed not have categorized space as a "form of the understanding," he "would have been surprised to hear that space was not held by him as a 'form of thought'" (132). Huxley countered on 14 January, citing six passages in which Kant *did* deny that space was a form of thought (132–133). The squabble escalated from there, drawing in correspondents both from within and outside academe.[84] By 29 January, *Nature*'s editor, Norman Lockyer, took matters into his own hands. He followed yet another letter in the exchange with the comment "This correspondence must now cease" (147). Lockyer certainly wanted his new publication venture to spark interest, activity, and interchange within the British scientific community, but enough was enough.

Sylvester's tenaciousness here, as in the correspondence with "Armed Science," reflected both a key aspect of his personality—his sense of right at all costs—and the pressure he was under in 1870 as he waited to see how the Royal Commission's recommendations would be implemented. Both of these exchanges show that he was more than capable of making a mountain out of a mole hill, though never wavering in his self-righteousness. Both also reveal his insecurity—hence his overwhelming need to *be* right—in the face of an uncertain future.

Life after Woolwich

The immediate future soon became quite clear. On or just after his fifty-sixth birthday on 3 September 1870, Sylvester was informed that his tenure at the

Royal Military Academy was over effective immediately and that he would receive a pension of £278.1 a year; he would not even be allowed to stay the six extra weeks that would have brought his period of service to a full fifteen years and thus qualified him for what would have been the usual civil service pension due him of a *minimum* of £338.[85] Sylvester, who had been making £550 a year in the 1860s with a house included, suddenly found himself homeless and with an income cut by some 50%.

Moving in to London and spending much of his time at the Athenæum, Sylvester tried once more to "recover [his] footing in the world's slippery path."[86] He turned first to two tried and true equilibrators—poetry and good works— dusting off his translation from German into English of the ballad of Sir John de Courcy and presenting it with extensive, colorful, and timely commentary to the working classes at a Penny Reading on 11 April 1871.[87] In particular, the irony of the fact that he was highlighting German poetry on an English theme just months after Bismarck's victory in the Franco-Prussian War was not lost on Sylvester. He commented wryly to his audience on how "singular" it was "that a narrative . . . so much to the honor and glory of England should come to us through a German channel." "It is not without some feeling of melancholy," he added, "that we can bring before our minds the very different tone in which England is now-a-days referred to by our German Hussites."[88]

If relations between Germany and Great Britain were strained in the wake of Prussia's war with France, relations were no better between Sylvester and the authorities at the War Office. Sylvester also sought to regain his footing in the spring of 1871 by fighting back against a system and a decision he viewed as unjust. The day before his Penny Reading, he wrote to his long-time friend, Benjamin Smith, soliciting his "aid . . . in bringing pressure to bear on the Government from their supporters."[89] Sylvester had been "advised to get as many parliamentary friends" as he could to argue on his behalf behind closed doors and, if need be, on the floor of Parliament, since "they will only yield if at all to party considerations."[90] By August, these efforts had borne fruit.

Sylvester had journeyed up to Scotland to serve as a Vice President of Section A of the BAAS at its annual meeting in Edinburgh, to give a brief note on partition theory to those assembled there,[91] and to receive an honorary LL.D. from Edinburgh University. He was enjoying the hospitality of the Edinburgh chemist, Alexander Crum Brown, when the *Times* reported extensively on his case and on its successful outcome. The edition of 17 August reported that Sir Francis Goldsmid, son of Sir Isaac Goldsmith, who had lobbied so hard earlier in the century for Jewish rights, had brought the matter up on the floor of the House of Commons and had threatened to "move an Address to the Crown that the Professor might receive as his pension 'two-thirds of the salary and emoluments enjoyed by him at the time of his removal, and humbly to assure Her Majesty that this House will make good the same.' "[92] After citing a host of examples where pensions had been liberally granted, the writer for the *Times* continued reproachfully that "it seemed to be a necessary conclusion that the men who manage these matters in the Government can be liberal, and even generous, at the expense of charity funds or of distant dependencies, but have a different measure to mete with in the case of a man of science who has given his ripe intellect to the service of this wealthy nation."

Not surprisingly perhaps, the company around Sylvester at Crum Brown's concurred. One of that number, the distinguished German physicist, Hermann von Helmholtz, recounted the story in brief in a letter home to his wife. "On Sunday," he wrote, "we had dinner with Crum Brown, with whom is staying a great mathematician from London, Sylvester, in aspect extremely Jewish, but otherwise an important and presentable person. . . . Mr. Sylvester has been treated as badly as could have happened at the hands of a Prussian Cultusminister—or even worse; and there was great indignation expressed by the company."[93]

Fortunately, right won out. At the end of Parliament's session, Gladstone "rectif[ied] . . . a lamentable departmental error" by conceding Sylvester's full fifteen years of service and thereby making him eligible under the seventh clause of the Superannuation Act for the more liberal pension Goldsmid had called for.[94] This decision thus corrected what Sylvester termed "an official oversight," that is, the result of some unthinking bureaucrat's blind adherence to the letter of the law.[95] It gave Sylvester an income of about £400 a year on which to live— an amount that moved him down from his former standard of living while still establishing him as solidly middle class—but it did not fill the void left by his premature retirement.[96]

By March 1872, Sylvester had once again turned to education in his efforts meaningfully to occupy his time. The seat for Marylebone on the London School Board that had been occupied by Thomas Huxley was coming open, and Sylvester decided to run for it. Addressing the Marylebone residents assembled in the Central Committee Room on Great Quebec Street in the district, Sylvester promised that "if you send me to the London School Board, I shall be prepared, while looking forward to the gradual adoption of a National system of Education, to adhere to that wise and moderate compromise by which, without violation of principle, you may obtain the use of existing machinery."[97] His bid was supported in writing by more than 40 men of science, among them, Norman Lockyer, who published Sylvester's brief address in full together with the names and credentials of Sylvester's supporters in *Nature* along with the journal's endorsement. It was important, in *Nature*'s view, to have "at least one representative of Science on the Board."[98] On 28 March, a week after *Nature*'s endorsement had appeared and on the very day of the Marylebone election, the *Times* ran a very curious paragraph entitled "A Little Mistake." It read: "The following paragraph appeared in Monday's *Record*:—'A paragraph from a respondent in

Wednesday's *Record* recommending Dr. Sylvester as a candidate for Marylebone was inserted in ignorance of the fact that Dr. Sylvester is a Jew.' "[99] The blatant sarcasm of the header more than suggested that the *Times*—unlike *Nature*— found Sylvester's faith an issue in his candidacy. In its coverage of the outcome the next day, the *Times* was able to report that the Jew had not won: "The successful candidate was the Rev. John Llewelyn Davies, who obtained 2,254 votes. The votes obtained by each of the other candidates were as follow:— Cremer, 1,038; Sylvester, 526; and Dyason, 15."[100] It had not even been close.

The weeks after his defeat found Sylvester testy. On 20 April 1872, he wrote in no uncertain terms to Stokes as Secretary of the Royal Society, once again stating categorically his unwillingness to referee for the Society. "The last was the *third* time of this having occurred and I have been subjected to much, unnecessary trouble on account of it," Sylvester railed. "I should have been more persuaded if . . . you had more effectually satisfied my mind that my quiet should not again be invaded by some slight expression of regard for the obliviousness on your part which has been the cause of this to me very painful correspondence."[101] This was certainly no way to address a colleague, but the sting of the School Board loss, together with Sylvester's own estrangement from mathematics, combined to make the thought of reading and evaluating the fruits of someone else's mathematical research labors unbearable.

That spring, Sylvester occupied himself not with research but as an Examiner in Mathematics for the University of London, and in the late summer, he left London and its trials altogether for a restorative trip to the Continent that took him to Poland, Russia, and Sweden by the end of September. Writing from Prague in mid-September to his long-time acquaintance, the Russian probabilist Pafnuti Chebyshev, Sylvester informed him of his intention to go to Moscow to view various exhibitions and asked for letters of introduction, since he knew no one in Moscow personally.[102] From Moscow, Sylvester journeyed to St. Petersburg to visit Chebyshev, only to find the mathematician out on the day of his call.[103] Back in London by November, Sylvester reported to his niece that "I feel much better . . . than I have for a long time past and think the doctor's advice is likely to be very useful. . . . Already I think I begin to feel in quite different spirits in consequence of following it."[104] Still, it was with a tone of resignation that he continued, "until I have settled in my plans which will be about the beginning of next year, music (and getting my books and letters in order) will be my chief occupations: it is quite as interesting to me now as Algebra used to be when I was a small boy."[105]

Like poetry, music—and specifically the piano and singing—had become a diversion for Sylvester in the post-Woolwich, nonmathematical years, and his niece, Edith, encouraged him to join her choral club. As she knew, it would help distract him from thoughts of his new and unwanted circumstances by occupying his time and exercising the tenor voice that had apparently benefitted from lessons with the French composer, Charles Gounod. Sylvester was eager, but unsure of himself. He did not feel strong enough vocally to carry his part alone but would gladly join in with others in the tenor section.

If his confidence in his musical talent was shaky in November, he soon had cause for confidence in the extent of his international reputation. On receiving in December a parcel from Chebyshev addressed to him as "Corresponding Member of the St. Petersburg Academy," he learned of his November election to that body.[106] The Americans followed four months later in March 1873, electing him "Foreign Honorary Member" of the American Academy of Arts and Sciences to replace the recently deceased French astronomer, Charles Delauney. Sylvester proudly asked the Assistant Secretary of the Royal Society "to add this title to the others in the list against my name in the Royal Society Catalogue."[107] These signs of recognition were clearly welcome salves for Sylvester's bruised ego.

New Possibilities on the Horizon?

Whether or not these international kudos rekindled his intellectual and mathematical spark, by 23 January 1874, Sylvester was lecturing anew, this time to the prestigious Friday evening meeting of London's Royal Institution in Albemarle Street. Founded in 1799 by Sir Benjamin Thompson (later Count Rumford) "for diffusing and facilitating the general introduction of useful mechanical inventions and improvements and for teaching by courses of philosophical lectures and experiments, the application of science to the common purposes of life," the Royal Institution had nurtured such eminent scientists and experimenters as Humphry Davy and Michael Faraday and had long supported a program of lectures on a wide range of mostly scientific topics.[108] Its subscribers were "highly-fashionable" members of society, who, only after recommendations from four members, paid a hefty £10 in their first year of membership and £5 thereafter to use the ample library, listen to any lecture delivered, and attend the weekly evening meetings.[109] In his colorful 1874 survey of "scientific London," Bernard Becker described the latter, in particular, as "a wonderful combination of science and society, of physics and fashion" (44). It was to that well-heeled audience that Sylvester had been invited to speak, following in the footsteps of such distinguished scientists as Huxley, John Tyndall, Lord Kelvin, and James Clerk Maxwell.[110]

Sylvester chose as his topic the mechanical conversion of circular into rectilinear motion, and while not a subject for everyone, according to Becker, it caused "those of a higher and drier turn of mind" to "experience ineffable delight."[111] Certainly a topic that fit well with the Royal Institution's statement of purpose, it was, nevertheless, an odd topic and an odd presentation for Sylvester. In fact, it harkened back to his unhappy days as Professor of

Natural Philosophy at University College when he was forced to carry out physics demonstrations in front of his classes. This wintry Friday, and of his own accord, however, he stood before his Royal Institution audience holding a curious device made of wooden sticks connected at their ends by pivots, and, while talking through the principles behind it, he manipulated it to show how it perfectly transformed one kind of motion into another. (See figs. C and D.)

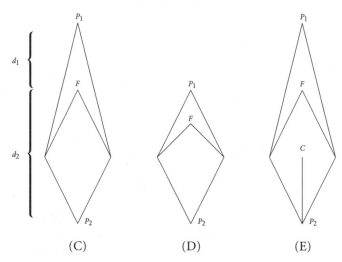

(C)   (D)   (E)

FIGS. C–E.   *Three types of Peaucellier cells:* Take four equal rods and arrange them in a diamond pattern. Attach them at the vertices of the diamond in such a way that they pivot. Now affix two more rods of equal length (but not necessarily equal in length to the other four) to either pair of opposite vertices in the original diamond, again in such a way that they pivot and attach their free ends with a final pivot. Call that final pivot F the *fulcrum*. The free pair of opposite vertices of the diamond P_1 and P_2 (called the *poles*) will then always be in a straight line with the fulcrum F, regardless of how the device is manipulated. If the two additional rods are longer than the original four, the cell is called *positive* (fig. C); if less, *negative* (fig. D). These are termed 6-bar linkworks.

Calling the distance d_1 and d_2 of each of these poles from the fulcrum the *arms*, it is easy to show that the product of the arms is invariant, that is, $d_1 d_2 = k$, some constant. This mathematical property gives the Peaucellier cell its surprising mechanical property.

Finally, consider the case (fig. E) where the fulcrum is fixed and where one of the poles is attached to a center C around which it can revolve, such that the center is equidistant from the pole and the fulcrum. This 7-bar linkwork with two fixed centers is the one that effects the true conversion of circular into rectilinear motion.

Sylvester had learned of the mechanical principles behind this device—what he called the Peaucellier cell—during a visit that Chebyshev had made to London in 1873 in conjunction with his participation in France at a meeting of the Association française pour l'avancement des sciences.[112] Sylvester, well aware of Chebyshev's long-standing interest in the theory of mechanisms, chanced to ask him about "the progress of the disproof of the impossibility of the exact conversion of circular into rectilinear motion" only to be told that it had been proven possible in 1864 by Charles Peaucellier, an engineering officer in the French Army, as well as independently by a freshman student in Chebyshev's own class.[113] Chebyshev then made a quick sketch of the simple device for Sylvester. With characteristic exuberance, Sylvester showed the drawing to a friend, the noted singing teacher and inventor of the laryngoscope, Manuel Garcia, who promptly went home and fashioned a rough model out of wood and nails. Sylvester could not have been more delighted when Garcia presented him with the contraption. He took it with him to dinner at the Philosophical Club of the Royal Society, where it was presented along with dessert to the ohs and ahs of "some of the most distinguished members of the . . . Club . . . not mathematicians, but naturalists, geologists, chemists, and physicists," as well as to the Athenæum, where Kelvin "nursed it as if it had been his own child, and when a motion was made to relieve him of it, replied, 'No! I have not had nearly enough of it—it is the most beautiful thing I have ever seen in my life.' "[114] Sylvester also talked about it with another friend, the Professor of Mathematics at University College, Olaus Henrici, who had a finer working model executed in zinc that he demonstrated after Sylvester's lecture.

The device and the theory behind it did have potentially far-reaching ramifications. As Sylvester noted in his historical introduction:

> Before Peaucellier's time all so-called parallel motions were imperfect, and gave merely approximate rectilinear motion; in substance they will be without exception found to be merely modifications of Watt's original construction, and to depend on the motion of a point in, or rigidly connected with, a bar joining the extremities of two other bars rotating around fixed centres, which may be described briefly as three-bar motion. Peaucellier's exact parallel motion depends on a link-work of seven bars moving like Watt's [Figure e], and the other imperfect parallel motions of the same class, round two fixed centres (8–9).

Thus, unlike the piston configuration that James Watt had developed in the last quarter of the eighteenth century and that had revolutionized the steam engine, Peaucellier's device involved absolutely no wobble. There was no imperfection in its translation of circular to rectilinear motion. The engineering and everyday ramifications were manifold: water pumps, water closets, "steam-engines, planing and grinding machines, the construction of maps by stereographic projection, millwrights' work, laying out of railway curves, dioptric apparatus for lighthouses, ornamental tracery, pendulum suspension to effect motion in a practically exact cycloidal arc, &c., &c."[115] An adaptation of the device could also have improved "certain machinery connected with some new apparatus for ventilation and filtration of the air of the Houses of Parliament, [then] under course of construction" and, according to the authorities at Woolwich, could have "saved several weeks' work, inconvenience, and expense in cutting out the fish-bellied torpedo casings [then] recently constructed in the laboratory department at the Royal Arsenal there."[116] Moreover, in suitable combination, Peaucellier cells could even serve as the underlying mechanism of calculating machines, and Sylvester demonstrated a cube root calculator composed of three such cells in the appropriate relative sizes and combination.

Sylvester's talk received quite a bit of press. It was reported on immediately in the *Examiner* and elsewhere, and it appeared later *in extenso* in printed form in Britain's *Proceedings of the Royal Institution of Great Britain*, France's *La Revue scientifique*, and North America's *Van Nostrand's Eclectic Engineering Magazine*.[117] It also attracted the attention of several readers, who then contacted Sylvester with questions and with their own ideas for combining Peaucellier cells mechanically and for analyzing them mathematically. Among these correspondents were Francis Penrose, author, astronomer, and architect and surveyor of London's St. Paul's Cathedral; Francis William Newman, brother of Cardinal John Henry Newman and former chair of Latin at University College London; and Harry Hart, an instructor at the Royal Military Academy who had been shown a copy of Sylvester's lecture by Morgan Crofton, Sylvester's successor and former colleague there.[118] By January 1875, the mathematization of these so-called linkworks had come to dominate Sylvester's thought as well, and he fired off a volley of letters to Cayley in January and February describing his combinatorial and algebraic findings.[119] Although this enthusiasm resulted in two short, largely nonmathematical papers—one in *Nature* and one presented to a meeting of the London Mathematical Society on 11 February—Sylvester's supposed mathematization ultimately went nowhere.[120] Still, linkworks immersed him once more in mathematical ideas after a drought that had lasted effectively since 1868 and the work on cyclodes, but it was too soon to tell whether he had really succeeded in regaining his mathematical "footing in the world's slippery path."[121]

Although his thoughts had turned once again to mathematics, he did not neglect his commitments to working class education. As he had told Charles Graves back in May 1869, "I have always taken a deep interest in the cause of education and have frequently examined for the College of Preceptors and the Society of Arts and have even personally instructed a class in Algebra at the Working's Man [sic] College in G[rea]t. Ormond Street."[122] In June 1874, he thus felt obliged to pass up a visit to the Sussex home of his old friend and former flame, Barbara Smith Bodichon, to attend "a meeting connected with a society for promoting education among the working classes," even though he "would [have] much prefer[red] to have been walking in a Sussex lane [with her] instead."[123]

Two months later on 12 August, he was again writing to Mrs. Bodichon telling her of his imminent plans to depart on a European tour that would take him first to Boulogne, then to Lille, the host of that year's meeting of the Association française pour l'avancement des sciences, where he would participate as a guest of the city, and finally to Athens and Constantinople via Italy before returning to England by way of Vienna. He thought to link up with Mrs. Bodichon's sister in Venice on his way through, and he was in no way pressed for time. "I am free to stay abroad if not (too much ennuyé) until the end of October," he told her.[124] Life thus went on after Woolwich; Sylvester was recovering his equilibrium.

Still, he longed for gainful employment. He was too young and too vital not to be doing something more productive with his life. In mid-February 1875, at the most fevered pitch of his researches on linkworks, an announcement in the *Times* caught his eye. The professorship of mathematics at the University of Melbourne had come open on the early death of its incumbent, William Wilson, and John Couch Adams, Cambridge's Lowndean Professor of Astronomy and Geometry and one of Sylvester's professional friends, had been appointed as the selection committee to fill the vacancy. Writing to Cayley on 15 February, Sylvester confessed to being "more than half inclined to go out to the Antipodes rather than remain unemployed and *living upon charity* in England" and asked Cayley to sound Adams out as to Sylvester's suitability for the post.[125] Cayley must have been reluctant to do this because, four days later, Sylvester was again writing, thanking him for his "desire to save me from the Antipodes."[126]

By late August, it was not the Antipodes but the United States that offered an intriguing possibility. A new university, the Johns Hopkins University, was being organized in Baltimore, and its first President, Daniel Coit Gilman, was hard at work putting together what would be his first faculty when the school opened in the fall of 1876. Gilman's efforts were highly publicized in the American press, and news of his faculty-finding trip to England in the fall of 1875 quickly reached members of England's intellectual élite, including, of course, Sylvester. The mathematician was tempted and immediately sent off letters to friends on both sides of the Atlantic soliciting their support. Letters poured in to Gilman from Joseph Henry in Washington, from Benjamin Peirce in Cambridge, Massachusetts, from Sir Joseph Hooker at Kew Gardens outside of London. All strongly endorsed Sylvester as a candidate; all echoed Henry's description of him as "one of the very first living mathematicians"[127]; all let Gilman know that Sylvester would seriously consider an offer were one to be tendered.

The moment of truth finally came at the end of September. Hooker had arranged a face-to-face meeting in London between Gilman and Sylvester, and

Sylvester felt the enormity of the occasion. Writing to Barbara Bodichon on 19 September, he allowed that "the result of our interview may be momentous for me."[128] The meeting ultimately went well; Sylvester and Gilman entered into a correspondence that lasted well into February of 1876.[129] In the end, the Johns Hopkins University had itself a mathematician, but only after hard bargaining from Sylvester on the issue of salary. Having been burned once by the authorities in the War Office and no doubt aware of the economic depression in the United States occasioned by the Panic of 1873, Sylvester would take no less than $5,000 *in gold*, and he wanted a house to boot. On 17 February, he finally accepted the Hopkins offer—$5,000 in gold annually as well as an annual housing allowance, also in gold, of $1,000. He was officially appointed by the Hopkins Board of Trustees on 6 March.[130] After some thirty years, Sylvester would return to the United States to teach, but even if the Johns Hopkins University of the 1870s was not the University of Virginia of the 1840s, it was not clear whether the "uneasy years" would soon be over or whether, in some sense, they were only just beginning. It was a gamble that Sylvester was nevertheless willing to take.

James Joseph Sylvester circa 1840. A painting by George Patten in the
possession of Alain Enthoven.

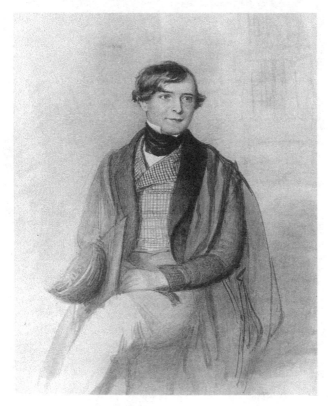

Watercolor graduation portrait of Arthur Cayley as the senior wrangler, 1842;
Trinity College, Cambridge.

SKETCHED NOVEMBER 1853 AT
THE UNIVERSITY of VIRGINIA

THE STUDENT

It gives me great pleasure to say that, although the vivacity of these blooded colts at the University frequently leads them into all sorts of deviltries and excesses, they have almost invariably the manners of gentlemen. R.C. 1853

Sylvester experienced in the 1840s the irreverence and arrogance of the "typical" University of Virginia student captured in this 1853 sketch.
Courtesy of Special Collections, University of Virginia Library

INDEX TO DEFINITIONS.

With the exception of the ~~lower~~ words "Eliminant" and "Quantic" all the above terms are of Mr Sylvester's creation; he mentions this as indicative of his having had a much larger share in the formation of the New Analysis than might be inferred from the author's preface ... of himself & his splendid ... M. Hermite equally a ... friend ... Mr Cayley has said that Mr ...

From Sylvester's copy of George Salmon's *Lessons Introductory to the Modern Higher Algebra* in the possession of Paul Garcia. Sylvester has written (with punctuation errors preserved): "With the exception of the words 'Eliminant' and 'Quantic' all the above terms are of Mr Sylvester's creation; he mentions this as indicative of his having had a much larger share in the formation of the

LESSONS ON HIGHER ALGEBRA.

LESSON I.

DETERMINANTS.—PRELIMINARY ILLUSTRATIONS.

1. If we are given n homogeneous equations of the first degree between n variables, we can eliminate the variables, and obtain a result involving the coefficients only, which is called the *determinant* of those equations. We shall, in what follows, give rules for the formation of these determinants, and shall state some of their principal properties; but we think that the general theory will be better understood if we first give illustrations of its application to the simplest examples.

Let us commence, then, with two equations between two variables—

$$a_1x + b_1y = 0, \qquad a_2x + b_2y = 0.$$

The variables are eliminated by adding the first equation multiplied by b_2 to the second multiplied by $-b_1$, when we get $a_1b_2 - a_2b_1 = 0$, which is the determinant required. The ordinary notation for this determinant is

$$\begin{vmatrix} a_1, & b_1 \\ a_2, & b_2 \end{vmatrix}$$

We shall, however, often, for brevity, write (a_1b_2) to express this determinant, leaving the reader to supply the term with the negative sign; and in this notation it is obvious that $(a_1b_2) = -(a_2b_1)$. The coefficients $a_1, b_1,$ &c., which enter into the expression of a determinant, are called the *constituents* of that determinant, and the products $a_1b_2,$ &c., are called the *elements* of the determinant.

B

"and in the right of all the world is the founder of the New Analysis" J.J.S.

New Analysis than might be inferred from the Author's preface. M. Hermite (equally a friend of himself & his splendidly old friend Mr Cayley has said that Mr S 'to the knowledge and in the sight of all the world is the founder of the New Analysis' J.J.S."

Daniel Coit Gilman, first president of the Johns Hopkins University.
Courtesy of the Ferdinand Hamburger Archives of
The Johns Hopkins University.

The interior (circa 1890) of the journal-, book-, and modeled-lined Mathematical
Seminary founded at Johns Hopkins University by Sylvester in the 1870s.
Courtesy of the Ferdinand Hamburger Archives of
The Johns Hopkins University.

James Joseph Sylvester sometime after his arrival in Oxford in 1884.
An engraving by G. J. Stodart from a photograph by Messers I. Stillard & Co.,
Oxford in the possession of the author. This image accompanied the
article on Sylvester in the series of "Scientific Worthies"
published in *Nature* 39 (1889).

CHAPTER 9

Exploring Familiar Ground on Unfamiliar Territory

How unlike America is to Europe! much further from us English than are the Italians, Germans, or Russians although they speak the same language and outwardly follow the same habits. I do not think it possible that I can ever regard America as a home.
—J. J. Sylvester to Barbara Bodichon, 21 August 1876

*I have just made a hit in the theory of Invariants . . . . I find that the tremendously difficult machinery of Clebsch and Gordan may be entirely dispensed with and that my theory of Partitions . . . is competent to solve the question much more (*infinitely *more) easily and completely than the method employed by C and G.*
—J. J. Sylvester to William Spottiswoode, 18 November 1876

Send-off parties and dinners hosted by his friends at the Athenæum, together with the inevitable delays involved in trying to arrange his affairs, forced Sylvester to postpone his departure for Baltimore several times in the month of April 1876. By the twenty-ninth, though, he had finally set sail. The Johns Hopkins University would open early in October, and he was anxious both to get settled in his new surroundings and to begin his association with his new colleagues. Most importantly, he had to reengage in his mathematical researches. His new post, unlike the position at Woolwich, *required* him not only to train future researchers but also to produce and publish original research. It reflected his long-held ideal of the professional mathematician, but was he up to it now in his early sixties? The work on linkages in 1874 and 1875 had never really gone anywhere mathematically, and the papers on cyclodes had come out in a flurry way back in 1868 and 1869. There had been a research drought in between and after. Where was he to turn for inspiration?

To complicate matters further, Gilman had big plans for Hopkins. To foster research and to provide suitable outlets for its publication, he proposed that his professors consider launching a journal in their respective fields. Sylvester, editor of the *Quarterly Journal of Pure and Applied Mathematics* back in England since its inception in 1855, was certainly no stranger to the world of journal publication, but where would he get the submissions to fill a research-level journal in the United States? Gilman's idea was that much of the material would come from the pens of his professors and from their students, but this only put more pressure on Sylvester to find his way—and quickly—to a fruitful research program.

It is no exaggeration to say that the sixty-one-year-old Sylvester faced real challenges when he returned to American shores in the spring of 1876, and they were not just of a professional nature. He had left the familiar and comfortable surroundings of London—its scientific societies, the Athenæum, his network of friends and acquaintances—for a strange city in which he knew essentially no one and a country in which he had failed miserably to find his place some three decades earlier. Sylvester was in decidedly unfamiliar territory. He realized that his second American venture would only succeed if he could revive his mathematical powers, flourish in the graduate classroom, and find a congenial social niche. But at least he had a job, a good job, and he was part of a new experiment in American higher education. The prospects were daunting, but exciting.

First Impressions

After his arrival in New York, Sylvester boarded the train southward for Baltimore, stopping for a number of days in Philadelphia to visit his brother, Frederick, and perhaps to take in the opening days of the Centennial Exhibition in honor of the one-hundredth birthday of the United States. Laid out on 450 acres of fairgrounds in the northwest of Philadelphia, the exhibition opened on 10 May amid intermittent showers and much fanfare. President Ulysses S. Grant and his wife were on hand to kick off the six-month-long event with its pavilions dedicated to the different states and to various nations, its gargantuan buildings devoted to art, invention, horticulture, manufacture, and agriculture, and its exotic foods and entertainments.[1] Not just Philadelphia but the whole country was abuzz with the centennial, and news of the exhibition filled the papers nationwide.

Toward the end of May, Sylvester left the hubbub of Philadelphia behind as he traveled the hundred miles farther south to Baltimore. There, he encountered another city bustling with activity—albeit of a different kind—and growing in all directions.[2] A port city with a broad natural harbor 14 miles up the Patapsco River from the upper end of the Chesapeake Bay, Baltimore had a population of some 300,000 people in 1876 and was often termed the "Liverpool" of the East Coast. Whether that association with his family's roots brought comfort or consternation, Sylvester knew firsthand the advantages and the disadvantages of a port city. On the negative side, there were certain areas to be avoided; in fact, one of Baltimore's other nicknames was "Mobtown" on account of its checkered history of riots and rowdiness. On the positive side, port cities enjoyed goods and services in abundance and, in general, had the wealth to partake of them. In Baltimore's case, the port was the number one market in the world for flour and grain,[3] at the same time that it dealt in a variety of other key commodities and put them to use in its extensive manufactures. In retrospect, the Baltimore of the decades from 1875 to 1895 was in transition into a "modern city." This was the Baltimore Sylvester experienced.[4]

The Johns Hopkins University he found was, architecturally at least, some-what at odds with this picture of progress and prosperity. Despite the $3,500,000 bequest of financier Johns Hopkins for the founding of a university, the Hopkins Trustees were bound and determined to spend the money not on expensive buildings but on scholars.[5] They thus bought two rather plain, Victorian, brick, mansarded residences on North Howard Street just north of the main com-mercial area, cobbled them together, and supplemented them with a new, but basic building that extended down Little Ross Street toward Eutaw Street. The unpresupposing complex was virtually lost among the fashionable shops, theaters, and host of small businesses that made up the genteel North Howard Street district. In particular, Baltimore's Academy of Music, nearby on North Howard, served not only as the focal point of Baltimore's musical, theatrical, and social scenes but also as the University's auditorium for major lectures and events.

The University was also a short walk from posh Mount Vernon Place with its imposing Washington Monument and stylish city homes of such wealthy Baltimoreans as John W. Garrett, President of the Baltimore and Ohio Railroad and one of Johns Hopkins's handpicked University Trustees. It was equally close to the stately 60,000-volume public library that entrepreneur and philan-thropist, George Peabody, had built in the 1860s also in Mount Vernon Place and that was then currently being expanded to accommodate as many as 500,000 volumes. Gilman and the Trustees realized that it would take them much time and money to build their own research library, so the Peabody was tailor-made for their immediate needs, even though it was a reading and not a lending library. The University itself may have come up short architecturally, but it was situated in the cultural, intellectual, and social heart of the city, a perfect location for partaking of all the finer things the city had to offer.

Sylvester took rooms at the elegant Mount Vernon Hotel right on Monument Street and a stone's throw from both the University and Mount Vernon Place. He immediately set to work. By 29 May, he had already written up a report for Gilman on the merits of the various candidates for the two, and possibly three, fellowships that would go to mathematics, but by 2 June, he was off for New York City.[6] There were two problems. First, a box of manuscripts that he had brought with him from England had failed to arrive, had then been due to come in on a later ship, but had apparently still not materialized. Distraught, Sylvester told Gilman that "the loss will be irreparable if the box should have permanently disappeared and send a cloud over all the remaining years of my life," so he needed to go personally to New York to try to track it down.[7] The other problem was timing. Sylvester had arrived at the moment when the cultured Baltimoreans with whom he would likely have associated were fleeing either northward or to the countryside to avoid their fair city's brutal summer weather. Gilman had forewarned him of this in a letter on 27 April and counseled that "your pleasure will be promoted by delay . . . . We do not expect the classes till Oct. 3."[8] Sylvester had not heeded this advice, and the saga of the missing box combined with increasingly oppressive temperatures and humidity drove him first to New York and then onward to Cambridge, Massachusetts, and the hospitality of his longtime friend, Benjamin Peirce. Sylvester planned to spend the remainder of the summer in the cooler Northeast. He would begin to settle in in Baltimore come fall and the start of classes.

Yet even in Massachusetts, Sylvester was at work for his new university.[9] He sounded Peirce out on some potential candidates for the Hopkins associateship in mathematics. The Associate, who would be responsible for the bulk of the undergraduate teaching in addition to supplementing Sylvester's graduate-level instruction, needed to have solid credentials in both teaching and research. At least two Harvard men had been mentioned in connection with the post, William Byerly and William Story, but Peirce gave the edge to Story on account of his further training in Germany and his Leipzig doctoral degree. Sylvester made a point of meeting Story and apparently liked what he saw, but Story also had to pass muster with Gilman. The personal interviews apparently went well enough, even if some sparks had flown on paper between Story and Gilman about the expectations for the Hopkins faculty. Story was hired in August.[10]

Sylvester's stay in Massachusetts contrasted with his brief exposure to Maryland. In Cambridge and Boston, he found cooler weather and friends with whom he could socialize and who could introduce him to society; in Baltimore, he knew only Gilman, who was busy traveling up and down the East Coast promoting and doing the business of the University in the final months before its opening. Moreover, although Sylvester had met essentially no Baltimoreans in his brief sojourn there, he found the Bostonians "more agreeable and suave in their manners and more like English people than most other Americans." Also of great importance, Peirce had arranged for Sylvester to have privileges at Boston's Somerset Club on brahmin Beacon Hill, a gentlemen's club "as good and comfortable . . . as any [Sylvester had] ever known in

England or elsewhere."[11] With Peirce as a surrogate Cayley, with the Somerset more than a viable substitute for the Athenæum, and in society not so unlike that of England, "Bosbridge" felt more like the home he had left than did the pulsing port town of Baltimore.

Ultimately, however, not even this was enough to smooth the transition. On 21 July, Sylvester wrote Gilman of his decision to return to England the next day for the remainder of the summer. The abnormal heat and what he described to Gilman as "depression" of spirits prompted the move, but Sylvester assured Gilman that he would "be anxious to carry out any directions with which you may favor me if I can do anything for the good of the University during my absence in the Old World."[12]

In fact, Sylvester did do something for the "good of the University" while in England. His spirits restored by his old, familiar surroundings, he was back at his mathematics and making new discoveries. By the time he set sail for his return to Baltimore on Tuesday, 29 August, he had sent a new paper to the *Philosophical Magazine* in which he showed how to simplify—and thereby render more intrinsic—some of the calculations that had come up in James Clerk Maxwell's classic treatise on electricity and magnetism. Where Maxwell had seen integrals, Sylvester found "purely formal or algebraical" arguments, and he chided the physicist in classic Sylvesterian style.[13] "To deduce purely analytical properties of spherical harmonics, as [Maxwell] has done, from 'Green's theorem' and the 'principle of potential energy,'" Sylvester wrote, "seems to me a proceeding at variance with sound method, and of the same kind and as reasonable as if one should set about to deduce the binomial theorem from the law of virtual velocities, or make the rule for the extraction of the square root flow as a consequence from Archimedes' law of floating bodies."[14] It was too soon to tell whether this paper marked a return of the mathematical muse, but it was at least an encouraging sign.

Getting Started

By the time classes at the Johns Hopkins began on 3 October, the new university had already attracted significant attention by inviting the ever-controversial Huxley to deliver an opening address in the Academy of Music on 12 September. Gilman had met Huxley in London in 1875 during the course of his mission to Europe in search of both faculty and insights into how universities operated abroad. Learning that Huxley would be in the United States in 1876 to visit his sister, Lizzie, Gilman invited him to give a course of public lectures at Hopkins during its inaugural semester. Huxley had to decline that longer-term commitment, but agreed to give a single lecture.[15] His stay in America would only last six weeks in August and September, and he had already committed himself to so many lectures that he hardly had time for the family visit.[16] The invitation to Baltimore, however, would allow him not only to get a sense of Gilman's new institution but also to see his former protégé, Gilman's new Professor of Biology, Henry Newell Martin, and to meet up again with Sylvester.

The newspapers had been full of the tour. After all, Huxley had become the avatar of materialism and atheism in some circles after his celebrated exchange with Bishop Wilberforce over Darwin's theory of natural selection in 1860. He was also seen by some as the physical embodiment of what Andrew D. White would later call the "warfare of science with theology in Christendom,"[17] so the Baltimoreans packed the auditorium to hear what he had to say.

From the outset, Johns Hopkins, the Hopkins Trustees, and Gilman had all been adamant that the new university be free from sectarian concerns. In his own inaugural address as Hopkins President earlier in the year, Gilman had told his audience that "religion has nothing to fear from science, and science need not be afraid of religion. Religion claims to interpret the word of God, and science to reveal the laws of God."[18] The invitation to Huxley drove home the University's commitment to this principle, and the lecture generated much press and publicity for the new university as a forum for informed intellectual debate (70–71).

By October, the excitement generated by Huxley's visit had been replaced by the ferment of the classroom. In mathematics, Sylvester oversaw the work of eight graduate students. Thomas Craig, had earned a degree in civil engineering at Pennsylvania's Lafayette College in 1875, had won Gilman over early on through his correspondence and through personal meetings, and had secured one of the three coveted fellowships that went to mathematics that first year.[19] George Bruce Halsted, another fellowship recipient, had also taken his undergraduate degree in 1875 but at Princeton, whereas the third fellow, Joshua Gore, had an 1875 civil engineering degree from the University of Virginia. Craig, Halsted, and Gore were joined by two fellows in engineering, Daniel Hering (Yale Ph.B., 1872) and Erasmus Preston (Cornell B.C.E., 1875), two more civil engineering graduates, Maxwell Hudgins (Washington and Lee University, 1876) and George La Tour Smith (Cornell University, 1874), and Fabian Franklin, who had earned his bachelor's degree from George Washington University in Washington, DC, in 1869.[20] Together with Story, the mathematical scientists thus numbered ten strong.

That first semester, Sylvester offered no formal classes, instead talking with the students individually and trying to keep up his own research momentum. Sylvester had interpreted the philosophy Gilman had defined for the University liberally and perhaps too literally. As he explained to Barbara Bodichon back in England, "my work lies exclusively with a few of the 'Fellows' and 'Post-graduates' i.e., young men who have already taken their degrees at other universities and is exactly what I like to make it as regards quantity, quality, and times of giving lectures.... I need not give a single lecture unless

I please." "In fact," he marveled, "it is impossible to imagine a more generous and appreciative mode of treatment by any institution of its professors than that which we experience here."[21] Compared with the Royal Military Academy, the Johns Hopkins *was* a scholar's paradise, but as Sylvester soon realized, regularly scheduled courses were not only expected and anticipated by the students, but they also represented a more efficient way both to convey material and to create an *esprit de corps*. Still, it had probably been a wise decision to offer no organized course that first fall. Sylvester had to acclimate to his new environment, but, more importantly, he had to produce new research results. That was most definitely a part of the Hopkins philosophy that he had in no way misunderstood, and the pressure was on.

The month of October breezed by, with Sylvester getting to know his students and settling in to a life in Baltimore. Back in August, he had precipitously declared to Barbara Bodichon that "I do not think it possible that I can ever regard America as a home,"[22] but by early November he was changing his tune. For one thing, he had found a new club, the élite Maryland Club to which all of Baltimore's finest gentlemen belonged. At the corner of Franklin and Cathedral Streets, it was centrally located three blocks from his office and an even shorter walk from the Mount Vernon Hotel, and "in its material arrangements [was] almost a compensation for the Athenæum Club in London." Moreover, "the cooking is *much better* than at that Club."[23] Sylvester also had to confess that "now I know the ways of the people I can live here much better and cheaper than I could in London" (164).

His renewed sense of well-being must only have been enhanced by 3 November, when the University hosted a welcoming dinner in his honor. Gilman did it right, inviting Sylvester's scientific friends like Henry from Washington and Benjamin Peirce from Cambridge as well as introducing him to others like astronomer Simon Newcomb at the Naval Observatory in Washington and geophysicist Julius Hilgard at the U.S. Coast Survey also in Washington.[24] The dinner was supplemented by the first in a series of twenty public lectures on the history of astronomy by Newcomb in Hopkins Hall, the University's own auditorium. Gilman had conceived of these short lecture series as a way of reaching out to the public and of advertising and highlighting the work of the University. Both Newcomb and Hilgard participated in the highly effective scheme in that first academic year.[25]

The Sylvester dinner provided more than the opportunity to launch a public lecture series, though. By bringing Newcomb together with Sylvester, Story, and their colleague, the physicist Henry Rowland, Gilman engineered a meeting of minds that resulted five days later in a call to the mathematical scientists of America to offer their support for "a periodical of a high class . . . to be called the American Journal of Pure and Applied Mathematics" and to be underwritten by the Johns Hopkins University.[26] It was a bold move. Up to that time, no specialized mathematical journal launched in the United States had lasted more than eight years, and most had died within a year or two of their founding.[27] In the fall of 1876, however, with Sylvester on American shores, with a scientific community united around institutions such as the American Association for the Advancement of Science and the National Academy of Sciences, and with the Johns Hopkins University and its mandate for the production of original research, perhaps the time was right to try again.

On Friday, 10 November, Sylvester made the short train trip to Washington to visit Henry for the weekend and to speak to him personally about the idea for the new journal. Although Gilman and Sylvester and his collaborators conceived of a purely research-oriented journal that would encourage "the propagation and advancement of mathematical knowledge" in the United States, Henry felt that expository articles were also needed to bridge the gap between the research level and the usual level of mathematical coursework. This idea apparently left Sylvester cold.[28] Henry may have been right about the level of mathematical instruction at most of America's colleges and universities in 1876, but Sylvester and his colleagues had a new model, the Hopkins, in mind. The journal should be a success with the cooperation of those in America already engaged in original mathematical research and with the contributions of those researchers at and to be trained at Hopkins. They should not have to sacrifice their standards to accommodate the lowest common denominator.

Whether the promise of the new journal—or the pressures of providing new material for it (should it come into existence)—had stimulated his creative juices, the weekend after his return from Washington found Sylvester writing his friend back in London, William Spottiswoode, about "a hit" he had just made "in the theory of invariants."[29] Sylvester had been revisiting some of his long-neglected work in the theories of both invariants and partitions in the context of the reading course he was doing in modern algebra with Halsted.[30] In a classic example of the fruitful interplay between teaching and research, Sylvester had dived back into invariant theory after a decade-long hiatus at precisely the point where the methods of the British school he and Cayley had animated had fallen short, in contrast to those of the German school led by Alfred Clebsch and Paul Gordan.[31]

In 1868, Gordan had proven, contrary to the conclusion Cayley had drawn in 1856 in his "Second Memoir upon Quantics," that a minimum generating set for a binary quintic is finite. (Cayley had mistakenly stated in the "Second Memoir" that the number of covariants in a minimum generating set of a binary quintic was infinite.) This result had both established Gordan's reputation and called into question the techniques of the British school. It thus became incumbent upon the British to vindicate their techniques by showing that they, too, were strong enough to provide a proof of Gordan's theorem. Moreover, because Gordan's methods did not lend themselves to efficient, direct computation of minimum generating sets—the prime objective of the British school—the British were still very much in the game.

Cayley had begun the rehabilitation process in 1871 with his paper, "A Ninth Memoir on Quantics." There, he used Gordan's theorem in the context of the British approach to calculate explicitly the covariants in a minimum generating set of a binary quintic. In that paper, too, he expressed his "hope that a more simple proof of Professor Gordan's theorem will be obtained—a theorem the importance of which, in reference to the whole theory of forms, it is impossible to estimate too highly."[32] Sylvester, largely out of the mathematical fray in the late 1860s and early 1870s, took up this challenge in earnest beginning in November 1876. The "hit" about which he wrote to Spottiswoode in a series of almost daily letters from 18 through 29 November was a purely combinatorial technique for determining the number of invariants of given degrees in a minimum generating set for a given binary quantic, and it involved precisely the same kind of partition-theoretic results he had obtained in 1858 on the so-called "problem of the virgins."[33]

Typical of Sylvester's working style and of the way of doing mathematics he had detailed in his 1869 response to Huxley, the letters to Spottiswoode reflected an inductive approach to the problem as well as ideas in flux. On 18 November, Sylvester had verified his new combinatorial technique, one involving generating functions, in the cases of the binary quartic, quintic, and sextic. Having gathered these data, that is, having performed these mathematical "experiments,"[34] he had then used them to conjecture a general formula for the associated generating function for any binary quantic.[35] The next day, he had completed "some complex calculations" of the generating function for the binary octavic. These confirmed his conjecture to the point "that there can now be no doubt as to the truth of the theorem."[36]

Still, he continued to calculate. On the twenty-fourth, "sitting in an easy chair" in the "very comfortable" Maryland Club, "after a moderate but well ordered [?] dinner washed down with a beaker of excellent rum-punch followed by a meditative cigar," he tackled the binary septic.[37] The next day he turned to the binary nonic. The combinatorial techniques he had originally derived and presented in 1859 were bearing totally unexpected fruit. "Little did I think," he wrote Spottiswoode, "that when I fell upon the theorem for reducing Compound to Simple Denumeration (in the course of the Lectures on Partitions of Numbers at King's College where I was honored by your attendance) that I held in my hand the Key to the unlocking of the deepest mysteries of the New Algebra!"[38] By the twenty-sixth and twenty-ninth, however, he was revising some of his earlier claims about the nonic and the octavic, respectively; he had made some errors in those "complex calculations," but they affected only the details—and not the substance—of his new and exciting method.[39]

This research success occasioned an announcement to the *Comptes rendus* of the Paris Academy in January 1877.[40] As with the work on Newton's rule back in the mid-1860s, Sylvester exercised his right as a foreign corresponding member of the Academy to open access to its proceedings. By presenting his new ideas to the Continental audience there, he staked his claim on the results in a venue noted for its speed of publication. The fuller exposition could follow later, and, in this case, Sylvester told Spottiswoode, he "hope[d] to have an article on the new theory in the first number of our projected Mathematical Journal 'The American Quarterly Journal of Pure and Applied Mathematics.'"[41]

Although that hope was not realized (the first number of the new journal did not appear until 1878), Sylvester bombarded the *Comptes rendus* with papers in 1877 and 1878.[42] It was even more critical than usual that his new work get out quickly and that it be recognized by the Europeans. Nothing less than England's national honor was at stake, and Sylvester minced no words in private on this score. "I see a splendid vista of investigations open to me on this subject," he told Spottiswoode, "destined I believe to reduce to annihilation all that the school of Clebsch and Gordan, by aid of methods borrowed by the Germans without acknowledgement from Cayley and myself, have attempted in this subject."[43] In his early sixties, Sylvester, confidence renewed, was definitely back in mathematical form. The Hopkins environment with its emphasis on research, publication, and the training of future researchers had nurtured him. There, he was "treated with confidence and *kindness*" as he had never really been before.[44]

Sylvester gave public expression to these feelings of regeneration as well as to his impressions of the new university in a speech delivered at Hopkins's Commemoration Day festivities on 22 February 1877. Standing in Hopkins Hall before an audience of colleagues, students, and assembled dignitaries, he first thanked them for "the many proofs of kind and generous feeling which, both within and without the walls of this University, have been so widely and unequivocally accorded me."[45] His first impression, back in the late spring and early summer, of the Baltimoreans as less than hospitable had clearly been wrong.

Moreover, the University itself—its working environment and students—completely diverged from all of his prior experiences and totally surpassed his expectations. Never before had he taught students who wanted to be in his classroom simply for the sake of learning. Never before had he been given free rein to pursue his own researches as part of his academic mission. Never before had his teaching and his research been mutually reinforcing. In short, never before had his mission as a professor been his ideal of the professional mathematician: to produce original research and to train future researchers. Johns Hopkins was unique on American shores in its adoption of these German ideals.[46]

Hopkins was also remarkable, in Sylvester's view, for its nonsectarian principles. Unlike the ancient English universities, which "until so late a period" have been "the monopoly of a party and the appanage of a sect,"[47] Hopkins closed no one out on the basis of religion.[48] "If I speak with some warmth on this subject," he told his audience, "it is because it is one that comes home to me—because I feel what irreparable loss of facilities for domestic and foreign study, for full mental development and the growth of productive power, I have suffered, what opportunities for usefulness been cut off from, under the effect of this oppressive monopoly, this baneful system of protection of such old standing and inveterate tenacity of existence."[49] Sylvester thus used the Hopkins podium to decry the religious prejudice that he, as a Jew, had suffered both as a student at Cambridge and in his efforts to secure a position and to make a career for himself in England. Until 1871, he could not have held a post at Oxbridge; in 1872, his Jewish heritage had been a factor in his race for the London School Board; in 1877, in Baltimore and at the Johns Hopkins University, he felt no religious disability whatsoever. "Happy the young men gathered under our wing, who, unfettered and untrammeled by any other test than that of diligence and attainments, have here afforded to them an opportunity of filling up a complete scheme of education," he intoned.[50] If only it had been like that for him back in England in the 1830s. The Johns Hopkins University truly offered unprecedented opportunities for those associated with it, and the new Professor of Mathematics was no exception.

Thriving in the Graduate Classroom

Just three days before he delivered his Commemoration Day address, Sylvester had begun teaching his first graduate-level course. Entitled "Determinants and Higher Algebra" and ostensibly drawn from Salmon's *Lessons Introductory to the Modern Higher Algebra*, the course, in fact, should have been called "My Current Researches." Meeting with seven auditors twice a week on Tuesdays and Fridays at noon, Sylvester aimed to introduce not only some of his newly discovered invariant-theoretic wonders but also, he hoped, new results inspired by his lecturing. This was surely not a course in any conventional sense. The students did not work through the text or even a series of readings or circulated notes. Instead, in class, they sat and listened as Sylvester, "with his white beard and few locks of gray hair, his forehead wrinkled over," presented his latest thoughts at the board.[51] In some cases, he went over calculations that he had written out previously and that he knew to be correct; in others, he pushed through ideas that he believed should yield new results, that he believed (but did not know) were in the "right" direction; in still others, he challenged his listeners to fill in the gaps in his arguments, to work out calculations he had glossed over or omitted altogether, to *do* mathematics themselves.

After class, Sylvester's students had to try to make sense of what they had just witnessed. They had to seek out books either at the Peabody or in their own "mathematical seminarium," a book- and journal-lined study room where they could meet and work; they had to put pen to paper and interpret their notes; they had to figure out how to prove those points that had been left for them. Sylvester threw them into the deep end of the pool, and they were forced either to sink or swim. Of that entering class, neither Hudgins nor Smith survived the spring semester; Gore, Hering, and Preston remained one more year but left without degrees; Craig, Halsted, and Franklin all ultimately made it to the Ph.D. Sylvester's teaching style was clearly not for everyone, but those with talent and drive found themselves in a kind of mathematical laboratory in which experiments conducted sometimes succeeded and sometimes failed, but in which new mathematics was being created.[52]

In the spring semester of 1877, Sylvester took his students on an inductive tour of invariant theory centering on the question of minimum generating sets—the number of elements in them and the exact form of those elements. He first presented the combinatorial work from the late fall on the number of *in*variants in minimum generating sets of binary quantics of degrees up to eight. This very naturally extended, however, to the questions: how many *co*variants are there in a minimum generating set of single binary quantics of successive degrees? what are they explicitly? and what about the number and explicit form of the in- and covariants in generating sets for *systems* of binary quantics? By 9 April, Sylvester was reporting to Spottiswoode on the relatively systematic, inductive assault he had been conducting in his class on these questions, and he had sent him a printed copy of his Commemoration Day address that included an appendix outlining his method for determining the in- and covariants of any given degree and order in a minimum generating set of a binary sextic.[53] Proceeding case by case, Sylvester had also worked out the in- and covariants in minimum generating sets of a binary quintic, a binary octavic, and those of the system of a binary quadratic and binary quartic. So confident was he in his new method that he further stated his intention "to address a letter to you officially asking for a grant from the Royal Society to enable me to extend these tables as the computations are very laborious and one *single* error makes the whole result worthless. I aspire to get the Fundamental Table for 3 cubics, for 3 quartics, for 3 quadratics, for a Quarto-Hextic and so on."[54] The calculational vistas were literally endless. As George Salmon put it, Sylvester "had rehabilitated Cayley's method," a method that had "naturally [been] dismissed in disgrace after it seemed to have broken down in [Cayley's] hands."[55]

Key to this resurrection of the British techniques were the students in Sylvester's class in the spring of 1877. He put them to work, and especially Fabian Franklin, as human calculators, working out these various special cases of ever-increasing calculational complexity,[56] and they uncovered several mistakes in the conclusions of the German school.[57] These successes piqued both Sylvester's competitiveness and his sense of national rivalry. In one of his papers in the *Comptes rendus*, he threw down the gauntlet, openly challenging "the disciples of M. Gordan, so numerous and so widely disseminated in Germany, Italy and elsewhere, to work together to do the work necessary to confirm or deny, by his method, the enumeration that I have recently published in the *Comptes rendus*."[58] Sylvester's students thus experienced not only the satisfaction of the successful mathematical hunt but also the emotion that could accompany the research enterprise.

They also must have felt the electricity of history in the making.[59] Already in his letter of 9 April to Spottiswoode, Sylvester had made a dramatic claim, which if true, would not only have rehabilitated the British methods but would also have largely supplanted the German techniques. On the basis of his limited experimental evidence, which did match up with known results, Sylvester declared to Spottiswoode that "I have at last completely solved this great problem of finding the Basic In- and Covariants for Binary Quantics and Systems of Binary Quantics."[60] Drawing from just a handful of examples of single binary quantics and one system of such quantics, he claimed to have solved the problem in general—and in a purely constructive, algebraic way—in the case of a form or system of forms in two variables.

Sylvester's correspondence that April was full of this claim. He wrote to John Tyndall on 20 April, explaining that he had sent an account of his finding the day before to the Paris Academy and describing how the result had stemmed from "the leisure and stimulus to work" that he "owe[d] entirely to the Johns Hopkins University."[61] Three days later Cayley received the news, although now there was a caveat. Sylvester had come to realize "that *anterior to all* verification this method *could not give* superfluous forms—but it is metaphysically conceivable that it might give *too few* grundformen."[62] In other words, in the three intervening days, Sylvester had recognized that the method still had some wrinkles that needed ironing out, but it nevertheless gave "a sure means of proceeding *step by step* from the lowest forms to higher ones until all were exhausted."[63]

By the end of the month, Sylvester thought he had even more. He dashed off a telegram to Hermite, asking him to "inform the Academy that, since my last communication, I have resolved the problem of finding the complete groundforms for any form in *n* variables."[64] This stunning claim—the solution of the problem of finding minimum generating sets not just for *binary* forms as Gordan had done but in total generality for *n-ary* forms—immediately prompted a letter to Sylvester from Camille Jordan. Jordan, who had tackled the problem of ternary forms without great success in 1876, implored Sylvester to publish a complete and detailed account of his method as soon as possible.[65] Jordan never got his wish. Sylvester did not have this most general result, and had he, it would have been a constructive proof in 1877 of the celebrated and ground-breaking finite basis theorem that David Hilbert would only prove in 1893.[66] It had been a frenzied April in which weeks had "passed like a shadow," and his students had witnessed every twist and turn.[67]

The month of May found Sylvester finishing up his first semester of true graduate-level teaching and writing up a couple more notes for the *Comptes rendus* on his spring's researches.[68] It had been an excellent term, and he was pleased with his students and their performance. In his final piece of departmental administration before leaving for a summer in England, Sylvester reported on the fellows and made recommendations for fellowship candidates for the 1877–1878 academic year. He recommended Craig and Halsted for renewal without qualification, although he was less sure about Gore, who, although hard working, seemed less talented. Among the new applicants, Franklin received an unequivocal endorsement with Sylvester "consider[ing] him to be peculiarly well fitted to become a Fellow of this University" and predicting "that his connexion therewith will be likely to be productive of benefit both to himself and to the University."[69]

His report off, Sylvester boarded a steamer back across the Atlantic. He planned to take "an Invariantive tour in England, France, and Germany," but, ultimately, the summer proved unsettled.[70] He was literally between two worlds. Back in the United States, his new scientific compatriots recognized him on 20 July as one of *their* élite by bestowing on him one of their highest honors, election to America's oldest scientific society, the American Philosophical Society in Philadelphia, while he was busy participating in the London scientific scene as one of its élite. Neither permanently in England nor permanently away from Baltimore, Sylvester felt torn and had trouble concentrating on his research.[71] Still, he persevered, and just before classes started in October, he was able to report to Cayley that he had "discovered *at last*" some of the combinatorial results he had been pursuing relative to his work on the theory of binary forms.[72] As before, he both wrote them up for quick exposure in the *Comptes rendus* and used the new work to begin his class on modern algebra.[73]

That fall, Sylvester worked with five students. Craig, Franklin, Gore, and Halsted were joined by newcomer Irving Stringham, an 1877 Harvard graduate. The classroom "formula" that had seemed so successful in the spring was back in place; Sylvester presented his evolving work from meeting to meeting. Once again, this proved to be a fruitful strategy for Sylvester's research, for just two weeks into the semester, he was writing to Newcomb telling him that "the new theory gives me constant occupation and takes almost exclusive occupation of my thoughts—there is so much to be done on it both in the way of extension and of filling up."[74] Sylvester could not have known how truly prophetic those words were. By 6 November, he had, in fact, "filled up" another major gap in the British approach to invariant theory.

In his "Second Memoir" of 1856, Cayley had actually dealt with two separate but related questions about a given binary quantic: first, and the question that had been engaging Sylvester and his students in the spring and fall semesters of 1877, what are the elements in—and so how large is—a minimum generating set, and, second, what is the maximal number of linearly independent covariants of any given degree and order?[75] Whereas Cayley's conclusion relative to the first question in the case of the binary quintic had been proven wrong in 1868, his work on the second, the so-called "counting problem," had seemingly stood the test of time.[76] In 1876, however, a new book by the Italian mathematician, Francesco Faà di Bruno, called even that part of Cayley's analysis into doubt.[77]

George Salmon had tipped Sylvester off to Faà di Bruno's book as early as April 1877 in response to Sylvester's request for information on the latest publications dealing with the issue of minimum generating sets.[78] Sylvester, notoriously bad both at keeping up with and at reading the current literature, did not become aware of the book's contents until Halsted questioned him on the aspersions cast there on Cayley's counting theorem.[79] Faà di Bruno claimed that Gordan's 1868 theorem implied that a certain set of equations assumed to be linearly independent by Cayley in his proof of the counting theorem were, in fact, not always so.[80] All Sylvester could say to Halsted at the time was that "the extrinsic evidence in support of the independence of the equations which had been impugned rendered it in my mind as certain as any fact in nature could be, but that to reduce it to an exact demonstration transcended . . . the powers of human understanding."[81] By early November, however, he reported jubilantly to Cayley that he had discovered "a *rigorous proof*"[82] of the theorem that not only completely settled the matter of linear independence but also was "so simple that I shall be able to give it to my class."[83] On 15 November, in fact, his students found the following note in Sylvester's hand posted near their room: "The attendance of all members of the class of Higher Algebra is especially requested on Friday the 16th at one o'clock as Prof[r] Sylvester will on that occasion give his newly-discovered Proof of the hitherto Undemonstrated Fundamental Theorem of Invariants."[84] Once again, the interplay between teaching and research had yielded a major result.

Sylvester continued to mine this vein of invariant theory in his graduate class in the spring of 1878,[85] but this research no longer occupied his thoughts as exclusively as it had in the fall. By January 1878, a new publication avenue was opening up that would significantly affect both Sylvester and research-level mathematics in the United States.

Launching the *American Journal of Mathematics*

In May 1877 as Sylvester was hard at work writing up the last of the results from that most productive first spring in Baltimore, other duties of his professorship had called. Gilman was anxious to get the groundwork laid for the proposed mathematics journal and had been trying to hammer out some of the business arrangements with the Trustees. In particular, they had to decide just what it would mean for the University to underwrite the publication. Because they quite naturally wanted to minimize their risk, they had proposed that Sylvester assume the financial responsibility for the proposed venture, in the sense that he would be responsible for assuring its financial solvency through adequate subscription.

Reporting on these discussions, Gilman had sounded Sylvester out. He wanted to get a clear sense both of Sylvester's thoughts on the business aspects of the journal and of his conception of his duties as editor. In a letter to Gilman on 9 May 1877, Sylvester made his thoughts crystal clear. "My desire," he stated, "is to be as useful as I possibly can to the undertaking *quâ* mathematician and to have my mind undisturbed by being mixed up in any way with [the journal's] financial arrangements."[86] He explained that in the case of the *Quarterly Journal of Pure and Applied Mathematics*, the journal he had actively edited since 1855, the publisher "carries on the Journal without any guarantee or subvention entirely at his own risk and has probably not a much larger public to appeal to than that which our projected journal will command."[87] In his experience, what the Trustees had in mind from a financial point of view was "altogether unusual on the part of a scientific Editor," and he would have nothing whatsoever to do with it.[88]

Diplomatically adept, Gilman navigated the stormy waters that had whipped up between Sylvester and the Trustees, and matters were settled in Sylvester's favor. The Trustees would subsidize the publication of each volume[89]; Sylvester, as editor-in-chief, would be the scientific editor entrusted with the journal's contents; William Story would be engaged as the associate editor in charge of seeing each number of the journal into and through production. Moreover, there would be three cooperating editors in the fields of mechanics, astronomy, and physics, namely, Benjamin Peirce at Harvard, Simon Newcomb at the Naval Observatory in Washington, and Henry Rowland, Sylvester's Hopkins counterpart in physics, respectively. It was a strong editorial team, and they hoped to put out their first number early in 1878.

By mid-January, the first number was in press, but printing problems delayed its appearance until March. When it finally did come out, it reflected the concerted efforts of the editorial team and contained an impressive array of papers: Newcomb on a class of transformations of surfaces in space of more than three dimensions, George William Hill of the U.S. Nautical Almanac Office on lunar theory, Henry Eddy of the University of Cincinnati on a theorem of three moments, Guido Weichold of Saxony on solving cubic equations, Cayley on group theory, Rowland on the theory of electric absorption, Charles Peirce on the method of least squares as propounded by the head of the geodetic division of the Italian Survey, and Sylvester on an application of the atomic theory of chemistry to the theory of invariants. Over a hundred quarto pages in length, the first number was true to the philosophy that the editors had laid out in their prefatory "Notice to the Reader." The journal carried contributions from abroad alongside those from American mathematical scientists, and it included at least one account of difficult-to-obtain foreign work.[90] The first number also fulfilled one of the journal's other express purposes. It served as a publication outlet for mathematical research issuing from the Johns Hopkins University.

Sylvester opted not to publish any of his new invariant-theoretic work in the *American Journal*'s inaugural number. In light of the timing, it might have given the new journal quite a bit of cachet to have carried his (the first complete) proof of Cayley's counting theorem, but he offered that instead to the *Philosophical Magazine* back in England. For the *American Journal,* Sylvester worked up some ideas he had presented in the fall of 1877 before the Johns Hopkins Scientific Association, a group of practicing scientists and scientists-in-training that met monthly to hear about new discoveries in mathematics and the physical sciences. In agreeing to speak, Sylvester faced more than the perennial problem of the mathematician, namely, how to communicate technical mathematical advances to a nonmathematical audience. Simple and rough analogies might suffice to convey a certain flavor of mathematical research to the layperson, but an audience of scientific peers would require more. Sylvester wanted to talk about invariant theory, but how could he drive home its import unencumbered by masses of calculations and by the highly technical jargon he had done so much to create?

The answer to this conundrum came to him one night as he tossed in his bed.[91] He perceived a tight analogy between the atomic theory of bonds and valences as developed particularly by the English chemist, Edward Frankland, and the combinatorial aspects of the algebraic theory of invariants. In short, Sylvester identified the invariants associated with any given system of binary forms with a "saturated system of atoms in which the rays of all the atoms are connected into bonds."[92] Thus, chemical graphs of molecules in terms of their component atoms and bonds provided a way of actually visualizing those mathematical abstractions—invariants—associated with systems of binary quantics. Moreover, the combinatorial properties of the chemical graphs translated into combinatorial properties of invariants and vice versa. It was a "chemistrization" of invariant theory at the same time that it was a mathematization of the atomic theory and, in keeping with Sylvester's broader positivistic philosophy of science, represented a welcome synthesis of the two seemingly disparate intellectual realms of mathematics and chemistry. As he explained in a letter to Newcomb, "I have also ventured upon some purely chemical suggestions which seem to me to make the 'Bond' or 'Valence' theory more consistent with itself— without affecting in any way the existing chemical constructions."[93] Still, he had to admit that he felt "anxious" about the work's reception, because "it will be thought by many strained and over-fanciful. It is more a 'reverie' than a regular mathematical paper."[94] It was, indeed, an odd choice for his first paper in the new research-oriented journal, but perhaps he thought that mathematicians in America, like his scientific audience at Hopkins, needed a less technical, more visual introduction to the intricacies of invariant theory.

Sylvester's paper did serve one key purpose for the new journal: publicity. From England, the chemist, Edward Frankland, and the mathematician, William Clifford, wrote to Sylvester about his ideas, and extracts of their letters appeared in subsequent numbers of the journal's first volume.[95] From Saxony, Guido Weichold contributed a note on some "Historical Data Concerning the Discovery of the Law of Valence," and from the University of Virginia, the chemist, John Mallet, argued that one of the chemical insights Sylvester only mentioned in passing "deserved careful consideration and more prominence than [Sylvester had] deemed it worthy of."[96] Finally, from the Johns Hopkins University, Sylvester's student Franklin picked up on some of the paper's combinatorial aspects and gave a partition-theoretic interpretation of them in "On a Problem of Isomerism."[97] Two things were clear. The new journal was being read in the United States, in England, and on the Continent. It was also sparking new research from the mathematical team at Hopkins.[98] These had been among the explicit goals of Gilman and the editorial staff. It was an auspicious start.

The journal, however, had at least one *implicit* goal. As Joseph Henry had noted in his diary after the Sylvester dinner on 3 November 1876, it aimed to prompt Sylvester "to finish and publish the results of his previous investigations not fully completed and which have not yet been given to the world."[99] By 1878 and 1879, that meant his systematic work on calculating minimum generating sets of binary quantics of successive degrees, only part of which had seen the light of day in the *Comptes rendus*. Sylvester's next substantive contribution to the journal appeared in the fourth number of volume one and was in precisely this vein. There, he gave "A Synoptical Table of the Irreducible Invariants and Covariants to a Binary Quintic" that foreshadowed a string of papers in 1879.[100] This work reached its climax in 1879 in "Tables of the Generating Functions and Groundforms for the Binary Quantics of the First Ten Orders," a paper completed with Franklin's expert calculational help and published in the *American Journal*.[101]

Old habits proved hard to break, however. Just as before the advent of the *American Journal*, Sylvester announced many of his results first in the *Comptes rendus* and even gave a lengthy paper on the technical underpinnings of the calculations to Carl Borchardt for the *Journal für die reine und angewandte Mathematik*.[102] Sylvester apparently felt that the *American Journal* was not yet well enough known or widely enough read to get the word out effectively and efficiently to the international mathematical audience.

Sylvester continued to publish invariant-theoretic work in the *American Journal* intermittently through 1882, pushing the calculations begun in the 1877–1878 academic year to systems of pairs of binary quantics of the first four orders in 1879, to the binary duodecimic, again with Franklin's help and supported by a grant from the BAAS, in 1881, to selected systems of triples of binary quantics, with the help of two new graduate students, William Durfee and George Ely, in 1882.[103] This later work, however, was done largely outside the context of Sylvester's graduate classroom as he shifted in the 1879–1880 academic year to mathematical topics other than invariant theory for his lecture courses (see chapter 10). Still, the case of Sylvester, invariant theory in the graduate classroom, and the launching of the *American Journal* underscores the wisdom of Gilman's three-pronged strategy of simultaneously supporting teaching, research, and the publication of original work.

Sustaining the *American Journal of Mathematics*

Sylvester, however, could not sustain the research content of the new publication singlehandedly, even though his contributions were critical to its success. As the journal's editor-in-chief, he also had to work to secure high-quality contributions and, in general, to talk up the venture in the right circles. A lot depended on his international connections; a lot depended on his international reputation. Sylvester used both to good advantage for the University and its *Journal.*

Back in England after the end of the busy but melancholy spring semester in 1878—one of his oldest American friends, Joseph Henry, had died on 13 May— Sylvester was gratified to learn firsthand of the reception of the journal's first number. Writing to Gilman on 7 June, he announced with pride and some hyperbole that "all the world speaks admiringly of the Journal. There appears to be a brisk demand for it in Germany which is a very good symptom of its possessing some sterling claim to regard."[104] This favorable buzz could only help in his efforts to line up submissions from mathematicians in Britain and on the Continent.

His summer travels took him first to Cambridge for a short visit with Cayley, then to France for several weeks from mid-July to early-August to meet up with mathematical friends and to view the latest art on display at the French Exposition, and finally to Dublin for the meeting of the BAAS in mid-August. By early September, he was able to report that he had managed to secure "the promise of a valuable paper from Hermite" as well as one from Rudolf Lipschitz at the University of Bonn, and he had a paper in hand from Clifford in addition to further commitments from him.[105] Although the papers by Lipschitz and Clifford appeared in later numbers of the first volume, neither Hermite's "valuable paper" nor further contributions from Clifford were forthcoming.[106] Still, papers and notes did come in from Cayley, Thomas Muir, and James Hammond in Britain, from Édouard Lucas in France, and from several American contributors in addition to Hopkins students, Craig and Franklin. As projected in the opening "Notice to the Reader," the first volume of the *American Journal* did, indeed, contain about 384 quarto pages. It exceeded that expectation, in fact, by a full three pages.

As Sylvester plunged headlong into work on the journal's second volume early in 1879, as he taught his graduate course in invariant theory, as he generated new invariant-theoretic results, he was also moved to return to an old passion: poetry. His new routine had energized him; he was confident in his renewed mathematical powers; he once again felt the tug of the muse. In February 1879, he gave a public reading in the hall of the Peabody Institute of his poem "Rosalind." Originally composed in March 1876 in the afterglow of his successful negotiations with Gilman for the Hopkins professorship, "Rosalind" was what Sylvester termed a "study in monochrome," that is, a poem in which every line sounded the same rhyme.[107] In the case of "Rosalind" that monochromatic rhyme was with the name of its title Shakespearean character, with the short or the long sound of the vowel "i" allowed. On Tuesday, 18 February, Sylvester mounted the podium before a packed crowd that "expected to find much interest or amusement in listening to this unique experiment in verse."[108]

The crowd knew Sylvester. It found "the great bald dome rising out of an abundant halo of gray hair, which stood out about its equator like the rings on Saturn, the impetuous delivery, that magnetism which attends 'the gentleness of greatness,' and the delicious confusion of manuscript which was almost certain to occur before he got through . . . irresistible."[109] They were not disappointed. Instead of hearing a reading of the actual poem, they heard a Sylvester completely enthralled in the numerous glosses he had made to his lines of verse; he insisted on reading and elaborating on those before actually getting to the business of reciting the poem. His student, Franklin was in the audience and later described the unfolding scene this way: "When he had dispatched the last of the notes, he looked up at the clock, and was horrified to find that he had kept the audience an hour and a half before beginning to read the poem they had all come to hear. The astonishment on his face was answered by a burst of good-humored laughter from the audience; and then, after begging all his hearers to feel at perfect liberty to leave if they had engagements, he read the Rosalind poem."[110] It had been a vintage Sylvesterian performance!

On a bright spring morning a month later, a chance observation of a pair of ladies in conversation on the sidewalks of Charles Street prompted another flight of monochromatic fancy, this one entitled "Spring's Début" and composed of more than 200 lines all rhyming with "een" or "in."[111] Sylvester had himself been walking, in conversation with Robert Garrett, son of John Garrett and heir to the family's fortune in the Baltimore and Ohio Railroad, when the two women caught his eye. As Sylvester described it, the poem "evolved out of an improvised epigram which, as he wended his way home that morning, formed itself in the Author's mind, intoxicated with the bright sun shining over head, the balmy air, the song of the birds and the new-come-out virgin Spring just beginning to peep over Old Father Winter's reverend shoulder."[112] Sylvester was clearly in fine form and in high spirits in the spring of 1879. Baltimore had become a comfortable home; his research was going well; the journal was up and running.

Work on the second volume was already well under way when Sylvester returned to England for the summer of 1879. Cayley continued to write up short notes and full-fledged papers at Sylvester's urging. Sylvester tapped his personal correspondence for material from, for example, Morgan Crofton, his former protégé and later successor at the Royal Military Academy. Other of Sylvester's mathematical correspondents—like Julius Petersen in Denmark and Lucas in France—sent in the fruits of their research for publication. But Sylvester also had his ear to the ground for new results and managed to snare Alfred Kempe's startling proof of "the theorem long known to mapmakers but never yet demonstrated that 4 colours are sufficient for marking off the determinations of different countries."[113] The publication of that particular work brought to light some of the difficulties inherent in the division of labor between Sylvester as the journal's scientific editor and Story as its business manager.

Kempe's paper arrived that summer while Sylvester was still in England. Thus, it fell to Story to see it into galleys and successfully into print. Reading the paper carefully, however, Story realized that there was a gap in the reasoning. Kempe had reduced the problem to an ingenious analysis of various cases, but Story recognized the possibility of other cases. He was in a bind. Since Sylvester had already accepted the paper for publication, it had to appear, but, in all good conscience, Story could not let the gap go unnoted. With production deadlines looming and with Kempe's paper as the lead article in the third number of volume two, Story exercised his prerogative as managing editor and decided to add his own "Note on the Preceding Paper." There, he pointed out some additional cases and tried—although unsuccessfully—to patch the proof.[114] When Sylvester saw what Story had done, he was livid, accusing Story of "*mala fides . . .* in his treatment of Mr. Kempe's valuable memoir."[115] Things only went from bad to worse the following year, again while Sylvester was in England for the summer.

Sylvester left Baltimore early in May 1880, having been given permission by the Hopkins Trustees to leave even earlier than in previous years in order "to attend the meetings of the learned societies and meet the Savants before they are on the wing for their summer vacation."[116] He needed to make contacts and round up new research for the *Journal.* Busy at work while on board the steamer, he posted an addition to his paper "On Certain Ternary Cubic-Form

Equations" when the boat put in in Queenstown, Ireland. By late July, he was corresponding with Gilman about the journal and about what he viewed as Story's irresponsible behavior. After writing to Story for acknowledgment of receipt of the addendum and after repeated requests for information on when the second number of volume three would appear, Sylvester had received little satisfaction. Fuming to Gilman on 22 July, he asked, "if he treats me this way, how is he likely to act towards other contributors?"[117] But it was not just a matter of neglected correspondence. Sylvester also complained that Story had "allowed Rowland to exceed the limits of the journal by 20 pages in flat disobedience to my directions and without referring the matter to me for my opinion" and, in so doing, had caused a lengthy and unnecessary delay in the appearance of the first number.[118] "Every one (persons of the highest position that I could name) says that this delay and irregularity are doing *immense injury* to the Journal," he sniped. Under these conditions, Sylvester had concluded "that it is inexpedient that [he and Story] should continue to act together in carrying on the Journal and as I am primarily responsible to the Public, to the Trustees and the World of Science for its success I formally request that arrangements may be made for dissolving the present connection of Story with the Journal and myself as I can no longer work satisfactorily with or feel any confidence in him—for I consider that his conduct has proved him to be wanting in loyalty and trustworthiness" (196).

This was really the rub. Sylvester, as editor-in-chief, was naturally assumed to know what was going on, but in light of his understanding of his role as scientific editor, he had been content largely to ignore the production side of things and to let Story shoulder those burdens. Now, when things were going wrong, and he was facing questions from his friends in England, he was in the dark precisely because Story was doing his best to maintain the division of labor upon which Sylvester had so strongly insisted.

Another letter to Gilman followed two days later in which Sylvester, still angry, reiterated his complaints against Story. Gilman, on holiday himself in Princeton, Massachusetts, worked to get to the bottom of things. After learning of the problems Story had been encountering in trying to get the journal out, he counseled him to send Sylvester "a frank explanation . . . of the serious difficulties you have encountered & an apology for any delay on your part to answer his telegram & letter." He added that "I should be truly sorry to have you lose his confidence and good will—& I think they are possessions which you will not lightly forfeit."[119]

Although Gilman's diplomacy prevailed in the short term, by February 1881 the situation had come to a head. On Thursday, 17 February, Sylvester stormed into Gilman's office and railed against Story. Two days later, Gilman reprimanded Sylvester in no uncertain terms. "I have always intended to treat you with the respect due to an honored colleague," he wrote, "but I must refuse to be again exposed to such a scene as occurred in my office on Thursday for no business can be transacted wisely when either party is excited." He continued by laying the cruel truth of the situation out on the table:

> I think it may appear when I look into the matter that you were absent from the country some weeks before the close of the last academic year & did not return until some time after the beginning of this academic year & that you were also absent during the progress of the recent examinations, so that Dr. Story in his desire to sustain the order of the university assumed responsibilities which would otherwise have fallen on you. . . . I must call your attention to the fact that during that period you were absent from your post for more than a week at a distance from Baltimore & that neither before you left nor after you returned did you say anything to me in respect to your absence.[120]

It was a strong rebuke, but Sylvester had let his temper get the best of him, and that was unacceptable. In the end, Sylvester and Story remained on cordial professional terms, but Story stepped down as associate editor in charge of the journal.

A number of other changes went into effect as a result of this flap. As Sylvester explained in a letter to Cayley that June, the University assumed responsibility for all of the business details as well as for the correspondence associated with the journal, thus lessening the clerical burden on the editorial staff. Moreover, Fabian Franklin, who had earned his Ph.D. in the spring of 1880 and who had been kept on as an Assistant in Mathematics, was lined up as "the (*anonymous*) assistant Editor" charged with keeping the journal running day by day and especially during Sylvester's annual summer absences.[121] Finally, and this had been at the root of the problems in 1880, the printing of the journal would henceforth be done not in Boston but in Baltimore, where communication between the printers and the editors could effectively be instantaneous. Gilman's hand is clearly visible in these solutions, since they required a kind of shuttle diplomacy between the Trustees and the Department of Mathematics. He had successfully averted the crisis.

As this account of the *American Journal*'s early years makes clear, it was not easy to sustain a research-level journal in mathematics even under the best of circumstances. First, there was the matter of personalities. Sylvester's limited sense of his obligations to the journal undoubtedly created problems. He wanted to be perceived as "in charge" without taking on the full responsibilities that would have entailed. He happily left the dirty work to Story, and that worked well enough as long as Sylvester was in residence and could be consulted. But when hard decisions had to be made while he was away, he seemed never to agree with them and complained bitterly that it was *his* reputation that was at stake when things were seen to go awry. There is no denying that Sylvester could be a difficult personality, especially when things did not go his way, but in all fairness, he had not really wanted to take on the responsibilities of a journal when he arrived at Hopkins, even though he had ultimately agreed to do it.

There was also the matter of the journal itself. The finances caused headaches; the mechanics of production presented difficulties; the scientific content had to be maintained at a high level and in sufficient amounts. Sylvester worked hard at his own research and filled many pages of those early numbers. His students and colleagues at Hopkins also consistently supplied quality material. In fact, almost half of the work published in the first six volumes of the *American Journal,* under Sylvester as editor-in-chief, came from those who were or had been associated with the Johns Hopkins University. Another quarter came from abroad and largely from Sylvester's many foreign contacts and summertime efforts, and the final quarter came from the mathematical scientists of the United States. Given this numerical profile, it seems clear that the journal's sustenance depended critically on Sylvester's efforts as a researcher, as a teacher, and as a prominent figure in mathematics internationally, even if his prickly personality sometimes created problems. Still, the *American Journal of Mathematics* owed its ultimate success not only to Sylvester but also to Story, to the Johns Hopkins University, and to the emerging community of research-level mathematicians in the United States. It was by no means a one-man show.

The journal, however, was not Sylvester's only or even his main concern at Hopkins. Teaching, animating a graduate-level program in mathematics, training future mathematical researchers, these were his primary missions. He had never taught at this level before, but the first, experimental year had gone well enough, and the second and third years had seen his teaching inspire even more new results. Yet these three years had, in some sense, been more about him than about his students as he strove both to settle in to new routines and to rekindle and sustain his own research agenda. Craig and Halsted had pursued their research interests in applied mathematics and geometry, respectively, largely independently of Sylvester's work; only Franklin had been drawn into Sylvester's special field of invariant theory. Moreover, Sylvester could not expect to teach invariant theory semester after semester after semester. That familiar territory may have served successfully to reengage him in research at a critical juncture in his career, but now he needed to think more about his students and less about himself. He needed to branch out in order to provide his researchers-in-training with a broader mathematical base. He needed to move on to other things.

CHAPTER 10

Tackling New Challenges
in a Home Away from Home

*I feel that my connexion with the Johns Hopkins University has given
me a new lease o[n] mathematical life and activity.*
—J. J. Sylvester to Sarah Hunt Mills (Mrs. Benjamin) Peirce, 25 March 1880

If research and publication had, in some sense, taken precedence over the training of future reseachers in the Department of Mathematics at Hopkins from the fall of 1876 through the spring of 1879 and the appearance of the first full volume of the *American Journal of Mathematics*, Sylvester began to focus more on the classroom beginning in the 1879–1880 academic year. After all, his own research was on a fast track, and the journal had gotten off to a strong start, so it was time—high time—to turn to the students. In April 1877, George Salmon had asked him, "what security have you that the men you endow will research?"[1] The short answer was "none," but from 1879 on, Sylvester was both teaching a wider variety of courses—number theory, combinatorics, the theory of substitutions, matrix theory—and working more actively to draw all of his students into the material and into the research endeavor. Perhaps the security Salmon wanted would come by engaging the Hopkins students as truer partners in research and not merely as human calculators.

Sylvester and His Early Students

If the numbers had been small in Sylvester's classes during his first two years at Hopkins—eight students in the first year and five in the second, of whom only one was new—four of those five returning students earned their doctorates under Sylvester's guidance. The highly self-motivated Thomas Craig was the first, taking his degree in 1878 for a thesis on "The Representations of One Surface upon Another, and Some Points in the History of the Curvature of

Surfaces." George Halsted followed a year later with work on a "Basis for a Dual Logic," while 1880 saw the completed work of both Irving Stringham on "Regular Figures in n-Dimensional Space" and Fabian Franklin on "Bipunctual Coordinates." These four dissertations—on topics in algebraic geometry in the cases of Craig, Stringham, and Franklin and in logic in the case of Halsted—provide an interesting glimpse into the early years of the Hopkins graduate program.[2] They suggest that mathematics students at Hopkins had the flexibility to explore areas other than those immediately engaging Sylvester. They also reflect the influence of the geometer Story, who, although charged with essentially all the undergraduate teaching, also played a key role in the graduate program. Finally, they indicate that Sylvester's courses in invariant theory were largely failing to draw his students into his own special area of interest. Sylvester may have been providing his students with a sense of the excitement of doing mathematics, but he was not enticing them to delve into invariant theory.

Fabian Franklin was the exception. The geometrical dissertation aside, his research interests primarily lay in invariant theory, where he had already proved instrumental as a human calculator in Sylvester's grand project of determining a minimum generating set of covariants for binary quantics up to and including degree ten.[3] Sylvester had also lured Franklin into the mathematical aspects of the analogy between chemistry and invariant theory as well as into invariant theory's more purely combinatorial components.[4] True to the objectives of the *American Journal,* moreover, Franklin's research, like that of his fellow students, regularly appeared on the journal's pages.

Franklin and Craig were also exceptional in that they remained at Hopkins after the completion of their doctoral work as members of the teaching staff, each ultimately rising to the rank of full professor.[5] In Franklin's case, this meant continued participation in the algebraic environment Sylvester animated as well as graduate-level teaching in modern algebra, determinants, and even statics and analytic mechanics by way of augmenting the departmental offerings. Researchwise, Franklin continued his work in invariant theory, publishing "On the Calculation of the Generating Functions and Tables of Groundforms for Binary Quantics,"[6] at the same time that he followed Sylvester into new investigations that grew out of Sylvester's lecture courses (see "From Invariant Theory to the Theory of Numbers" and "A Troubled Transition to the Theory of Partitions"). Craig, on the other hand, pursued his separate interests in the theory of functions, differential equations, mechanics, and hydrodynamics largely in the context of his graduate teaching. Like Franklin, he used the *American Journal* as the principal outlet for his new work. In the five years from 1879 through 1884, the journal carried twelve of his papers, ranging from results "On the Motion of an Ellipsoid in a Fluid" to "Some Elliptic Function Formulæ" to research

"On Quadruple Theta-Functions."[7] Franklin and Craig clearly flourished in the mathematical environment centered on Sylvester at Hopkins.

Sylvester's other two early Ph.D.s, Halsted and Stringham, made their ways in the world outside Baltimore. Halsted's case proved the most problematic. With no small ego, the twenty-two-year-old Halsted had sung his own praises in a tightly penned, four-page letter to Gilman as early as 10 April 1876. Halsted opened unctuously by expressing an "overmastering anxiety to be a partaker in [Gilman's] rich feast of learning" and begging Gilman to bear with him while he "hurr[ied] through what to [him] is a most painful process—the enumeration of [his] own claims" on one of the then-recently announced fellowships for Hopkins's opening year.[8] After confessing to a love of algebra from his high school days in Newark, New Jersey, he went on to detail an undergraduate career at Princeton, where he quickly "devoured with the greatest eagerness everything on [mathematics] that fell [his] way." "This naturally brought [him] very soon to the end of all the ordinary American textbooks, and [he] only too gladly took up those splendid works which have made Cambridge and Dublin so proud." Still, he continued, "this tells nothing as to Thoroughness—the Quality of my work—which in mathematics is all-important." No matter! He went on to detail all the first prizes he had taken throughout his undergraduate years and all the additional honors that had been heaped upon him not just in mathematics but in "Rhetoric, Bible, Eng. Literature and Chemistry" as well as in "Logic, Metaphysics, Physiology, Civil Government, Psychology, Crystalography, Oratory, and Physics." In closing, he wrote, "I must humbly crave your pardon for what must seem an exceedingly egotistic and harsh letter." Gilman was impressed, but perhaps not in the way Halsted had intended. Writing in response to a letter on Halsted from his scientific adviser, Simon Newcomb, Gilman confided that "Halsted wrote so to me, 'and more so.' I told him in reply that I could not promise a fellowship,—but I thought it clear he was one of the young men we are in search of, & I begged him to plan to come here some-how, next year."[9]

Halsted clearly did come, actually as a fellow, and his first year went well enough. He had met with Sylvester privately during that first semester and had helped rekindle Sylvester's researches in invariant theory through his many questions. The next year, however, his lack of tact created problems between him and his professor.

In the spring semester, Halsted had planned to give a short course of lectures on logic, a subject in which he had become interested during his undergraduate days at Princeton under the tutelage of James McCosh and on which he was writing his doctoral thesis. He had intended to include one entire lecture on the writings of Charles Peirce, son of Benjamin Peirce, and a rising figure in American intellectual circles.[10] The younger Peirce, in fact, was angling for some sort of position at Johns Hopkins, so this kind of publicity could have been advantageous for his cause.[11] On 14 March 1878, however, Halsted wrote to

Peirce that his plans had changed. As he intemperately put it, "the opinion was expressed here to me that I had better not do it and that you had exhibited in your writings a tendency to undervalue everybody and everything mentioned. Besides this I was particularly discouraged by Prof. Sylvester's adding that your articles in the Popular Science Monthly were pretentious without being at all profound and that any body could have written them."[12] This prompted Peirce to write Sylvester five days later. "I was surprized [sic] to learn from the enclosed letter that you are acting against my being invited to the Johns Hopkins University," Peirce admitted. "I thought you had given me to understand that you would be friendly to me in this matter." "I cannot say that young Mr. Halsted appears to me remarkably distinguished for *savoir-vivre*," he continued. "He is no way entitled to address personal comments to me. Unless you indicated to him your willingness to have your observation repeated to me, his action would seem to have been somewhat indiscreet."[13]

In dismay, Sylvester took the matter—and Halsted's letter enclosed by Peirce—immediately to Gilman. The President wrote to Peirce on 20 March to assure him that Sylvester "in his conversations with me, so far from appearing unfriendly to you . . . has constantly referred to you in terms of high appreciation and has left upon me the abiding impression that in his opinion you would bring to the J. H. University all manner of intellectual powers [?] which would be most valuable."[14] Sylvester was so upset by the incident that 29 March found him writing to Gilman to inform him that he would be leaving for New York City the next day, Saturday, to visit the younger Peirce and would return the following weekend.[15] Gilman's reassurances and Sylvester's visit apparently served to undo the damage done by Halsted's letter; cordial relations between Peirce and Sylvester were restored, at least temporarily.[16] The matter did not end on such a positive note for Halsted, however.

Halsted gave his spring course of lectures on logic, but they were not well received. In particular, his pronouncements on syllogism—the topic to which he was devoting the written thesis he would submit for the doctoral degree later that spring—were deemed incorrect. Halsted had failed his examination for the degree, and Gilman apparently gave Halsted the option of writing a new thesis. On 7 June 1878, Sylvester, writing from London in response to a letter just received from Gilman, allowed that "I do not see how you could have acted otherwise than you have done in the case of Halsted. He too will I hope ultimately himself be benefitted by this temporary check."[17]

September found Halsted in Princeton, not in Baltimore. On the third, he sent Gilman an update on his work: "I have written a new Thesis, retaining the notation I used in my lectures, but not saying a word about the Syllogism, to a Statement and Reduction of which you will remember my last thesis was devoted." He added that "I will send on to you this new Thesis as soon as I receive a line from you to let me know that you have returned to Baltimore and are ready to receive it. Any hint you can give me in regard to my Examination will also be gratefully received."[18]

Six weeks later, Halsted was writing Gilman again, this time to alert the Hopkins authorities to a favorable mention some of his work in logic had received in the *Zeitschrift für Philosophie.* "As you have now somewhat to decide whether my ideas and my claims to be able to found a Dual Logic are valid or not," Halsted baited, "I thought you might desire to have your attention directed to that short notice which you now have on your Library Tables." He closed with one final jab: "Of my article on the syllogism I corrected the proof some time ago and it will appear in the next number of the Journal of Speculative Philosophy, to be judged of by the world at large."[19]

Halsted ultimately received his Hopkins Ph.D. in 1879, but Gilman, not Sylvester, seems to have been the intermediary. The events of the spring of 1878 had seemingly estranged student and professor, but they did not prevent Halsted from finding gainful employment at Princeton as a Tutor in Mathematics from 1879 to 1881 and then as an Instructor in Mathematics there from 1881 to 1894.[20]

Sylvester enjoyed much better relations with Stringham, who arrived in the program's second year of 1877. Although Story's geometrical leanings played the greater role in shaping Stringham's mathematical interests, the young mathematician-in-training faithfully attended Sylvester's graduate class on invariant theory in the 1877–1878 and 1878–1879 academic years and followed him as his graduate teaching branched out the following year into number theory. Unlike Halsted, however, Stringham had success in the series of lectures he gave on his research in the spring of 1880. Between January and May, he spoke twice before the more intimate and mathematically intense Mathematical Seminary and twice before the more general Scientific Association to good effect. He was thus able to write a very solid annual report in which he detailed his coursework as well as his four lectures. "The chief results of these [latter] investigations will be contained in a paper entitled 'Regular Figures in N-Dimensional Space,' to be published in the forthcoming number of the *American Journal of Mathematics,*" he told Gilman proudly.[21]

It was on the basis of his spring performance and this publication that Stringham earned his 1880 Ph.D. He followed this work immediately with a foreign study tour to Leipzig, where Felix Klein had just joined the faculty. In that first year, Klein directed a seminar on geometric function theory, in which Stringham was an active participant. During the winter semester of 1881, in fact, he lectured on the research he had done for his Hopkins Ph.D.[22] After one more year in Germany, Stringham returned to the United States to take up a professorship at the University of California, Berkeley, largely arranged for him by former Berkeley President, Daniel Coit Gilman. Writing to Gilman from Germany late in April 1882 about the recently received offer, he expressed the hope that he would "have the pleasure of meeting Prof. Cayley and Prof. Sylvester in England" that summer.[23] If Sylvester had exerted perhaps less influence on Stringham in three years than Klein had in two, Stringham still had fond regards for his former professor.

Craig, Halsted, Franklin, and Stringham were the first products of the Hopkins experiment in mathematics. They pursued their graduate studies at a time when Sylvester was striving to regain his former mathematical powers, and although they had the opportunity to draw inspiration from his lectures, they were somewhat more on their own mathematically than perhaps they should have been. Their successors, who earned their Ph.D.s in the years from 1882 through 1884, however, encountered at Hopkins a Sylvester with "a new lease o[n] mathematical life and activity."[24] The Sylvester they experienced was much more actively interested in teaching them a fuller range of mathematical topics and in engaging them as true partners in research.

From Invariant Theory to the Theory of Numbers

The fall of 1878 saw a significant upswing in graduate admissions into the Hopkins Department of Mathematics. After admitting only one new student in the fall of 1877, the program had six new faces in 1878. Of these, two—Oscar Mitchell and Christine Ladd—ultimately earned their Ph.D.s; both finished four years later in 1882. Although admissions remained strong—six new students in 1879, four in 1880, seven in 1881, five in 1882, six in 1883—only three more went on to earn Ph.D.s under Sylvester: William Durfee and George Ely in 1883 and Ellery Davis in 1884.[25] The work that these students did both in conjunction with their graduate studies and for their doctoral degrees reflected the broadened scope of the Hopkins program. Mitchell and Ely worked in number theory, Durfee focused on partition theory and its ramifications, Ladd concentrated on logic, and Davis delved into algebraic geometry. Of these later students, Ladd represents a particularly interesting case.

Christine Ladd had graduated from Vassar College in 1869 at the age of twenty-one.[26] Founded in 1861 by businessman, Matthew Vassar, for the higher education of women, the college had opened in 1865, the year before Ladd entered. Despite the fact that her curriculum had included little mathematics, Ladd developed an active interest in the field as she pursued a career in teaching. In the early 1870s, she contributed to both the question and the answer sections of England's *Educational Times* and to *The Analyst*, an American mathematical periodical begun in 1874. She also pursued her formal study of mathematics at males-only Harvard thanks to special arrangements with William Byerly and James Mills Peirce, son of Benjamin and brother of Charles. Perhaps emboldened by her success in gaining access—if only unofficially—to America's oldest institution of higher education, she ventured in March 1878 to seek admission to one of the then newest universities, the trendsetting Johns Hopkins. As she knew full well, though, Hopkins, like Harvard, was not coeducational.

Ladd addressed her inquiry directly to Sylvester, not neglecting to lay out her credentials. "It is my desire," she wrote, "to listen next year to such of your mathematical lectures as I may be able to comprehend. Will you kindly tell me whether the Johns Hopkins University will refuse to permit it on account of my sex? I am a graduate of Vassar College and I have attended mathematical lectures at Harvard University."[27] Sylvester, who like his friend, Barbara Smith Bodichon, had long been sympathetic to the claims of women to higher education, immediately instructed Franklin to do a literature search to see what, if any, publications Ladd may have had.[28] If he was going to make a case for her before Gilman and the Trustees, he needed not only to be certain of her qualifications but also to have put together as compelling a case as possible.

Ladd's letter arrived at the height of Sylvester's worries about Charles Peirce in the wake of Halsted's indiscretion. In fact, Sylvester was on the train to New York to see Peirce three days after Ladd wrote to him. Her case, however, intrigued him. On the train northward, Sylvester happened to run into two of the more liberally minded Hopkins Trustees, Dr. James Carey Thomas and Francis T. King, and he took the opportunity to sound them out on Ladd's petition. Writing to Gilman from New York on 2 April, Sylvester recounted that "they seemed to favor the notion and to be inclined to give her every facility for carrying out her wishes on the subject."[29] As for his own views, Sylvester felt that Ladd "would be a source of additional strength to the University." He added cajolingly that "I cannot but think that with your fertility of resource you would hit upon some plan of utilizing her for the purposes of the University."[30]

Sylvester had set the wheels in motion. Even before his return from New York, the Trustees were discussing not only Ladd's particular case but also the much broader issue of coeducation. They deliberated throughout the month of April; they were profoundly divided. On the twenty-sixth, their decision was conveyed to Ladd through Gilman. She would not be admitted to the University, but she would be permitted to sit in on Sylvester's courses. "As this is an exceptional recognition of your mathematical scholarship," Gilman concluded, "no charge will be made for tuition & your name will not be enrolled on the annual Register."[31] They had given Ladd leave to be an "invisible student" among them.[32] Sylvester and the Trustees who favored coeducation had won the battle but lost the war. Ladd could continue her higher education under Sylvester, but the Johns Hopkins would not be a coeducational institution.

Ladd and Mitchell both entered the program in the fall of 1878, and both participated in Sylvester's year-long course on invariant theory. The next fall, however, Sylvester took his graduate class into new mathematical territory, number theory, while Charles Peirce had finally joined the Hopkins faculty as Lecturer in Logic. Although Mitchell supplemented his purely mathematical work with courses in logic from Peirce, Sylvester lured him into number theory, the subject that ultimately inspired his dissertation research.[33] Ladd, on the other hand, continued to sit in on Sylvester's courses, but found the topic of her disseration research with Peirce. Thanks to Sylvester's interventions with Gilman, she had not only gained access to Peirce's classes but also won one of the coveted Hopkins fellowships. A note discreetly but publicly placed in the first volume of the *Johns Hopkins University Circulars* in December 1879 announced, in fact, that "Miss Christine Ladd, a graduate of Vassar College, whose mathematical attainments had been commended by the Faculty as worthy of the holder of a Fellowship in Mathematics, was invited by the Trustees, June 2 1879, to continue her studies here for a year, and the usual stipend of a Fellowship was voted to her."[34] The language here was very careful. She was "worthy" of a fellowship, and the stipend "was voted to her," but she was not actually *given* the title of "fellow." If she was no longer totally invisible, she was still unofficial.

In his new course, Sylvester guided eight students—among them Franklin and Stringham in addition to Mitchell and Ladd—on a tour of number theory that included, not surprisingly, some of his own original work. On his passage back to England earlier in June, he had been thinking about number theory in preparation for his fall course and in reaction to an article received for the *American Journal* shortly before his departure. Writing from Paris in May, Édouard Lucas had submitted a paper in which he revisited work that had engaged Sylvester as early as 1847 on the general problem of solving (over $\mathbb{Z}$ or $\mathbb{Q}$) ternary cubics of the form $Ax^3 + By^3 + Cz^3 = Dxyz$, where A, B, C, $D \in \mathbb{Z}$.[35] In particular, Lucas considered the special case of integral solutions of

$$x^3 + y^3 = Az^3. \tag{1}$$

As he noted, Euler and Legendre had shown that this equation has no integral solutions when $A = 1, 2, 3, 4, 5$, and Sylvester had asserted in 1856 that there were at least six more general classes of instances in which equation (1) is insoluble.[36] In 1879, Lucas succeeded in proving Sylvester's six general cases in the context of one general theorem, namely, "if p and q denote primes of the forms $18n + 5$ and $18n + 11$, respectively, it is impossible to decompose any of the following numbers into two cubes, either integral or rational: p, $2p$, $4p^2$; q^2, $2q^2$, $4q$."[37] Moreover, in a letter that apparently accompanied his submission, Lucas also informed Sylvester of work of another French mathematician, Théophile Pepin, that provided 10 additional general classes in which equation (1) is insoluble. It was precisely this constellation of ideas that occupied Sylvester on the return voyage to England in June 1879. By month's end, he had sent off a brief reaction to the work of Lucas and Pepin that formed the first part of what would ultimately be a mammoth, eighty-page paper published in four installments over two volumes of the *American Journal* in 1879 and 1880.[38]

Just as with the interplay between his graduate lecturing and his invariant-theoretic research, the ideas Sylvester developed in this major paper were largely what he presented in successive stages to his class. As he explained in the official published announcement, the course approached number theory "by various methods, many of which are new and original," at the same time that it "embrace[d] a statement of the result of independent investigations by the lecturer, among which may be specified . . . applications to a new theory of cubic-form equations."[39] Sylvester verified that his results were, indeed, new as he prepared his lectures. Among the books he assigned for the course were Dirichlet's *Zahlentheorie*, Legendre's *Théorie des nombres*, and Bachmann's *Theorie der Kreistheilung*,[40] and he, as well as his students, consulted and drew inspiration from these masterworks.

Mitchell, for example, quickly immersed himself in the material, and especially in Dirichlet's book, publishing his first paper in the *American Journal* in 1880 "On Binomial Congruences; Comprising an Extension of Fermat's and Wilson's Theorems, and a Theorem of Which Both Are Special Cases."[41] Sylvester's influence on Mitchell's research was manifest. In a number of places in his text, Mitchell made reference to Sylvester's work and used, for example, some of the peculiar terminology his professor had introduced in the course of his lectures.[42] Mitchell continued mining this number-theoretic vein in the context of Sylvester's course the following year, publishing "Some Theorems in Numbers" in the spring of 1881.[43] It was the latter work that he ultimately defended for the Ph.D. in 1882 before leaving Baltimore to take a position as Professor of Mathematics at his *alma mater*, Marietta College in Ohio.

Things did not go as smoothly for Mitchell's classmate, Ladd. She, too, successfully completed the requirements for the Ph.D. in 1882 for a thesis written, like Mitchell's, in 1881, but under Peirce's direction.[44] Although Sylvester, Peirce, and others strongly endorsed her candidacy for the degree, the Board of Trustees refused to reverse its earlier stand on coeducation. Ladd did not become the first woman to receive a Ph.D. from the Johns Hopkins,[45] although she remained in Baltimore following her marriage in the summer of 1882 to Fabian Franklin and continued her efforts both in mathematics and in fighting for the rights of women.

A new student, George Ely, joined Mitchell and Ladd in what proved to be Sylvester's final course on number theory in the fall of 1881 and was lured into the subject by his professor's lectures on Bernoulli numbers and related concepts. Like any good scholar, Ely readied himself by compiling a bibliography of almost exclusively nineteenth-century works on the topic. Published in the *American Journal* in 1882, the bibliography included the work of Englishmen, John Couch Adams, Cayley, James W. L. Glaisher, and Sylvester; Frenchmen, Lucas and Hermite; Germans, Carl G. J. Jacobi, Leopold Kronecker, and Ernst Kummer; and Italians, Ernesto Cesaro and Angelo Genocchi, among others.[46]

In mastering these known results and in consultation with Sylvester, Ely hit upon a problem to consider. It was well-known that the series expansions of the trigonometric functions tan x, cot x, and csc x involve the Bernoulli numbers. Moreover, "differentiating the series for tan x and cot x we obtain the expansion of $(\sec x)^2$ and $(\csc x)^2$, and thence $(\tan x)^2$ and $(\cot x)^2$. The problem was therefore suggested . . . to find expansions for the nth powers of the trigonometric functions."[47] While this proved to be little more than a clever exercise, it, together with his other work at Hopkins, sufficed in 1883 to earn him both the Ph.D. after only two years in residence and a professorship of mathematics at Buchtel College in Ohio.[48] Sylvester had clearly adopted the criterion of "original research" for his program at Hopkins, but as some of the early Ph.D.s in mathematics exemplify, that research may not always have been too deep.

Students like Mitchell and Ely were not alone in having been drawn into new mathematics as a result of Sylvester's lectures on number theory between 1879 and the end of 1881. Indicative both of the cohesiveness of the mathematical program at Hopkins and of the *American Journal* as a stimulus for new research, Story and Franklin were both attracted by the ideas Sylvester was presenting and the research he was publishing. (Franklin had been appointed to the Hopkins mathematical staff as an Assistant in Mathematics in the fall of 1879 even before he had earned his Ph.D. in the spring of 1880.)

As a student in the 1879–1880 course, Franklin apparently accepted Sylvester's classroom challenges to prove results by providing a new proof of at least one of the results that ultimately appeared in Sylvester's paper, "On Certain Ternary Cubic-Form Equations."[49] In the fall of 1880, however, he bested his previous year's performance; Franklin, combining the computational expertise and facility he had gained as Sylvester's human computer in invariant theory with the new number-theoretic ideas he was encountering in Sylvester's course, came up with a surprising and elegant new proof of Euler's classic, pentagonal number theorem.[50]

In his monumental 1748 text, *Introductio in analysin infinitorum*, Leonhard Euler had given an analytic proof of the following product-sum identity:

$$\prod_{n=1}^{\infty}(1 - x^n) = \sum_{n=-\infty}^{\infty} (-1)^n x^{\frac{n(3n-1)}{2}}.$$

(Notice that, for n ranging from 1 to ∞, the powers that occur on the right-hand side are just the pentagonal numbers 1, 5, 12, 22, . . .) Legendre revisited this result in 1830 in his *Théorie des nombres* and established the link between the pentagonal numbers and partition theory. He noted that, for any $m \in \mathbb{Z}^+$, the difference between the number of partitions of m into an even number of parts and the number of partitions of m into an odd number of parts was $(-1)^n$ if

$m = [n(3n \pm 1)]/2$ and zero otherwise.[51] In 1881, and in a note published in no less prominent a place than the *Comptes rendus* of the Paris Academy of Sciences, Franklin gave a purely graphical proof of Legendre's result that hinged on an argument attributed to Norman Ferrers.[52] Ferrers had shown that the number of partitions of n into at most m parts equals the number of partitions of n in which no part exceeds m, and his proof was essentially "by picture."

Consider, for example, the partition $4 + 3 + 3 + 2 + 1 + 1$ of 14. This partition can be represented graphically as

```
*   *   *   *
*   *   *
*   *   *
*   *
*
*
```

Note that this is a partition of 14 in which no part exceeds 4, but note, too, that interchanging the rows and columns in this array gives the new partition $6 + 4 + 3 + 1$ of 14 that has

```
*   *   *   *   *   *
*   *   *   *
*   *   *
*
```

as its graphical representation. The latter partition of 14 has at most four parts. Clearly, this flipping process ensures the one-to-one correspondence Ferrers asserted.[53] Franklin used a similarly spirited graphical demonstration—involving rearranging graphical representations—in his proof of the pentagonal number theorem.[54]

Whereas Franklin drew his inspiration from Sylvester's lectures, Story, as production editor for the *American Journal*, had carefully to go over all the papers that appeared in print. This meant, in particular, that he read in detail all Sylvester's work, and the paper "On Certain Ternary Cubic-Form Equations" apparently piqued his research interest. By the time the last number of volume three was ready to go to press, Story had extended the ideas Sylvester had published in that volume's first number on the so-called theory of rational derivation on a cubic curve, that is, on the theory of points on a cubic curve that can be expressed as rational functions of an arbitrary initial point on the curve.[55] Story's paper came out in the last issue for which he was managing editor and just days before Sylvester had his unfortunate blow-up in Gilman's office. (Recall chapter 9.) The tensions between them over the journal did not blind Sylvester to the value of Story's research, however. Writing to Cayley on 19 January 1881, he acknowledged that "Story was doing good work . . . on connecting my Derivation Theory with Elliptic Functions."[56]

In this same letter, Sylvester also provided a tantalizing snapshot of the Hopkins program aimed at convincing Cayley to come for an extended teaching stay in Baltimore. "I wish 1000 times over that you would come over here," Sylvester coaxed, "where you could more than double your emoluments as Professor and where you could command a class of 10 or 20 enthusiastic hearers and followers and really found a school." As for the students,

> You will see Franklin's resumé of the G[enerating] F[unction] method in No. 2 of the Journal; it seems masterful to me.
>
> We have two men Mitchell and another [Ely] *researching* entirely on their work in the Theory of Numbers.
>
> Mitchell I think will do great work. Miss Ladd has a first rate article coming out on the connection and infinite extension of the Algebraical processes.

And as if these guarantees of serious, talented students were not enough, Sylvester added that "besides founding a school your social relations here could be very agreeable: They are a great church-going and very friendly people." He was clearly at ease in Baltimore; he was clearly pleased with his students and his program; but he just as clearly missed his regular interactions with Cayley. He had never had a mathematical equal in Baltimore, and, since 6 October 1880, he no longer had one in all of the United States. Benjamin Peirce had died. Three days later, in the context of a "beautiful and simple service," Sylvester, along with Harvard President Charles Eliot, Simon Newcomb, and Oliver Wendell Holmes, among others, had carried the casket of their longtime friend to its final resting place.[57] A sense of loss and increasing exile had been coming over Sylvester since then, and these were only strengthened by the fact that he had just learned that, back in England, he had been awarded the Copley Medal by the Royal Society of London, that body's highest award for scientific achievement. Sylvester longed for the mental and mathematical stimulation that the other side of the Atlantic had for so long provided him, but he had to make do in Baltimore with his research and teaching in number theory.[58]

Flirting with melancholy in January 1881, Sylvester turned once again to poetry for solace. On the thirteenth, he had written Newcomb of a public lecture he had delivered the week before in Hopkins Hall on his laws of verse as well as of his efforts to polish "some old sets of verses."[59] By 20 February, in the immediate aftermath of the blow-up in Gilman's office over Story's handling of the *American Journal*, Sylvester was writing to another acquaintance, Harvard professor Charles Eliot Norton, about what had become a fixation with poetry. "What could I do to get rid of this perilous [?] stuff which weighs upon the brain, so as to have done with them for good and all?," he queried rhetorically.[60] Sylvester was also considering the publication of a new, expanded, American edition of his *Laws of Verse* and asked Norton his thoughts on the possibility of securing a publisher. As in the wake of the events that had led to his premature retirement from Woolwich, Sylvester let poetry distract him in the winter of 1881 from the various stresses in his life.

These poetic diversions were further supplemented by an occasion of grand proportions, a visit to Baltimore on Saturday, 12 February of lame-duck President Rutherford B. Hayes. Gilman, Baltimore Mayor Ferdinand Latrobe, and industrialist, William Keyser, met Hayes at the train station in the morning and escorted him to Gilman's home. From there, Hopkins Trustee, Ladd supporter, and hospital advocate, Francis T. King, escorted Hayes on a tour of the as yet unfinished Johns Hopkins Hospital as well as of the Peabody Library, before returning at 1:00 for a luncheon in Hayes's honor hosted by Gilman. Gilman used the opportunity to highlight the talent he had assembled in Baltimore. He had invited "thirty or forty of the best people," among them Sylvester, and Hayes found his visit to Baltimore "altogether a happy occasion."[61]

Sylvester's temper and his outlook improved by March. On the twenty-third, as he taught his course in number theory and pursued his own number-theoretic researches, he wrote to Cayley in the belief that he had finally found a British-style proof of Gordan's finiteness theorem.[62] Three days later, however, he had to confess that "the theorem which I sent you the day before yesterday is *wrong*—it proceeded from a certain new principle which is correct—but the application made of the principle was erroneous."[63] This time, invariant theory distracted him only briefly. Unable to push the "principle" through, he refocused on number theory for the remainder of the semester and into the summer.

Sylvester published one more notable number-theoretic paper in the *American Journal*. In the fall of 1881, and in conjunction with what proved to be his last course at Hopkins in that area, he considered the method that his friend Chebyshev had devised in 1852 for producing upper and lower bounds on $\pi(x)$, the number of primes less than a given number x. Chebyshev had shown that

$$A_1 < \frac{\pi(x)}{x/\log x} < A_2,$$

where $0.922 < A_1 < 1$ and $1 < A_2 < 1.105$.[64] In his work, Sylvester improved somewhat on these bounds but encountered in the process a certain system of linear equations that turned his attention from number theory to the theory of "universal algebra" or what would today be called matrix theory. It was to this new set of ideas that he exposed his graduate students in the spring semester of 1882 and to which he turned his own researches for much of the rest of that year.[65]

One of the new graduate students, William Durfee, graphically described what it was like to be in Sylvester's graduate classroom at the time of this sea change. In the fall of 1881, he explained, "Sylvester began to lecture on the Theory of Numbers, and promised to follow Lejeune-Dirichlet's book; he did so for, perhaps, six or eight lectures, when some discussion which came up led him off, and he interpolated lectures on the subject of the frequency [of prime numbers], and after some weeks interpolated something else in the midst of these."[66] "After some further interpolations," Durfee continued, Sylvester "was led to the consideration of his Universal Algebra, and never finished any of the previous subjects." Thus, in Durfee's first year in the program, 1881–1882, the students "never received a systematic course of lectures on any subject," but they "had been led to take a living interest in several subjects." To Durfee's way of thinking, they "were greatly the gainers thereby."

From Number Theory to "Universal Algebra"

Already in January 1882, before classes had begun for the spring semester, Sylvester was at work on the new results he would present to his students. Although he credited the shift in his research to ideas encountered in the course of the Chebyshev paper,[67] he was undoubtedly influenced also by the publication in the second number of the 1881 volume of the *American Journal* of Benjamin Peirce's ground-breaking study on "Linear Associative Algebra."[68] Peirce had first presented the ideas in this paper before a largely bewildered audience at the National Academy of Sciences in 1870 and had had one hundred lithographed copies made for private distribution. The work had thus never really seen the light of day, and Sylvester resolved both to rectify this and to honor the memory of his friend by seeing it into print.

In his memoir, Peirce had adopted a very abstract, structural point of view to the study of algebras. Taking his lead in some sense from Sir William Rowan Hamilton's abandonment of commutativity in his discovery of the quaternions in 1843, Peirce considered algebras of dimensions one through six over the complex numbers and completely classified them in terms of their multiplication tables. He laid the groundwork for his systematic study by developing a theory of algebras, defining, for example, the notion of a zero divisor (he called them nilfactors), that is, a nonzero element a in an algebra that satisfied the equation $ab = 0$ or $ba = 0$ for some nonzero element b in the algebra. He also defined the concepts of nilpotent elements, those which when raised to some power vanish, and idempotent elements, those which when raised to the second or higher powers give the element back again. Zero divisors and nontrivial nilpotents and idempotents did not exist in the usual arithmetic of the real numbers, but Peirce—like Hamilton—was willing to break with that standard and consider a whole new class of mathematical objects: algebras.[69]

In preparing his father's memoir for the press, Charles added numerous notes and explanatory glosses. In particular, he showed how a notation he had developed in the context of his own research on the logic of relatives could be used to express his father's algebras in terms of what would today be called a basis of matrix units, and he systematically recast the algebras in this notation in a series of accompanying footnotes.[70]

Sylvester may well have had these glosses in mind when he wrote to Peirce on 30 December 1881 about his evolving ideas. In his reply of 4 January 1882, Peirce suggested that Sylvester might find it "useful to write $(a)_{ij}$" for "the quantity which stands in the ith column and j^{th} line of the matrix denoted by a," and, after laying out some additional notation, he closed with the question "will these suggestions taken from the logic of relatives have any useful application to your vast new algebra?"[71] Sylvester must have responded immediately either in person or by letter, for Peirce was writing him the next day with further clarifications, seemingly sparked by a direct question, as to the "precise relationship of [Sylvester's] algebra of matrices to [Peirce's] algebra of relatives."[72] After a bit of explanation, Peirce concluded that "it, thus, appears to me just to say that the two algebras are identical, except that mine also extends to triple & other relatives which transcend two dimensions." Still, he added, "I have not considered any problem similar to those which seem to have occupied you, my studies having been chiefly logical—i.e., relating to the form of the algebra, the different operations necessary, etc.—and not to any great extent mathematical." The next day brought another letter from Peirce. Clearly, Sylvester had not liked Peirce's suggestion that his ideas were anything but totally original, but Peirce retorted that "I lay no more claim to your umbral [matrix] notation than I do to the conception of a square block of quantities! What I lay claim to is the mode of multiplication by which as it appears to me this system of algebras is characterized. *This* claim I am quite sure that your own sense of justice will compel you sooner or later to acknowledge." He closed, saying "I am sorry you seem to be vexed with me."[73]

Relations between Sylvester and Peirce were once again frosty, but Sylvester proceeded to develop his ideas on matrix algebra in his graduate lectures and in notes especially to the *Comptes rendus*. In particular, he first concerned himself with the characteristic equation associated with an $n \times n$ matrix M, namely, the equation $\det(M - xI) = 0$, for I the $n \times n$ identity matrix. He was interested in those values x over $\mathbb{R}$ that satisfied this equation as well as in determining the properties of those roots; his preliminary findings appeared in the *Comptes rendus* in January.[74] The next month found another paper by Sylvester on matrix equations in the *Comptes rendus*. This time he was considering the question, when is an $n \times n$ matrix M over $\mathbb{R}$ unitary, that is, in his sense of the term, when is it the case that $M^i = I$, for a given positive integer i?[75] Typical of his inductive, experimental approach to mathematics, however, these early questions were all directed toward a much larger problem. As he put it in the February paper, "I have already established a general functional theory of matrices, and . . . I no longer consider these [matrices] as schemata of elements, but as communities, or, if you will as complex quantities."[76] Thus, Sylvester laid claim to having established a theory of matrices *per se*. He viewed matrices as mathematical objects with their own distinct theory; he conceived of the possibility of matrix algebra.[77]

The spring of 1882 brought more to Hopkins than Sylvester's lectures on matrix algebra, however. After repeated entreaties from Sylvester by letter during the troubled spring of 1881 and after a personal visit to Cambridge in early August, Sylvester had finally won Cayley over to the idea of coming to Baltimore for a semester. By October, he had successfully brokered the deal between Gilman, the Hopkins Trustees, and Cayley. The Cayleys would come for the spring semester of 1882. They arrived late in December of 1881, just in time for Sylvester to bounce his new ideas on matrices off of Cayley and to seek solace in the wake of his altercation with Peirce. It was a surprised— but characteristically unembarrassed—Sylvester whom Cayley reminded of Cayley's own paper on matrices published almost a quarter-century earlier in 1858 in a no less conspicuous place than the *Philosophical Transactions of the Royal Society!*[78] Still, when Sylvester wrote to Charles Hermite on 15 January 1882 asking him to present the paper on unitary matrices to the next meeting of the Paris Academy, he unabashedly confessed to the situation, while maintaining the novelty of his results. "Mr. Cayley has arrived," he wrote Hermite, "and seems convinced that the theory I gave in the previous note [of January 1882] is completely new. When I wrote it, I did not know of his Memoirs on Matrices Phil. Tr. 1858, 1866; he traced the foundations of the theory in the first of these, but he will be the first to admit that I have given an extension of and completely new and unexpected solutions of this same theory."[79]

More lectures and more results on matrix algebra followed. Describing his Baltimore routine in a letter to Thomas Hirst back in England, Cayley captured a sense of the dynamic. "I have my lecture twice a week to a class of about 12 . . . on the Abelian and theta functions . . . , on the other two days I hear Sylvester, who began a course of 'three or at most four' lectures on multiple algebra, which he has gone on with since the beginning of the year, & they are not finished, nor likely to be."[80] Sylvester was in fine form, and Durfee, a student in that class, described his professor's frequent fits of exuberance in vivid terms. "I remember distinctly," he wrote, "an incident that occurred when [Sylvester] was at work on his Universal Algebra. He had jumped to a conclusion which he was unable to prove by logical deduction. He stated this fact to us in the lecture, and then went on, 'GENTLEMEN' [here he raised himself on his toes],

'I am *certain* that my conclusion is correct. I will WAGER a hundred pounds to *one*; yes, I will WAGER my *life* on it.' The capitals indicate when he rose on his toes and the italics when he rocked back on his heels." "In such bursts as these," Durfee continued, "he always held his hands tightly clenched and close to his side, while his elbows stuck out in the plane of his body, so that his bended arm made an angle of about 140°."[81] Sylvester was deeply into the material, and the results flowed.

A stream of short papers issued from his lectures, but unlike the earlier results that went principally to the *Comptes rendus* for publication, many of these stayed at home on the pages of the *Johns Hopkins University Circulars*. The *Circulars* had been launched in December 1879 in an effort to familiarize the wider Hopkins community with the broad range of research being conducted within the University's walls, but they were also abstracted, for example, by the Royal Society in its mammoth effort to catalog all scientific literature in the nineteenth century. As Sylvester later described them to Cayley, the *Circulars* served as "a sort of record of progress in connection with the work and personality of the Johns Hopkins."[82] One manifestation of his own progress in the theory of matrix algebra—first presented in his lecture course and then published in the *Circulars*—was his discovery of a nine-dimensional matrix algebra analogous to the quaternions.

As early as 18 January, Sylvester, Cayley, and Peirce had given a special series of lectures before the Mathematical Seminary in celebration of Cayley's arrival, with Peirce presenting a matrix representation of the quaternions in terms of his language of relatives.[83] Sylvester had recast this result in his own notational scheme in his lecture course. As is well-known, the quaternions are a four-dimensional algebra (over $\mathbb{R}$) with basis 1, i, j, k satisfying the relations

$$ij = -ji = k, \quad jk = -kj - i, \quad ki = \quad ik = j, \tag{2}$$

$$i^2 = j^2 = k^2 = -1. \tag{3}$$

Put more generally, for u, $v \in \{i, j, k\}$, the relations in (2) may be expressed as $uv + vu = 0$, and Sylvester noted that for $\theta = \sqrt{-1}$, the matrices

$$u = \begin{pmatrix} 0 & 1 \\ -1 & 0 \end{pmatrix} \quad \text{and} \quad v = \begin{pmatrix} 0 & \theta \\ \theta & 0 \end{pmatrix}$$

satisfy the relations in (2) and (3) so that the matrices

$$\begin{pmatrix} 1 & 0 \\ 0 & 1 \end{pmatrix}, \begin{pmatrix} 0 & -1 \\ -1 & 0 \end{pmatrix}, \begin{pmatrix} 0 & \theta \\ \theta & 0 \end{pmatrix}, \begin{pmatrix} \theta & 0 \\ 0 & \theta \end{pmatrix}$$

form a basis of 2×2 matrices for the quaternions.

In keeping with his inductive approach to mathematics, Sylvester then naturally asked whether an analogous algebra could be constructed in terms of a basis of 3 × 3 matrices. The answer was "yes." By taking

$$u = \begin{pmatrix} 0 & 0 & 1 \\ \rho & 0 & 0 \\ 0 & \rho^2 & 0 \end{pmatrix} \quad \text{and} \quad v = \begin{pmatrix} 0 & 0 & 1 \\ \rho^2 & 0 & 0 \\ 0 & \rho & 0 \end{pmatrix},$$

for ρ a primitive cube root of unity, he showed that the matrices I, u, v, u^2, uv, v^2, u^2v, uv^2, u^2v^2 form a basis of a nine-dimensional algebra analogous to the quaternions. Sylvester dubbed this algebra the "nonions," and it clearly pleased him.[84]

Unlikely as it may seem, this little work on the nonions resulted in a complete and permanent break between Charles Peirce and Sylvester. It was only months later, at the end of 1882 or early in 1883, that Sylvester apparently read the published version of his note, and when he did, he saw red. As he recounted things, he had given Peirce leave "to supply in my final copy for the press, such reference as he might think called for,"[85] but what Peirce had added, immediately after the presentation of the nine basis elements of the nonions, was the sentence "these forms can be derived from an algebra given by Mr. Charles S. Peirce (*Logic of Relatives*, 1870)."[86] Sylvester insisted that this sentence should have read "*Mr. C. S. Peirce informs me that these forms can be derived from his Logic of Relatives*, 1870." He went on to announce that "I know nothing whatever of the fact of my own personal knowledge. I have not read the paper referred to, and am not acquainted with its contents."[87]

Peirce, feeling unjustly accused, complained to Gilman on 18 February 1883 and sent a rejoinder with his own account of the facts that he asked Gilman to read and then publish in the next *Circular*. A month later, the matter had still not been settled. Gilman had tried to let things calm down and had not yet acted, but on 26 March, Peirce had sent him another formal text on the issue for publication. When Gilman responded that he would proceed by showing Peirce's text to Sylvester, Peirce replied that "it is not quite fair, because his was not shown to me. Still, if it were possible to avoid further dispute, I should heartily consent."[88] Peirce did not hold out much hope of Sylvester's acquiescence, however. He added that "I do not think . . . that it is possible; because he is blinded by arrogance & I shall not go without having the imputations that have been made upon my conduct, completely refuted." Peirce's rebuttal as well as another remark by Sylvester appeared in the April issue of the *Circulars*. There, Peirce recounted the prior discussions he and Sylvester had had about matrices and the logic of relatives and rightly questioned why Sylvester had not consulted Peirce's earlier writings to determine for himself the relation between Peirce's work and his own.[89] For his part, Sylvester lamely reiterated that he had "not read [Peirce's] Logic of Relatives and [was] not acquainted with its contents."[90]

Sylvester was clearly in the wrong on several levels here. His life-long reluctance to read the work of others, his overwhelming need for priority, his sense of the importance of his work, his impatience and inattentiveness with proof-sheets, all contributed to this unnecessary and childish blow-up. Yet, just as with his outburst in Gilman's office over Story's handling of the journal, for Sylvester's part, all was soon forgotten.[91] By the spring of 1883, he had moved on to other things.

A Troubled Transition to the Theory of Partitions

The productive spring of 1882 was only enhanced by the Cayleys' presence in Baltimore. Together they had enjoyed what Sylvester described as Baltimore's "very agreeable" society in addition to each other's company both within the University's walls and on long walks together in the "beautifully wooded" countryside around Baltimore.[92] It was thus a melancholy Sylvester who accompanied them northward to New York City for their departure back to England on 10 June.

He stayed behind, for the first time since he had taken the position at Hopkins, not going to England for the summer. From New York City, he made a short excursion to Niagara Falls, spending nine days viewing their "sublime but desecrated beauties" before returning to Baltimore.[93] "Business connected with the Journal added to [his] weariness of solitude and the desire of being near [his] books and the libraries" had brought him back. As with Story, Sylvester had had "a serious difference" with his "anonymous" managing editor, Franklin, before he had left for New York and had given him notice that his services would no longer be needed after mid-July. This was another first. For the first time, Sylvester would have to deal with the actual mechanics of the journal, unless he could enlist someone else's help. Fortunately, he soon lined up his recently graduated student, Oscar Mitchell, and then his former student and current Associate in Mathematics, Thomas Craig, after Mitchell left to take up his new teaching postion in Ohio.

Still, after a month, things were becoming just too much for him. He had barely been able to write up the report he owed the British Association on the grant that he, Cayley, and Salmon had been awarded to calculate tables of fundamental invariants of binary quantics. He could not concentrate on his research on matrices in the debilitating summer heat and humidity. The real problem, however, was that he missed the Cayleys and began dwelling on what he increasingly viewed as his exile in America. In the letter he wrote to Cayley on 3 August accompanying the report, he allowed that "it was *a great mistake* my not going over to England this summer," and he longed for "an opening . . . *at home* in England!"[94]

Sylvester's mental state only went from bad to worse over the course of the next week. Earlier in the spring, he had suggested Sir William Thomson to Gilman as a candidate for a visiting Hopkins lecturership to give a two-week, short course of lectures. Sylvester had known Thomson at least since 1845 when, as a fresh Cambridge graduate and, like Sylvester, a second wrangler, Thomson had assumed the editorship of the *Cambridge and Dublin Mathematical Journal.* In the interim, Thomson had become England's premier mathematical physicist and so was a natural choice for a distinguished Hopkins lecturer. Gilman had liked the idea and had asked Sylvester to approach Thomson, extending an invitation for any time between October 1882 and June 1883. When Thomson wrote back proposing to come early in the summer of 1883, however, Gilman found the timing problematic and asked Sylvester to write Thomson again. Sylvester felt that such a follow-up letter would put him in an awkward position, especially since Thomson had already chosen a date within the original bounds Gilman had set, and so he left the matter in Gilman's hands.

This was the last Sylvester heard about the matter until August, when Gilman once again asked him to write. Sylvester snapped. Writing to Gilman on 10 August, he minced no words. "I feel very grieved and mortified by the course which you have adopted in respect to the invitation tendered to Sir William Thomson," he declared. "Notwithstanding your promise to write to him you again ask me to undertake the duty which properly devolves on yourself of explaining to him why his acceptance although coming within the terms of the offer made to him has been declined."[95] This letter contained more than just this rebuke, though. It contained a bomb shell. "I am wearied and dispirited and feel no longer equal to the discharge of what I consider to be the duties of my office in a manner satisfactory to myself or conducive to the best interests of the University," Sylvester continued. "I write therefore (after much anxious deliberation) to request that you will take the first opportunity to lay my resignation before the Board of Trustees to take effect as soon after the 1st of October next as may be found convenient" (209–210).

Gilman was stunned. Away at his summer house in Massachusetts, he wrote back to Sylvester on the thirteenth, apologizing for his handling of Thomson's invitation and assuring him that a letter was in the post to Thomson that very day exonerating Sylvester of any blame in the situation.[96] "As to the other matter referred to, I will write you after a little reflection," Gilman said. "I cannot think with composure of your withdrawal from the University where you have been so useful in counsel, in instruction and in investigation—& hope to persuade you to reconsider the whole subject."[97]

It took Gilman three days to compose himself and a suitable letter to Sylvester. Despite the fact that Sylvester could be difficult at times, Gilman viewed him as a personal friend, a valued adviser, and a major reason for the success thus far of the Johns Hopkins University.[98] He did not want to lose him in any of these capacities. The letter he finally sent was at the same time businesslike, understanding, and cajoling. He opened in his capacity as the University's President, explaining to Sylvester the timetable according to which his resignation, should he still insist on tendering it, would be brought before the Trustees but then quickly shifted into his role as an understanding and concerned friend. "I hope . . . you will find that you have been laboring under a temporary depression in respect to your professorship," Gilman wrote, "& with the cooler autumn weather, & the return of your pupils and friends to Baltimore, that you will feel your usual enthusiasm for the work which you have prosecuted with such great ability & success during the six years of your residence among us."[99] To lift Sylvester's spirits and undoubtedly to soothe and stroke his ego, Gilman next enumerated the mathematician's many achievements at Hopkins: "The honorable positions which have been offered to Craig, Franklin, Stringham & Mitchell; the remarkable papers which they & others have been led under your guidance to prepare; the high reputation of the Am[erican] J[ournal] of Mathematics; your own contributions to its pages; the articles which you have elicited from mathematicians at a distance; the intellectual quality of the students whom you have drawn around you,— are indications to my mind, of the strongest character, that you have found a useful & profitable field of labor" (210–211). Gilman was not exaggerating. He recognized and appreciated all that Sylvester had done for the University. He had, indeed, fulfilled all of the high hopes that Gilman had had for his first faculty, and he had done it with a Sylvesterian flare. Gilman knew that in Sylvester he had a "personality," but when all was said and done, it was a "personality" he truly valued for all its flaws and quirks. "Whatever may be your decision (and I hope it will be to withdraw your last letter) be sure," Gilman closed, "that I have a most grateful appreciation of all your wise counsel & hearty cooperation in the past & that I am with undiminished respect & regard Yours sincerely" (211).

Sylvester replied a week later. "The very kind and considerate tone and tenor of [Gilman's] two letters," together with a break in the heat, had markedly improved his frame of mind and eased his "*extreme depression of spirits*."[100] He withdrew his resignation but asked to be relieved of his teaching duties in the fall so that he could concentrate on the research he needed to produce for the journal. As he explained to Gilman, "I had hoped to have filled the vacant space in the number on hand with a portion of a memoir on the subject of my lectures during the last *semester* but have been unable to collect my ideas sufficiently to write a single line—so that the press is at a stand-still for the want of matter."[101] He went on to assure Gilman that "the provisional cessation from lecturing . . . will make no difference in my relations to our students to whom I shall always remain accessible as I have no desire—(nor do I think it would be of any use in reestablishing my mental vigor) to leave the city where the environment is more favorable for quiet study than any other place in the union would be to me in my present condition" (213).

Although he was sympathetic to the request, Gilman hesitated to bring it before the Trustees. After all, he had come to understand Sylvester's moods, and he knew that it was quite possible that Sylvester would see things differently as the start of the semester drew nearer. "Perhaps when our company reassembles," he wrote, "you will feel once more, as you have always felt hitherto, a fresh enthusiasm in the guidance of your followers."[102] He was right, but mathematical success prior to the return of society contributed fundamentally to the turnabout.

The month of September found Sylvester reengaged in research, but not in the research on matrix algebras that had dominated his thoughts in the spring, rather in the invariant theory that had proved so successful in 1876 in resuscitating him mathematically. In fact, invariant theory had never been far from his thoughts after he made the shift to other topics in his graduate teaching in the fall of 1879. He had returned to it briefly in that unsettled March of 1881 when he thought Gordan's finiteness theorem had finally fallen to his techniques.[103] In September of 1882, it was that same gap in the British approach that he once again seemed to have bridged.

Writing to Cayley on the sixth, he was back in form and brimming with enthusiasm about his new line of investigation, destined for the *American Journal*. After sketching his argument and acknowledging that it hinged on the limit of a certain sequence of rational numbers going to infinity, he averred that "I can prove Gordan's theorem." Moreover, of the limit going to infinity, he had "little or no doubt and the proof (however difficult it may be) belongs to the province of ordinary algebra or the theory of partitions."[104]

The month of September flew by, with Sylvester so thoroughly engrossed in pushing these ideas that he had inadvertently let a much welcomed letter from Cayley go unanswered. Finally replying on 6 October, he brought Cayley up to date on the results he had obtained during the past month, and although he had again to confess that his "supposed proof of Gordan's theorem was a *Delusion*," he was pleased to report that his latest paper "originally designed as a Note to fill up a gap of two or three pages in the journal—now extended to 7 sections likely to occupy about 40 pages."[105] When that was finished, he continued, "I hope to fall back on what I originally meant to occupy the summer vacation viz. the subject of Matrices or Multiple Algebra so that there is plenty of work before me for the next two or three months to come." And as if his letter did not make it patently obvious to his long-time friend, he added that "I am all right again."[106]

The return of less oppressive weather had brought the return of the mathematical muse, but it had also occasioned the gradual return of Baltimore society.

The Gilmans were back as were Franklin and his new bride, Christine Ladd Franklin; Sylvester's fellow Athenæum member, Herbert Spencer, had come to town on the evening of 28 September and had taken a tour of the University with Sylvester before retreating for a week to nearby Montebello, the country estate of John and Rachel Garrett; the noted theologian, William Carpenter, had been fêted in the context of his Hopkins Hall lecture to mark the start of the new academic year.[107] Sylvester took part in all of these social affairs. The three long, debilitating months of solitude were finally over.

The students were back, too, and, as Gilman had predicted, Sylvester was with them in the classroom. How could he refuse, when Craig asked him to give a course on the theory of substitutions out of Eugen Netto's brand-new and path-breaking book (even if, as he confessed to Cayley, he knew "hardly anything" about the topic and found "it very hard to learn out of books")?[108] Nine attended the late afternoon lectures each Tuesday and Friday—among them, staff members, Story, Craig, and Franklin and students, Ely, Durfee, and Davis—yet, like the preceding fall, the attempt to follow a set text was doomed. "We all got the text," Davis recounted. Sylvester "lectured about three times, following the text closely and stopping sharp at the end of the hour. Then he began to think about matrices again. 'I must give one lecture a week on those,' he said. He could not confine himself to the hour, nor to the one lecture a week. Two weeks were passed, and Netto was forgotten entirely and never mentioned again."[109]

Nor did that fall's class continue long with matrices. By the end of October, Sylvester had bounced from a passage in Netto's book on forming tables of symmetric functions to Cayley's work on the same topic to Sylvester's own calculational technique. Durfee pursued precisely this line of investigation and earned his doctorate for his efforts the next semester. These ideas also brought Sylvester once again to the same theory of partitions that had cropped up in his own work on Gordan's theorem in September and that had briefly engaged Franklin in 1881.[110] By 15 November, Sylvester was speaking on partitions to the assembled Mathematical Society.[111] He was hooked, and he plunged from the paper he was writing in installments for the *Journal* on September's invariant-theoretic results to the theory of partitions. The paper on invariant theory, which he had predicted would run to 40 journal pages, in fact ran to almost 60 with a 17-page closing "excursus" on partitions,[112] and on Christmas Day, he was reporting to Cayley that he had "just discovered and proved" another "very beautiful theorem in partitions."[113] There was no doubt now that partitions would be the subject of his lectures come January.

One newcomer—"invisible" Christine Ladd Franklin—joined the same nine students from the fall to participate in what proved to be Sylvester's most successful graduate-level course. No pretense was made of following a book. Like the earlier courses on invariant theory, this one was on Sylvester's evolving research, but there was a key difference. Although Sylvester had given that series of lectures on partitions at King's College London in 1859, he had hardly thought about them since. Thus, whereas the students in his invariant theory classes were competing with one of the founders of the area, those in the spring of 1883 found themselves on a more even playing field with their professor. This class ultimately resulted in a truly joint research effort on what Sylvester dubbed a "constructive theory" of partitions.

As with the previous spring's course on matrix algebra, the underlying assumption of this spring's assault on partitions was "modern." Just as Sylvester had treated matrices as mathematical objects and had sought to develop a matrix *theory*, so he "consider[ed] a partition as a *definite thing*" and aimed to develop an intrinsic theory of these objects *per se*.[114] The constructive theory then hinged on the systematic numerical representation of partitions in terms of their parts arranged by (usually descending) order of magnitude and on their corresponding graphical representation as an array of nodes. (Recall the previous discussion of Ferrers's Theorem in "From Invariant Theory to the Theory of Numbers.") The constructive theory was thus juxtaposed with the analytical approach of Euler or Legendre. It would rely primarily on the manipulation of graphical representations rather than on that of infinite sums and products. Franklin's proof of the pentagonal number theorem back in 1881 had actually been an example of a constructive approach, but Sylvester and his students— Franklin among them—only succeeded in turning that approach into a *theory* in their work in 1883.

The paper that resulted, "A Constructive Theory of Partitions, Arranged in Three Acts, an Interact, and an Exodion," bore only Sylvester's name under the title, but inside, results and even entire sections were explicitly credited to individual members of the class.[115] More than ever before, Sylvester's graduate classroom crackled with mathematical energy generated at the blackboard as well as at the desks. As Sylvester described it in a letter to Cayley on 16 March 1883, the dynamic was more one of friendly but fierce competitors than of teacher and students. "I sent to the Comptes Rendus two or three days ago my proof of the wonderful theorem (discovered by observation) of the *class-to-class* as well as the *one-to-one* correspondence between partitions of *n* into odd numbers and its partitions into unequal numbers," he bragged. "Franklin, Mrs. Franklin, Story, Hathaway, Ely, and Durfee? were all at work trying to find the proof—but I was fortunately beforehand with the theory and the only one in at the death."[116] In this course, Sylvester had finally come into his own in the graduate classroom. By working with—rather than lecturing to—his students, he had led them over the hurdle that separates the student of mathematics from the mathematical researcher.

Sylvester supplemented and advertised this effectively joint work with a stream of short papers and notes of his own on partition theory in the *Johns*

Hopkins University Circulars and in the *Comptes rendus* in 1883. After such a traumatic summer, it had been a rewarding fall, winter, and spring. The winter and spring of 1883, however, were full of more than partition theory. As the announcement in 1875 of the founding of the Johns Hopkins University had presented an undreamed of possibility for the then unemployed Sylvester, so the winter of 1883 brought a tragedy that sparked the hope of a life's dream come true.

Henry Smith Is Dead

On 12 February 1883, Cayley wrote to tell Sylvester of the "grievous loss we have had by the death of Prof. Smith."[117] Just three days before, Henry Smith, the Savilian Professor of Geometry at Oxford since 1860, had died, leaving a vacancy in one of England's most prestigious chairs.

When Sylvester took the professorship of natural philosophy at University College London late in 1837, he, as a Jew, had been ineligible for fellowships—much less professorships—at England's ancient universities. The spirit of reform that had been afoot in the Cambridge of Sylvester's student days to remove disabilities for Jews had, from the mid-1850s to the early-1870s, finally brought many of the desired changes.[118] In the wake of this new legislation, Sylvester's *alma mater*, St. John's, had already acknowledged his student and subsequent achievements by awarding him his long overdue B.A. and M.A. *honoris causa* in 1872. They had acknowledged him even more completely—as a member of their actual fellowship—in June 1880 when they elected him an honorary fellow. Perhaps not to be outdone, Oxford awarded him an honorary D.C.L. also in June 1880. As all these changes imply, in 1883, it was actually possible for Sylvester, as a professing although not practicing Jew, to hold the Savilian professorship of geometry, but did he dare try to push the point? After all, he was sixty-eight years old. Would that be considered too old? Would he have the fortitude to bear a rejection?

These and other thoughts weighed on him through the remainder of February and into March. By 16 March, though, he was close to making a decision. "I shall probably offer myself as a Candidate for the succession to [Smith's] professorship," he confided to Cayley, "but await the result of inquiries before coming to an absolute decision."[119] He then immediately made one of those inquiries, asking "do you think I am likely to be appointed?," before laying out some of the pros and cons. "If the chances are considerably against me," he argued, "it would be impolitic to offer." Still, he continued, "perhaps even if impolitic it would be right on my part to do so by way of testing what I consider—although you may not agree with me—an important principle," namely, the religious freedom guaranteed by the Universities Tests Act of 1871 and the University's willingness actually to open its faculty to Jews. Then there was the financial aspect of the post to consider. The salary of £700 was significantly less than the $6,000 Sylvester was then making at Hopkins,[120] but, on the other hand, "the appointment would suit me," he frankly admitted, "and I should . . . have enough to live upon—especially as I have been habitually spending only a portion of my income and laying by the rest as a reserve, to give me time to turn about in the event of going up or losing from any cause my present appointment." The lesson of Woolwich had been learned. Sylvester never wanted to be in that kind of financial situation again and had planned accordingly.

As his professional thoughts focused the rest of the month on the new work in partition theory and on the escalating dispute with Charles Peirce over the nonions and their exposition in the *Circulars*, his private thoughts turned repeatedly to the opportunity the Savilian vacancy presented. By 10 April, he had made his decision. He had submitted his formal application for the chair through his friend and one of the members of the selection committee, William Spottiswoode.[121] Now came the wait.

At least he had much to occupy him. The lectures on partitions had really shifted into high gear by April, and results for the paper on the constructive theory were pouring forth. By the end of the month, he had also learned that, together with Cayley and Joseph Bertrand, he had been elected a Foreign Associate of the National Academy of Sciences, that body's highest honor to foreign scientists.[122] His ego enlarged perhaps by this fresh manifestation of the high regard in which he was held, he grew prickly again in May over university politics. The faculty was scheduled to meet on Wednesday, 23 May to discuss, among other things, fellowship allocations for the 1883–1884 academic year. On Saturday, Sylvester had exploded at Henry Rowland, the Professor of Physics, over what he viewed as aspersions cast on teaching in the Mathematics Department. The next day found Sylvester writing to Gilman to ask his leave not to attend. "The laboratory departments work as a unit together—the same is true of Latin and Greek," he stated. "Mathematics has only a chance of receiving the full measure of consideration to which it is entitled when it receives assistance from your support. It is everywhere an unpopular department of science except to the initiated."[123] By the day of the meeting, though, Sylvester had realized that he had misinterpreted Rowland and had once again behaved badly. After the faculty voted to make the award to Sylvester's fellowship candidate, he felt even more ashamed of himself.[124] It had been a spring of ups and downs by the time he left Baltimore to sail on the White Star Line's *Adriatic* on the thirty-first. The question was, would he return?

The issue of who would succeed Smith in the Savilian chair was due to be decided no later than 7 July. Once back in England, Sylvester waited as he busied himself in his usual summer routine—working on his research, attending meetings of the London Mathematical Society, participating in the daily life of the Athenæum. Gilman was also scheduled to spend part of the summer in London, and Sylvester had put the necessary wheels in motion to secure for him an honorary membership at the Athenæum during his stay. Writing with this news on 19 June, Sylvester also updated Gilman on the Savilian. "I have heard nothing about the Oxford business except that Spottiswoode who as P[resident of the] R[oyal] S[ociety] is one of the electors is ill (I regret much to say) with typhoid fever," he explained. "His friends were at one time (as also his physician) in great alarm about him but the malady has happily assumed a mild form and it is hoped that the worst is past."[125] The worst was not passed, however. Sylvester attended his long-time friend's interment in Westminster Abbey on 5 July. Just before Spottiswoode's death, moreover, Sylvester had learned that the election of the Savilian chair had been postponed indefinitely. Sylvester now expected the worst, but still he had to wait.

He wrote to Cayley, another of the electors, on 10 July asking if he knew when the election process would recommence, whether previous candidates would have to reapply formally, and when the successful candidate would be expected to assume his duties. Everything was up in the air. Should he try to get a leave of absence from Hopkins to remain in England until the matter was decided? Should he plan to return to Baltimore?[126] By the eighteenth, it had been decided that the election would take place on 4 December, and Sylvester had asked Gilman to put his request before the Trustees for a leave of absence without pay for the fall term. Still, he was by no means sanguine about his chances of being named to the position, and he realized that he could be jeopardizing his good relations with Hopkins. "Possibly I may withdraw my candidature to make room for younger men—and to avoid the mortification of a refusal," he wrote to Simon Newcomb. "The trustees of Johns Hopkins will have a perfect right to put before me the alternative of returning to Baltimore *animo morandi* or to resign."[127]

In the midst of all this turmoil, Sylvester was trying to get some research done, as always, to help fill the pages of the *American Journal*. He had spent some time in Cambridge in July and then returned at the first of August, hoping that the change of scenery and Cayley's company would inspire him. It had been to no avail. Back in London by the eighth, he could not concentrate; he could not write. It made him "very sad and despondent."[128] Unlike during the summer of 1882, however, he managed not to let this despondency turn into depression. On the twenty-second, he was once again writing enthusiastically to Cayley of research under way. He had gone back to the theory of matrices, and he was making progress.[129]

Matrices distracted him from the election throughout the fall. In September, he had decided to return to Baltimore, but he had also tendered his resignation effective 1 January 1884, regardless of the outcome in Oxford. As he told Cayley shortly before he set sail, "if I am not elected my present purpose is either to buy a life annuity with my savings of the last 7 years or to go in for some business partnership."[130]

Back at Hopkins, he led what could well have been his last class through his new matrix-theoretic results, as he prepared short notes for the *Circulars* and for the *Comptes rendus*. It was by now the usual routine, but he was really biding his time until December. As early as September, he had mentioned the possibility to Gilman of Felix Klein as his replacement in mathematics. Klein, a dynamic young mathematician then at Leipzig, was approached by Cayley as an interested third party, and discussions proceeded throughout the fall.[131] They went into full swing after 5 December, when Sylvester learned by telegram from Cayley that he had been elected to the Savilian chair.[132]

The telegram arrived for Sylvester at the St. James Hotel where he had taken rooms for the fall, and before he could swear the hotel's manager to secrecy (he most definitely wanted Gilman to be the first to know), word had spread like wildfire.[133] Just two days before and in anticipation of the news of the election (whether ultimately in Sylvester's favor or not), the Board of Trustees had presented him with an offer that would have still tied him to the University even after his resignation took effect on 1 January. They proposed that he teach occasionally as he desired, that he serve as an agent for the *American Journal* in securing quality submissions, and that he receive for these services an honorarium of $1,000 for each of the next five years providing he accepted no position at another university. It was an unequivocal testament of the Board's esteem, and it deeply gratified Sylvester.[134]

Still euphoric on the thirteenth, he wrote to Cayley to detail all of the happenings of the preceding week. As the reality of the situation had begun to sink in, he also began to feel nostalgic about the University and Baltimore, his students, colleagues, and friends. "There are really an unusual proportion of truly charming people—meaning especially women—in this city to which I just as I am on the point of leaving it am beginning to feel attached," he told Cayley, "but it will be nice to be back in old England."[135]

There was still much to do before his departure on the twenty-second. He was hard at work writing up his fall lectures on matrix algebra for the *Journal*.[136] He had to tie up his business affairs and pack his books and possessions. He had to say good-bye to all his friends, and they all wanted to fête his success. In particular, on the twentieth, the University hosted a farewell reception in his honor in Hopkins Hall.

The whole University community turned out for the event. After all, Sylvester had been a member of the University's first faculty, and he was the first of those original professors to leave. Moreover, despite his not infrequent tantrums and fits of paranoia over priority, Sylvester had endeared himself at Hopkins through his contagious exuberance, his theatricality, his erudition, his commitment, his counsel, and his hard work. It was thus an occasion of much warmth and sentiment. The many speeches recording Sylvester's accomplishments at Hopkins and testifying to his membership in the Hopkins family were followed by Sylvester's own remarks of gratitude and pride at having been a part of Gilman's great educational experiment. "I shall always remain in spirit in the sacred precincts of this great University," he told the crowd in Hopkins Hall, "and always have a grateful recollection of the kind feeling that you have evinced towards me, and the favor and indulgence which I have experienced on all sides, during the seven years I have passed in this city."[137] He could be difficult, and even he realized it. In Baltimore, though, he had been treated kindly; he had been indulged; he had been understood. On 22 December 1883, he left all this behind to return to England. He would be making yet another fresh start, this time at the age of sixty-nine, but he was going home triumphant to one of the most prestigious chairs in mathematics at one of the most distinguished universities in the land, acknowledged for his accomplishments at long last.

CHAPTER 11

A Bittersweet Victory

She is a good dear mother our University here and stretches out her arms with impartial fondness to take all her children to her bosom even those whom she has not reared at her breast.
—J. J. Sylvester to Arthur Cayley, 8 November 1884

Entre nous *this University except as a school of taste and elegant light literature is a magnificent sham. It seems to me that Mathematical science here is doomed and must eventually fall off like a withered branch from a Tree which derives no nutriment from its roots.*
—J. J. Sylvester to Daniel Coit Gilman, 11 March 1887

Sylvester's appointment to Oxford's Savilian professorship of geometry was a dream come true. As a young student at Cambridge, he had hoped for the chance to earn his living in England through his mathematics, but his Jewishness had effectively prevented that in the late 1830s. The social and political reforms effected from the 1850s through the 1870s, however, had finally made it possible. Sylvester expressed his sense of the enormity of the appointment in his farewell address to his friends, students, and colleagues at Hopkins. "I want nothing more," he told them. "I do not want to be Prime Minister—I would not care to be, nor Lord Chancellor, nor Lord Chief Justice. [The Savilian chair] is a sure haven, an honored position, but what I value most of all, an opportunity for uninterrupted thought, and evolution of one's ideas. I do, undoubtedly, feel it a very great honor and distinction, and I am grateful to my friends for considering me worthy of being appointed to that honor of the chair which does deservedly rank among the most honored chairs existing in any old English University."[1]

Still, on returning to England to take up this post, Sylvester was once again entering unfamiliar territory. Oxford was not Cambridge, and Oxford was certainly not the Johns Hopkins. It was an ancient university with objectives different from those, in particular, of the new, forward-looking, modern, research university he had left behind in Baltimore. He would not be able simply to pick up where he had left off. He would have to adjust his sense of "the professor" and the "professional mathematician" to the realities of Oxford with its undergraduate thrust and its well-entrenched system of college tutors and set examinations. He would have to accept that he would no longer control his own department, for departments in the Hopkins sense did not exist in Oxford. He would have to acclimate to life within New College, the college to which the Savilian Professors of Geometry and Astronomy were attached. There, he would be a new member of an established fellowship, in which leaders were already defined and where real policy-making was in the hands of a very few. The kind of relationship he had enjoyed with Gilman at Hopkins was unlikely to be replicated within New College's walls. England may have been home, but Oxford and life as a professor in a cloistered college were foreign. Sylvester faced yet another challenge to fit in.

An Oxford in Transition

Sylvester made his initial visit to Oxford and New College early in January 1884 after an uneventful crossing followed by several days of recuperation at the Athenæum in London. Although he scarcely realized it then, he had entered an Oxford in the throes of a difficult transition from a collection of essentially ecclesiastical colleges into a modern university. That transition had begun in 1850 when Lord John Russell, the then Prime Minister, convened a Royal Commission to look into the University's educational, financial, and other affairs. The so-called Oxford University Commission met with stiff opposition within college walls, with virtually no college heads or other college officials agreeing to sit on the commission or to cooperate in any way with its inquiries. As a result, the commission report which ensued in 1852 represented solely the critics' views.

The Commission called for a reorientation of the pecking order within the University and a revision of the instructional roles of the tutors, the other teaching fellows, and the actual professors. In so doing, it effectively attempted to create a new profession—that of the academic—and so fit squarely into other midcentury efforts to define the class of professional men. According to the commission's scheme, tutors and other teaching fellows would devote themselves to drilling students, whereas the professors would be in charge of lecturing in their respective subjects, setting examinations, and evaluating student performance. This would have strengthened markedly the position of the professors, who, before midcentury, were more adjuncts to than key participants in the teaching mission of the colleges. At the same time, it would have weakened significantly the influence of the tutors and teaching fellows, who had taught the variety of examination subjects, set the examinations, and done the marking and grading. The professoriate, as the Commission conceived of it, would represent a viable career objective for the academically talented fellow, rather than having college fellowships be, as they had traditionally been, mere stepping stones into the hierarchy of the Church of England.[2]

Not surprisingly, the tutors were less than pleased with this proposed change in their status, although in many ways they were in agreement with the Commission's underlying objective of creating an academic profession. Shortly after the publication of the Commission report, they organized themselves into the Tutors' Association and began a pamphlet-writing campaign aimed at articulating their ideas for reform. In their view, the tutor should no longer teach a wide range of subjects but rather should be encouraged to specialize and deepen his knowledge of the particular fields—the classics, in particular—encompassed within the examination schools. This would address the ideal of higher scholarship underlying the Commission's proposals at the same time that it would maintain, what the tutors viewed as, the pedagogically superior system of tutorials as opposed to larger lectures. With the role of the tutor thus reconceptualized, they argued, "tutor" could become a professional career objective—especially if the numbers of students were increased to provide for more jobs—without prohibiting the strengthening of the professoriate as an extracollegiate resource and as a viable career move up for the particularly gifted tutor (335–341).

When William Gladstone, the Member of Parliament for Oxford and an Oxford graduate sympathetic to the tutors' position, drafted the Oxford University Act in 1854, he basically ignored the Commission's more university-oriented recommendations in favor of the tutors' college-centered views. The bill that resulted largely maintained tradition, although it did make possible an open, merit-based competition for a number of fellowships at each college at the same time that it made it officially possible for dissenting students to enter and take their B.A.s. It also reformed university governance, giving the tutors a greater voice, even if it did not address the tutors' call for increased opportunities for specialization.[3]

Perhaps the main result of this variety of efforts in the first half of the 1850s was to define the different constituencies at Oxford—in particular, the tutors and the professors—and to raise awareness of the issues that would confront a modern as opposed to an ancient university. Among those issues was that of more specialized teaching in areas appropriate for undergraduate examinations. Traditionally, classical studies had formed the basis of the examination school in *Literæ Humaniores*, but just before the formation of the Royal Commission, new examination schools in the natural sciences (1849) and in law combined with modern history (1850) had been created. The questions naturally arose, should there be examination schools in other subjects as well, and, if so, how should the colleges—unwilling to decrease the number of their fellowships, especially in the classics—provide for instruction in those new fields?[4] These and other questions were addressed by a new commission convened in 1877.

Relative to more specialized teaching, the commission of 1877 recommended the creation of so-called "Boards of Faculties" to be made up of representatives from throughout the university in a particular field who would give, in addition to their tutorials within their respective colleges, lectures in their own more specialized field to any and all interested students regardless of their college affiliation. These boards would then serve as the examining bodies for those students. At the same time, they represented a first step toward the institutionalization of specialization at Oxford. Regarding staffing, the commission also looked into the finances of the various Oxford colleges and found vast resources idiosyncratically allocated and unevenly distributed between the colleges as well as between the colleges and the University. Clearly, one way to fund more fellowships was to reallocate and redistribute these resources more equitably. Although this would also help to make a teaching fellowship a more viable academic career, the celibacy requirement for fellows still represented a major roadblock to that ultimate goal. The commission thus called for the creation of the new category of "official fellow" for those who wanted to marry *and* pursue a career in academe. Finally, the commission tried to strengthen the professoriate by augmenting professors' salaries with funds from the colleges to which they would be associated and by making them *ex officio* chairs of the different Boards of Faculties. Although the first of these initiatives was somewhat successful, the latter failed owing, once again, to lobbying by the tutors who did not want to see their power undermined. The compromise that was reached found the professors as *ex officio* members of the boards but only the tutors with voting rights to elect new members.[5] The Oxford Sylvester encountered in 1884 was one largely under the sway of these 1877 reforms.

As Savilian Professor of Geometry, moreover, Sylvester was attached to New College, which, despite its name, had been founded in 1379 by William of Wykeham. Intimately linked to Wykeham's school for boys, Winchester College, New College had traditionally been a closed foundation limited to some 70 fellows. Its fellowships, granted probationally at the time of matriculation, were limited to those who had held scholarships at Winchester. Because Winchester scholarships were given not on the basis of academic prowess, but rather on that of patronage, and because the Winchester education of the early nineteenth century was not of a particularly high caliber, these students were not necessarily the most intellectually inclined. Before the midcentury reforms, then, New College was a highly inbred institution, the intellectual life of which "was at a low ebb."[6]

Still, like the university of which it was a part, New College was fundamentally affected by the atmosphere of reform. Described as "a sort of fossil" before 1850, the New College of the last half of the nineteenth century became one of the most preeminent colleges at Oxford (72). That transition owed largely to several key players within the College's walls who shared and worked to implement many of the reformers' ideals. It also owed to one influential who at least did little or nothing to prevent those ideals from being realized. The latter was James Sewell, Warden of New College from 1860 to 1903.

Sewell, dubbed by the students "The Shirt," was a man of the old school. He had come to New College as a student in 1827 and, except for a period of a few months in the 1830s, had lived in college ever since. His mindset was very much that of the *status quo*; he saw no need to expand the curriculum or the student body; he saw a need neither for specialization nor for an academic profession; he was perfectly happy with the College's finances and statutes; he had little thirst personally for academic attainments. Still, as his colleagues increasingly lobbied for change, Sewell did not officially oppose or attempt to block their efforts. One of the first of these initiatives was the change in the constituency of both the student body and the fellows.

The modest reforms that issued from Gladstone's 1854 bill were already having a visible impact on New College just as Sewell became Warden. In 1857 new ordinances were adopted that, when they went into effect in 1860, replaced the College's 70 fellowships by 30 fellowships and 30 scholarships. Of the fellowships, 15 were to be assigned on the basis of open competition with 15 reserved for those from either Winchester or New College. As for the scholarships, they were no longer limited to scholars of Winchester but could be held by any qualified Winchester student. Moreover, should Winchester fail to provide adequate candidates, the scholarships, too, would be awarded through competition. This particular reform thus opened what had been a totally closed, Wykehamist fellowship to men with other intellectual backgrounds, while another reform required that the costs of the Savilian Professors of Astronomy and Mathematics be charged to New College's accounts.[7] These changes—coupled with the facts that teaching had improved at Winchester, that several fellows resigned their fellowships in the 1850s to be replaced by new blood, and that Sewell took a fundamentally *laissez-faire* attitude during his over forty-year tenure as Warden—help explain the transformation New College was able to undergo after 1860.[8]

Prior to the Commission of 1877, New College had already abandoned the celibacy requirement, and it had undertaken a massive expansion, building the immense but architecturally uninspired Holywell Building to house a student body that had increased in size from 30 in 1860 to 100 in 1870 to 200 in 1880.[9] After the commission issued its report and New College's statutes were changed once again in 1882, the college fellowship increased to 36 of whom five were professorial fellows (among them the two Savilian professors), a maximum of 10 were tutorial fellows, and a maximum of four were research fellows elected on the basis of their scholarly attainments. The other fellows held ordinary, seven-year, nonrenewable fellowships, which went alternately to Wykehamists and non-Wykehamists, all on the basis of competitive examination.[10]

As for college governance, that changed, too. The Warden and College officers dealt with the day-to-day business, whereas the Committees on Tuition (1870) and Estates (1871) were formed with membership from the senior fellowship to oversee the College's teaching mission and its various properties, respectively. "The establishment of these two major committees . . . marked the college's recognition that teaching was its main task, and financing its own expansion as an institution of learning its main problem" (92). With its increased physical plant, its growing student body, its enlarged and more professionalized fellowship, and its more structured and shared governance, this was Sylvester's New College.

In addition to Sewell, who by the time of Sylvester's arrival was in his seventy-third year and increasingly content to pass his time in the Warden's Lodge "carefully annotating the statutes laid down by the Founder, while his colleagues got on with running the college under the statutes that had replaced them" (85). Sylvester's new senior colleagues included the politically active Hereford George, tutor and a modern who viewed college teaching and research as a viable career option; William Spooner, tutor and dean in charge of student discipline; and Alfred Robinson, Senior Bursar, intercollegiate lecturer, and one of the major forces for reform within the college and university. To varying degrees, these men were responsible for New College's transition into the modern age. These were among the men Sylvester would have to work with and get to know. He was the newcomer; they were well entrenched.

Hereford George was born in 1838 the year after Sylvester had sat the Tripos in Cambridge and the year in which he had begun teaching as Professor of Natural Philosophy at University College London. A burly man with a square auburn beard, George had proceeded to New College from Winchester in 1856, had taken his B.A. in 1860 with a second class finish in both the classical and mathematical schools, had become M.A. in 1862, and had been called to the Bar at the Inner Temple in 1864. After three years in practice, he returned to New College as a tutor in the combined school of law and modern history, before moving into the position of tutor of modern history in 1872 after those schools were separated. He held the latter post until 1891 but retained his fellowship and continued to work actively until his death in 1910.[11]

In his long association with New College and with Oxford, George "was both an exemplary and an influential figure, and one of the most important of the benign forces for change."[12] He completely embraced the dual objective of teaching and research, teaching modern history and doing original work especially in military history and in the interrelationship between history and geography, at the same time that he helped define the new profession of "academic" at Oxford. The first married fellow in Oxford (he married in 1870), George also argued forcefully before the Commission of 1877 for the institutionalization of research seminars to be taught by the professoriate for the benefit of the teaching fellows, in particular, and was active in the governance of the reformed New College, serving at various times as Sub-Warden and on the powerful Tuition Committee.[13]

William Spooner, six years younger than George, came along just after the reforms of 1857 had gone into effect in 1860. In 1862, he became the first non-Wykehamist to be admitted as a scholar at New College and, in 1903, on Sewell's death, the first non-Wykehamist to be elected Warden.[14] He was thus the first drop of that new blood that infused the College from the 1860s on, becoming a fellow in 1867 and serving both as a tutor in ancient history and philosophy beginning in 1869 and as dean from 1876 to 1889.

A rather small man who suffered from albinism and so even as a boy had white hair, a pink complexion, and very short-sighted eyes, Spooner was known at New College for "his sweet temper, keen wit, and ingratiating appearance, . . . his capacity for dealing humourously with youthful folly, his sanity and Christian charity, and the perennial freshness of his interests and conversation."[15] Although not a particularly strong scholar himself, he nevertheless appreciated the new research and scholarly ideals, while devoting himself to his mission as a teacher. He also worked hard within New College's walls to see to its effective governance, serving in a variety of capacities— Sub-Warden, member of the critical Estates Committee, and others—before becoming Warden.[16]

Alfred Robinson was new blood of another type. He had studied at University College London before winning, in 1864, one of the new open fellowship competitions at New College and taking up teaching duties there. His expertise lay in mathematics and political economy, and although he, like Spooner who had attended his lectures while a student, was not that strong a scholar, he channeled his energies into the governance and reform of the College and the University.

A "formidable North-Country man with a square beard," Robinson was one of the driving forces, along with George, of the success of the intercollegiate lecture system institutionalized by the 1877 reforms.[17] He also embraced the philosophy behind the fiscal reforms proposed in 1877 and, in fact, had been at work on the College's finances on assuming the powerful post of Senior Bursar in 1875. In that position, Robinson not only controlled the College's purse strings but also oversaw—and to a large extent made possible through shrewd financial management—the College's new building project in Holywell Road, its increased student body, and its greater financial contribution to the purposes of the University.[18] Spooner made it clear that Robinson ruled New College with an iron fist when he wrote that "some found the ever present pressure of such a personality, exerting itself as time went on with ever increasing force, as irksome, or even in a few cases, intolerable: but most of us lived under his predominance with a sense of security and confidence which it would have been very difficult to disturb."[19] Robinson had a clear idea of the direction in which New College and the University should go—the direction indicated by the reforms of 1877—and he was confident that he knew exactly how to get them there.[20]

On his arrival at New College in 1884, unlike when he had arrived at the Johns Hopkins eight years earlier, Sylvester entered a fellowship in which established leaders had been working for more than a decade to effect change and to bring their college and their university into the modern era. They had made significant strides; their strategies were working; they knew how to proceed. Given this environment, it would be unlikely that Sylvester would reprise the role of trusted adviser he had played so well at Hopkins. What then would his role be? How would he fit in? Would he fit in? His first months as Savilian Professor found him unclear as to the answers to these questions.

Between Two Worlds

Sylvester made his way to Oxford from London on Friday, 4 January to be introduced, as he explained proprietarily in a letter to Gilman, "to my new College."[21] It was only a weekend trip, though. He did not yet have rooms in college, and the university, still in the recess between Michaelmas and Hilary terms, would not reconvene for another two weeks or so.[22] Nothing yet tied him to Oxford.

Back at the Athenæum by Monday, he spent the week in London before making the return trip. On Tuesday, the fifteenth, he was admitted into the fellowship at New College. The rooms he had chosen the week before had been furnished temporarily until his books and furniture arrived from Baltimore. His residency in Oxford had begun. Not surprisingly, he occupied himself with letter writing, composing by his own count about 30 letters over the two days from 20 to 21 January and sending his "autographs . . . flying all over the world" with word of his new situation.[23]

One of those autographs flew to Cambridge and the box of Charles Taylor, Master of St. John's College. Taylor had written back in late December to congratulate the mathematician on his new appointment, but his letter had missed Sylvester in Baltimore, failed to find him in London, returned to Baltimore, and finally arrived in Oxford. In thanking him for his warm wishes, Sylvester allowed that "I . . . find every body friendly and disposed to make me feel at home in my new surroundings." He also wondered if "I may perhaps be able to serve some useful purpose as a common term (to speak in Algebraical phrase) between our two great Universities and the Universities with which I have been affiliated on the other side of the Atlantic."[24] Gilman had consciously encouraged and benefitted from Sylvester as a mathematical link between Hopkins, England, and the Continent. In seeking to define his new niche, Sylvester thought that perhaps he could now play the role in reverse.

The upbeat tone of the letter to Taylor masked, however, the loneliness that Sylvester had already begun to feel in Oxford. Writing to Cayley on the twenty-second, he enthused that "this is a delightful college to live in and I take part in all the college business" but then asked "will you come and pay me a visit here if and when I can find a bedroom for you in College?"[25] Just as in his efforts to lure Cayley to Baltimore, he immediately followed the invitation with a description of the College's attributes. "There are two sitting rooms in my quarters so that you can have a room to yourself to work in. I will at once inquire (if you say *yes*) about a vacant bedroom. If you could come some Saturday and stay over Sunday you could attend the beautiful service in the College chapel. . . . If you could come you would be able to show me my way about Oxford (as you did at Baltimore) and find out the walks to take in the neighbourhood."[26] Two days later, Sylvester was back at the Athenæum, having gone to London "for a day or two change of air and hav[ing] just succeeded in getting out of cough and bronchitis, thanks to this and the reappearance of the *sun*."[27] He had decided to exercise his right to a "grace term" from teaching and so would not take up his teaching obligations until the Trinity term in late April.[28]

This decision certainly did not help his acclimation to Oxford and life at New College. His restlessness manifested itself over the months from January through April in an almost constant back-and-forth between Oxford and elsewhere. There were frequent sojourns in London sometimes coinciding with meetings of the London Mathematical Society, a trip to Cambridge in early March to see Cayley and other old friends like Ferrers, and a journey to Edinburgh in April as the representative of the Johns Hopkins University on the occasion of the tercentennial celebrations at Edinburgh University. Between two worlds, Sylvester was seeking comfortable companionship and diversion and avoiding the hard adjustment to his new life.

This avoidance was facilitated by the fact that his mind was still very much back in Baltimore. In search of a new Professor of Mathematics, Gilman repeatedly consulted Sylvester for advice during the winter and spring, asking for his help in the negotiations and briefing him on how things were progressing. Sylvester had thought at the time of his departure that his successor would likely be Felix Klein. After all, he had proposed Klein's name as early as 5 September 1883, and the initial overtures in the fall of 1883, carried out through Cayley, had been favorably received. By December, Gilman had offered Klein the job, and the two men had exchanged correspondence regarding some of the specifics of the appointment such as salary and pension provisions. Klein had followed up with a set of queries addressed personally to Sylvester about the nature of both the position and Sylvester's reasons for leaving it, and Sylvester had answered candidly and positively on 17 January 1884. Unfortunately, Gilman had not ultimately been able to convince the Hopkins Trustees to agree to Klein's two deal-breaking conditions: a salary equal to Sylvester's terminal salary of $6,000 and the establishment of an annuity to provide for Klein's family in case of his death.[29]

By early February, Sylvester had learned from Gilman that the negotiations with Klein had failed. Writing to Cayley on the third, he stated plainly that "the J.H.U. has made a mistake" and explained that "Gilman said to me that if they made Klein's stipend $6,000 all the other professors would require their salaries to be raised to a like amount. I did all I possibly could to insure that the offer should be raised to $6,000 but Gilman as you are aware is not absolute and has the Trustees to reckon with."[30] In that same letter, Sylvester also replied to Cayley's inquiry about Ferdinand Lindemann as a possible candidate. Lindemann had made a splash in the mathematical world just two years earlier with his proof of the transcendence of π and was certainly a candidate with some cachet. Still, Sylvester wrote, "I can offer no opinion as to Lindermann [sic]. He does not bring the same prestige as Klein would have done to *compensate* for his being a German i.e. a foreigner with imperfect command of English. But possibly Lindermann [sic] may know English better than Klein."[31] A letter from Gilman four days later, in fact, announced that he intended to go next not after Lindemann but after Cayley, and enlisted Sylvester's help. "It is almost too much to expect that you could urge his coming," Gilman wrote, "but we are confident that you will do all that you can to second our overtures."[32]

Although it is not clear how hard—if at all—Sylvester lobbied Cayley to take the Hopkins post, Cayley was not interested in leaving Cambridge. Gilman next fixed on his early scientific adviser, the mathematical astronomer Simon Newcomb, and asked Sylvester his opinion of Newcomb as a successor. The mathematician and the astronomer had gotten to know and respect each other during Sylvester's time at Hopkins—Newcomb had been the motive force behind Sylvester's selection as Foreign Associate of the National Academy of Sciences—so it is not surprising that Sylvester had nothing but praise for this latest candidate. "I can only say," he told Gilman, "that in my opinion if you can secure Newcomb it will be a very great hand indeed."[33] Sylvester gave Gilman more than just his opinion, though. He had also put the question of Newcomb's reputation to his colleague, Charles Pritchard, the Savilian Professor of Astronomy, and had gotten a glowing review.[34]

As the matter of his successor continued to bind Sylvester to Baltimore so his matrix-theoretic research was a constant reminder of his last course of lectures there. During the fall of 1883, Sylvester had continued to write up his evolving ideas on matrices and their properties. As usual, the results were worked out first in his Hopkins classroom and then came out in preliminary announcements in both the *Johns Hopkins University Circulars* and the *Comptes rendus*. One of the last mathematical projects he had completed before setting sail for England, in fact, had been an article for the *Circulars* on a set of "three laws of motion" that he had formulated in the matrix-theoretic context and that he conceived of as parallel in some sense to Newton's three laws of motion. The three laws of motion, so called because "as motion is operation in the world of pure space, so operation is motion in the world of pure order," dealt with certain properties of the eigenvalues (or "latent roots" as Sylvester called them) of a matrix, the behavior of matrix multiplication when an $n \times n$ matrix is multiplied successively by the powers of another $n \times n$ matrix, and certain properties of the product of two $n \times n$ matrices.[35] These, in Sylvester's mind at least, were as fundamental to universal algebra (that is, the theory of matrices) as Newton's laws were to physics. The corollary as to Sylvester's perceived importance of his contribution to this area of mathematics is clear!

Sylvester's correspondence in January was full of exuberant overstatement about the value of these three laws as well as prescient pronouncements about the import of universal algebra. Just after he reached London in January 1884, he made one of the overstatements in predicting to Gilman that the three laws of motion "ought to make a sensation in the philosophical as much as in the mathematical world."[36] Two weeks later, he waxed prescient as he extolled to the Master of St. John's College, Cambridge "the vast subject of Universal Algebra which swallows up Quaternions as Aaron's rod swallowed up those of the Egyptian magi." "I wonder," he asked rhetorically, "what your antepredecessor, the Master in my time Dr Wood (whose Algebra was our textbook in those days) would have thought of this the *real* new Algebra, or as I sometimes term it Algebra the 2^d, called to the throne after the dynasty of Algebra the 1st dating from Harriott's posthumous work had held rule for almost exactly a quarter of a Millenium! The new Algebra and the new Electricity—these it seems to me will be one of the principal occupations of the next century to familiarize and develop."[37]

By the end of the month, Sylvester was back actually doing mathematics rather than merely writing about work he had already done. He focused, in particular, on the matrix algebra of nonions and on questions such as, given two 3×3 matrices M and N in the algebra of nonions, under what conditions is it the case that M and N commute, namely, when is the equation $MN = NM$ satisfied?[38] His attack on these and similarly spirited questions—often cast in general terms of $n \times n$ matrices—centered on the direct computation and solution of a variety of matrix equations, among them the characteristic equation.[39] This then naturally led him to an analysis of the eigenvalues or characteristic roots, that is, the roots of the characteristic equation.

All this work culminated later in 1884 in what was to have been the first in a series of "Lectures on the Principles of Universal Algebra" published in the *American Journal*. There, Sylvester, as systematically as his mathematical style allowed, specified and gave the elementary properties of the operations on matrices—multiplication, addition, scalar multiplication—defined some of the basic notions like inverses, and discussed key constructs like the characteristic equation and its associated roots. The examples that he worked out, however, were all 2×2 or 3×3, and he provided no general proofs. Still, he recognized the possibility of a general theory, even if his approach through explicit matrix calculations did not seem amenable to one. He closed his paper, noting "that quantities of every order admit of being represented in the mode strictly analogous to that in which quantity of the second order is represented by quaternions."[40] Although he took it no further, he was certainly proven right that a general theory was both possible and highly fruitful.

In addition to the matrix-theoretic work carried over from the Hopkins years, March 1884, in particular, also found Sylvester refocused on some of the invariant-theoretic results he had obtained while in Baltimore. In the latter case, it was another matter of priority. Percy MacMahon, a Captain in the Royal Artillery and an instructor in mathematics at Sylvester's former institution, the Royal Military Academy, had just published a paper in the *American Journal* entitled "Seminvariants and Symmetric Functions," in which he sketched the proof of a theorem to which Sylvester laid claim.[41] The so-called Correspondence Theorem represented a key computational and conceptual development of the British approach to invariant theory.

Recall that in his "Second Memoir upon Quantics" of 1856, Cayley had given a combinatorial formula for determining the number of linearly independent covariants of a given degree and order associated with a particular binary form, in addition to tackling the related counting problem of the number of covariants in a minimum generating set of that form. The latter problem had especially engaged Sylvester and his Hopkins students from 1877 through 1879 as they performed the horrendous calculations that allowed them to solve it systematically for binary forms of successive degrees. The key difficulty in this process, of course, was the determination of the algebraic dependence relations, that is, the syzygies, between the covariants. In trying to deal with these problems, Sylvester had developed in 1882 a theory of what he termed "subinvariants" or polynomials in the coefficients a, b, c, ..., k, l of a binary form that are annihilated by the differential operator $\Omega = a\partial_b + 2b\partial_c + 3c\partial_d + \cdots + ik\partial_l$.[42] Without explicitly naming them, Cayley had already recognized in 1856 that the problem of determining the covariants associated with a given binary form reduced to that of finding the subinvariants of that form, because the leading coefficient of a covariant is a subinvariant and because every covariant is uniquely determined by its leading coefficient. In his 1882 work, Sylvester had realized that any syzygy between subinvariants implied a syzygy between the corresponding covariants, so this provided a new way of trying to isolate syzygies among elements in a minimum generating set. MacMahon reduced the problem once again in 1884 by proving the Correspondence Theorem, namely, that the subinvariants were, in fact, equivalent to special kinds of symmetric functions.[43] The import of MacMahon's result was that the long-studied and well-known theory of symmetric functions could now be brought to bear on the invariant-theoretic problem of determining minimum generating sets.

In a letter to Cayley on 2 March 1884, Sylvester commented that "MacMahon can well afford to forego the credit of 'discovering' the correspondence theorem to which he has not a shadow of a claim. It is Brioschi's and my remark."[44] In Sylvester's view, Francesco Brioschi had proven the key facts about symmetric functions, and Sylvester had seen how to apply them in the subinvariant calculations in his 1882 paper. As the month of March went on, Sylvester stewed more and more over MacMahon's work. By the thirtieth, he had actually read MacMahon's paper and taken serious exception to his proof. Writing again to Cayley, he enumerated the proof's faults and concluded that "on these grounds giving MacMahon all possible credit for the happy idea of looking to the roots of the associated quantic (which be it observed I had drawn attention to in a previous paper) I have now no hesitation in speaking of the . . . theorem of Correspondence as my theorem or at the very least as the Sylvester–MacMahon theorem."[45]

He proceeded to stake his priority claims publicly both abroad in the *Comptes rendus* and at home in the *Messenger of Mathematics*.[46] At issue, in his view, was the credit for reducing invariant theory—with its strongly geometrical overtones—to algebra or combinatorics. This had, after all, been one of Sylvester's lifelong goals as a mathematician. As he put it in the *Comptes rendus*, "it is clear that the same principle can be applied to invariants of all kinds, such that, thanks to Mr. MacMahon's beautiful discovery, with the generalization (almost intuitive in a way) that I gave, one can today treat the most difficult and the most essential parts of the theory of algebraic forms . . . by abstracting out, so to speak, all question of substance (of matter contained in the forms), and by limiting oneself to a purely combinatorial calculation."[47] Having had his say, by the end of April, this matter of priority was forgotten, and Sylvester was back at his matrix theory.

He was also back at something else—teaching—but it was not the kind of teaching to which he had become accustomed and in which he had excelled at Hopkins. It was low-level, undergraduate instruction in the prescribed subject of geometry—never one of his favorite topics and one that he had systematically striven to eliminate from invariant theory—and he had begun it in the Hilary term. As Sylvester had explained to Cayley earlier in January, "I am bound by usage and by the wants of the undergraduates reading for *mods* to give a course or courses of lectures on *pure* synthetic geometry."[48] Moderations, or "mods" as they were familiarly called, were examinations taken after four or six terms at Oxford and represented the students' first major hurdle toward the degree. It was thus important that they get the "right" preparation, and Sylvester was clueless as to even what textbook to use. His anxiety over this came out clearly in another letter to Cayley in late January of 1884. "What do you think would be the best or at all events a good book to use as a text book for lectures on Synthetic Geometry?," he asked plaintively. "Smith did not use any but it would be much better (I am persuaded) for me and my class to adopt one. Would Townsend be a good book? . . . What other books could you name for me to have for my private use alongside of Townsend or other text book that I might decide upon?"[49]

Early February still found him worrying—and still to Cayley—about how he was going to do undergraduate teaching in geometry. "Under the 'new statutes' my province is more strictly defined than was the case with my predecessor," he explained, "and I shall have enough to do to qualify myself for this without extravagating in my lectures into any outside topics. There will be plenty of work for me to do in concentrating all my efforts upon duly qualifying myself as a lecturer of geometry 'pure and analytical.' "[50] Sylvester recognized that, at Oxford, he was not going to enjoy the liberties he had had at Hopkins. No longer would he be able to rove in his lectures where his immediate research interests took him. In fact, his immediate research interests would ostensibly play no role in his Oxford classroom. Moreover, he was going to have to set and keep to a text at the same time that he got himself up to speed with the material. A realist who recognized his own foibles, Sylvester was rightly concerned about whether he would succeed in disciplining himself to this kind of focused teaching.

By May, he was well into his course, giving three lectures a week on projective geometry and using Luigi Cremona's book, *Éléments de géométrie projective*, for his personal preparation.[51] After a somewhat fitful start, he was able to report to Cayley on 20 May that "I am getting on now without trouble in my lectures" and "manage to draw on the board sufficiently well for the purposes of instruction all the geometrical constructions required."[52] With the teaching going well, he was feeling more comfortable in his new position and in his new surroundings, all the more so because he had begun to entertain friends in college. Cayley had come in March, but in May, Cremona had stayed with him for a couple of days just as he was immersing himself in his Italian friend's text, while Leopold Kronecker, his wife, and daughter spent "a happy day" with him, dining as his guests in the Common Room in the new Holywell Buildings.[53]

Just as he seemed to be settling into a new routine in Oxford, however, a poignant reminder of Baltimore drew his thoughts back to the other side of the Atlantic. His former student, Thomas Craig, visited him at New College in June, bearing a most unexpected gift—a gold medal commemorating his association with the Johns Hopkins University, together with a letter of appreciation signed by Gilman and numerous of Sylvester's former colleagues and friends. He hardly knew how to express his deeply felt emotions on receiving this testimonial. "Words fail me to give adequate utterance to the gratitude which I feel for the kindness which has inspired the preceding and which I trust I shall continue to entertain unto the latest hour of my existence," he wrote to Gilman. "Let me hope to live still in your recollections as one who cherishes the remembrance of your almost too great kindness and whose dearest aspirations would be gratified if he could but feel himself more worthy of being the object of it." He closed his letter "with the sincerest sentiments not merely of collective but of individual and personal regard and esteem," and signed off "your obliged and attached, Servant and Friend."[54] Sylvester had, indeed, left behind true friends in Baltimore. Their expressions of friendship and esteem only drove home the fact that he had yet to make any new friends among the Oxford fellowships.

On 24 June, just days after Craig's visit, Sylvester left Oxford for London and a summer of work on matrices at the Athenæum. The club may have felt more like home than New College, even after a term performing the duties of the Savilian chair, but this flight from Oxford only exacerbated the adjustment to his new life. He returned on 1 August, but by the fifth, he bemoaned to Cayley that "for the last 3 or 4 days" he had been "without a soul to communicate with—every fellow and undergraduate having left the College which I am the sole tenant of except for the Warden and Subwarden who live in their own houses and have taken no notice of me."[55] This kind of isolation—like he had experienced in Baltimore in the summer of 1882—was never good for Sylvester. This time, though, depression did not set in. He was leaving for London again that very afternoon and proposed to Cayley "if it be agreeable to you to join you at Grassmere."

Sylvester spent most of August and the first week of September in the Lake District, at Grassmere near the Cayleys and also in Keswick where he met up with other friends and was even entertained by the local vicar. He occupied himself, too, with long rambles through the picturesque countryside and with work on the inaugural lecture that he was obliged to give as a condition of his chair.[56] The prospects of that lecture had been weighing on him since January. In fact, it was probably one of the reasons he had opted to take the grace term from teaching. Technically, he was supposed to have given the lecture on beginning his teaching duties, but he had managed to postpone it yet again.[57]

His anxiety stemmed from the facts that while he had no idea for a topic, he wanted to make the best possible impression. By 20 May, Robert Clifton, who was Professor of Experimental Physics and Director of the Clarendon Laboratory at Oxford, had suggested that he speak on Descartes as a geometer, an idea Sylvester found "admirable" since "Descartes' name is very much up just now in Oxford as a Philosopher."[58] Still, Sylvester was no historian, and he agonized over how to gather the material necessary to put together such a lecture. In July, he rued "the dreadful incubus of a Public Lecture which the University Statutes require[d] of [him] next term," especially since it was keeping him from expending his full energies on his matrix-theoretic research.[59] By September, he simply had to bite the bullet. "I have laid aside mathematical work for the present," he told Cayley, "and am reading up all I can about Descartes—with a view to compose the dreaded lecture to which I am pledged—a sad waste of time!"[60]

He left the Lake District for London on Tuesday, 8 September, and aside from a few days back in Oxford, spent essentially the rest of September in London trying unsuccessfully to keep his mind on Descartes and off of matrices. On the twenty-eighth, he was on board the *Invicta* from Dover to Calais, "under the vivifying and genial rays of a bright and benignant sun," en route to Paris, ever seeking companionship.[61] From Paris's Hotel Normandy, he wrote Cayley on the thirtieth that there was "*nobody in Paris.*"[62] He had already gone to the Academy of Sciences, where he had met up with a few colleagues, but "they are all staying out of town." Nevertheless, even a scientifically empty Paris was apparently better than Oxford. Sylvester stayed there into the third full week of October, missing Hermite but seeing Émile Picard, Gaston Darboux, Camille Jordan, Joseph Bertrand, and others, and working on matrices and the lecture.

The research on matrices, in particular, had a number of consequences: a lengthy technical note by Sylvester in the *Comptes rendus*, Sylvester's short heuristic account in *Nature* of how the results in that note evolved, and a brief but ultimately influential paper in the *Comptes rendus* by Henri Poincaré highlighting the links between Sylvester's matrix-theoretic researches and work being done on the Continent by Sophus Lie and others on Lie groups as well as on hypercomplex number systems, that is, algebras over $\mathbb{C}$ or $\mathbb{R}$.[63] It had been a productive and encouraging month in Paris, but Sylvester could avoid the inevitable no longer. He returned to Oxford just days before he had to start teaching on 29 October.

A Rocky Year

The 1884–1885 academic year was Sylvester's first full year in an Oxford classroom, and he was of necessity more tied down than he had been thus far. His ongoing indecision about the inaugural lecture and its eventual postponement got the autumn off to a rocky start, but he was at least initially optimistic about his teaching. He was lecturing three times a week on analytic geometry "to a numerous class" and had taken as his texts "three Chapters out of Salmon and Fiedler—viz On abridged notation, On binary invariants (from Fiedler) and on The invariants to ternary quadratic forms as applied to Conics."[64] While this may have been "a much more interesting subject to [him] than Cremona on projections," it was certainly not the sort of material that students would expect to be tested on in mods.[65] It was invariant theory, and not even a particularly gentle introduction thereto, coming from a textbook in German. Sylvester's choice of subject matter might have made sense for his former graduate classes at Hopkins, but it was not a very wise choice for Oxford undergraduates operating within a system of set examinations.

By the end of his first full week of lectures, Sylvester was already feeling the pressure of trying to balance non-research-oriented teaching with his own work, and looking ahead to the end of the term. "Alas!," he sighed to Cayley. "I shall be for some time occupied in preparing my lectures. . . . I . . . am obliged to postpone other matters to endeavoring to do justice to my auditors. In another month term will be up and in future I only propose to lecture twice in the week."[66] Still, he seemed upbeat about Oxford in general. "She is a dear good mother our University here and stretches out her arms with impartial fondness to take all her children to her bosom even those whom," like himself, "she has not reared at her breast." Moreover, he had gotten new rooms in College in October, and they were "a great improvement" on his earlier quarters. "You ought to feel pleased," he continued, "in having taken so important a part in bringing me back to my native country and enabling me to pass the evening of my days in so delightful an environment!"

Despite his professed pleasure in living in Oxford and the acknowledged burden of his teaching preparations, Sylvester slipped back over to Paris just after penning this letter to Cayley. Presumably he went for a meeting of the Academy of Sciences, but while there he suffered an attack of bronchitis that detained him for more than a week at the Hotel Normandy and caused him to miss that week's worth of classes back in Oxford.[67] By January 1885, however, he was reporting to Gilman that "I am now pretty well acclimated again to our English damp and gloom and thank God! enjoy perfect health."[68]

He had also been back at his poetry, giving readings in London and Oxford of four new compositions and distributing printed versions of them to the Gilmans in Baltimore, to his niece and her family in Florence, and to Warden Sewell in Oxford, among others. One of those compositions, a sonnet, described "some verses" that he had "lately recited before a company in New College, Oxford."[69] That he had felt a welcome member of the fellowship came through poignantly despite the belabored lines:

> *Where glows by Wykeham's fane the sacred fire*
> *And generous hearts the Muses' precincts throng*
> *A murmur spread that I should wield the lyre*
> *And strike the chords of my untutored song.*
> *Welcome acclaim did the glad notes prolong*
> *Pure as hosannahs of the shrill-voiced choir,*
> *Waking amaze that could my thoughts not wrong*
> *While some were moved with wonder to enquire*
> *"Truant from alien fields! where didst thou learn*
> *To paint the peerless subject of thy praise?"*
> *"The screen that pictures whom my soul prefers*
> *Freely" I answered "to the light I turn:*
> *Mine is the prism which unbinds the rays*
> *She is the sun, colours all are hers."*[70]

His colleagues' "generous hearts," their request that he "wield the lyre," and their "welcome acclaim," all clearly moved the "truant from alien fields," Sylvester. Unfortunately, with the new term about to begin, he realized that his "verse making w[ould] be at an end for some long time to come."[71] The duties of his office called; he had to prepare his lectures.

He was once again lecturing on analytic geometry from Salmon's text. Initially, he found it "quite interesting," but by the end of the term, he was "in a most wretched mental condition," having begun "to think that [he] was unequal to [his] post and to weigh whether [he] might not in good faith to resign the professorship."[72] The spring brought a brighter mood, more lecturing on analytic geometry to "a fairly numerous class," and a supplemental, intercollegiate course on matrices at the request of a number of the Oxford tutors.[73] The latter was just the kind of course Hereford George had argued in favor of before the 1877 commission, and it was much more similar in spirit to the kind of courses Sylvester had taught at Hopkins than his undergraduate lectures in geometry. Nevertheless, unlike at Hopkins, the course did little to stimulate Sylvester's mathematical creativity. Aside from four short and derivative notes, he accomplished little in the way of research during the 1884–1885 academic year.

There was a bit of politicking. His priority snit of the previous spring forgotten, he sounded Cayley out in April on MacMahon as a possible candidate for a Royal Medal from the Royal Society of London on the basis of his "colossal investigation of the Sextic Perpetuant" on the pages of the *American Journal*.[74] A direct outgrowth of the work on subinvariants that Sylvester had published in the *American Journal* in 1882, MacMahon's research focused on perpetuants— subinvariants that cannot be expressed as a polynomial in other subinvariants—and extended not only the British-style calculations from the case of the binary quintic to the binary sextic form but also the British theory. Recall that in his 1882 paper, Sylvester had begun laying out the theory for determining minimum generating sets of irreducible covariants in terms of subinvariant calculations. The principle he employed for detecting syzygies, however, what he called the fundamental postulate, was more a heuristic than a proven fact. Posing a prize question for proof or disproof in the *American Journal* about this problem of syzygies, he received a paper from James Hammond, a thirty-two-year-old English mathematician, with a disproof by way of a counterexample in the case of the binary septimic.[75] As Sylvester saw it, "my perpetuant paper began the ball. Then Hammond comes out with his syzygies provoked by the A.J.M. prize question. Then [Cayley] appl[ies] this to correct my result for the 5^c perpetuants. . . . Lastly comes MacMahon and . . . finds the *per*petuants themselves in *substance* as well as in *number* or say the *substance* and not merely the shadow!"[76] To this way of thinking, then, a Royal Medal to MacMahon would represent further acknowledgment of Sylvester's ongoing contributions to mathematics at its most sophisticated level.

After receiving Cayley's assurance in April that he would support MacMahon's candidacy, Sylvester officially put his name forward in May. The medalist would not be named until the following year. It was not clear who the competition might be. Sylvester would just have to wait and see.[77]

The end of the term in June again brought with it Sylvester's almost immediate departure from Oxford for London, where the Athenæum again served as the home base for his summer work and travels. Trips back and forth to Oxford were punctuated by a sojourn in Paris in July, a country retreat in August at the home of his friend, Clement Ingleby, the writer and Shakespearean critic, and a turn around the Continent in September and early October including a trip to the Netherlands, before his return to England in time for the start of classes. In London at the end of August, he took time out to answer a letter Gilman had sent him back in July and to report on his first full academic year in Oxford.

Sylvester opened with nothing but praise for Gilman's accomplishments at Hopkins—a new physical laboratory, the success of William Thomson's lectures, and the "truly admirable" quality of the *American Journal*.[78] "The Johns Hopkins . . . I often tell people here and at Oxford is the first University in the world for what it can do for its students and [for] the true university spirit which animates the place." Oxford was another matter. "If you miss me as you are good enough to say at the Johns Hopkins I miss you no less," he confessed. "For the last year I regret to say (or at all events since last October) I have done scarcely anything in Mathematics—the stimulus and excitement to work having been absent. I make it now my main business to qualify myself in the subject which I have to teach and to which I feel it my duty to devote myself." Still, he added, "I am well treated at Oxford and have no grievance but my life there is not particularly cheerful owing to the want of companionship with persons of suitable age and kindred tastes in science. Also the climate does not agree with me—although at the present time I am enjoying good health." These confessions betray the bittersweetness of Sylvester's Oxford victory. He held a prestigious English post, but it constrained him. It gave him no intellectual stimulation, stifled his mathematical research, and provided no natural confidants. Little wonder that he had taken every possible opportunity to seek companionship elsewhere since his appointment. If only Oxford could be more like Hopkins, his letter seemed to say, maybe then he could settle into a more comfortable and satisfying life there.

Hopkins in Oxfordshire?

At Hopkins, Sylvester had enjoyed three key things: the ear of the University's President, a strong research agenda, and graduate students. It would take at least two of these for Oxford to become more of a Hopkins for him. The influence he had had with Gilman and so on the broader affairs of the University, while gratifying, had been perhaps the least important ingredient in the Hopkins mix and was, in fact, impossible at Oxford given its decentralized system of governance. However, Sylvester *would* need to be actively and successfully engaged in a research program, and he *would* need an appreciative audience of advanced students before whom he could present his ideas as they emerged. The fall of 1885 brought with it the hope that he might actually be able to recapture these two components of his earlier success.

Although, as he told Gilman, he had done almost no research during the 1884–1885 academic year, early July of 1885 had found him in London thinking fleetingly about a new line of mathematical inquiry and agonizing over both his lectures for the fall and a referee's report he owed Stokes on a paper Alfred Kempe had submitted to the *Philosophical Transactions*. As usual, this kind of situation frustrated him. The new mathematical ideas centered on what Sylvester called reciprocants, that is, polynomial expressions in successive derivatives of y with respect to x that remain invariant when x and y are interchanged. This, he had decided, would form the topic of his long-postponed inaugural lecture, and he had forced his own hand by "giv[ing] notice in the Oxf^d Univ^y Circular of a Public Lecture on Reciprocants for the 12th of Decr next."[79] Still, on 4 July 1885, he was "obliged to postpone all work of this kind," convinced that "the Herculean task of reporting on [Kempe's] memoir . . . and the lectures together [were] enough to crush all the life out of [him]."[80]

Indeed, Sylvester gave seemingly little thought to reciprocants until October, when, as a guest of Cornelius Grinwis, the Professor of Mathematical Physics and Mechanics at the University of Utrecht, he learned about Andrew Forsyth's new book, *A Treatise on Differential Equations*.[81] Ordering a copy of his own, and actually *reading it* after his return to England, Sylvester became immersed in the new theory of reciprocants—a special theory, after all, of differential equations—and rapidly proved new results. By 9 November, he was writing Cayley that "the Theory of Reciprocants has made great advances."[82] By 23 November, the first installment of his announcement "Sur une nouvelle théorie de formes algébriques" had been read before the Paris Academy. By 28 November, he was writing to Gilman about his "great and unlooked for *revelation* which will alter the whole face of Analytical Geometry and also produce no less effect in the theory of Differential Equations and Transcendental Functions."[83] On 12 December, the day of the inaugural lecture, the Sylvester of Hopkins days was definitely back, and he was extolling the virtues of his latest work with characteristic hyperbole.

A large audience turned out that Saturday to hear Sylvester speak, with friends like Pieter Schoute, Professor of Mathematics at the University of Groningen, journeying all the way from the Netherlands, others like Percy MacMahon riding up from London, and still others coming from the various Oxford colleges.[84] Sylvester felt obliged to open with an apology for and an explanation of the tardiness of the lecture. "I have waited, before address-ing a public audience," he told them, "until I felt prompted to do so by the spirit within me craving to find utterance, and by the consciousness of having something of real and more than ordinary weight to impart."[85] The lecture on Descartes that he had contemplated and agonized over simply had not moved him. It would not have been a lecture of real "weight," but rather something perfunctory worked up merely to satisfy an obligation. Now, though, he had hit upon something exciting, something "far bigger and greater, and of infinitely more importance to the progress of mathematical science" than the topics of any of the public lectures he had ever given.[86] Whereas "no subject during the last thirty years has more occupied the minds of mathematicians, or lent itself to a greater variety of applications, than the great theory of Invariants," the the-ory of reciprocants he was about to describe "infinitely transcends in the extent of its subject-matter, and in the range of its applications" ordinary invariant theory.

To give the flavor of his subject, Sylvester presented the example that had so energized him in October, namely,

$$\left(\frac{d^3y}{dx^3} \div \frac{dy}{dx}\right) - \frac{3}{2}\left(\frac{d^2y}{dx^2} \div \frac{dy}{dx}\right)^2. \tag{1}$$

"If in this expression the x and y be interchanged," he told his audience, "its value, barring a factor consisting of a power of the first derivative, remains unaltered, or, to speak more strictly, merely undergoes a change of algebraical sign."[87] The reciprocant in (1) is an example of what Sylvester dubbed a "mixed reciprocant," because it involves the first derivative; a reciprocant is "pure" oth-erwise. Geometrically, "every pure reciprocant corresponds to, and indicates, some singularity or characteristic of a curve, and vice versâ every such singular-ity of a general nature and of a descriptive (although not necessarily projective) kind, points to a pure reciprocant." In this light, the algebraic isolation of pure reciprocants provides geometric information about the underlying curve. Pure reciprocants thus represent another means toward Sylvester's lifelong quest of algebrizing geometry. Little wonder that he was so taken with this new theory.[88]

After sketching for his audience the theory of pure reciprocants—and show-ing them how it closely paralleled the usual theory of invariants—he moved on to do the same for the more challenging theory of mixed reciprocants. A brief poetic interlude inspired by his new researches and addressed "To a Missing Member of a Family Group of Terms in an Algebraical Formula" preceded "a sort of after-course" of "general reflections arising naturally" from his subject matter.[89] One of those general reflections centered on Oxford, his perception of its educational mission, and his view of his role there. It involved a confession.

"During the past period of my professorship here, imperfectly acquainted with the usages and needs of the University," he acknowledged, "I do not think that my labours have been directed so profitably as they might have been either as regards the prosecution of my own work or the good of my hearers: my attention has been distracted between theories waiting to be ushered into existence and providing for the daily bread of class-teaching."[90] His solution was, not surprisingly, for him to abandon his prescribed undergraduate teaching and, as he had at Hopkins, offer courses solely on the research in which he was immediately engaged. "Thus, by example," he would "give lessons in the difficult art of mathematical thinking and reasoning—how to follow out familiar suggestions of analogy till they broaden and deepen into a fertilising stream of thought— how to discover errors and to repair them, guided by faith in the existence and unity of that intellectual world which exists within us, and is at least as real as that with which we are environed." This, he told his listeners, was what he had done at the Johns Hopkins, and it had resulted in much new work produced in conjunction and in cooperation with his students. "It was frequently a chase, in which I started the fox, in which we all took a common interest, and in which it was a matter of eager emulation between my hearers and myself to try which could be first in at the death" (297). Introducing yet another metaphor, like the painting master, he hoped "to induce [the members of his classes] to take the palette and brush and contribute with their own hands to the work to be done upon the canvas" (297).

In short, Sylvester sounded the call, at least in mathematics, to make Oxford a Hopkins, a university with the dual mission of research and the training of future researchers. To this end, he had already announced that his Hilary term course would be on the theory of reciprocants. By offering high-level courses and with the aid of a few more men of demonstrated mathematical talent, Sylvester declared, he, together with his "brother Professors and the Tutorial Staff of the University . . . could create such a School of Mathematics as might go some way at least to revive the old scientific renown of Oxford, and to light such a candle in England as, with God's grace, should never be put out" (300).

The sentiments Sylvester expressed in his inaugural were as self-serving as they were bold. He had been miserable—especially in his teaching—since his arrival at Oxford, and, at the same time, he recognized how successful the Hopkins dynamic had been for his mathematical research. What better way to improve his own lot than to take matters into his own hands and publicly redefine his professorship? If the experiment worked, moreover, he would have succeeded not only in creating a better life for himself but also in making Oxford more of a competitor in the international mathematical arena. It could be a win-win situation.

High on the reception of his lecture, Sylvester left Oxford to spend the inter-term break in London, where reciprocants dominated his thoughts. Back in college by 18 January 1886 to attend a "more than ordinarily important 'slated quarterly meeting'" of the College fellowship, he seemed eager to begin his new experiment.[91] Unfortunately, duties setting papers for the university's examination for mathematical scholarships distracted him initially, but by 18 February, he was writing exuberantly to Cayley that "I have a class of 14 or 15 comprising several (5 or 6) of our college tutors to whom I lecture twice a week on Reciprocants."[92] Among his auditors was Leonard Rogers, who had taken first-class honors in mathematics at Balliol College in 1884, who remained at Oxford as a tutor until 1888, and who, according to Sylvester, had just "made a brilliant discovery regarding Projective [that is, differential] Reciprocants." Because pure (differential) reciprocants were analogous to the subinvariants that Sylvester had studied so earnestly in 1882 and because subinvariants had seemed so promising relative to yielding a British-style proof of Gordan's theorem, Sylvester now held out great hope that reciprocants would finally succeed in the task. Rogers's result looked like a key step in that direction and nourished in Sylvester "the undying hope that . . . we shall be able to prove the finitude of the ground-forms of Invariants and Reciprocants by some simple process of reasoning."

Sylvester now had talented auditors like Rogers to keep him on his toes, but he also had something more. The same James Hammond who had so impressed him in 1882 with his counterexample to the prize question posed in the *American Journal of Mathematics* had moved to Oxford, and he and Sylvester were meeting daily to talk about their mutual mathematical interests.[93] Hammond had taken up the reciprocant banner, and he and Sylvester had established a Sylvester–Cayley type of sounding board relationship as Sylvester worked up the results he presented to his class. Hammond, thirty-six years Sylvester's junior, was also attending the lectures on reciprocants and helping Sylvester to systematize, polish, and write them up in a form suitable for publication.

Sylvester continued the class in the spring and fall terms of 1886, delivering some 33 lectures in all which appeared in print over three volumes of the *American Journal*.[94] There, he fleshed out the theory he had only given in a skeletal form in his inaugural lecture. It was his hope to provide through his lectures "a practical introduction to an enlarged theory of Algebraical Forms" and "to rouse an interest in the subject" first among his auditors and later among his readers.[95] He proudly announced that at least among the former there had already been much success. "Since the delivery of [the] public lecture in December last," Sylvester wrote in 1886, "papers have been contributed on the subject to the *Proceedings of the Mathematical Society of London* by Messrs Hammond, MacMahon, Elliott, Leudesdorf and Rogers."[96]

Indeed, Sylvester had sparked new research by a number of gifted college tutors and mathematical lecturers. In addition to Rogers, Edwin Bailey Elliott had taken a first class in the final mathematical schools in 1873 and was a tutor at Queen's and mathematical lecturer at Corpus Christi, whereas Charles Leudesdorf had taken his Oxford M.A. in 1876 and was a fellow and mathematical lecturer at Pembroke. Sylvester seemed to be realizing not only his personal goals of stimulating his own research program and of animating a mathematical school at Oxford but also Hereford George's ideal of research seminars given by the professoriate to nurture the research agendas of the teaching fellows. As he had bragged to Cayley in January 1881 of his classes at Hopkins, so Sylvester bragged to Gilman in the summer of 1886 of his most recent classes at Oxford. The theory of reciprocants "has taken root in Europe as to my knowledge memoirs have already been published or are in the course of publication relating to it," he recounted. "Three or four of the writers are men who have attended my lectures—and some of them or indeed I may say all of them have made really valuable additions to the subject."[97] Perhaps Oxford could be a Hopkins after all.

Sylvester spent a busy summer of 1886 working up reciprocant-theoretic results, but this did not preclude travel to the Continent. Late September and early October found him in Stockholm enjoying the hospitality and the mathematical conversation of Gösta Mittag-Leffler. The forty-year-old Mittag-Leffler had studied with Hermite in Paris and with Weierstraß in Berlin, had been Professor of Mathematics in Stockholm since 1881, and, in 1882, had begun a new, international mathematics journal, *Acta Mathematica*.[98] During Sylvester's stay, he and Mittag-Leffler discussed reciprocants as well as some of Georges Halphen's work on differential equations. Sylvester also made the acquaintance of Mittag-Leffler's colleague, Sonja Kovalevskaya, whom he thanked "for the pleasure and honor she conferred upon [him] by her society" in what he described as such a "lovely city."[99] It had been a good summer.

Sylvester returned to Oxford just in time for the celebration of the twenty-fifth anniversary of Sewell's wardenship. With much fanfare, it was commemorated by a portrait of Sewell presented to the College Hall before what the *Times* described as "a large gathering" that included such worthies as "Lord Selborne and the Bishops of London, Southwell, and Salisbury."[100] The good summer was followed by an auspicious start to the new academic year. Sylvester was churning out new results; he had a strong and engaged class; he had Hammond as a mathematical confidant and amanuensis; and he had new international connections in Sweden.

In a buoyant mood, he was also enjoying Oxford society and the return of the poetic muse. On 14 November, he met "at a friend's house a lady visitor to Oxford who was to sing that evening at one of the hebdomadal concerts in Balliol College, and the conversation happening to turn on the gifted mathematical lady Professor in the University of Stockholm," he was moved to compose a sonnet paying homage to both women that he then distributed far and wide, on the pages of *Nature*, to Sweden, and to friends like Tyndall and

Francis Galton in England.[101] Maybe he could be happy and fulfilled in Oxford after all.

With the subject of reciprocants exhausted after three consecutive terms of lectures, Sylvester made the fateful decision to turn back to a nonresearch-inspired topic for his course in the winter term of 1887—surfaces of the second order as illustrated by the new set of models he had acquired for the University from the German firm of Ludwig Brill.[102] The course was not a success. Unlike the lectures on reciprocants that drew in the tutors because of their potential to stimulate original research, this course, although it was something Sylvester wanted to think about, covered well-known material. Moreover, since the University's regulations required that questions on the mathematical examinations come not just from elementary geometry but actually from Euclid, there was absolutely no incentive for the undergraduates to attend. The Association for the Improvement of Geometrical Teaching of which Sylvester was a participating member was, in fact, trying to get those regulations changed at exactly the time Sylvester was teaching his course, but their efforts were too late to affect the enrollment in his Hilary term class.[103]

After his teaching triumphs of 1886, this experience proved devastating. On 11 March 1887, he confided miserably to Gilman that "I am out of heart in regard to my Professorial work in this University in which the real power of influencing the studies of the place lies in the hands of the College Tutors and in which I can see no prospect of doing any real good."[104] "It depends exclusively on the Tutors," he went on, "whether a Professor can get undergraduates to attend his lectures and their attendance on such lectures may be as irregular as they please. Few of the Tutors recommend such attendance and many not merely discourage but actually prohibit it to those under their control on the ground that it will not pay in the examinations in the schools." Sylvester had known this, of course. It had been the reason why the attendance in his low-level classes had fallen steadily since his arrival at Oxford, and it was the source of the confession he had made in his inaugural lecture. He had naïvely thought that he could declare it best for him to teach what interested him at the moment, and the students would come. It had worked with the research seminars of 1886, but it could not work at the undergraduate level given the power the tutors had acquired in the wake of the University reforms and given the well-entrenched system of examinations that was only very slowly changing during Sylvester's tenure at Oxford. He felt betrayed, declaring that "this University except as a school of taste and elegant light literature is a magnificent sham. It seems to me that Mathematical science here is doomed and must eventually fall off like a withered branch from a Tree which derives no nutriment from its roots." So distraught was he by this dawning certainty that he was driven to ask Gilman whether it would be possible for him to return—accompanied by Hammond—to his Hopkins professorship. "I think that possibly my presence there with Hammond who is my *alter ego* in Mathematical work . . . might not be without some correspondent advantages in keeping alive the spirit of the Mathematical research in your great University and in contributing to the maintenance of the high place it already occupies in the eyes of the world," he wrote. It was a low ebb. Oxford was never going to be a Hopkins for Sylvester, and now he knew it.

CHAPTER 12

The Final Transition

*. . . Robert Browning . . . stopped me to say that he had been
looking at my portrait in the Academy—and thought it one of the
best portraits he had ever seen and congratulated the artist and
myself on the success. . . . Of course when I look at it (which is
seldom) I think of photographs taken a quarter of a century ago
and murmur* Quantum mutatus ab illo.
—J. J. Sylvester to Robert Forsyth Scott, 14 July 1889

The tough realization in the spring of 1887 that he would not succeed in
recreating at Oxford what he had had at Hopkins proved almost cathar-
tic for Sylvester. The tutors may not have recommended his courses to their
charges, and the students may not have cared for what he chose to impart, but
what difference did any of that really make? He was the Savilian Professor of
Geometry regardless, so he might as well just perform the duties of his chair as
required and stop worrying about his classes and their enrollments. He was also
never going to have the influence at Oxford that he had had at Hopkins, but,
again, what difference did that make? He would serve the fellowship of New
College in whatever ways he was asked and try to be content. As for research,
the University did not care whether he produced original work, so why not
release some of the pressure he had for so long exerted on himself and let the
mathematical muse visit him whenever she saw fit? Why not, in short, just get
on with life as he wanted to live it—thinking about mathematics, composing
verses, participating in things that mattered to him, being with friends, traveling, enjoying the fruits of his accomplishments—and stop trying to make his
Oxford position something that it could not then be?

The remainder of the spring and summer of 1887 found Sylvester doing just
that. In March, he made the short trip to London to attend and make remarks
at a festival dinner held by the Anglo-Jewish Association at the Grosvenor
Gallery. The Association, founded in 1871 "not only to assist in safeguarding

the interests of oppressed Jewish populations, but more particularly to raise the social, intellectual, and moral status of the Jews in semi-civilized countries," had then recently helped "to establish schools in handicrafts in great centres of Jewish population in the East," and its efforts had been publicly applauded by members of both the House of Commons and the House of Lords.[1] Sylvester's support of these Anglo-Jewish initiatives reflected his continuing humanitarian concerns for oppressed adherents of his faith.[2] In College, he quietly finished out the spring term class he was giving on the theory of numbers from Dirichlet's *Vorlesungen über Zahlentheorie*. He also served on the committee that oversaw the library and participated in other day-to-day business,[3] while he successfully used his College connections and position to secure for his great nephew, Reginald Edward Enthoven, one of the positions reserved there for matriculants intending to prepare themselves for the India Civil Service.[4] In London again at the end of the academic year, he played the role of a grand old man of English mathematics, entertaining, for example, Mittag-Leffler and accompanying him to the Royal Society's soirée on the evening of 8 June "to introduce [him] to our English savants."[5] Finally, he did not neglect his research. Together with Hammond, he wrote the only joint paper that actually carried a name other than his own after its title. It was a spin-off of a piece he had published earlier in the year in Crelle's *Journal* on Tschirnhausen transformations, that is, special transformations that effect a removal of lower powers of the variable from a polynomial of a given degree.[6] He further advertised these new results when he traveled to Toulouse to speak about them in the international forum of a meeting of the Association française pour l'avancement des sciences.[7] By the time he returned to New College from the Continent in the fall, he was finally ready to settle into the routine that New College and his life at Oxford could offer.

Increasing Signs of Age

The establishment of that routine was delayed somewhat in October by the news that the Council of the London Mathematical Society had awarded him its highest honor, the De Morgan Medal for outstanding contributions to mathematics; the presentation would take place at the Society's meeting on 10 November. An honor bestowed every three years, the De Morgan Medal had been first awarded in 1884 to Cayley, after the names of Sylvester, MacMahon, and Hermann Schubert had been discussed and after the names of Cayley and Schubert had officially gone to the Council for its deliberation and final choice. The second medal of 1887, however, went to Sylvester with no other candidates even considered or proposed. His colleagues were unanimous that the medal of 1887—the second medal—belonged to Sylvester and Sylvester alone.[8] A month would have to elapse before he would make the proud trip to London for the presentation. In the meantime, he would have to get his Michaelmas term teaching under way with his announced class on elementary number theory.

Although the subject matter held no particular challenges for him, the course did get him thinking once again about one of his lifelong interests, mathematical nomenclature. In a short "Note on a Proposed Addition to the Vocabulary of Ordinary Arithmetic" published in *Nature*, he "without immodesty [laid] claim to the appellation of the Mathematical Adam," having given, he believed, "more names (passed into general circulation) to the creatures of the mathematical reason than all the other mathematicians of the age combined."[9] In light of the fact that almost all of the terminology of invariant theory—like invariant and covariant—and some of the terminology from matrix theory—like matrix and nullity—issued from his linguistically fertile mind, his claims here may not have been exaggerated. Apparently in presenting the elementary theory of divisibility to the students in his fall class, he had hit upon the words "manifoldness" or "multiplicity" to signify the number of distinct primes dividing a given number and proposed to call "a number whose Manifoldness is n . . . an n-fold number."[10] Developing a new vocabulary around this term, he proceeded to translate a number of well-known results into his new language. The paper was what Sylvester might have called a mathematico-literary diversion.

Although the course sparked some new results on prime numbers, perfect numbers, and other aspects of number theory,[11] this work was seriously interrupted in January and February. In early November, a fall and ensuing leg injury had prevented Sylvester from going to London to accept his De Morgan Medal on the tenth. He had sent Hammond in his stead, but how gratified he would have been to have heard the Society's President, Sir James Cockle, speak of "those vast attainments and discoveries which will make [Sylvester's] name immortal."[12] Less than a week after the award ceremony, Sylvester was reporting to Cayley that the leg was improving but that he had been confined to his bed.[13] By January, however, more drastic action had been required. He had had to have an operation, and William Esson, mathematical lecturer and tutor at Merton College, had been appointed as a deputy to execute the teaching duties of the Savilian chair during the Hilary term.

Sylvester spent the winter and early spring in London and at the seashore in Devonshire, revising some of his poetry and continuing to push his number-theoretic results.[14] By 26 February 1888, he was writing to Gilman of his ordeal but assuring him that he was now well, that he "[had] not been quite idle since his accident occurred," having "recently fired off a few papers for Nature and the Comptes rendus," and that he was even "contemplat[ing] running over to Algiers" to spend the month of March.[15] He had been invited to speak in Oran at another meeting of the Association française pour l'avancement des sciences, and he had decided that the change of both scenery and air would do him good.

In Oran, he spoke on some new number-theoretic work on what he called Dirichlet's Theorem or Dirichlet's Principle, namely, "to show that the number of primes of a given form is infinite, . . . construct an infinite progression of integers mutually relatively prime and which each contain [as a factor] (at least) one prime number of the given form."[16] He effected such a construction to prove that, given any integer m, there are infinitely many primes of the form $mx + 1$. The lecture, however, might have been too much for him, given the delicate state of his health. On 14 April, he wrote Mittag-Leffler from the Hôtel de la Régence in Algiers that he had been "detained here by a severe cough which ha[d] confined [him] to the house for the last two days."[17] The cough was apparently not so debilitating that it prevented him from doing his mathematics, though. He told Mittag-Leffler that "I have obtained the proof of Dirichlet's theorem for 12 i.e. $12x + 1, 5, 7, 11$." He presented this and the similar result for an arithmetic progression of the form $8x + j$, for $j = 1, 3, 5, 7$ in the published version he presented to the Paris Academy of Sciences later that month.

By the end of April, Sylvester was finally back at New College and back at his teaching, this time a more advanced course on number theory in which he gave Gauss's proof of the constructability of the 17-gon, Lindemann's demonstration of the transcendence of π, and a proof of the fact first shown by Niels Henrik Abel that general polynomial equations of degree 5 are not resolvable by radicals. He was also back at one of the duties that increasingly defined the "mathematical citizen" of the last half of the nineteenth century—refereeing—although there was apparently still no universal agreement on how that process should work.

Sometime during his absence from Oxford, Lord Rayleigh, who had succeeded Stokes as Secretary of the Royal Society in 1885, had sent Sylvester a paper to referee by Cambridge's Andrew Forsyth, entitled "A Class of Functional Invariants." By 28 April, Sylvester had read it and declared that, despite its title, it was actually "upon *Ternary Reciprocants*, a subject already treated of under that name at considerable length and with much success by Mr. E. B. Elliott of Queen's College, Oxford."[18] This was surely trouble, especially since "no reference is made throughout the paper to the work of Mr Elliott, or of M. Halphen, or of Captain MacMahon, or of myself and others in the subject of *Reciprocants* although the nomenclature and the modes of reasoning, and even in some cases the actual formulæ are borrowed or adapted." Forsyth had definitely stepped on some toes. Sylvester viewed him as appropriating for himself much of the work that Sylvester and his entourage had done on reciprocants in 1885 and 1886 and of which Sylvester was so proud. He was blunt in his assessment to Rayleigh. "Mr Forsyth's mode of composition appears to me," he stated, "to be similar to that of some astute conveyancer who draws up conditions of sale in such a manner as to cover the imperfections of title without making any positively false averment which would lay his client open to an action for damages." For this reason, Sylvester could not "recommend the paper for insertion in the Transactions of the Royal Society. It would in [his] opinion be setting a very bad precedent to publish a paper constructed on the principle of studious concealment of the links of connexion between it and all that has been previously published on the subject."

The next day, Sylvester, still thinking about Forsyth's paper, was prompted to write Rayleigh once again on the matter. Having recognized Cayley's "well-known and unmistakable handwriting" on the manuscript copy he had been sent, he knew that the paper had already gone to his friend for refereeing before coming to him, and he wanted to give further justification of the concluding paragraph of his report in which he had made it clear that the paper should definitely be sent to a third referee.[19] In his follow-up letter, he suggested guidelines that Rayleigh might follow in taking this course of action. "It would be desirable that such remarks [Cayley's] should be preserved in a copy but be obliterated from the MS. Or else his [Cayley's] report along with mine should be brought under the notice of the arbiter—supposing (as I infer from internal evidence) that Prof.r Cayley recommends the publication of the paper." This was a matter of both principle and standards. Cayley, not an expert on recip-rocants, recommended a paper by one of his junior colleagues; Sylvester, one of the creators of the theory, most assuredly did *not* recommend it on account of its lack of proper attribution. Even though Sylvester himself was notoriously careless about citing the literature, it could not be condoned in this case. The matter clearly needed to be settled by some third party. "There are occasions when it is necessary to speak out," Sylvester closed.

By 12 June, the matter had been settled, and Sylvester was aghast. "I cannot suppose," he wrote in reply to Rayleigh, "that the Council of the Royal Society will finally consent to publish Forsyth's paper without taking the opinion of some 3.d referee. I have never known such a course pursued in any other case and a departure from it on the present occasion would in my opinion be allowed with very grave consequences."[20] He went on to suggest Alfred Greenhill, Professor of Mathematics at the Artillery College in Woolwich, as a possible third referee, since he, too, had done work on the theory of reciprocants. Sylvester ended, assuring Rayleigh that "I have had no communication with Greenhill relative to the paper in question" and that Greenhill's "high moral character as well as his great and extensive knowledge make him out as preeminently fit to act as arbiter in the question at issue." Although Sylvester's suggestion was not followed, Cayley did write to Forsyth regarding the addition of suitable references, and Forsyth complied. His paper appeared in 1889, and Sylvester did very little further refereeing for the Royal Society.[21] Sylvester had never found refereeing an agreeable task, but if it was a necessary evil to ensure the steady professionalization of the field of mathematics, it should at least be conducted in such a way that the true expert's opinion carried the most weight and differences of opinion were fairly and not capriciously adjudicated by the editors.

In the midst of this minor skirmish over professionalization at the national level, Sylvester was also engaged in an experiment to try to raise mathematical awareness and competence locally. On 29 May 1888, he gathered together Oxford's mathematical lecturers in his rooms in New College and "plead[ed] for the formation" of a mathematical society in Oxford that would provide a venue for the discussion of nascent mathematical ideas, ideas that were perhaps not ready for the floor of a meeting of the London Mathematical Society but that could be honed in the friendlier, more relaxed atmosphere of regular, local get-togethers.[22] With all present in agreement with his plan, Sylvester had his young colleague and fellow reciprocant theorist, Edwin Elliott, send out a circular to all mathematics graduates in Oxford inviting them to attend an organizational meeting on Saturday, 9 June. A gratifying three dozen people answered the call to constitute the original membership of the Society. At that first official meeting, they elected their officers—Sylvester as President, William Esson and Bartholomew Price, the Sedleian Professor of Natural Philosophy and a fellow of Pembroke College, as Vice Presidents, and Elliott as Secretary—and set their rules. Foremost among the rules was the prohibition on any sort of a society publication. The rationale for this "was two-fold. On the one hand slight or tentative papers, which their authors could not think of as enough developed for publications were welcomed, and on the other more ambitious efforts could be talked about without prejudice to their admissibility by, say, the London Mathematical Society" (12–13).

That Sylvester had accurately taken the pulse of Oxford mathematics in 1888 seems clear from the fact that during the Society's first five years some 23 different contributors presented an average of just over three papers at each of the six annual meetings for a total of just under 100 papers. Of the contributors, not surprisingly, Sylvester was the most prolific with 17 presentations, but Hammond, Elliott, Leudesdorf, and others also took advantage of the forum to test new results on an audience of mathematical peers. If Oxford was not yet ready in the 1880s for the active training of future researchers in its classrooms, it apparently *was* ready for the extracurricular support system that the Oxford Mathematical Society provided for those who nevertheless wanted to contribute to mathematics at the research level. At least in this small way, Sylvester did help realize the goal of fostering research and a research ethos in an Oxford resistant to the transformation into a modern university.

Sylvester undoubtedly left Oxford for the summer of 1888 with a sense of professional accomplishment, but his mathematical activism of the late spring did not deter him from indulging in his poetical avocation over the summer. Back at the Athenæum, he was waxing lyrical over an incident that had occurred before his departure from New College. The sonnet over which he labored, his seventh by his own count, was "To a young Contralto who sang in my room overlooking the College garden without accompaniment."[23] The poetic nature of the scene comes easily to the mind's eye. The bald-pated, white-bearded professor seated in an overstuffed chair gazed into the lush late-spring garden, lost in the pure, a cappella strains of the attractive young woman beside him. The old and the young, the intimacy of the moment, the beauty of the music and the spring, all combined to create a *moment privilégié* that Sylvester longed to capture in perfect sonnet form, but it was difficult to get right. In one undated, scratched through, and marked up version, the poem ran:

> *The jealous ambushed Nightingale's despair*
> *To match those notes so tender sweet and low*
> *That pure as in night's hour new fallen snow*
> *Vouch thou art good e'en as thou art fair*
> *What need hast thou of gems to deck thy hair*
> *Of all the wealth Golconda's mines bestow*
> *Rubies or pearls rash divers seek below*
> *Thou canst in nobler wise thy worth declare*
> *Oft shall thy votary in his ivied cell*
> *Tired with vain search for reason's guiding clue,*
> *Pause to bid Memory with her magic spell*
> *Recall that innocent childlike form to view*
> *And in fond fancy hear thy voice anew*
> *Till life to gladness breathes its last farewell.*

Despite his ongoing revisions, on 18 June, Sylvester had sent the sonnet to poet and fellow Athenæum member, Frederick Locker, soliciting his help in securing a publisher for it. When ten days passed with no word, Sylvester, in agony, continued to revise. The reply that finally came was "kind and sympathetic."[24] "Your approval in itself," Sylvester wrote, "is a rich reward for the trouble the sonnet has cost me." Still, the state of doubt provoked by the ten-day silence, prompted Sylvester to ask for advice.

> Would it be troubling you too much to ask if you think
> > *[Oft shall the tenant of this quiet cell]*
> would be an improvement or the contrary?
> > Also in the 3[d] from end
> > > *Restore that lissome youthful form to view,*
> > I would very much like to be favored with y[r] opinion if
> > > *[heav'n-lit]* for *[youthful]*
> would not be an improvement . . . .

If only he could perfect this sonnet and get it into print, he told Locker, "I should then be encourage[d] to bring out a complete edition of all my verses in the most perfect form to which I have been able to bring them."[25]

Poetry over the summer followed by teaching in the fall left little time for "the votary in his ivied cell" to "search for reason's guiding clue" to new mathematical results. This search was further impeded by physical limitations. Sylvester's eyes had been causing him much trouble, and early February 1889 found him in Wiesbaden, Germany seeking treatment for them from a noted oculist.[26]

On a more positive note, the days around the first of February also brought word of a new and totally unexpected manifestation both of the esteem in which he was held and of how far he had come. On 31 January 1889, Robert Scott, the Master of Sylvester's *alma mater*, St. John's College, Cambridge, wrote to inform him that "a number of your friends and admirers in College have long been of opinion that we ought to have your portrait in College. The thing is now taking shape and I write on behalf of those acting with me to ask whether you would kindly give sittings for such a portrait."[27] His portrait in College! A college in a university from which, as a Jew, he did not take a degree in 1837, this college now proposed to honor him by hanging *his* portrait among those of the College worthies who had stared down upon him during his student days. It was "an honor," he later wrote to Scott, "which I regard as second to none that could befall any living man."[28]

The College had approached Alfred Emslie, "a very rising artist," to do the portrait, and Emslie had accepted the commission, especially because he had heard that Sylvester was "picturesque."[29] During the month of February, the mathematician gave the painter "3 long sittings at his (very cold) studio," and, while the portrait was progressing, Sylvester prepared Scott for what might be a less than satisfactory result: "I fear from the state of my eyes and the general depression of spirits I am laboring under the portrait may not be all you and my other kind friends might desire."[30] By April, the work was finished, and it had received widespread approval, even if "the expression might have been made more agreeable."[31] "When I look at" the portrait, he later confessed to Scott, "I think of photographs taken a quarter of a century ago and murmur *Quantum mutatis ab illo*.[32]

If he was both gratified and sobered by the experience of sitting for the portrait, he was unquestionably buoyed that spring by a new research hit concerning what is known as Buffon's needle problem, namely, "suppose that in a room, in which the parquet is simply divided by parallel joints, one tosses a rod in the air and that one of the players bets that the rod will not cross any of the parquet's parallels while the other bets, on the contrary, that the rod will cross several of the parallels. What are the odds for these two players?"[33] The Comte de Buffon had proposed this problem in so-called geometric probability as early as 1777, and a number of mathematicians—among them Joseph-Émile Barbier,

Joseph Bertrand, Sylvester, and Sylvester's protégé from Woolwich days, Morgan Crofton—had pushed through results in the area in the decade from 1860 to 1870 that involved tossing not rods but rigid rectifiable curves and even pairs of figures in rigid relation to one another.[34] In the spring of 1889, Sylvester realized how to generalize these earlier French and English results to $1 + n$ $(n > 0)$ figures rigidly related to one another, and he wrote at length of his finding to Mittag-Leffler on 3 May, wondering if a paper on the subject might interest his Swedish colleague for his journal, *Acta Mathematica*.[35] Sylvester also publicized his new results at home. With Hammond as his intermediary, the paper was delivered to the London Mathematical Society at its meeting on 13 June.[36]

If the work on geometric probability had brought momentary relief from the research drought that had followed the 1888 work on Dirichlet's Principle and if the portrait for St. John's College had represented a testament to the regard of his peers, the winter and spring of 1889 had also brought the real fear of permanently failing health. As Sylvester told Gilman in a letter dated 30 July 1889, "I have been troubled considerably about my eyes and in other ways during the last half year and more."[37] The fall and resultant leg injury in 1887, the surgery and respiratory problems in 1888, and the eye troubles that began to worsen markedly in 1889, all of these were increasing signs of age and decline, and Sylvester realized it. He closed his letter to Gilman on a note of resignation: "The air of Oxford does not suit many people, and I am one of them but I do my best to keep on working."

It was hard to work, however. A month after penning this letter, Sylvester was in Sweden for an extended stay in Jönköping, where he was being treated by yet another oculist. Once again, the initial prognosis seemed optimistic. He wrote to Mittag-Leffler that the doctor "thinks he can cure me: and that he can arrest or even dissipate the opacity in the least affected eye in a short time."[38] The problem, however, was cataracts, and, unfortunately, both eyes were involved.

Sylvester remained in Sweden into September, but it was hardly a pleasant sojourn. Although there were more than a dozen other English patients seeking eye treatments—among them, William Gladstone's daughter and her entourage—"the life here is very dreary except for Mr. Gladstone's relatives," he told Mittag-Leffler.[39] "Do you happen to know," he asked plaintively, "any people in Jönköping who speak English, French or German to whom you could introduce me?" Sylvester, out of his element and with no congenial companions, was miserable.

Fortunately, by 8 September he was in Göteborg after a detour to Norway and before continuing on to England via Denmark and the Netherlands. In Norway, he used Mittag-Leffler's introductions to meet the professors at the university in Christiania (now Oslo) where he "pass[ed] 5 or 6 days . . . *very agreeably.*"[40] Then in Copenhagen, he visited with both the mathematician, Julius Petersen, and the noted historian of mathematics, Hieronymus Zeuthen. It was Petersen who pointed out to Sylvester problems with an argument recently published by Cayley in Felix Klein's *Mathematische Annalen*.[41] This chance revelation set Sylvester off in a fresh research direction and into an intense international competition with Petersen that occupied him well into 1890.

An International Competition

Nine months earlier in January 1889, Cayley had sent a paper "On the Finite Number of Covariants of a Binary Quantic" to Klein for the *Annalen* in response to a paper that had just appeared there by David Hilbert.[42] The issue at hand was that same issue that had for so long dogged the British invariant theorists, namely, a British-style proof of Gordan's finiteness theorem for binary quantics. In his paper, Hilbert gave a simpler, more modern proof of Gordan's result; in his, Cayley thought he had greatly simplified Hilbert's argument along British lines. It could have been that long-sought vindication of British techniques, except for one thing. It was wrong.

Interestingly, Klein, the editor, knew it was wrong; his two referees, Hilbert and Gordan, had both pointed out the error. Still, Klein gave Cayley the option "to withdraw the note or to have it printed nevertheless."[43] In the latter case, though, Klein would follow the paper immediately with a rejoinder. As he recounted to Hilbert, "with the stubbornness of an old gentleman Cayley has chosen the second alternative." When Klein asked Hilbert if he would write the promised rejoinder, however, Hilbert declined, wanting neither to seem immodest nor to hurt Cayley's feelings. In his view, "as soon as [Cayley] has the printed text before him, he will see himself that he is wrong. Should this fail to happen and should nobody else clear up the error, then the best occasion for correction arises when I myself publish something new on the subject. . . . Cayley's publication has at least the merit that it shows the attentive reader why every step in my proof is necessary."[44] One who was an attentive listener as well as reader was the Dane, Julius Petersen.

While in Copenhagen, Sylvester had apparently mentioned to Petersen a letter he had received from Cayley in which he had sketched a simplification of a proof published by Hilbert. When Sylvester recounted the argument, one he had presented earlier to the Oxford Mathematical Society,[45] Petersen quickly saw the flaw and found that Cayley's erroneous argument had actually been published. On 20 September, shortly after Sylvester's departure, Petersen wrote to Klein about the matter, stating that "I cannot but notice that this proof is wrong" and proceeding to show where Cayley had gone astray.[46] Here was the person Hilbert had foreseen who could "clear up the error," and Klein duly asked Petersen if he would write the rejoinder.[47]

Sylvester's interest in Cayley's error, but especially in Hilbert's proof, was also piqued by Petersen's observation. Sylvester marveled that "so great a Genius and so practiced a Veteran as Cayley had allowed himself to be betrayed as if he had been some novice for the first time grappling with an arduous question."[48] What really interested him, though, was Hilbert's correct proof. At one point in it, Hilbert needed to determine when a polynomial of the form $p(x_1, \ldots, x_n) = \prod_{i<j}(x_i - x_j)^{e_{ij}}$, for unknowns e_{ij}, has the same degree in each of its variables x_k. He realized that this question reduced to solving a Diophantine system of $n - 1$ homogeneous, linear equations in the $[n(n - 1)]/2$ unknowns e_{ij}, for $n \in \mathbb{Z}^+$, and he knew, thanks to a theorem of Gordan's, that the system was solvable.[49] Although it had been precisely this system of equations that Cayley had thought his proof avoided, it was also precisely this system of equations that intrigued Sylvester and Petersen. In his proof, Hilbert just needed to know that a solution existed; Sylvester and Petersen wanted actually to effect that solution explicitly.

Back at New College, Sylvester had begun tackling the problem case by case by 1 October. For the binary quartic, the explicit solution of the Diophantine system was "extremely simple," but the case of the binary quintic proved more difficult.[50] Petersen independently approached the problem in the same inductive way. By Friday, the eleventh, Petersen had sent Sylvester his solutions for the quartic and quintic, and Sylvester not only stated that he had "been in possession" of both solutions "for many days past" but also acknowledged that his solutions agreed with Petersen's and claimed to have a proof in the general case of a binary n-ic form.[51]

Sylvester spent the weekend deeply engrossed in this work. On Saturday, he penned two letters to Petersen, confident that he had further generalized his solution from the case of a single binary n-ic to any system of binary forms. Sunday must have brought doubts, for on Monday, he was forced to admit that, although his solution for a single binary n-ic was solid, his more general claims had been erroneous.[52] Undeterred, he wrote to Klein on Tuesday, announcing his discovery of "the general solution of Hilbert's linear Equations" for a single binary n-ic and his "possession of a strict proof of the same" and inquiring whether Klein might like his result for publication in the *Nachrichten* of the Göttingen Academy of Sciences.[53] The paper "would occupy 3 or 4 pages of the Journal," he added. "I prove my theorem without algebra by a sort of method of Valence." This "method of valence" was what would today be called graph theory, and it harkened back to the earlier work "On an Application of the New Atomic Theory to the Graphical Representation of the Invariants and Covariants of Binary Quantics" that Sylvester had published in 1878 in the first volume of the *American Journal*.[54]

Sylvester and Petersen exchanged numerous letters throughout the months of October and November of 1889 as they labored to develop this new theory.[55] Although Sylvester quickly realized that his claim to Klein had been premature, by 24 November, he thought that he at least had the result in the case of binary n-ics, for n odd, and he proposed to come over to Göttingen to present it in person at the upcoming meeting of the Göttingen Academy on 4 December.[56] The timing was right. On Saturday, 29 November, Sylvester would be finishing up the Michaelmas term course that he had been giving on partition theory in parallel with and informed by his new research. Moreover, he was feeling energized both by the mathematics and by the give-and-take with Petersen. It had been a good fall after a difficult winter, spring, and summer. What better way to celebrate it than to make a triumphal appearance before one of the premier mathematical audiences in Europe?

Although Sylvester did not finally make the journey to Göttingen, he did continue to work hard at graph theory throughout the month of December and into January of 1890. Letters to Klein on 12 December sketched the results he thought he had obtained thus far, while a letter to Petersen on the same day carefully laid out a history of how the theory had developed between them.[57] In the latter, Sylvester, ever conscious of matters of priority, wanted to lay claim to *his* methods of proof at the same time that he freely acknowledged Petersen's. They were both working on the same question, but their techniques, while similar, were not identical. Sylvester thus proposed officially "to join forces with [Petersen] in the further prosecution of this great and momentous question," and he invited him to come to Oxford so that they could work in concert with an aim of "furnish[ing] Klein with a joint memoir." If such a memoir resulted, Sylvester continued, "your name would be attached to your proof, and my name to mine—the contrast between the two methods is in itself a very wonderful feature of the subject."

By the end of December, Petersen was in England, and he and Sylvester were hard at work on the theory. Also by the end of December, Klein was trying to assess Sylvester's work. The letters Sylvester had sent on the twelfth had been so illegible that Klein had had to have them copied in order to be able to read them. By 25 December, he had sent the copies to Hilbert for his opinion, but "to tell the truth," he confided, "I have little desire to pursue this matter further: Sylvester is too irregular, altogether a 'genius'; in matters of business there is no way of getting anywhere with him."[58] Hilbert's reply came four days later. While he found it "certainly of interest . . . to see . . . in which sense Sylvester is working on the diophantine equations," he "[could] not convince [him]self from reading the letter that in spite of [Sylvester's] emphatic claims of their importance, Sylvester's results are really deep and will appeal also to those who are not content with the joy of purely formal developments."[59] This was a clash between the old and the new. Sylvester, the grand old man of seventy-five, had always pursued mathematics from an inductive, enumerative point of view; construction and explicit exhibition had always been his goals. Hilbert, the young Turk just shy of his twenty-eighth year, was a harbinger of so-called modern mathematics; fully general existence theorems were of much greater interest to him than messy calculation case by case.

While Hilbert was passing his preliminary judgment on Sylvester's claims, Petersen was in England where he remained for some two weeks. On 4 January 1890, Sylvester was once again writing Klein about their work. "We have come to the agreement," he explained, "that I shall send you a paper on our Graphs—and that he shall follow this with another, also for the Annalen."[60] Although Sylvester predicted that his paper would follow in one or two weeks, 19 January found him writing Klein once again "to state how the matter stands. My previous proof," he explained, "labored under an imperfection which I have only recently succeeded in removing and in so doing discovered a new and more general theorem than that which I thought I had established when I wrote you last."[61] The problem now was that the Hilary term had begun, and Sylvester was pressed for time to devote his research. "I must beg you," he wrote in desperation to Klein, "to grant me a little more time, say two or three weeks, to write out my investigation in a form sufficiently clear to be suited for the Annalen."

A week later, Sylvester still thought he had the result. Writing to John Couch Adams on 25 January, he boasted that "I have just completed a very difficult but very beautiful investigation concerning Chemical Graphs in which [Petersen] had been running me very hard, but now unless there is some illusion in my mind on the subject, I have far outstripped him. . . . My work will I hope soon appear in the Math. Annalen, but my labors are subject to frequent interruption (as at the moment) from eye troubles and general nervous prostration."[62]

Ultimately, Sylvester did have "some illusion in [his] mind." He never succeeded in getting his proof. Following his visit to England, Petersen sensed that this would, in fact, be the case. He explained in a letter to Klein on 26 January 1890 that "Sylvester thought twenty times that he had found a proof . . . , but every time the proof was insufficient. . . . This repeated itself quite often during the two weeks that we spent together, and by the time I left he had made no progress. Now he writes that he is sure to have solved not only the even case, but also the far more difficult odd one."[63] Petersen seriously doubted this claim, but, at the same time, he did not want to see Sylvester embarrassed. In the event that Sylvester actually submitted a paper to the *Annalen*, Petersen asked Klein "in all confidentiality either to have a very close look at his paper yourself or, should you be unwilling to do so, to leave the refereeing to me."

Sylvester seemingly never submitted and definitely never published any of the fruits of his graph-theoretic labors, although Petersen's paper did finally appear in 1891, marking the first major contribution to the mathematical theory of graphs.[64] Sylvester had lost the competition. It had simply been too much for the aging mathematician. As he explained to his Danish competitor in mid-June 1890, "the trouble in my eyes and certain causes of mental inquietude have incapacitated me for the last five months and more for all mathematical work of investigation—and had it not been for Mr. Hammond's aid, I doubt whether I could even have prepared the ordinary lectures for my class."[65] At least, he continued, "my lectures are now at an end which is the element of mental disturbance removed—and I am more reconciled to the grievous trouble with my eyes: therefore with the blessing of Providence I hope to be able to resume real work before long."

A Steady Decline

The winter and spring had, indeed, been difficult for the ever-competitive Sylvester. He did not like to lose, but he had never failed so completely to meet a mathematical challenge. It had been a debilitating blow that he had tried to blunt in February 1890 with two sonnets, one inspired by the departure of one of his New College colleagues, the other by a visit to Oxford by former (and soon to be again) Prime Minister, William Gladstone.[66] According to his colleague, Spooner, "the sonnets [Sylvester] used to re-edit & polish up in the most painstaking & sometimes embarrassing fashion, producing in the same day three or four editions of them differing only in some small & to inattentive eye or ear, unnoticable particular; but in the course of these essays he evolved some curious & . . . not wholly valueless canons of prosody & accent, and ended in producing a result which, though not a work of genius & generally obscure up to its end was still serious and in the grand manner."[67]

Trying to lose himself in these poetic reveries and revisions and following four months of intense mathematical activity, Sylvester had also largely neglected his correspondence. On 9 March, his faithful friend, Hermite, wrote to him, concerned by his long silence. The last note Hermite had received had contained news of Sylvester's trip to Germany in search of relief for his eyes, but that had been a year ago. Now, Hermite wrote, "I act on the authority of my anxiety and of our old friendship, to know how you are doing. You will know, I hope, that I am not only a friend in good times, and not at all doubting that I would want to know of your interest were I myself ill, I ask for several words from you to let me know what is happening with you, whether good or bad."[68] Sylvester replied promptly, but the news was not good. He had been suffering again from bronchitis; his eyes were no better; his work, motivated by Hilbert's paper, had not been going well. Still, he told Hermite, Hilbert's main result *was* stunning in its simplicity. "How surprised we would have been, thirty odd years ago," Hermite replied, "if the future had revealed itself [and] we had seen all the fruit of our long and great efforts collapse . . . destroyed by the repeated blows of Messrs Gordan and Hilbert!"[69] It was hardly a sentiment Sylvester needed to see expressed at such a vulnerable juncture in his life, but it had about it a certain element of truth.

Sylvester also took the time to send a letter westward across the Irish Sea to Dublin and Robert Graves, the nineteenth-century biographer of Sir William Rowan Hamilton, the brother of John and Charles Graves, and a friend of

Sylvester's since the late 1830s.[70] In a retrospective mood, Sylvester had taken the opportunity presented by his latest poetic flights to send Robert a copy of his pair of sonnets, and this had elicited an appreciative reply in which Graves had told Sylvester of the recent celebration of his eightieth birthday and of the completion of his three-volume biography of Sir William Rowan Hamilton. "I congratulate you on passing your 80th birthday so recently," Sylvester answered, "in the full possession of so much vigor & freshness as your letter to me gives evidence of. And what a comfort to you to have brought to a successful end your monument to the memory of your (and I may add my) illustrious friend."[71] Sylvester, however, feeling anything but in the "full possession of . . . vigor & freshness," also wrote almost wistfully that "I hope you have daughters or other female relations to take care of you." He very much felt the weight of his own failing health, advancing age, and loneliness, even as he tried to take pleasure in the achievements of others.

By May, his thoughts had turned to yet another old friend, Daniel Gilman, who along with his wife and daughters were due to visit Oxford toward the end of their year-long, recuperative leave of absence from Baltimore. In 1889, Gilman had finally succeeded in opening the Johns Hopkins University Hospital, the other half of Johns Hopkins's original $7,000,000 bequest, and the end of that ordeal had left him on the verge of nervous collapse. Recognizing this and in full appreciation of his herculean efforts, the Board of Trustees had granted him a year's paid leave, and he had decided to spend it on a trip to the Orient that brought him through England in the late spring of 1890.[72] Sylvester wanted to show the Gilmans the best of times, proposing to take them on 25 and 26 May to the "two [Boat] Race Suppers (which are in great request and attended by all the élite of Oxford) in the Hall followed by music, our choristers' part songs, and sometimes dramatic or other recitations" as well as to the College's Commemoration Day celebrations in early June.[73] Still, he forewarned Gilman that "I have had much trouble and must expect more with my eyes—and I expect you will find me much altered—and my eye troubles make it difficult for me to do much hard thinking—but I sometimes amuse myself with a little work in making verses—and send you 'a sample of some' herewith."[74] He was self-conscious of his debilitation, and he was sad, but he could not pass up the chance to share with his dearest Baltimore friend "a real poet's May," with the "gardens . . . now looking their loveliest," and "to hear all that is doing at Baltimore or is to be done."

If the Gilmans' visit was one bright spot in the otherwise melancholy and introspective year of 1890, there were at least three more. In the winter, Sylvester had learned that the President of the French Republic had conferred upon both him and Cayley, at the instigation of the President and members of the Paris Académie des Sciences, the "Décoration d'Officier de la Légion d'Honneur," one of the country's highest civilian honors.[75] Then on 19 March, the fellowship of New College had voted to increase the salaries of the Savilian Professors of Astronomy and Geometry by £150 effective retroactively from the beginning of the year, an increase of some 20%.[76] Finally, on 10 June, Sylvester received an honorary D.Sc. degree from his *alma mater*, Cambridge University, in recognition of his high mathematical attainments.[77] Much appreciated though these gestures were, they served in mid-1890 more to underscore, in Sylvester's mind, what had been rather than what he fervently hoped—but increasingly doubted—might still be. Such accolades, once so keenly desired, were now bittersweet.

A quiet summer preceded an uneventful fall. Sylvester was once again teaching the course on surfaces of the second order as illustrated by the collection of Brill models, and he continued with his college committee work. He had served on the Library Committee since October 1885; two years later, he was also appointed to the influential Tuition Committee; and then in 1889, he picked up a third assignment on the equally powerful Estates Committee. He was thus heavily involved in the College's governance. In the fall of 1890, in particular, he served with colleagues Hereford George, Charles Pritchard, and William Spooner, among others, in tending to the various aspects of the business at hand.[78] Spooner, however, later recalled Sylvester's efforts as a working member of the College in less than flattering terms. Sylvester, he wrote, "was not a man on whom judgement in practical matters much reliance could be placed. A quick temper, an extreme sensitiveness & readiness to take offence even when no offence was intended made him rather difficult as a member of a corporate body."[79] Although his former colleagues at Hopkins would certainly have recognized the description of the temper and the sensitiveness, they, at least, had relied in key ways on Sylvester's judgment in all manner of practical matters. Perhaps Sylvester just had more of a feel for the new universty than for the five hundred-year-old, tradition-bound college.

Regardless, by early December all committee work was forgotten. Sylvester was back in Paris being treated once again for his eyes, this time undergoing a series of minor surgeries on his tear ducts. The cataracts, however, were still of major concern. As he reported to Simon Newcomb in early January 1891, "I have lost the use of one eye by Cataract and the other is affected with an incipient form of the same affection—for which there is no help but the extraction of the lens when the proper time arrives."[80] Despite this disheartening prognosis, Sylvester was managing to do a little in the way of mathematics. On 9 December 1890, his brief "Proof That π Cannot Be a Root of an Algebraic Equation with Integer Coefficients" was read before the Paris Academy of Sciences, and he was also announcing a public lecture in Oxford for early in the Hilary term of 1891 on the related "Story of the Quadrature of the Circle."[81] This lecture was intended as a teaser for his announced course on what he described as "the application of the higher arithmetic to the division of the circle."[82]

More number theory lectures followed—on elementary aspects of the subject—in the Trinity term of 1891. As it had in the Michaelmas term of 1887, this topic sparked fresh insights on prime numbers, but now those insights related to the work he had done while at Hopkins in 1881 on Chebyshev's upper and lower bounds on the number of primes less than a given number.[83] Writing to Cayley about these new results on 17 and 19 April 1891, he sounded almost like the old Sylvester. "It is not difficult," he wrote on the seventeenth, "to obtain a superior limit to the number of primes not surpassing p," yet before he got the letter into the mail, he had added the postscript that "there is a *flaw* in my supposed proof."[84] Two days later, though, all was well. He reported that the theorem "is now proved," and "the proof D. V. is to appear in the Messenger."[85]

The proof—and more—did, indeed, appear in two lengthy installments in consecutive numbers of the *Messenger of Mathematics*. In the first part, Sylvester gave what, by his own admission, was a "wearisome proof" of the fact that "whatever n and whatever i may be, provided that m is relatively prime to i and not less than n, the product $(m + i)(m + 2i) \cdots (m + ni)$ must contain some prime number by which $2 \cdot 3 \cdots n$ is not divisible."[86] "It will not surprise the author of it," Sylvester continued, "if his work should sooner or later be superceded by one of a less piece-meal character—but he has sought in vain for any more compendious proof." Before the paper went to print in May, however, Sylvester had a mathematical epiphany while "taking a walk on the Banbury road (which leads out of Oxford)." He realized how, from earlier results of Stirling and Chebyshev, "all [his own] results are made to flow," and this new development formed the paper's second part over which he labored throughout the month of June. If graph theory had ultimately bested him in the fall and winter of 1889–1890, such was not the case with number theory in the spring of 1891.

The term ended with both this research hit and a visit from Europe's brightest mathematical star, Henri Poincaré. Sylvester had met and talked with the French prodigy on more than one occasion during his frequent visits to Paris, and the two, while not exactly friends, were cordial professional acquaintances. In June of 1891, the thirty-seven-year-old Poincaré was in England to observe the English universities, and Sylvester hosted him at New College, providing him rooms, entertaining him, and introducing him to College and University society. Still, although Sylvester found the Frenchman "as simple and modest as he is eminent," he confessed to Cayley the day before Poincaré's arrival in Oxford that "I rather dread the encounter as there is so little in the theory of Mathematics upon which I can hope to talk to him."[87] The elderly algebraist, Sylvester, had been struggling mathematically, while the young analyst, Poincaré, had stunned the mathematical world as recently as 1889 with his prize-winning work on the three-body problem.[88] Once again, Sylvester was feeling his years.

Perhaps owing to Poincaré's visit, the end of the term did not bring with it Sylvester's immediate flight from Oxford. He did attend the Royal Society's gala *soirée* in London on 17 June, an evening open to gentlemen *and* ladies on this occasion on which Burlington House was uncharacteristically decorated "with a profusion of greenery and flowers" and where a wide range of exhibits and demonstrations provided "a great deal of entertainment and instruction," but he returned immediately to Oxford where he remained into the month of July.[89] Prospects of a new project occupied him. On 11 June, he had confided to Cayley that "I have consented perhaps unwisely to undertake the revision of my collected papers for the [Cambridge] University [Press]. I fear," he continued, "that many of them will be unworthy of republication and the task of correction will be formidable."[90] Cayley had been in the process of compiling his own collected works since the late 1880s, with the first two volumes appearing in 1889, the third in 1890, and the fourth in 1891, so he knew well the magnitude of the task. Moreover, he had already gone through the negotiation process with the press, so he could potentially provide valuable advice on how to deal with the practical issues involved in such a project.

Two weeks later, Sylvester was inquiring about just such an issue. In discussions with the press about the matter of print run, he realized that he really had no idea what was appropriate. "Will you please inform me (if you are at liberty to do so) how many copies are printed by the University Press of your collected works?," he asked Cayley. "The information if you desire it shall be treated as confidential."[91] A month later, all these deliberations seemed moot. "As to the Republication of my Omnia Opera," he told Cayley with a flash of testiness, "there is a slight hitch in the matter owing to the *gaucherie* of the secretary to the Syndicate in his way of wording his Communications to me. I therefore rest on my oars—and wait to be further applied to on the subject. . . . Please accept this in *confidence*."[92] He did not want his collected works published so badly that he would accept anything less than the respect due him from the functionaries of the press, and he certainly would not approach *them* on the matter. This, at least, was the overt message, but Cayley knew better. Sylvester would have been devastated had the project ultimately fallen through. As it was, though, he did not undertake it that summer.[93]

In fact, the summer was spent largely at New College in quiet contemplation. He had considered a trip to France, this time to the south and Marseille for the annual meeting, beginning 17 September, of the Association française pour l'avancement des sciences and had anticipated a "splendid reception . . . by the princely merchants of the city" as well as "excursions arranged to visit Tunis and Algeria after the meeting."[94] Instead, he remained in and near his rooms, passing his time with issues of the conservative *Quarterly Review* and the more liberal *Edinburgh Review* in a kind of political point-counterpoint.[95] At the end of September, the contemplative summer of 1891 came to a close as he grappled with mathematical reform at Oxford as part of his duties as President of the Association for the Improvement of Geometrical Teaching (AIGT).

The AIGT had been formed in 1871 and had been bolstered by that BAAS committee—seated in 1869 and on which Sylvester had served—to study ways to improve elementary geometrical teaching in England. (Recall chapter 8.) Twenty years later, at the beginning of 1891, Sylvester had been elected to the AIGT presidency, but his health and other commitments had kept him from playing a very active role in its affairs. A letter late in September from E. M. Langley, mathematics teacher at the Bedford Modern School and an honorary secretary of the AIGT, forced him, however, to confront some of the AIGT's concerns. The letter contained a draft of an AIGT document to be presented to the Board of the Faculty of Natural Science at Oxford. At issue was Oxford's handling of its low-level or "pass" examination in geometry, and the AIGT, as part of its expanded sense of mission, wanted to make specific recommendations for improvement.[96] Sylvester felt overwhelmed by this communication on a number of levels.

First, although he was sympathetic to the AIGT's position, it nevertheless placed him in a delicate situation. Following the implementation of the reforms of 1877, the Savilian Professors were *ex officio* members of the Board of the Faculty of Natural Science. Thus, "as a member of that Board," Sylvester explained to Langley, "I should be placing myself in an anomalous position in signing the Memorial."[97] Moreover, in Sylvester's view, the document "will not . . . produce any immediate effect if presented," although "if it came before the Board and I were referred to on the subject, I should of course express the opinion I entertain of the desirability of what your address recommends being adopted—but there are I am aware good grounds for believing that no measure would have a chance of being adopted which might appear to increase the difficulty of the existing pass examinations" (152–153).

Second, Sylvester was perplexed as to how the memorial had even come into being. As far as he could recall, it had not come up at the only general meeting at which he had been present, so either he had not been sufficiently attentive, or there had been a meeting of the Council that he, as President, did not know about, or there had been private discussions to which he had not been privy. Whatever the case might have been, he did not feel sufficiently in control. "Please feel assured," he told Langley, "that I have the best feelings towards the Association but at the same time feel that I am quite unable to perform the duties which ought to attach to the office of its president" (153). He was having a hard enough time dealing with his health and his teaching obligations. He just could not deal with the extracurricular demands of the AIGT, too.

In the fall of 1891, those teaching obligations, as they had the previous fall, centered on the now familiar lectures on the Brill models. Sylvester had planned to follow that in the winter of 1892 by reprising the course in elementary number theory from the preceding spring—and perhaps, once again, make some new and welcome research hits—but his health, both physical and mental, took another turn for the worse. For the second time during Sylvester's tenure, a deputy was appointed to assume the teaching duties of the Savilian chair. This time John Griffiths, an 1862 Oxford M.A. and fellow of Jesus College, taught in Sylvester's stead from January through June.

Sylvester was despondent. He had been unable to attend the interment of his friend and fellow Johnian, John Couch Adams, at the end of January, "owing to the inclemency of the season and the state of [his] health."[98] When asked to serve on the committee being formed by the Master of St. John's to place a bust of Adams in Westminster Abbey, he agreed, but "did not feel equal to taking the train to Cambridge so as to attend the meeting." As he explained self-consciously to Scott,

> I have been suffering so much of late and for some time past from mental depression, the state of my eyes, insomnia and other afflictions that . . . for *days altogether* do not find energy to stir out of college. This I mention in order that you and my other friends at St. John's may acquit me of apathy and indifference in any matter which concerns one in whom the College has so deep an interest as in our late loved and honored colleague. That would be shameful in me who feels indebted to you and other members of the college for repeated manifestations of regard so far beyond my deserts.

It was clearly a difficult time made even more so by word of Cayley's ever-worsening health. Sylvester tried to distract himself with regular reading in the library's periodicals; his mathematics was largely confined to a handful of questions to the *Mathematical Questions . . . from the Educational Times*.

By the summer of 1892, he was feeling well enough to make some trips back and forth to London and even farther afield to Brighton. He had been befriended by the 1888 New College graduate and then teaching fellow in modern history, Herbert Fisher, who had invited him for a restorative visit to the southern coast. "My mother is anxious that I should write to you to enforce the merits of Brighton as a health resort," he wrote. "She wishes me to say that if you are contemplating a change, she will get you some comfortable lodgings and that she hopes you will come to our house as often as you please."[99] For his part, Fisher could "answer for the climate [t]here. It has been most deliciously cool ever since the thunderstorm of Tuesday and a fresh breeze from the sea has made the air delightful." "I hope," he added, "that you may see your way to spending some part of the vacation in Brighton."

By early August, Sylvester had traded the seaside for London.[100] The Athenæ-um was undergoing renovations, so he was staying at the United Service Club just across from it on the corner of Pall Mall and Waterloo Place. On Sunday the seventh, his former student, Fabian Franklin, had cheered him there with a visit and had filled him in on all of the happenings at the Johns Hopkins and in Baltimore. The following Saturday, though, Sylvester had made the trip back to Oxford and, in so doing, had missed seeing Gilman in London by two short days. When Gilman's letter of the fifteenth reached him at New College on the seventeenth, he was once again dispirited. He could not see how to arrange a meeting, and yet he would have done almost anything to have had the pleasure of Gilman's company and conversation. Except for one other person who would be leaving the next day, he was all alone "in this vast college" and did not think he could "face a second time this awful solitude."[101] Gilman had heard this before in that terrible summer of 1882, so he knew well his friend's mental state. Still, as in 1882, he also knew that what he had referred to then as "the return of your pupils and friends" would restore Sylvester's spirits.[102]

By 21 October, Sylvester was, indeed, back at his lecturing, offering a course on "invariants and covariants of systems of conics." He was using Salmon's *Treatise on Conic Sections*, and he intended to focus for the entire term on its thirty-four-page, eighteenth chapter on that topic.[103] As he explained in his opening lecture, "I propose . . . to follow the order of the book, elucidating passages in the text in cases where the author's meaning seems to be obscure, and working out the examples as they occur, with the exception of some few or more of them which possess no real importance."[104] This would be a very basic introduction to invariant theory, a course that would not put undue strain on him. Still, by the end of the term, Sylvester knew that he could no longer go on teaching.

The course on invariant theory in the Michaelmas term of 1892 was his last. In March 1893, it was announced that a temporary deputy would be appointed, although it would take a year before the elector for the permanent deputy, Sylvester's New College colleague, Edward Hayes, would be named.[105] In the meantime, Sylvester passed his time in College and in London. His mental state was clearly reflected in a letter he penned at the Athenæum in response to Cayley on 30 May 1893. "I now rejoice," he wrote, "to receive so good an account of yourself and Mrs. Cayley, almost beyond which one ventured to anticipate. I cannot say much of the same complexion about myself. Please do *not* send me the draft, unless it is absolutely necessary that I should see it."[106] Sylvester could not concentrate, and he could not bear the thought of trying to attend to any serious business, mathematical or otherwise. Other than a few more questions posed in the *Mathematical Questions . . . from the Educational Times* and a brief note in the *Messenger of Mathematics* on a variant of the fifteen school girls problem, he, in fact, did no mathematics in 1893.[107]

As his melancholy deepened, he took less and less interest in the correspondence which had at times in his life so thoroughly consumed and defined him. The post did bring some welcome news—such as word of his election on

28 May 1893 as an honorary member of the University of Kasan on account of his "tireless work and important merits in the domain of the mathematical sciences"—but he was little inclined to write to others.[108]

His many friends and supporters within the Jewish community thus learned of his retirement through an announcement in *The Jewish Chronicle*. Sylvester had long been a source of pride for that community, with his achievements regularly reported on the *Chronicle*'s pages, so news of his disability was particularly unwelcome. After seconding the hopes publicly articulated at Oxford University that Sylvester "might in time be restored to the studies to which he was so keenly devoted," the *Chronicle*'s writers offered their own wishes and special sense of appreciation. "We take this opportunity," they wrote, "of expressing our sympathy with our veteran coreligionist in his illness. The name of Professor Sylvester is one which commands the universal regard of the Jewish community. Foremost amongst the existing mathematicians of the century, the veteran Professor has shed lustre on his race. . . . We trust that he may be spared to enjoy some years of health, as he has been enabled to win the fame of profound scholarship and original research."[109]

Gilman and Sylvester's former colleagues in Baltimore also heard the news from the press. "It is long since we have heard from you directly," Gilman wrote on 24 February 1894, "& we hardly know how to interpret the announcement made in a recent English paper that you are to be relieved of active duties."[110] Gilman also communicated an official statement passed unanimously by the Hopkins Board of Trustees at their annual Founder's Day meeting. It expressed "the regret with which they [had] heard of his continued ill health" and offered him "their best wishes for the restoration of his strength, & their grateful remembrance of his long continued services, here & elsewhere, in the advancement of mathematical science."

Later in 1894, other friends wrote—again on account not of word from Sylvester but of an article in the newspaper—to congratulate him on being named one of the twelve foreign members of the prestigious Italian Società dei Quaranti.[111] The Anglo-Jewish community, like the Italian intellectual community, also honored him that year. In May, he accepted an honorary membership in the Maccabæans, "a body," as characterized by *The Jewish Chronicle*, "formed for the precise purpose of paying homage to those qualities in our race that are of a purely intellectual character, as distinguished from others which betoken material success. It is the function of the Maccabæans," the article went on, "to pick out from our race those members of it and those only who are remarkable for their achievements in the realm of science and in arts and in letters. . . . The existence of honorary membership for this purpose was a wise provision on the part of the founders. . . . And it is quite certain that in no single instance could the selection for the compliment be more fitting than in the case of the distinguished mathematician who has just accepted it."[112] As Sylvester successively withdrew into himself, those around him and those who knew and respected him gratifyingly persisted in reminding him of their esteem.

The Final Years

After his retirement, Sylvester spent most of his time in London, living with friends as well as in apartments in the well-heeled area of Mayfair, but not infrequently seeking relief from his maladies and from his loneliness—especially after Cayley's death in January 1895—at the Spa Hotel in Kent's fashionable Tunbridge Wells.[113] As it always had, however, the Athenæum served as his social focal point. He sometimes met friends there, like Spooner from New College, who, in town on 15 March 1894, made it a point to visit.[114] Mainly, though, Sylvester spent hours sitting in a corner "with his green eye shade over his eyes" working on his poetry.[115] Beginning in 1895, Latin verses particularly captivated him, although there he was "less successful."[116] "His earliest attempts," at least in Spooner's view, "were almost ludicrous, full of bad mistakes, both in grammar & prosody, but he persevered & ended in producing verses which[,] if they often counted that elegance & correctness which the verses of a well-instructed school boy would seldom be without[,] had sometimes a fire & force about them which the verses of even a well-instructed school boy would often lack."

Among those verses was the collection,"Corolla Versuum," inspired largely by an extended trip to Scotland in August and September of 1895 and revised for private distribution at least twice after their first appearance in October.[117] Sylvester's excursion took him to the towns of St. Andrews, Pitlochry, and Aberfeldy in addition to "the palatial residence" of Sir William Armstrong at Cragside in Northumberland.[118] Armstrong, founder of the lucrative Elswick Ordnance Company and inventor of the so-called Armstrong water pressure wheel that had dramatically improved hydraulic machinery, had held the post of Engineer of Rifled Ordnance at Woolwich when Sylvester was on the faculty there. The mathematician had run into his old acquaintance while both were guests at Castle Menzies just outside of Aberfeldy in Perthshire, Scotland, and Armstrong had extended "a pressing and repeated invitation to pay him a visit . . . on [Sylvester's] way south."

Sylvester enjoyed himself immensely while at Cragside, admiring the "magnificently situated" house "full of art treasures" such as Sir John Everett Millais's "Chill October" and enjoying the company of Armstrong, his nephew, and his nephew's wife. Sylvester had also passed "a fascinating week" at the home of Sir Andrew Noble, the friend who had helped him so many years before to prepare the lectures on partitions he had given at King's College and now long Armstrong's partner at the Ordnance Company. (Recall chapter 6.) Sylvester's fascination had been sparked primarily by the Noble's "eldest and sole unmarried daughter," who, as he recounted to his niece, Edith, "has a divine soprano voice, sketches admirably and is most intellectual." In fact, Sylvester, just days shy of his eighty-first birthday, seemed more than somewhat smitten. He waxed lyrical first in Latin and then in his own English translation "To a Lady of Scotch Extraction":[119]

"When I sent her a present of sweetmeats"

> *Lillias! sweetest maiden, I beseech thee eat these sweets,*
> *And from sweet mouth Enchantress! give us back sweet song.*

and then again *"when I saw a scrap of her handwriting"*

> *As Orpheüs, Eurydice, with song and lyre did once rouse up [to earth]*
> *So dost thou with voice and hand rouse me up [from earth] to heaven.*

and yet again *"when she thought I was laying on too thick the compliments I paid to her singing"*

> *(a) For that I was too much enthralled, blind Chance hath played unfair,*
> *Praise past compare to thee let fall, pain that lives on, to me.*
> *(b) For that I was too much thy slave, chafed Fortune shows her spite,*
> *Darts tipped with praise at thee lets fly but arrows barbed at me.*

and, finally, *"when I was told she would in any case decline a gift from me under the circumstances"*

> *Why cautious over-much dost thou refuse from me a bride's gold wreath!*
> *The wreath I now set on thy brow, a laurel wreath, there fixed remains."*

Clearly, Miss Noble had touched him. At the same time, she had also done what only a very few had succeeded in for years: lifting his spirits.

Sylvester regretted having to bring this wonderfully diverting holiday to a close. On leaving Cragside, he had decided to postpone for just a bit the inevitable and "pass another week somewhere in the neighbourhood before returning to dangerous and unpleasant London."[120] On his return, his Caledonian euphoria was replaced by despondency so extreme that "early in 1896, his condition caused alarm to his friends."[121] The spring found him somewhat restored and back at his poetry, publishing Latin verses in *The National Observer* in honor of Lord Rayleigh and laboring over poetic lines in Italian.[122] Almost miraculously, though, by August "he quite suddenly became again interested in mathematical subjects, and this appeared to make him calmer and happier."[123] He returned to an old standby, number theory, and was absorbed in a problem no less notorious than the Goldbach conjecture: every even number can be written as the sum of two primes. In his last published paper "On the Goldbach-Euler Theorem Regarding Prime Numbers," he hoped that techniques he had successfully deployed especially in the context of the theory of partitions—namely, techniques for "determin[ing] the number of solutions in positive integers of any number of linear equations with any number of variables"—would prove fruitful in providing insight into the conjecture.[124] Not surprisingly, his approach was inductive, and he told his readers in *Nature* that he had "verified the new law for all the even numbers from 2 to 1000."[125] To the end, he favored what Hilbert had described as "enumerative methods" to more sweeping, general proofs.[126]

He was still at work pushing these ideas through, when, on 26 February 1897, he suffered a paralytic stroke. He died peacefully at 5 Hereford Street, Mayfair, in the early morning hours of 15 March and was buried four days later in the cemetery of the West London (Reform) Synagogue at Ball's Pond.[127] The ceremony was simple, but it included representatives of institutions that had played central roles in Sylvester's life: Anglo-Jewry, the group which had taken pride in his achievements and with which he had associated and identified himself throughout his life; the Royal Society, which had recognized his mathematical promise as early as 1839 by electing him a fellow; the London Mathematical Society, which he had served as its second President, and which ultimately proved instrumental both in raising the professional standards of mathematics in England and in creating the kind of high-level research environment that Sylvester himself had so desperately wanted to participate in; and New College, Oxford, the ancient college at the ancient university that had finally accepted him on the basis of his merits and achievements.[128] As Percy MacMahon put it in his obituary of Sylvester, "one of the giants of the Victorian era had been laid to rest."[129]

James Joseph Sylvester:
The Man and His Legacies

A mathematician . . . is not really known while he is alive;
he must wait for history to do him justice and estimate
his real worth and scientific position.
—Benjamin Peirce, 14 March 1880

"A mong the varied types of intellect which the Jewish race has produced in all ages, there are some which may be said to be historic, for they stand out in connection with individual lives that seem to be as landmarks in the history of our people."[1] This was how Oswald Simon prefaced the obituary he wrote of James Joseph Sylvester for *The Jewish Chronicle,* a key voice of Anglo-Jewry; Sylvester, in his view, was one of those "historic" individuals. The *Chronicle*'s staff writer concurred. "The death of Professor J. J. Sylvester, F.R.S., Hon. D.C.L., Savilian Professor of Geometry at Oxford," he wrote, "removes a personnage of great distinction from our midst." His "our" referred specifically to Anglo-Jewry; the list of honors and titles he attached to Sylvester's name reflected that community's unconcealed pride in its lost son.[2]

Sylvester had, in fact, long served that community as a standard bearer of sorts, an avatar of the high intellectual attainments of which Anglo-Jews were capable. As early as March 1849, as Lord Russell continued to pursue what would ultimately be the successful removal of the disability that barred Jews from holding seats in Parliament, a writer in the *Chronicle* underscored and countered some of the blatantly anti-Jewish sentiments then being aired in the English press. The weekly *Britannia* had, for example, asked the rhetorical questions: "Is there a man among them [the Jews] at this moment known as anything but a money-dealer? Where is the Jew holding a rank in any one of the bold, or learned, or intellectual or graceful pursuits of public honour, welfare, or ornament? Where is the eminent Jew soldier, or philosopher, or poet, or orator, or artist, in England, or in Europe, or on the face of the earth?"[3] The *Britannia*'s reply was immediate and categorical. "All are sunk in one slough of stock-broking; all swim in one unwholesome pond; all live and struggle and die in one mass of mediocrity. And this is the class," it concluded in disgust, "which is to reinforce the intellectual vigour or avert the moral decay of a British legislature." Not surprisingly, the *Chronicle*'s writer had a very different answer. He produced a long list of Jews at the Bar, in other learned pursuits, in painting, in music, in the Army, as orators, both at home *and* abroad. Of the three Anglo-Jews he singled out in the life of learning one was "Professor Sylvester . . . in Mathematics," styled "Professor" despite the fact that he had last held that title in 1842 and was then currently the actuary and secretary at the Equity Law and Life Assurance Society. But Sylvester had been a professor both in England and in the United States, a singular accomplishment for a midcentury Anglo-Jew and one once achieved not to be taken away.

The Anglo-Jewish press had continued to follow Sylvester's career over the remainder of its sixty-year course, noting his appointment to the professorship of mathematics at the Royal Military Academy in Woolwich in 1855, congrat-ulating him publicly on his election in 1863 as foreign corresponding member of the Paris Académie des Sciences, taking "great pleasure" in his "great dis-covery" of the proof of Newton's Rule in 1865, decrying his ill treatment at the hands of the government over his pension in 1871, rejoicing in the honorary D.C.L. Oxford conferred on him in 1880 and then again 10 years later when Cambridge followed suit, lamenting the ill health of "our veteran co-religionist" that forced the appointment of a deputy at Oxford in 1894 and hoping then that "he may be spared to enjoy some years of health, as he has been enabled to win the fame of profound scholarship and original research."[4] It had also touted him as an exemplar of and pathbreaker in the Anglo-Jewish cause. In 1857, it reminded its readers—when Arthur Cohen, the future distinguished jurist and Queen's Counsel, became the first Jew actually to take a Cambridge degree— that Sylvester's second place Tripos finish of 1837 had bested Cohen's fifth place finish of 1853 by sixteen years.[5] It reminded them of Sylvester's path-breaking feat once more, in 1869, when Numa Hartog broke his record of by then over thirty years' standing to become the first Jewish senior wrangler.[6] And it twice reported in 1883 on Sylvester's appointment as Oxford's Savilian Professor of Geometry, although, in quoting extensively from notices in both the *Times* and *Nature*, it failed to note that this made him the first Jew ever to hold an Oxbridge professorship.[7] For all these accomplishments, it was the case, as a writer in the *Chronicle* asserted in 1894, that "the name of Professor Sylvester is one which commands the universal regard of the Jewish community."[8] On his death, then, it comes perhaps as no surprise that the Anglo-Jewish community sought to establish a permanent memorial—a symbol of a lasting legacy—to a life that had reflected so brilliantly upon it.

It seemed, moreover, a critical time in which to highlight the accomplish-ments of Anglo-Jewry.[9] The 1880s had witnessed a large influx of Jews fleeing the pogroms in Russia that had followed in the wake of the assassination of Tsar Alexander II of Russia in 1881. These decidedly foreign, Yiddish-speaking, orthodox Jews contrasted sharply with the well-established, thoroughly accul-turated, and religiously relaxed Anglo-Jews. They were thus viewed as a threat to "the position that assimilated Jews had fought so hard to attain in English society" (79). These Russian Jewish refugees were not, however, the only threat to that hard-won acceptance; the highly publicized Dreyfus affair in France occasioned a strong anti-Semitic backlash on the Continent and in England in the 1890s. The end of this troubled decade was thus a propitious time to remind the non-Jewish public of the "right" image of the Jew in England.

In London, the Maccabæans were a group that sought to do just that. Founded in 1891 and composed of mostly young, cultured, Jewish professional men, they were dedicated to the "promot[ion] of the higher interests of the Jewish race," and in 1897, one way they saw to do that was by keeping alive the name and example of their nationally and internationally renowned honorary member, Sylvester.[10] Indicative of these broader motivating concerns, a chemist and not a mathematician was behind the initiative. Raphael Meldola, Professor of Chemistry at the Finsbury Technical College, recognized that "a high-profile memorial" to Sylvester "would provide a timely link between the Jewish and scientific communities, and demonstrate that Anglicized Jews were prominent bearers of English culture."[11] The proposal he brought before a meeting of the Maccabæans just two days after Sylvester's interment in the cemetery of the West London Synagogue was approved, and by 8 April 1897, he was presenting his idea before the Royal Society in his capacity as a new member of its govern-ing council. Six weeks later, following editorials and calls for subscribers on the pages of The Jewish Chronicle and elsewhere and after the anonymous donation to the fund (by Lord Nathan Meyer Rothschild) of £100, the Council voted to "accept the offer to found a medal to be associated with the name of the late Prof. Sylvester, and to be awarded triennially for the encouragement of pure mathematical research, irrespective of nationality."[12]

Meldola's efforts to raise the £1,000 desired for the medal's endowment redoubled in the months following the Council's action. On 4 June 1897, for example, he sounded a lengthy "Appeal to the Jewish Community" on the pages of the Chronicle, in which he reminded his readers that Victoria's reign had "witnessed the complete emancipation of our people." Since that reign, he continued, "has for us been made additionally lustrous by the world-wide fame of men such as Sylvester, it is our bounden duty to see that his name is kept alive throughout the ages by our individual efforts."[13] The Chronicle agreed. "True," it admitted, "it is a national rather than a Jewish memorial that is to be raised. But Prof. Sylvester's fame as a mathematician, his foremost position in

University circles, were important links in the history of Jewish emancipation in England."[14]

Ultimately, £880 was collected in denominations as small as £1 and as large as £50 and £100 from members of the Anglo-Jewish community, the British scientific community, and the scientific and mathematical communities in Britain, Europe, Russia, and the United States.[15] In 1901, the first Sylvester Medal was awarded to Henri Poincaré, that same young man who, ten years earlier, had visited the elder statesman of English mathematics in Oxford and had provoked dread in him because of the raw force of his mathematical talents.[16] It was a fitting choice. In 1901, Poincaré was undoubtedly one of the foremost mathematicians in the world. Moreover, given Sylvester's intimate ties with the French scientific community and its deep influence on his career, there was a kind of poetic justice that Sylvester would have appreciated in making the first award to a Frenchman. The Sylvester Medal honored Poincaré, but it also kept alive the memory—as Meldola and the Maccabæans had intended—of nineteenth-century Anglo-Jewry's brightest mathematical light. In so doing, it thus symbolized Sylvester's Jewish legacy. Equally—and perhaps more enduringly—it also bore testament to his mathematical legacy.

IN THE OBITUARY HE WROTE OF HIS MENTOR, Sylvester's former Hopkins student, Fabian Franklin, sagely identified the Englishman's major contributions to mathematics. "His influence upon the development of mathematical science," Franklin wrote, "rests chiefly . . . upon his work in the theory of invariants. . . . But his genius is quite as strikingly shown in researches of a more isolated character. Ten years before the date of his work in invariants, he wrote in quick succession several remarkable memoirs on algebraic subjects, especially on Sturm's functions and on elimination. His researches in the theory of partitions of numbers are among the most original and remarkable of his works. He made important and striking additions to the theory of matrices. In the theory of numbers, he was especially interested in ternary cubic forms."[17] And, as Franklin also pointed out, yet another "striking," "isolated" result was his proof of Newton's Rule for determining the complex roots of a polynomial equation (302). Sylvester did not venture into geometry *per se*; he did not, as Hermite once scolded him, embrace the analytic research so characteristic of the nineteenth-century tradition associated with the name of Karl Weierstraß;[18] he did not absorb the analytic number theory that had developed at the hands of Gauss, Jacobi, Dirichlet, and Eisenstein.[19] His approach to mathematics was unquestionably and self-consciously *algebraic* in nature. As his younger German contemporary, Max Noether, put it in the candid and thought-provoking assessment he wrote for the pages of the *Mathematische Annalen*, "Sylvester's [was] a truly *combinatorial-algebraic* genius."[20]

Sylvester unquestionably saw mathematics through an algebraic lens, seeking to algebracize all the mathematical ideas with which he came into contact whether they be invariant-theoretic or number-theoretic or partition-theoretic or applied mathematical or even chemical. From the 1850s on, he peppered his correspondence and his publications with references to what he viewed as the "New Algebra"[21] or "the algebra of the future"[22] and proclaimed that "all analysis must ultimately clothe itself under this form."[23] Central to his algebrization of mathematics, moreover, was the area that he had done so much to create: invariant theory. He saw it everywhere. "As all roads are said to lead to Rome," he once wrote, "so I find, in my own case at least, that all algebraical inquiries sooner or later end at that Capitol of Modern Algebra over whose shining portal is inscribed 'Theory of Invariants.' "[24]

But the "Modern Algebra" of which Sylvester spoke was not what the twentieth century would come to understand by that phrase and which mathematicians in Germany like Richard Dedekind and David Hilbert were already fashioning in the last decades of Sylvester's life. It was not the systematic study of mathematical entities defined by systems of well-crafted axioms and of a structural approach to those systems in which new mathematical objects are systematically analyzed after adding or subtracting axioms.[25] It was an organic outgrowth and fundamental extension of what Sylvester termed "Algebra the 1st," the algebra of the seventeenth century born, as he saw it, in the *Artis analyticæ praxis* of his countryman, Thomas Harriot, an algebra that concerned itself with the theory of equations.[26] Sylvester's *modern* algebra went beyond mere polynomials in one variable to homogeneous polynomials in two, three— he even conceived of n—variables at the same time that it went beyond the polynomials *per se* and explored their structural, inner workings, their "morphology" to use his word.[27] Yet, some of his work, especially during the Hopkins years of the 1880s, did show glimmers of the twentieth-century notion. For him, a partition, for example, was an object, a "*definite thing*" to be studied in and of itself.[28] Similarly, matrices, studied in and of themselves and subjected to definite laws of operation, constituted something new and different that he presciently recognized as "Algebra the 2nd," the reign of which would effectively succeed that of "Algebra the 1st."[29] With one foot in the past and one foot in the future, Sylvester, as Noether rightly put it, "glimpse[d]," from the lower summits he had attained, "the forms of mountains of previously unexpected heights," yet he was not quite able to reach those newly perceived heights himself.[30]

Moreover, in a career that effectively spanned Victoria's reign, Sylvester lived to see his mathematical style and approach supplanted. His was a world akin to that of the nineteenth-century British naturalist.[31] Just as the naturalist sought to name, classify, and describe the luxuriant plant and animal kingdoms opened up by British imperialism, so Sylvester sought to name, classify, and descibe the invariants that presented themselves before him in such mathematical profusion. His joy in the discovery of a new invariant was as complete as Darwin's in finding a new species of finch or barnacle. His approach, like that of the naturalist, involved a methodical search of the landscape, a gathering of specimens one by one, and a subsequent analysis of them to determine their morphological similarities and differences. It was an inductive approach, a systematic working up from the smallest example to the next largest to the next to the next, potentially *ad infinitum*. Along the way, generalizations were made and a theory was built just as Darwin generalized from his observations to craft the theory of evolution by natural selection.

Sylvester's presentational style, however, was not the more rigorous, deductive, theorem-proof style that had been developing especially in France and Germany. Rather, it was a looser, more narrative, almost stream-of-consciousness style that ultimately prompted George Salmon's efforts in 1859 to systematize invariant theory in his book, *Lessons Introductory to the Modern Higher Algebra*. Sylvester's was a mathematical style embedded in time and place. His version of invariant theory, shaped by broader currents within British science, was different from that that had developed concurrently on the Continent and especially in Germany at the hands of Aronhold, Clebsch, and Gordan. Hilbert recognized Sylvester and Cayley as mathematicians in at the very beginning of the field, involved in the initial fact-finding, data-gathering phase critical to any mathematical theory. He called it the "naïve" phase, and he contrasted it with the "formal" phase, of which he viewed his German predecessors as the representatives, and the "critical" phase that he himself had inaugurated in the late 1880s and early 1890s when he shifted the emphasis from the exhibition of and search for concrete examples in successively higher degrees and numbers of variables to the development of sweeping existence theorems that handled all of the specific examples in one fell swoop.[32] Sylvester and Cayley had the distinction of having laid a solid foundation for the theory of invariants. Others built upon it.

The mathematical achievement in which Sylvester and Cayley shared was reputation making. They founded and worked to develop a new and fruitful area *de novo*. Not many mathematicians in the history of the subject can make that claim. Sylvester's eulogists understood full well where this placed him in the pantheon of mathematicians past. "While it is certain," wrote MacMahon, "that he was one of the greatest mathematicians of all time, it may be doubted whether he will take a place amongst the small band who occupy absolutely the front line."[33] Franklin echoed this assessment. "Among these giants Sylvester has without question the right to be reckoned. In the history of mathematics, [however], his place will not be with the very greatest."[34] Great, but not the greatest. It is a mathematical legacy not many have shared.

∽ ∽ ∽

SYLVESTER LEFT YET ANOTHER LEGACY, what might be called a social legacy to the history of science and mathematics. Some of its components were born of his outsider status within English society. At least one followed from his reaction to broader currents of social change.

As a Jew, Sylvester had been forced to craft his own career as a mathematician. He had come to understand and appreciate the models of the Oxbridge mathematical tutor and the Oxbridge mathematical don while a student at St. John's, but these paths were both closed to him on account of his faith. He had then faced the questions: how could *he* be a mathematician? and what would that even mean outside the cloistered walls of Oxbridge? There had to be acceptable answers to these questions.

Looking to both France and Germany, he realized that being a mathematician would necessarily involve the pursuit and publication of original research, but it should also bring with it tangible rewards. It should not have to be an after-hours pursuit unsanctioned by all but the pursuer. He framed his rhetorical question this way in the essay on canonical forms he had privately printed in 1851: "The fortunate proclaimer of a new outlying planet has been justly rewarded by the offer of a baronetcy and a national pension, which the writer of this wishes him long life and health to enjoy," he stated. "In the meanwhile, what has been done in honour of the discoverer of a new inexhaustible region of exquisite analysis?"[35] Sylvester believed not only that mathematical research should be as recognized and appreciated within English society as astronomical discoveries but also that it should be highly enough valued to merit actual remuneration. The "research mathematician" should be able to take his place within the growing category of "professional men" in England.

When he accepted the Woolwich professorship in 1855, he thought he had achieved the latter objective in a three-day teaching load that would allow four days a week for research and quiet contemplation, but over the next 15 years, he recognized all too clearly that his conception of the post—of a research mathematician—was wildly at odds with that of the military authorities. After his move to the Johns Hopkins University, however, he realized his ideal. There, he was expected to teach, to carry on a high level of research, and to train future researchers. These were the three components of the job for which he was paid and for which he was greatly appreciated. This was what he had hoped for for so long. Unfortunately, on his return to England to take up the Savilian chair, he was ultimately unsuccessful in transplanting that ideal to Oxford. His was a conception of the research mathematician ahead of its time for Oxford in the 1880s and 1890s, but, as soon became clear, his was the way of the future. By his example, he had helped to shape a new professional category.

Sylvester's outsider status also sent him twice to the United States in search of a congenial university position. The second of those two sojourns had profound consequences for mathematics on the other side of the Atlantic. Throughout the nineteenth century, the United States had witnessed the progressive establishment of a professional scientific community with societies, journals, jobs, and limited but key support from the states as well as from the Federal government for the pursuit of science.[36] In the post–Civil War era of the 1870s, 1880s, and 1890s, the concept of the research university also evolved there, with Hopkins as its exemplar.[37] The graduate program that Sylvester set up at Hopkins marked the first step toward the professionalization of research-level mathematics—as distinct from the other sciences—on American shores. His was the country's first *bona fide* mathematical research department, setting an example that other universities key in the history of American higher education—the University of Chicago, Cornell, Harvard, Yale, among others—soon followed.[38] Moreover, the *American Journal of Mathematics*, which he founded at Hopkins in 1878, proved under his able editorship to be the first sustained, research-level mathematics journal on American shores.[39]

Sylvester's experience at Hopkins was emblematic as well of yet another aspect of his outsider status in English society and of his many efforts to overcome it: his international outlook. As early as the late-1830s, Sylvester had begun making contacts with mathematicians and scientists abroad—first in France and then in Germany, Italy, and elsewhere—both to learn of the mathematical research being done outside the British Isles and to make himself known on the Continent. Sylvester realized that the mathematical world was larger than England, that the Continent, not England, set the mathematical standard in the nineteenth century. It would be important to overcome the handicap he suffered at home because Oxbridge fellowships and professorships were denied him there, but it would be equally if not more important to compete successfully in the wider international arena where real mathematical reputations were made.

There was, however, a nationalistic component to Sylvester's international outlook. As an Englishman, he wanted to see his own mathematical research and that of his countrymen acknowledged on the international stage. Priority, after all, was a key means to reputation, and English work—especially his own—must not be overlooked. He also wanted to see England cast off its cloak of insularity and take a lead in nineteenth-century mathematics. In 1855 when he assumed the editorship with Norman Ferrers of the *Quarterly Journal of Pure and Applied Mathematics*, he had sounded this call loudly and clearly in the opening sentence of the new journal's first page. "At a period when Mathematical Science is putting forth new powers and induing itself with a more perfect and vigorous form of organisation," he wrote, "it would be little creditable to English Mathematicians that they should stand aloof from the general movement, or else remain indebted to the courtesy of the editors of foreign Journals, for the means of taking a part in the rapid circulation and interchange of ideas by which the present era is characterised."[40] By the end of the century, British mathematics may not yet have completely embraced the international vision that Sylvester and several others like Cayley shared, but it was well on its way to "trad[ing] its earlier reputation of insularity for that of a community unafraid and quite capable of operating on an international scale."[41]

As Sylvester endeavored in these various ways and on these different fronts to counterbalance the disabilities imposed upon him by his outsider status, he also confronted the inertia of mathematics as it had become institutionalized in England. This was not a confrontation born of his outsider status, however. It was one that all research-oriented, English mathematicians would ultimately have to face because it was at the heart of creating a modern, research-level mathematical community. Sylvester's strongly research-oriented outlook may have developed as a way for him to overcome his outsider status and to make a name for himself, but regardless of its origins, it starkly contrasted with the largely pedagogical, more purely scholarly outlook that for centuries had characterized mathematics at Oxbridge. When mathematics had for so long been seen in England primarily as a means for honing and disciplining a gentleman's mind, when it had been so firmly embedded at all levels in centuries-old curricula, how could the new research ethos that Sylvester and others embraced gain a foothold?

In a sense, Thomas Huxley faced the same question in his contemporaneous efforts to carve a niche in British society for the "new scientist," for the "new sciences," and for the "new teaching." Biology, in particular, for which he was England's most vocal proponent, was one of those "new" sciences. It had exploded into the mid-nineteenth-century consciousness with the publication of Darwin's *On the Origin of Species*. It was burgeoning as an area of research both in the field and increasingly in the laboratory. Yet, unlike mathematics, it did not have a niche in the established curricula. The problem that Huxley and his supporters faced was thus at least twofold: to find a way to expand the curricula to expose young minds to the area and, in so doing, to gain recruits for the future; and to secure funding and job opportunities to make it possible for those envisioned, new recruits actually to earn a living through their scientific research.[42] The problem they did not face, however   and one that very much plagued the research mathematicians—was that of breaking out of old stereotypes, of changing curricula so solidly built upon Euclid that *The Elements* were perceived "as second in sacredness to the Bible alone, and as one of the outposts of the British Constitution," and of escaping a long- and well-defined niche at Oxbridge by creating a new one.[43] This was a problem for all forward-looking, English mathematicians, not just outsiders like Sylvester.

Sylvester the researcher recognized that while the problems of the mathematicians and the biologists might be somewhat different, both groups were ultimately fighting for the same causes: the updating of the curricula to pave the way for the training of researchers and the establishment of the category of the "professional scientist." In his speech before Section A of the BAAS in 1869, Sylvester thus overtly allied himself with Huxley's efforts on behalf of the "new scientists" at the same time that he fought to break down the stereotype the biologist had embraced of mathematics as a largely calcified endeavor that "knows nothing of observation, nothing of experiment, nothing of induction, nothing of causation!"[44] That view, Sylvester recognized, if left unchallenged, threatened to exclude the mathematicians from Huxley's nationwide efforts to professionalize science in Britain. Moreover, Huxley's calls for an alteration of the curricula to include biology equally resonated with the forward-thinking among the mathematicians, who actively lobbied for a major overhaul of the mathematical curricula that would bring them into the late nineteenth century.[45] The fact that Sylvester was a Jew may initially have spurred him strongly to embrace research as early as the late-1830s; Huxley's "trajectory away from Anglicanism" may likewise have prepared the way for his willingness to confront authority and to establish new authorities and new structures for their support; but others, like Cayley, were very much within the Anglican fold and yet equally embraced the ideals of the researcher and of the professional scientist. The shared values of these three men—otherwise so different in their backgrounds and beliefs— reflected of those deeper currents within Victorian society that had resulted by the end of the nineteenth century in reform, in a strong middle class, and in a category of professional men within which the professional scientist ultimately took his (and later her) place.

∽ ∽ ∽

SYLVESTER ONCE TOLD BENJAMIN PEIRCE THAT "a mathematician is not really known while he is alive; he must wait for history to do him justice and estimate his real worth and scientific position."[46] That has been the purpose of this book.

Sylvester may not have been, as MacMahon said, in "the very top rank" of mathematicians of all time, but in his time and in his place, he was both a leader and a pathbreaker. A temperamental, strong-willed man with a vision for himself as well as for mathematics as a subject and as a profession, Sylvester fought many battles and overcame many obstacles to see those visions become reality. His life makes clear that mathematics is neither sterile nor created in a vacuum. It illustrates that the mathematician is not, as he himself once put it, "a sort of poor dumb visionary creature only capable of communicating by signs and symbols with the outer world."[47] It demonstrates, rather, that the mathematician is as fully and completely embedded in the world as the world is in him or her.

Notes

Introduction

Epigraph: Cajori, *Teaching and History of Mathematics*, 265.

1. "Washington's Birthday: Its Observance in Baltimore—Johns Hopkins University Celebration—Speeches, &c.," *Baltimore Sun*, 23 February 1877. The description in the next paragraph follows this source.

2. Halsted, "Sylvester at Hopkins," 178.

3. Halsted to Cajori, 25 December 1888, in Cajori, *Teaching and History of Mathematics*, 265.

4. "Washington's Birthday." For the text of the talk, see Sylvester, "Address on Commemoration Day" (1877), in *The Collected Mathematical Papers of James Joseph Sylvester*, (Cambridge, UK: Cambridge University Press, 1904–1912), 3:72–87 (hereinafter cited as *Math. Papers JJS*).

5. Sylvester, "Address on Commemoration Day" (1877), in *Math. Papers JJS*, 3:80.

6. Halsted, "James Joseph Sylvester," 295.

7. See Epigraph note.

8. For a bibliography of publications dealing with Sylvester's life and work up to 1936, see Archibald, "Unpublished Letters of James Joseph Sylvester," 90–95. Bell wrote famously (or infamously) about Sylvester the following year in his chapter "Invariant Twins" in *Men of Mathematics*, 378–405.

9. See Epigraph note.

10. Elimination theory was an area of nineteenth-century mathematics that involved, among other constructs, the determinant. It led in key ways to the development of the modern field of linear algebra.

11. Sylvester, "Address on Commemoration Day" (1877), in *Math. Papers JJS*, 3:72.

12. On the Cambridge tradition in mathematical physics, see Warwick, *Masters of Theory*.

13. Sylvester, *Laws of Verse* (1870).

14. For the reference to Gounod, see Baker, "Biographical Notice," in *Math. Papers JJS*, 4:xviii.

15. Desmond, *Huxley*, xiv, 617–618.

16. I have discussed the place of biography in the historiography of both science in general and mathematics in particular in Parshall, "Telling the Life of a Mathematician," 285–302. There, I also considered the strengths and weaknesses of biography as a methodology for understanding Sylvester's life and work. On the issue of scientific biography more generally, see Shortland and Yeo, *Telling Lives in Science*.

17. Parshall, *James Joseph Sylvester: Life and Work in Letters* (hereinafter cited as *JJS: Life and Work in Letters*).

Chapter 1

Epigraph: "Address on Commemoration Day" (1877), in *Math. Papers JJS*, 3:83. For the quote in the chapter title, see ibid., 81 (note ∗).

1. "Hamburgh Mail," *Times* (London), 3 September 1814. The next quote is also from this piece.

2. For the quotes in this paragraph, see "French Papers," *Times* (London), 3 September 1814.

3. "Extract of a letter from an Officer, dated Bermuda, 14 July 1814," *Times* (London), 3 September 1814. The following quote is also from this letter.

4. See Endelman, *Jews of Britain*, especially chapters 1 and 2 for a succinct overview of this 150-year period.

5. Simon Joseph's name is also sometimes rendered as Simeon Joseph. Zipporah's maiden name is unknown, nor is it known where in Germany Simon was born. Information on the Joseph family—including the English spellings of their names—comes from the "Register Book of the Jews of Liverpool," Liverpool Old Hebrew Congregation, 296 OHC/30/1, Liverpool Record Office.

6. From 1700 to 1750, some 6,000 Ashkenazim came to Britain; from 1750 to 1815, between 8,000 and 10,000 more followed. See Endelman, *Jews of Britain*, 41–42.

7. Chalklin, *Provincial Towns of Georgian England*, 38.

8. On provincial Anglo-Jewry in the eighteenth century, see Roth, *The Rise of Provincial Jewry*. On the extended Joseph family in Wakefield, see Wolfman, "Liverpool Jewry in the Eighteenth Century [Part 2]," 29.

9. Information contained in a letter from John Goodfield, archivist at the main library in Wakefield, to Joseph Wolfman, the honorary communal archivist of the Merseyside Jewish Representative Council. I thank Joe Wolfman not only for sharing his research with me but also for extremely helpful conversations on the early history of Liverpool's Jewish community.

10. Wolfman has conjectured that the bankruptcy in 1769 and subsequent departure for London of Simon's relatives, Samuel Joseph and Jonas Israel, may have prompted the move. See his "Liverpool Jewry in the Eighteenth Century [Part 2]," 29.

11. Waller, *Democracy and Sectarianism*, 1. Population estimates vary in different sources, with some placing Liverpool's population in 1800 at just under 80,000. Regardless of the precise figure, Liverpool was England's second largest city at the time.

12. For impressions of eighteenth- and early nineteenth-century Liverpool, compare Waller, *Democracy and Sectarianism*, 1–19; Brooke, *Liverpool As It Was during the Last Quarter of the Eighteenth Century*; Jarvis, *Liverpool Central Docks, 1799–1905*, 1–16; Lane, *Liverpool: Gateway of Empire*, 21–51; and see Chalklin, *Provincial Towns of Georgian England*, 98–112.

13. Wolfman, "Liverpool Jewry in the Eighteenth Century [Part 1]," 37–40. Wolfman's account differs from that in the traditionally accepted source on the history of Jews in Liverpool, namely, Benas, "Records of the Jews in Liverpool," 45–50. Wolfman's research, however, is based on numerous archival sources untapped by Benas. Benas's son extended some of his father's work, still unaware of those sources, in Benas, "Survey of the Jewish Institutional History," 23–38.

14. Again, the accounts of Wolfman and Benas differ in some of the particulars. I follow Wolfman. Compare Wolfman, "Liverpool Jewry in the Eighteenth Century [Part 1]," and Benas, "Records of the Jews in Liverpool," 50–51. Baron Benas estimated that by 1790 there were some fifty to seventy regular worshippers among the roughly 100 Jews in and around Liverpool. See Benas, 53.

15. Wolfman, "Liverpool's First Jewish Mayor," 60. In 1778 and again in 1782, Simon Joseph was named in legal documents related to the acquisition and maintenance of the Frederick Street synagogue. He and his two eldest sons, Elias and Samuel, were members of the nine-person synagogue committee. See Wolfman, "Liverpool Jewry in the Eighteenth Century [Part 2]," 28.

16. *Gore's Liverpool Directory* (1790). The first of these directories was published in 1766; subsequent editions came out irregularly thereafter. They recorded sufficiently wealthy householders and businessmen in Liverpool and their addresses and occupations. Members of the Joseph family first appear in *The Liverpool Directory* (n.p.: n.p., 1787) (on microfilm at the Liverpool Record Office); the next earlier directory was published in 1781.

17. Wolfman, "Liverpool's Jewish Mayor," 60–61. See also Benas, "Records of the Jews in Liverpool," 53.

18. Abraham actually appeared in *The Liverpool Directory* in 1787, but his listing dropped out until 1796. Simon and Zipporah's fifth son, Ellis, had died in 1792.

19. For Portsmouth and its naval connection, see Stapleton, "The Admiralty Connection," in *Population and Society in Western European Port-Cities*, 212–251.

20. On naval agents, see Green, *Royal Navy and Anglo-Jewry*, 120–124. It has been estimated that by the end of the Napoleonic Wars, a third of the some 400 naval agents operating in England were Jewish. See Lipman, *History of the Jews in Britain since 1858*, 5.

21. For details on how the prize was divided and disbursed, see Green, *Royal Navy and Anglo-Jewry*, 105–106.

22. Portsmouth, like Liverpool, had a significant Jewish community, in which the Woolf family played a key role. On the Jews of Portsmouth, see, among other sources, Roth, "Portsmouth Community," 157–187. On the Woolf family *per se*, Roth, "Portsmouth Community," 171 and 176–177.

23. Green, *Royal Navy and Anglo-Jewry*, 111.

24. Private merchantmen—privateers—were granted royal licences that effectively allowed them legally to plunder enemy ships. Records documenting Joseph Joseph's business dealings as a privateers' agent may be found in the Liverpool Record Office. See "Privateering Papers Belonging to Joseph Joseph, Silversmith of Liverpool, during 1802–1807," 380 MD 44, Liverpool Record Office, and the secondary account of Joseph Joseph's operation in Green, *Royal Navy and Anglo-Jewry*, 124–129.

25. See Sharples, *Liverpool*, 231. William Gladstone, the future Prime Minister, was born on Rodney Street. His father, John, was one of Liverpool's wealthiest citizens.

26. Endelman, *The Jews of Georgian England*, 119.

27. Endelman, *The Jews of Britain*, 65.

28. The following brief history of the Athenæum has been drawn from the following sources: Shaw, *History of the Athenæum, Liverpool, 1798–1898*, 37–49; Shaw, "The Athenæum and Its Place in Liverpool History"; and Carrick and Ashton, *The Athenæum Liverpool 1797–1997*, 1–6, 50–54. See also Wilson, "The Cultural Identity of Liverpool," 63–64.

29. Currie, *Memoir of the Life, Writings, and Correspondence of James Currie*, 487.

30. Wolfman has labeled the Josephs "the most English of the Liverpool Jews." See his "Liverpool Jewry in the Eighteenth Century [Part 2]," 38.

31. See "Subscription and Accompt Book for the Intended New Synagogue," Liverpool Old Hebrew Congregation, 296 OHC/1/1, Liverpool Record Office. The denomination of

£36 has a religious significance: the double *chai*. In Hebrew, the word both for "eighteen" and for "life" is *chai*. Hence, thirty-six is a propitious "double *chai*."

32. The second volume of the "Share Register" held in the library of the Athenæum records the transfer of Abraham Joseph's share to one Thomas Steele on 2 July 1808, which also points to a date around that time for the family's move to London. I thank the librarians of the Athenæum, John Rogers and Vin Roper, for allowing me to consult their archives and for showing me their historic collection. Coincidentally, the slave trade, which accounted for a significant percentage of Liverpool's commercial base in the eighteenth century, was also abolished in 1808. It has been estimated that average profits in the slave trade ranged between ten and fifteen percent. See Jarvis, *Liverpool Central Docks*, 2.

33. Abraham had also made arrangements to hold a seat in the synagogue for which annual rent was required, although he never paid and only settled his debt to the congregation after his move to London. It seems unlikely, however, that he actually had a financial reversal at this time that prevented him from paying his pledge. The annual financial ledgers of the Seel Street synagogue exist from 1808 on and served as a key archival source in Wolfman's research on the Joseph family.

34. Lipman, *Three Centuries of Anglo-Jewish History*, 70–71, and compare the maps on both the inside and outside endpapers.

35. Endelman, *The Jews of Britain*, 49. Beginning in the mid-1820s, wealthy Jews moved from the east to newer, upscale areas like Hyde Park to the west of the City. See also Lipman, *Social History of the Jews of England*, 12. From the surviving documents, it is not clear what the family's actual *residential* address was at this time.

36. The listing is actually recorded as "Abraham, Joseph." See "1808–1810 List of Members" (Great Synagogue), United Synagogue and Predecessors, ACC/2712/GTS/089, London Metropolitan Archives. I thank Justin Brummer for serving as my research assistant in several London archives related to Anglo-Jewry.

37. According to Wolfman, there may have been another son, George, born in London in light of financial records of the Seel Street Synagogue in Liverpool. On 22 May 1836, James and George are recorded as having each made a financial offering of five shillings, presumably to pray or to request that a prayer be said for their mother who was gravely ill. Miriam Joseph died on 28 May 1836. See chapter 2.

38. See Hyman, "Hyman Hurwitz: The First Anglo-Jewish Professor," 232–242, and Endelman, *Jews of Georgian England*, 159.

39. Hyman, "Hyman Hurwitz," 232 and 234.

40. Endelman, *The Jews of Britain*, 96.

41. Singer, "Jewish Education in the Mid-Nineteenth Century," 171 (note 33).

42. Endelman, *Jews of Georgian England*, 259–265.

43. See, for example, Sylvester, "On a Theory of the Syzygetic Relations of Two Rational Integral Functions" (1853), or *Math. Papers JJS*, 1:572–579.

44. See Sylvester, "Presidential Address: Mathematical and Physical Section" (1869), 1–9; reprinted as "Presidential Address to Section 'A' of the British Association," in *Math. Papers JJS*, 2:650 (note *). For more on the contents of this lecture, see chapter 8.

45. MacMahon, "James Joseph Sylvester," *Proceedings of the Royal Society of London*, ix–x. For the quotes that follow in this paragraph, see pp. ix and x, respectively.

46. Compare MacMahon, "James Joseph Sylvester," *Proceedings of the Royal Society of London*, x. Unfortunately, I have been unable to uncover any traces of this school. It is unknown, for example, whether it was a Jewish or a secular school. I have not, however,

found mention of it in the literature on nineteenth-century Anglo-Jewry, which may suggest that it was a secular school. Compare the discussion below.

47. Twenty-five-year-old Elias Joseph Sylvester may have accompanied his brother at this time; the peregrinations of the Sylvester brothers are complicated—if not actually impossible—to sort out. Two letters are held in the University College London Archives, dated 8[?] January 1829 and 25 February 1829, and signed "E. J. Sylvester" that most probably were from Elias. See University College London Archives (hereinafter cited as UCL Archives), College Correspondence, Sylvester, E. J. 1829: 1614 and 1615, and compare the discussion below. If so, this would imply that his emigration occurred later than that of Sylvester Joseph Sylvester.

Part of the confusion stems from the naturalization declaration discovered by Rosemary Flamion, a descendent of Elias Joseph Sylvester. The name recorded on it is that of Sylvester Joseph Sylvester. He then records that he was born in Liverpool on 28 December 1801 and was age twenty-five on his arrival on 27 August 1826. Sylvester Joseph was born on 28 December but in 1798; his brother Elias was born in 1801 but on 3 November.

That Sylvester Joseph Sylvester did, in fact, emigrate in 1826 is corroborated by the fact that his entry in the *Post Office London Directory* exists in the 1822–1826 edition but not in the edition of 1827. Moreover, the lottery and exchange office of N. & S. Sylvester was listed in New York directories at 130 Broadway between the years 1827 and 1830. See Archibald, "Unpublished Letters of James Joseph Sylvester," 101. He was surely well established in New York City by 1830 when he began publication of *Sylvester's Reporter*, subtitled "A Weekly Report of Lotteries, Bank Notes, Broken Banks, Stocks, &c." This periodical is held in the New York Public Library.

Elias Joseph Sylvester did emigrate, settling first in Philadelphia. He married Mary Drinker of Philadelphia most likely in the mid- to late-1830s, converted to Presbyterianism, and died of tuberculosis in Washington, DC in 1850. I thank Hugh Stewart for sharing with me Rosemary Flamion's genealogical work.

In *JJS: Life and Work in Letters* (4 5 [note 9]), I stated, based on Flamion's work and other genealogical records that I had, that Elias Sylvester emigrated in 1826. Based on the newly found evidence presented above, this would appear to have been a mistake. Moreover, I incorrectly offered the possibility that sisters Ellen (or Eleanor) or Elizabeth could have authored the letter, but they did not adopt the surname "Sylvester."

48. Susser, *Jews of South-West England*, 57.

49. As noted, Nathaniel Sylvester also eventually emigrated to the United States and joined his brother, Sylvester Sylvester, in the lottery and brokerage business.

50. Wolfman, "Liverpool Jewry in the Eighteenth Century [Part 2]," 37.

51. The case of Sir Francis Palgrave (1788–1861), barrister, constitutional historian, and first archivist of the Public Records Office in London, provides an extreme illustration of this point. He was born the son of Meyer Cohen and Rachel Levin Cohen but was baptized, married a non-Jew, and adopted his wife's mother's family name of Palgrave. See Endelman, *Jews of Georgian England*, 259. Palgrave would be one of the Directors of the Equity and Law Life Assurance Society at its founding in 1844. Sylvester would serve as that company's actuary from 1844 to 1855. See chapter 4.

52. Stephen and Lee, *Dictionary of National Biography*, s.v. "Gregory, Olinthus Gilbert" by Alexander Gordon (hereinafter cited as *DNB*).

53. Endelman, *Jews of Georgian England*, 260.

54. This is the address that appears on the letters from E. J. Sylvester to the authorities of the University of London dated 8[?] January and 25 February 1829 and referred to

above. From the mid-eighteenth through the mid-nineteenth century, Queen Square was the site of a girls' school nicknamed the "ladies' Eton." See Weinreb and Hibbert, *London Encyclopædia*, s.v. "Queen Square."

55. For the history of University College London and the ideals embraced by its founders, see Bellot, *University College London.*

56. For the story behind De Morgan's appointment, see Rice, "Inspiration or Desperation?" 257–274.

57. See *University of London. Register of Students*, vol. 1. I thank Adrian Rice for this information. From this point forward, I refer to James Joseph as Sylvester.

58. Bellot, *University College London*, 79.

59. *Second Statement by the Council of the University of London*, 19, as quoted in Rice, "Augustus De Morgan and the Development of University Mathematics in London in the Nineteenth Century," 64.

60. For the complete list of topics under each of these headings, see *Second Statement*, 42–45, as interpreted in Rice, "Augustus De Morgan," 75–76.

61. De Morgan, "An Introductory Lecture," UCL Archives, MS.ADD.3, f. 3, as quoted in Rice, "Augustus De Morgan," 76.

62. For the quotes, see UCL Archives, College Correspondence, No. AM/7, Committee Report on the Appointment to the Chair of Natural Philosophy, 18 November 1837, f. 3, as quoted in Rice, "Augustus De Morgan," 83.

63. UCL Archives, College Correspondence, Testimonial from De Morgan, [May 1841], as quoted in Rice, "Augustus De Morgan," 83.

64. E. J. Sylvester to the University authorities, 8[?] January 1829. The next quote is also from this letter. It is not clear why brother Elias and not father Abraham was engaged in this correspondence.

65. Bellot, *University College London*, 180.

66. E. J. Sylvester to the University authorities, 25 February 1829. In general, it has been reported that Sylvester was expelled from University College London. This letter makes it clear that he was withdrawn by his family, although perhaps just in time to forestall expulsion.

67. Ellen Joseph listed herself as Eleanor Joseph beginning in the 1829 edition of the renamed *Gore's Directory of Liverpool and Its Environs*. In the next edition in 1832, Abraham Joseph is listed as also residing at this address which by then housed the ladies' seminary as well.

68. Ormerod, *Liverpool Royal Institution*, 10.

69. Brown, *Royal Institution School Liverpool*, 21. The next two quotes are also from this page.

70. University of Liverpool, Archives of Liverpool Royal Institution and Liverpool Learned Societies at the University of Liverpool, The Liverpool Royal Institution Archive 1813–1942, Administrative Records and Correspondence, Minutes of the Committee and Sub-Committee 1822–1873, Index to Minute Books 1822–1970, LRI 1/2/2.

71. Brown, *Royal Institution School Liverpool*, 151. See also MacMahon, "James Joseph Sylvester," x, for an account of the kudos Sylvester received on winning the mathematics prize.

72. John Forshaw to A. Theodore Brown, 22 April 1897, in Brown, *Royal Institution School Liverpool*, 49.

73. William Leece Drinkwater to A. Theodore Brown, 26 June 1897, in Brown, *Royal Institution School Liverpool*, 53–54.

74. William Leece Drinkwater to A. Theodore Brown, 24 April 1897, in Brown, *Royal Institution School Liverpool*, 50–51. The three quotes in the next paragraph are also from this letter, pp. 51, 52, and 52, respectively.

75. Keatinge had married Harriet Joseph, the daughter of James's uncle, Samuel Joseph, in September 1814. Harriet had been born in 1792, the same year as James's eldest sister, Ellen, so the Keatinges would have been mature authority figures for the wayward Sylvester.

76. Baker, *Math. Papers JJS*, 4:xvii.

77. According to Percy MacMahon in his obituary of J. J. Sylvester, one of Sylvester Sylvester's friends, D. V. Gregory, had posed the problem of how best to package lottery numbers. See MacMahon, "James Joseph Sylvester," xi–xii.

78. For more on this story, see Ezell, *Fortune's Merry Wheel*, 168–170. Sylvester Joseph Sylvester's work as a lottery broker was not unrelated to the businesses of naval agent and privateersman engaged in, respectively, by his father and uncle. In the United States at least, lotteries and those involved in running them paved the way for what Ezell has called "the eminently respectable occupation of private banking and . . . the twentieth-century chain of stock brokerages" (82).

79. Baker, *Math. Papers JJS*, 4:xvii.

80. Endelman, *Jews of Georgian England*, 259.

81. Wolfman, "Liverpool's Jewish Mayor," 61.

82. Endelman, *Jews of Britain*, 101–103. Support for Jewish emancipation was by no means universal within the Jewish community, however. Many Jews were happy with their situation and reluctant to call attention to themselves and to their religion through political agitation. See Endelman, *Jews of Britain*, 104–105.

Chapter 2

Epigraph: "Art. IX: The Cambridge Controversy: Admission of Dissenters to Degrees," 467.

1. As quoted in Thomson, *England in the Nineteenth Century*, 170.

2. There were Conservative governments under Sir Robert Peel in 1834–1835 and again in 1841–1846.

3. For more details, consult Roth, *History of the Jews in England*, 248–256. Among the many published statements in support of the Jewish position, see Francis Henry Goldsmid's pamphlet, *Remarks on the Civil Disabilities of British Jews* and Thomas Babington Macaulay's anonymous essay, "Statement of the Civil Disabilities and Privations Affecting Jews in England." The latter, based on a speech he gave before Parliament, has been described as "a classic of English apologetics." See Roth, *History of the Jews in England*, 252.

4. Guicciardini, *Development of Newtonian Calculus in Britain*, 124–138.

5. Winstanley, *Early Victorian Cambridge*, 85.

6. Dyer, *Academic Unity*. See Garland, *Cambridge before Darwin*, 76.

7. Winstanley, *Early Victorian Cambridge*, 86.

8. Miller, *Portrait of a College*, 74.

9. Winstanley, *Early Victorian Cambridge*, 88–89.

10. Airy, *Autobiography*, 102, as quoted in Garland, *Cambridge before Darwin*, 76.

11. See Epigraph note. The situation at Cambridge had also been discussed in the liberal *Edinburgh Review*. See "Art. X: "English Corporations and Endowments," 496–498.

12. Winstanley, *Early Victorian Cambridge*, 73–78.

13. Winstanley, *Early Victorian Cambridge*, 95–96; Garland, *Cambridge before Darwin*, 77–78.

14. Bristed, *Five Years in an English University*, 121 (the first edition appeared in 1851). On the rivalry between Trinity and St. John's, see also Winstanley, *Early Victorian Cambridge*, 384–385.

15. The Mathematical Tripos, a culminating university-wide examination, was generally taken in the January of a student's fourth and final year. The students' papers were graded and ranked in an "order of merit" which then broke down into three honors categories and the nonhonors category of "poll men." Those with the highest scores were termed "wranglers," with the "senior wrangler" being at the top of the order of merit. The wranglers were followed by the "senior optimes" and finally by the "junior optimes."

16. These figures come from Tanner, *Historical Register of the University of Cambridge*, 483–504, 602–612.

17. Miller, *Portrait of a College*, 77.

18. In a testimonial for Sylvester written in 1837, Wilson acknowledged that "some years ago, when Mr. Sylvester became one of my pupils, he obtained a prize of £100, for the solution of a problem proposed in America." See "Testimonials Obtained by Professor Sylvester, on Occasion of Becoming a Candidate for the Chair of Natural Philosophy, in University College, London, (commonly called the University of London), in the Year 1837," (hereinafter cited as "Testimonials 1837"), p. 16, Daniel Coit Gilman Papers Ms. 1, Special Collections, Milton S. Eisenhower Library, the Johns Hopkins University (hereinafter cited as DCG Papers, JHU). Recall the discussion of the prize in chapter 1.

19. See *Admissions 1802 to 1835—Sharpley to Truell*, vol. 9, St. John's College Archives (hereinafter cited as SJC Archives), entry 3867, "Sylvester, James Joseph."

20. It was required that a candidate's name be officially entered on a college's rolls by the end of the Easter term of the year in which he intended to matriculate. This necessary formality could be accomplished by any Master of Arts of the University. Compare Bristed, *Five Years in an English University*, 26, and Quekett, "Life at St John's in 1821," 149–150.

21. *Admissions 1802 to 1835—Sharpley to Truell*, entry 3867, "Sylvester, James Joseph."

22. Wilson, *System of Plane and Spherical Trigonometry*.

23. "Testimonials 1837," 16.

24. These figures were culled from *The Cambridge Guide*. In the 1790s, St. John's had roughly 120 students in residence and admitted about 40 new students each year. By 1851, the College was admitting 90 new students annually and had 371 students in residence. See Miller, *Portrait of a College*, 72.

25. On the issue of the sizes of the colleges and the resultant differential in competition levels, see Bristed, *Five Years in an English University*, 78–79, 114, 131, 163–164.

26. Ibid., 15. Traditionally, sizars had performed various menial services for the fellows or for the fellow-commoners and had eaten leftovers from the fellows' table in exchange for the privilege of attending the college. By the end of the eighteenth century, although these outward manifestations of servility had largely vanished, class distinctions persisted. See below in this section.

27. Miller, *Portrait of a College*, 80.

28. *Cambridge University Calendar for the Year 1832*, 304. As Rothblatt remarked in *Revolution of the Dons*, "although the sizars were confessedly poor men, they did not come from the 'poor classes' of society: they were, like scholars or exhibitioners, mainly sons of professional men, especially the sons of clergymen" (76).

29. *Cambridge University Calendar for the Year 1831*, 172–173. Pensioners paid £15, but sizars paid £10.

30. Miller, *Portrait of a College*, 76.

31. Miller, *Portrait of a College*, 76–77, and Pugh, *Victoria History of the Counties of England*, 443.

32. As quoted in Miller, *Portrait of a College*, 81.

33. In 1863, for example, Sylvester took part in a by-invitation-only party in London honoring Colenso. See chapter 7.

34. Pugh, *Victoria History of the Counties of England*, 441.

35. Scott, *St. John's College Cambridge*, 102.

36. See Bristed, *Five Years in an English University*, 71–83 for a vivid description of such an evening.

37. As quoted in Miller, *Portrait of a College*, 81.

38. Brown, *Royal Institution School Liverpool*, 51. Drinkwater's letter to Brown was dated 24 April 1897, just over a month after Sylvester's death.

39. The Smith's Prizes were established in 1769 by Robert Smith, a former master of Trinity College. In the 1830s, they were worth £25 each and were given annually based on yet another examination. "The examination . . . takes place immediately after the result of the Mathematical Tripos is declared, and . . . serves to rectify or confirm the arrangement of the first three or four Wranglers." Bristed, *Five Years in an English University*, 126. See also *The Cambridge Guide*, 21–22, and compare with Barrow-Green, "'A Corrective to the Spirit,'" 271–316.

40. Biographical information on eighteenth- and nineteenth-century Cambridge students may be found in Venn, *Alumni Cantabrigiensis*.

41. Scott, *St. John's College Cambridge*, 78–80, and Miller, *Portrait of a College*, 73–74.

42. Bristed, *Five Years in an English University*, 25.

43. Miller, *Portrait of a College*, 79. Sylvester may have had an exemption—as a Jew—from chapel attendance.

44. On the day-to-day activities of the Cambridge undergraduate, see Bristed, *Five Years in an English University*, 33–38, and Wright, *Alma Mater*, 1:116–124.

45. Venn, *Early Collegiate Life*, 258.

46. Wright, *Alma Mater*, 1:126.

47. Bristed, *Five Years in an English University*, 96–97. Bristed was describing Trinity, but the scene at St. John's would have been much the same.

48. *Cambridge University Calendar for the Year 1833*, 291–297.

49. The Cambridge academic year was divided into three terms. The Michaelmas Term began on 10 October and ran until 16 December; the Lent Term started on 13 January and ended on the Friday before Palm Sunday; and the Easter Term opened on the second Wednesday after Easter Sunday and officially concluded on the Friday after the first Tuesday in July.

50. Bristed, *Five Years in an English University*, 122.

51. Compare Wright, *Alma Mater*, 1:206–207, and Bristed, *Five Years in an English University*, 114, 121, and 133.

52. Bristed, *Five Years in an English University*, 134.

53. The junior soph rankings appear in *Cambridge University Calendar for the Year 1834*, 245. For the placements in 1835, see Tanner, *Historical Register of the University of Cambridge*, 494.

54. *Admissions 1802 to 1835—Sharpley to Truell*, entry 3867, "Sylvester, James Joseph."

55. Percy MacMahon reported that "in June, 1833 [Sylvester] became seriously ill and had to remain at home till November. He then returned to the University, but again unfortunately became ill in February, 1834, and was obliged to remain at home for nearly two years, not rejoining his college till January, 1836." See MacMahon, "James Joseph Sylvester," *Proceedings of the Royal Society of London*, xii. Henry Baker was less categorical, stating that Sylvester "resided continuously till the end of the Michaelmas Term, 1833, though he seems to have been seriously ill in June of this year. For two years from the beginning of 1834 his name does not appear as a member of the College, and apparently he was at home on account of illness." See Baker, *Math. Papers JJS*, 4:xvii. It is true that Sylvester suffered throughout his life from bronchial complaints.

56. Bristed, *Five Years in an English University*, 183.

57. [J. J. Sylvester], *Collection of Examples on the Integral Calculus* (1835). A copy of this booklet may be found in the John T. Graves Collection, No. 1275, 6.d.18 in the Manuscripts and Rare Books Library, University College London. This copy has Sylvester's signature written in under the designation "A Member of the University." As noted below, Sylvester also had another early work published anonymously with this same label. Adrian Rice discovered this previously unknown 1835 work of Sylvester's in the course of his thorough mining of the nineteenth-century mathematical resources of University College.

58. Years later, a partially apochryphal story of this book appeared in *The Jewish Chronicle* (London), a major voice for the Anglo-Jewish community. The anonymous writer recounted that "Mr. Sylvester . . . wrote a book upon a mathematical subject, which Deighton undertook to publish; but when the manuscript was put into the printer's hands, he threw up his task in despair [on account of Sylvester's handwriting] and Mr. Sylvester's book never saw the light." See "Cambridge: Admission of a Jew to a Degree," *The Jewish Chronicle* (London), 24 July 1857, 1088.

59. See [Sylvester], *Collection of Examples on the Integral Calculus* (1835), p. 9 for the first example and p. 97 for example 98. Sylvester would have found the notational convention $\int_x$ in the writings of Sir George Biddell Airy.

60. *Cambridge University Calendar for the Year 1837*, 266.

61. *Admissions 1802 to 1835—Sharpley to Truell*, entry 3867, "Sylvester, James Joseph."

62. C. C., "Obituary: The Rev. H. H. Hughes," 209.

63. Cumming attested to Sylvester's double attendance in his testimonial of 9 October 1837. See "Testimonials 1837," 12. On Sylvester's ongoing and ultimately long-standing interest in chemistry, see chapter 9.

64. Cumming, "On the Connexion of Galvanism and Magnetism," 268–279, and "On the Application of Magnetism as a Measure of Electricity," 281–286. Compare Gillispie, *Dictionary of Scientific Biography*, s.v. "Cumming, James" by Bernard Finn (hereinafter cited as *DSB*).

65. *DNB*, s.v. "Cumming, James" by George Thomas Bettany.

66. Bristed, *Five Years in an English University*, 173.

67. *Cambridge University Calendar for the Year 1831*, 179. These appearances took place beginning in the Lent term of the third year of residence and continuing into the Michaelmas term of the fourth.

68. Rouse Ball, *History of the Study of Mathematics at Cambridge*, 181–182.

69. Newton, *Mathematical Principles of Natural Philosophy*, 1:32.

70. J. I. [sic] Sylvester, "A Supplement to Newton's First Section" (1836), v (his emphasis). A copy of this work—heavily annotated by Sylvester in August 1836 and inscribed (most likely subsequently) "With the author's respects for the Revd R. P. Graves"—may be found in the John T. Graves Collection, 18.h.27/20 in the Manuscripts and Rare Books

Library, University College London (hereinafter referred to as "Supplement to Newton's First Section" (1836), UCL). On the substance of some of these annotations, see below.

Robert Perceval Graves was the brother of John T. Graves, the latter the codiscoverer (with Arthur Cayley) of the octonions and holder of the chair of jurisprudence at University College after 1837. On R. P. Graves's copy, the middle initial of the author has been penned in to read "J." Sylvester most likely failed to proofread the pamphlet adequately. Once again, Adrian Rice uncovered this particular copy of the pamphlet.

Although MacMahon alluded to this early work in his obituary of Sylvester, it was essentially unknown until Raymond C. Archibald pointed out the existence of copies in the Boston Public Library and in the Newcastle-upon-Tyne Public Library. MacMahon, "James Joseph Sylvester," *Proceedings of the Royal Society of London*, xii; Archibald, "Unpublished Letters of James Joseph Sylvester," 95–97, 153–154.

71. Sylvester, "Supplement to Newton's First Section" (1836), UCL, v.

72. Ibid., viii (his emphasis). In the annotations Sylvester dated August 1836, he confessed at this point that "I am told this is Boscovich's. It is not worth quarreling about. . . . This little book was written to *patch* up the old system—and less for any ultimate utility than as a specimen of the argument from old definitions." The natural philosophical ideas of the Dubrovnik-born polymath, Rudjer J. Boškovič (1711–1787) (or Roger Joseph Boscovich), had been under discussion in England at least since the middle of the eighteenth century. See Feingold, "A Jesuit among Protestants," 511–526. That some of the ideas in the pamphlet might have been "worth quarreling about" was perhaps implicit in the remark and indicative of an early awareness of matters of priority and mathematical "ownership."

73. Sylvester, "Supplement to Newton's First Section" (1836), UCL, 14 (his emphasis).

74. A copy of the work in the Library of St. John's College, Cambridge, has a title page with title, publication information, and general layout identical with the University College version, the only difference being in the identification of the author. Hugh Stewart made the connection between this anonymous pamphlet at St. John's and the UCL version. I thank him for sharing his discovery with me.

75. MacMahon, "James Joseph Sylvester," *Proceedings of the Royal Society of London*, xii. Compare *Admissions 1802 to 1835—Sharpley to Truell*, entry 3867, "Sylvester, James Joseph."

76. On the practice of utilizing pseudonyms, see Smith and Wise, *Energy and Empire*, 176.

77. Sylvester, "Supplement to Newton's First Section" (1836), UCL, iii.

78. Ibid., final blank page.

79. Ibid., third from the last blank page (his emphasis).

80. Ibid., fourth blank page from the back, labeled "9" at the top (his emphasis).

81. Ibid., fifth blank page from the back labeled "8" at the top (his emphasis).

82. Compare Bristed, *Five Years in an English University*, 187.

83. Rouse Ball, *History of the Study of Mathematics at Cambridge*, 213.

84. Winstanley, *Early Victorian Cambridge*, 412; Bristed, *Five Years in an English University*, 192.

85. See Baker, *Math. Papers JJS*, 4:xxii.

86. "Testimonials 1837," 9.

87. Blondheim, "A Brilliant and Eccentric Mathematician," 120.

88. "Testimonials 1837," 10.

89. *DNB*, s.v. "Hopkins, William" by Robert Edward Anderson.

90. For one of many accounts of the private tutorial, see Forsyth, "Old Tripos Days at Cambridge," 166–167, 173–174.

91. Hopkins, *Remarks on Certain Proposed Regulations*, 10 (privately printed pamphlet), as quoted in Wilson, "The Educational Matrix," 16.

92. Much has been written on the nineteenth-century Cambridge Tripos. I have drawn the brief synthesis given here principally from Rouse Ball, *History of the Study of Mathematics at Cambridge*; Wilson; "The Educational Matrix"; Becher, "William Whewell and Cambridge Mathematics," 1–48; and Gascoigne, "Mathematics and Meritocracy," 547–584.

93. On the efforts of the Cambridge Analytical Society to reform mathematics education at Cambridge beginning in 1812, see the following works by Enros: "The Analytical Society: Mathematics at Cambridge in the Early Nineteenth Century," "The Analytical Society (1812–1813)," and "Cambridge University and the Adoption of Analytics in Early Nineteenth-Century England"; in addition to Wilkes, "Herschel, Peacock, Babbage, and the Development of the Cambridge Curriculum," Fisch, "The Problematic History of Nineteenth-Century British Algebra," and Durand-Richard, "L'École algébrique anglaise."

For a sense of the broader implications and perspectives, see the classic study by Cannon, *Science in Culture* and the more recent and politically and theologically oriented works by Becher, "Radicals, Whigs, and Conservatives" and Ashworth, "Memory, Efficiency, and Symbolic Analysis."

94. Wilson, "The Educational Matrix," 19. Of course, the few students, like Sylvester, who elected to hear Cumming's chemical lectures might know more about Oersted's ideas than the vast majority of their Cambridge cohort. In 1841, for example, Bristed was one of only three students attending Cumming's lectures. See Bristed, *Five Years in an English University*, 166.

95. The letter, dated 20 January 1820, is quoted in Rouse Ball, *Study of Mathematics at Cambridge*, 210.

96. For the timetable of the 1837 Tripos, see *Cambridge University Calendar for the Year 1837*, 142–144.

97. Compare Bristed, *Five Years in an English University*, 287–291. When Bristed took the Tripos in 1845, it had been lengthened from five to seven days, but in its generalities, his account holds true for Sylvester's year of 1837. Note that all students vying for honors had to take the Mathematical Tripos first; only those who wanted to try for honors in classics also took the Classical Tripos.

98. Bristed, *Five Years in an English University*, 290–291 (his emphasis). The Johnian, Stephen Parkinson, did win that year over the man from Peterhouse, the future mathematical physicist and gem in England's scientific crown, William Thomson (later Lord Kelvin). Compare Smith and Wise, *Energy and Empire*, 80–82.

99. Blondheim, "A Brilliant and Eccentric Mathematician," 120.

100. On this atmosphere, compare Wright's description of the 1824 Tripos in *Alma Mater*, 2:93–96; Bristed's descriptions of those in 1842 (when Arthur Cayley was senior wrangler) and 1845 in *Five Years in an English University*, 125–126 and 290–295; and the more general account given by Forsyth in "Old Tripos Days at Cambridge," 176–177.

101. Sylvester's friend, William Drinkwater, stayed with him in College in 1837 on the occasion of taking his own M.A. and reported much later that Sylvester "would have been Senior Wrangler, but some of his writing was undecipherable." See Brown, *Royal Institution School Liverpool*, 52. This story was also told twenty years after Sylvester's Tripos finish, when the first Jew, Arthur Cohen, actually took a Cambridge degree.

See "Admission of a Jew to Cambridge," *The Jewish Chronicle* (London), 24 July 1857. Anyone who has tried to read Sylvester's handwriting—much less his hand in the heat of passion—would not fail to find this story plausible! Other second wranglers who went on to make great scientific reputations for themselves were William Thomson in 1845 (as noted above), James Clerk Maxwell in 1854, William Kingdon Clifford in 1867, and J. J. Thompson in 1880.

102. For the most definitive biography of Green to date, see Cannell, *George Green, Mathematician and Physicist.*

Chapter 3

Epigraph: The letter, undated but from 22 May 1843, appeared in Archibald, "Unpublished Letters of James Joseph Sylvester," 123. Sylvester drew here from the fifth act of Shakespeare's "Midsummer Night's Dream."

1. Cannon, *Science in Culture*, 30.

2. Ashworth, "Memory, Efficiency, and Symbolic Analysis," 633–636.

3. See, for example, Schaffer, "Babbage's Intelligence"; Ashworth, "The Calculating Eye"; and especially, Desmond, *Politics of Evolution.*

4. Cannon, *Science in Culture*, 142.

5. Heyck, *Transformation of Intellectual Life in Victorian England*, 65.

6. Cannon, *Science in Culture*, 146.

7. J. J. Sylvester to Lord Brougham, 18 June 1861, UCL Archives, Brougham Correspondence, Sylvester, J. J.:20,248, published in Parshall, *JJS: Life and Work in Letters*, 103.

8. Jenkins and Jones, "Social Class of Cambridge University Alumni." The figures here and to follow are from this source. See Jenkins and Jones, 99. Compare, too, Reader, *Professional Men*, 11–15. One third of the fathers of Cambridge students during this period were Anglican clergymen, so the sons favored this career by a margin of two to one.

9. Reader, *Professional Men*, 150.

10. Rothblatt, *Revolution of the Dons*, 90.

11. Reader, *Professional Men*, 120–123.

12. *DNB*, s.v. "Ritchie, William" by George Stronach; and *Times* (London), 21 September 1837.

13. J. J. Sylvester to the "Secretary of University College &c. &c. &c. London" (Charles C. Atkinson), 23 October 1837, UCL Archives, College Correspondence, Sylvester, J. J., 1837:4,143, in Parshall, *JJS: Life and Work in Letters*, 4.

14. John Hymers and the Rev. Yate to whom it may concern, 16 October 1837, "Testimonials 1837," 14–15.

15. Edward Bushby to whom it may concern, 10 October 1837, in "Testimonials 1837," 15.

16. James Heaviside to whom it may concern, 9 October 1837, in "Testimonials 1837," 3 (his emphasis).

17. Samuel Earnshaw to whom it may concern, 16 October 1837, in "Testimonials 1837," 4.

18. John Hymers to whom it may concern, 10 October 1837, in "Testimonials 1837," 7.

19. Olinthus Gregory to the Right Honourable the President and Vice-President, and the Members of Council of University College, London, 31 October 1837, in "Testimonials 1837," 18–19.

20. *DNB*, s.v. "Webster, Thomas" by Henry Truman Wood.

21. Committee Report on the Appointment to the Chair of Natural Philosophy, 18 November 1837, UCL Archives, College Correspondence, AM/7. The direct quotes in this and the next paragraph are from this source.

22. Much has been written on the evolution of this conception of the professor in the German states, and especially in Prussia, as a result of the Humboldtian reforms in the opening decades of the nineteenth century. See, for example, Turner, "The Growth of Professional Research in Prussia."

23. J. J. Sylvester to Charles C. Atkinson, 21 November 1837, UCL Archives, College Correspondence, Sylvester, J. J., 1837:4,179.

24. *DNB*, s.v., 'Lardner, Dionysius" by James McMullen Rigg, and Hays, "The Rise and Fall of Dionysius Lardner."

25. See, in particular, the historical maps and drawings in Bellot, *University College London*, 34, 75, and 172–173, as well as Harte and North, *World of University College London 1828–1978*, 20–29.

26. Besant, *London North of the Thames*, 406–414. See also Weinreb and Hibbert, *London Encyclopedia*, s.v. "Doughty Street."

27. Desmond, *Politics of Evolution*, 27.

28. Compare Desmond, *Politics of Evolution*, 25–100, for a detailed and nuanced account of the political and social agendas underlying and driving University College in the 1830s.

29. See "The First Flaherty Scholarship," *Times* (London), 30 October 1838; and compare Brown, *Royal Institution School Liverpool*, 51.

30. Rice, "De Morgan and the Development of University Level Mathematics," 208, 228–229, and J. J. Sylvester to Henry Malden, Dean of the Faculty of Arts, 26 April 1839, UCL Archives, College Correspondence, Sylvester, J. J., 1839:4,533.

One of the schoolmasters in Sylvester's night class, Isaac Todhunter, earned his degree at University College in 1842 and proceeded, on De Morgan's advice, to St. John's College, Cambridge. Todhunter took his Cambridge B.A. in 1848 as senior wrangler and first Smith's Prizeman and went on to a successful career as a teacher and writer of scientific textbooks, historical studies on mathematics, and pedagogical tracts. See, among other references, Macfarlane, *Lectures on Ten British Mathematicians*, 134–146.

31. *University College, London, Session 1839–40: Faculty of Arts and Laws*, 8. See also Rice, "De Morgan and the Development of University Level Mathematics," 208.

32. Bellot, *University College London*, 175–179.

33. Compare Bellot, University College London, 179 for a sense of the salaries in the 1830s, and see Rice, "Sylvester and Friendship," unpublished lecture given at New College Oxford, 15 March 1997 at a celebration of the centennial of Sylvester's death.

34. Dr. Davison to Charles C. Atkinson, October 1839, UCL Archives, College Correspondence, 4683a.

35. J. J. Sylvester to Charles C. Atkinson, 26 November 1838, UCL Archives, College Correspondence, Sylvester, J. J., 1839:4,431.

36. J. J. Sylvester to Charles C. Atkinson, 22 April 1839, UCL Archives, College Correspondence, Sylvester, J. J., 1839:4,525.

37. Kelland, "On the Transmission of Light in Crystallized Media," 323–352, 353–360.

38. The synopsis that follows of early-nineteenth-century developments in this area is drawn from the excellent overview provided by Spearman's essay, "Mathematics and Theoretical Physics," 209–214.

39. Cross, "The Organization of Science in Dublin from 1785 to 1835," 232–235.

40. Sylvester, "Analytical Development of Fresnel's Optical Theory of Crystals" (1837–1838), or *Math. Papers JJS*, 1:1–27.

41. Sylvester, *Math. Papers JJS*, 1:1 (his emphasis).

42. Sylvester, "On the Motion and Rest of Fluids" (1838), or *Math. Papers JJS*, 1:28–32; "On the Motion and Rest of Rigid Bodies" (1839), or *Math. Papers JJS*, 1:33–35; and "On Definite Double Integration" (1839), or *Math. Papers JJS*, 1:36–38.

43. On this work and Sylvester's other research in applied mathematics, see Grattan-Guinness, "The Contributions of J. J. Sylvester, F.R.S., to Mechanics and Mathematical Physics." The discussion of these papers appears on pp. 256–258.

44. Compare the discussion of this work in Parshall, "The Mathematical Legacy of James Joseph Sylvester."

45. Sturm made the announcement before the Paris Académie des Sciences. He published it in Sturm, "Mémoire sur la résolution des équations algébriques." Compare the discussion of Sturm's work in Sinaceur, *Corps et modèles*, 33–144.

46. Compare Burnside and Panton, *Theory of Equations*, 198–203. See Burnside and Panton, *Theory of Equations*, 203–205 for the case of Sturm's Theorem when $f_n(x) = 0$, that is, for the case when $f(x) = 0$ has a multiple root.

47. Sinaceur, *Corps et modèles*, 126 (my translation; her emphasis). Here, Sinaceur is using the terms "algebraic" and "analytic" in their modern senses. The "analytic" approach championed by Babbage, Herschel, Peacock, and the Cambridge Analytical Society in the 1810s was, in fact, "algebraic" as opposed to "geometric." The terms "algebraic" and "analytic" are contrasted in the modern sense below.

48. See Sylvester, "A Method of Determining by Mere Inspection the Derivatives from Two Equations of Any Degree" (1840), or *Math. Papers JJS*, 1:54–57.

49. See Sylvester, "On Rational Derivation from Equations of Coexistence, That Is to Say, a New and Extended Theory of Elimination. Part I" (1839), or *Math. Papers JJS*, 1:44. The other paper in which Sylvester presented this early work is: "On Derivation of Coexistence. Part II. Being the Theory of Simultaneous Simple Homogeneous Equations" (1840), or *Math. Papers JJS*, 1:47–53.

Compare also the discussion of these papers and their results in Parshall, *JJS: Life and Work in Letters*, 6–8, and Parshall, "The Mathematical Legacy of James Joseph Sylvester," 247–267. For more on Sturm's work and Sylvester's extension of it *per se*, consult Sinaceur, *Corps et modèles*, 33–144.

50. Recall from chapter 2 that Sylvester was already musing on the language of mathematics in his 1836 pamphlet, "A Supplement to Newton's First Section."

51. Sylvester, "On Derivation of Coexistence. Part II" (1840), in *Math. Papers JJS*, 1:53 (his emphasis).

52. Sylvester, "On Rational Derivation of Coexistence. Part I" (1839), in *Math. Papers JJS*, 1:44.

53. Compare the analysis of this point in Parshall and Seneta, "Building an International Reputation," 210–222. That Sylvester rapidly succeeded in making himself known in Paris is exemplified by the fact that Joseph Liouville was actively and independently touting Sylvester's work at meetings of the Bureau des Longitudes by 1841. See Lützen, *Joseph Liouville 1809–1882*, 100.

54. Thomas Graham to Joseph Henry, 5 October 1841, "Incidental Testimonials," DCG Papers, JHU, 40. Graham's letter was intended as a letter of introduction for Sylvester to Joseph Henry, then Professor of Natural Philosophy at Princeton. Sylvester did, indeed, meet Henry during his sojourn in the United States. See "Struggling against an 'Adverse Tide of Affairs.'"

55. J. J. Sylvester to Charles Wheatstone, 8 July 1838, UCL, General Catalogue, S. R. MS MISC 2S; J. J. Sylvester to John Herschel, 27 May 1841, Royal Society of London, Herschel Letters, HS.17.160 (see also Parshall, *JJS: Life and Work in Letters*, 5–9); and J. J. Sylvester to Charles Babbage, 27 May 1841, British Library, 621-621v, 32301.

56. Royal Society of London, Certificates 1830–1840, EC/1839/24.

57. The presence of three physicians and of Robert Grant among Sylvester's recommenders indicates that Sylvester had established connections within the London medical community and may partly explain his fleeting idea of pursuing a career in medicine in 1838.

58. *DNB*, s.v. "Goldsmid, Sir Isaac Lyon" by Claude Goldsmid Montefiore.

59. Sylvester to Babbage, 27 May 1841.

60. Bruce, *History of the University of Virginia*, 3:73, and De Morgan to the Board of Visitors of the University of Virginia, 22 May 1841, "Testimonials Obtained by Professor Sylvester, on Occasion of Offering Himself as a Candidate for the Professorship of Mathematics in the University of Virginia in the year 1841," (hereinafter cited as "Testimonials 1841"), DCG Papers, JHU, 23–24.

61. De Morgan to the Visitors, 22 May 1841, in "Testimonials 1841," 23.

62. Sir John Herschel to J. J. Sylvester, 31 May 1841, "Testimonials 1841," 32; Charles Babbage to Andrew Stevenson, American Ambassador to the Court of St. James, 28 May 1841, in "Testimonials 1841," 33; James Ivory to J. J. Sylvester, undated, in "Testimonials 1841," 25; and John T. Graves to whom it may concern, 29 May 1841, in "Testimonials 1841," 31.

63. Robert G. Latham to the Board of Visitors of the University of Virginia, 22 May 1841, in "Testimonials, 1841," 26–27.

64. John Coombe to whom it may concern, 30 May 1841, in "Testimonials 1841," 34.

65. Robert Cullen and John Robson to whom it may concern, 25 May 1841, in "Testimonials 1841," 35.

66. Sylvester had, in fact, been acquainted with the Graves brothers since Cambridge days, when, on a reading holiday in the Lake District under Philip Kelland, he had met their other brother, Robert, then a young minister engaged there in the Lord's work. See J. J. Sylvester to Robert Graves, 24 June, 1890, Trinity College Dublin, MS 10047/47/150.

67. McDowell and Webb, *Trinity College Dublin 1592–1952*, 246–248.

68. I thank Brian O'Donnell, former librarian at the Hamilton Library of Trinity College Dublin, for this insight and for his help in tracking down the details of Sylvester's association with Trinity College.

69. Compare J. J. Sylvester to Charles Babbage, [1841], British Library, 258-259v, M32299, where Sylvester explained that he was to leave for Dublin that evening to take his Dublin M.A. Given that 6 July 1841 was a Tuesday, this letter dated only "Saturday Eve" was most likely written on Saturday, 3 July 1841.

The first Jew to have taken a degree at Trinity College Dublin, Nathan Lazarus Benmohel, did so five years earlier than Sylvester in 1836. In fact, Benmohel was the first Jew to take a university degree in the British Isles. See Roth, "Jews in English Universities," 102.

70. Six days after receiving his degrees, Sylvester presented the first part of his research on the dialytic method of elimination before the Royal Irish Academy. Sir William Rowan Hamilton was in the chair. Sylvester signed his paper "J. J. Sylvester, Esq. A.M., of Trinity College, Dublin, and Professor of Natural Philosophy in University College, London." See Sylvester, "Part I. of a 'Memoir on the Dialytic Method of Elimination'" (1840–1844), *Proceedings of the Royal Irish Academy*, 130.

Before his departure from Dublin, Sylvester actually visited personally with Hamilton at the Dunsink Observatory. See J. J. Sylvester to Sir William Rowan Hamilton, 20 September 1841, in Graves, *Life of Sir William Rowan Hamilton*, 2:348–349.

71. Charles Graves to J. J. Sylvester, 8 June 1841, in "Testimonials 1841," 28.

72. James MacCullagh to J. J. Sylvester, 5 June 1841, in "Testimonials 1841," 29–30.

73. Minutes of the Rector and Board of Visitors of the University of Virginia, vol. 3, 1837–1855, 437, University Archives, Special Collections, University of Virginia Library.

74. See Sylvester, "On the Relation of Sturm's Auxiliary Functions to the Roots of an Algebraic Equation" (1841), or *Math. Papers JJS*, 1:59–60.

75. Sylvester expressed precisely these sentiments in a letter to the University's Rector, Chapman Johnson. Compare Bruce, *History of the University of Virginia*, 3:74. The letter now unfortunately seems to be lost.

76. Andrew Stevenson to Joseph Carrington Cabell, 11 October 1841, Cabell Family Papers (#38-111), Box 32, Special Collections, University of Virginia Library.

77. Bruce, *History of the University of Virginia*, 2:169–174.

78. Minutes of the Rector and Board of Visitors of the University of Virginia, vol. 3, 1837–1855, 437, University Archives, Special Collections, University of Virginia Library.

79. Bruce, *History of the University of Virginia*, 2:298–299.

80. Bruce, *History of the University of Virginia*, 2:1–3. Objections to foreign professors persisted well into the 1840s. See "Professor of Mathematics at the University of Virginia."

81. See Bruce, *History of the University of Virginia*, 2:205–317 for a lengthy discussion of the numerous complaints and associated incidents.

82. "University of Virginia," *Watchman of the South*, 5 August 1841. This editorial was discovered and quoted by Feuer in "America's First Jewish Professor," 156.

Feuer provided a probing analysis in this article of Sylvester's brief association with the University of Virginia that aimed to set the record straight. Much had been written on the topic; stories abounded. Compare the accounts in Archibald, "Unpublished Letters of James Joseph Sylvester"; Baker, *Math. Papers JJS*, 4:xxiii; Blondheim, "James Joseph Sylvester," *The Jewish Comment* (Baltimore), 25 May 1906; Bruce, *History of the University of Virginia*, 3:73–77; Halsted, "James Joseph Sylvester," 295–298; Macfarlane, *Lectures on Ten British Mathematicians*; and Yates, "Sylvester at the University of Virginia," 194–201; among others. An even further fictionalized version of the story found its way into the short story, "August Blue," by the noted homosexual writer Guy Davenport. See Davenport, "August Blue" in *A Table of Green Fields*, 1–14.

The account here largely follows Feuer's carefully researched study, although it incorporates some new information and differs in interpretation on a few points. See "Professor of Mathematics at the University of Virginia."

83. Feuer, "America's First Jewish Professor," 156.

84. Bruce, *History of the University of Virginia*, 2:361–371.

85. Feuer interpreted this as evidence of antisemitic sentiments at work, although the prejudice would seem more broadly based than that.

86. Stevenson to Cabell, 11 October 1841.

87. Bruce, *History of the University of Virginia*, 2:63.

88. Minutes of the Faculty, University of Virginia, vols. 4–5, 2 July 1834–25 April 1842, University Archives, Special Collections, University of Virginia Library. This source is unpaginated but is in chronological order.

89. Bruce, *History of the University of Virginia*, 2:3.

90. On the practice of illuminations, see Bruce, *History of the University of Virginia*, 2:273–274. On Sylvester's reception *per se*, compare Robert Dabney to his mother,

Elizabeth Dabney, 15 December 1841 in Johnson, *Life and Letters of Robert Lewis Dabney*, 52, as quoted in Feuer, "America's First Jewish Professor," 154.

91. William Barton Rogers to "his brothers in Philadelphia," in *Life and Letters of William Barton Rogers*, 1:201. Feuer quoted the supporting passage in his article on p. 154.

92. Robert Dabney to Elizabeth Dabney, 15 December 1841 in Johnson, *Life and Letters of Robert Lewis Dabney*, 52–53 and quoted in part in Feuer, "America's First Jewish Professor," 157. Actually, Sylvester had turned twenty-seven in September 1841. Dabney was six years Sylvester's junior.

93. At 47 the School of Mathematics was the fourth largest of the nine Schools at the University in 1841. Chemistry was the largest with 75, followed by law with 54, and modern languages with 48. After mathematics, moral philosophy had 46, anatomy and surgery 42, natural philosophy 41, and medicine and ancient languages 39. See *Catalogue of the Officers and Students of the University of Virginia: Session of 1841–42*, 2.

94. The fees were actually calculated on the basis of the number of Schools a student attended. If one, the student paid $50.00 for one professor; if two, $30.00 for each professor; and if three, $25.00 for each professor. Almost all students attended three Schools, so the figures given in note 95 are calculated on the basis of $25.00 per faculty member per student.

95. For the salary figures for the academic years 1835–1836 through 1837–1838, see Bruce, *History of the University of Virginia*, 2:182. The fees for these and all years up through 1841–1842 can be calculated based on the enrollment figures given in *Catalogue: Session of 1841–42*, 2.

The Professor of Chemistry was the University of Virginia's highest paid faculty member in 1841–1842, making $2,175. Comparatively speaking, the highest paid Harvard professor that year made $2,000. See *Sixteenth Annual Report of the President of Harvard University to the Overseers, on the State of the Institution for the Academical Year 1840–41*, 8. I thank my colleague, Joseph Kett, for this reference.

In his printed testimonials, Sylvester greatly overestimated the salary when he wrote that "Mr. Sylvester succeeded in obtaining this appointment, the annual emoluments of which appeared, from the Proctor's Books, to have averaged, for the last four years, $3,900 per annum, exclusive of house and land." See "Testimonials 1841," 35. In fact, the average for the four years preceeding Sylvester's arrival was roughly $2,388.

96. On the mathematics curriculum in American colleges in the mid-nineteenth century, see Cajori, *Teaching and History of Mathematics in the United States*, and compare the discussion in Parshall and Rowe, *Emergence of the American Mathematical Research Community*, 1–23.

97. *Catalogue: Session of 1841–42*, 13. The English translation of volume one of Laplace's work had been completed by the American, Nathaniel Bowditch, in 1829. See "Struggling against an 'Adverse Tide of Affairs.'" Unfortunately, the surviving records do not permit the determinations of the precise editions and translations of the other texts used.

98. The details of this incident and of the faculty's meeting on it may be found in the Minutes of the Faculty, University of Virginia, vols. 4–5, 2 July 1834–25 April 1842. The direct quotes that follow in this and the next paragraph are from this source.

99. Ibid.

100. Bruce, *History of the University of Virginia*, 2:291.

101. Robert Dabney to Elizabeth Dabney, 15 December 1841 in Johnson, *Life and Letters of Robert Lewis Dabney*, 52 as quoted in Feuer, "America's First Jewish Professor," 157.

102. Willis Woodley to Lucian Minor, 5 January 1842, Minor Family Papers (#3750), Special Collections, University of Virginia Library.

103. Minutes of the Faculty, University of Virginia, vols. 4–5, 2 July 1834–25 April 1842.

104. As noted, students generally enrolled in three Schools. Wade's third School was that of ancient languages.

105. Minutes of the Faculty, University of Virginia, vols. 4–5, 2 July 1834–25 April 1842. The direct quotes in the next three paragraphs all come from this source.

106. Charles Coleman Wall, Jr. provided an insightful analysis of the role of honor in student behavior at the University of Virginia prior to the Civil War in "Students and Student Life at the University of Virginia, 1825 to 1861," especially pp. 90–111. Wall treated the case of Ballard and Sylvester in addition to a number of other incidents. Feuer's argument did not explicitly take the Southern sense of a gentleman's honor into account. Compare Feuer, "America's First Jewish Professor," 165–166.

107. Minutes of the Faculty, University of Virginia, vols. 4–5, 2 July 1834–25 April 1842 (my emphasis).

108. Minutes of the Faculty, University of Virginia, vols. 4–5, 2 July 1834–25 April 1842. The problems with Ballard did not let up, however. After being reported for excessive absences and receiving a reprimand from the faculty at the end of April, Ballard was expelled from the University on 5 July 1842. When his petition for readmission in October was denied, he moved on to Princeton. See Minutes of the Faculty, University of Virginia, vols. 6–7, 30 April 1842–26 September 1856.

109. Minutes of the Rector and Board of Visitors of the University of Virginia, vol. 3, 1837–1855, 446.

110. In a letter dated 9 April 1842 to Lucian Minor, the University's proctor, Willis Woodley indicated that Sylvester had left without even making provisions for the packing or transport of his belongings. He also indicated that Sylvester was planning "to return in a few days." It is not clear whether Sylvester came back to Charlottesville or, if so, how long he remained. See Minor Family Papers (#3750), Special Collections, University of Virginia Library.

111. As Feuer argued, the brothers were most likely William and Alfred Weeks of Louisiana. William Weeks had been one of the main witnesses for Ballard, and Alfred Weeks had dropped mathematics on Sylvester's request. See Feuer, "America's First Jewish Professor," 172.

112. Sylvester's student at Johns Hopkins, George Bruce Halsted, was the first to tell this story in print, although his version has an overly dramatic ring to it. See Halsted, "James Joseph Sylvester," 296. Halsted's account was promulgated by his colleague at the University of Texas, Alexander Macfarlane, in the widely read Lectures on Ten British Mathematicians, 108: "Sylvester drew his sword-cane, and pierced the young man just over the heart; who fell back into his brother's arms, calling out 'I am killed.' A spectator, coming up, urged Sylvester away from the spot. Without waiting to pack his books the professor left for New York, and took the earliest possible passage for England. The student was not seriously hurt; fortunately the point of the sword had struck fair against a rib." That there was such an assault was then disputed in the literature. See, for example, Archibald, "Unpublished Letters of James Joseph Sylvester," 97–100.

Feuer provided previously unknown corroborating evidence—the autobiography of one of Sylvester's Virginia colleagues, George Tucker—that there was, indeed, an assault. See Feuer, "America's First Jewish Professor," 170–172.

113. Many of the standard accounts of Sylvester's life, for example, those by Baker and MacMahon, give Sylvester's aversion to slavery as the reason for his departure from the University of Virginia. There seems to be no evidence to support this claim. Indeed, if slavery had been foremost in Sylvester's mind, it is doubtful that he would have left

London for a slave state in the first place. Still, his experiences in Virginia may well have helped form the negative opinion of slavery he may well have articulated later in life to colleagues like MacMahon and Baker.

114. Bruce, *History of the University of Virginia*, 2:144–147, 298–299.

115. Robert Dabney to Elizabeth Dabney, 15 December 1841 in Johnson, *Life and Letters of Robert Lewis Dabney*, 53 and quoted in part in Feuer, "America's First Jewish Professor," 157.

116. This is the very same book that Sylvester would later adopt in his senior mathematics class at Virginia.

117. For Bowditch and Peirce in the context of the development of research-level mathematics in the United States, see Parshall and Rowe, *Emergence of the American Mathematical Research Community*, 2–20.

118. J. J. Sylvester to Benjamin Peirce, 5 September 1842, in Archibald, "Unpublished Letters of James Joseph Sylvester," 116.

119. Sylvester's letter to Peirce of 28 February 1843 suggested that Peirce thought he could line up a two- or three-year appointment for his new friend, but this came to naught. See Archibald, "Unpublished Letters of James Joseph Sylvester," 119–120.

120. Oehser, *Sons of Science: The Story of the Smithsonian Institution and Its Leaders*, 12. The wording is from Smithson's will.

121. Much has been written on the history of the Smithsonian Institution. See Reingold, *Science in Nineteenth-Century America*, 152–154, for a brief overview and Moyer, *Joseph Henry: The Rise of an American Scientist* for a fuller account. The matter was finally decided in 1846 when the Smithsonian Institution was founded as a research facility, natural history collection, and library under the directorship of the Princeton physicist, Joseph Henry.

122. Sylvester wrote to Peirce in 1843 that "I feel grateful to your friends, see plenty of Sam Ward in New York. He knows all my affairs and is kindly disposed." See Elliott, *Uncle Sam Ward and His Circle*, 365.

123. Sylvester to Peirce, 28 February 1843, in Archibald, "Unpublished Letters of James Joseph Sylvester," 119 (Sylvester's emphasis).

124. J. J. Sylvester to Benjamin Peirce, 19 May 1843, in Archibald, "Unpublished Letters of James Joseph Sylvester," 121 (Sylvester's emphasis).

125. In his only book, *Laws of Verse* (1870), 75, Sylvester mentioned that he had discovered his poetic bent during his sojourn in New York City. Moreover, in a letter to Sylvester dated much later, Joseph Henry recalled Sylvester's time in the United States and mentioned that he had recently come upon "a copy of the poetry which [Sylvester] had breathed forth under the inspiration of the tender passion." See Joseph Henry to J. J. Sylvester, 8 April 1869, Record Unit 7001, Joseph Henry Collection, 1808, 1825–1878, and related papers to circa 1903, Collection Division 4, Outgoing Letters of Joseph Henry in Letterpress Books, 1865–1878 and undated, Smithsonian Institution Archives, in Parshall, *JJS: Life and Work in Letters*, 134. For more on Sylvester and poetry, see chapters 8–12.

126. J. J. Sylvester to Benjamin Peirce, dated "Monday morning" (and most likely 22 May 1843), in Archibald, "Unpublished Letters of James Joseph Sylvester," 122. Archibald managed to identify Miss Marston thanks to a passage from the autobiography of the Egyptologist, Flinders Petrie. Referring to a friend of his mother's, Miss Marston, Petrie wrote that "that good lady's life had been crippled by a deep attachment to the mathematician Sylvester; she felt the difference in religion prevented their marriage, and

they parted, he to follow a meteoric career of bachelor wanderings from one university to another, she to decline upon a lady companion, lap-dogs, charitable work and anti-vivisection." See Archibald, "Material Concerning James Joseph Sylvester," 216.

127. Sylvester to Peirce, dated "Monday morning" (and most likely 22 May 1843), in Archibald, "Unpublished Letters of James Joseph Sylvester," 122.

128. J. J. Sylvester to Joseph Henry, 6 June 1843, Record Unit 7001, Joseph Henry Collection, 1808, 1825–1878, and related papers to circa 1903, Collection Division 6, Incoming and Outgoing Correspondence, 1800–1878 and undated, Smithsonian Institution Archives, in Parshall, *JJS: Life and Work in Letters*, 10–11, and Reingold and Rothenberg, *The Papers of Joseph Henry*, 5:355–357 (hereinafter cited as *Henry Papers*).

129. Thomas Graham to Joseph Henry, 5 October 1841, "Incidental Testimonials," DCG Papers, JHU, 40, and Sylvester to Henry, 6 June 1843, in Parshall, *JJS: Life and Work in Letters*, 11.

130. J. J. Sylvester to Joseph Henry, 10 June 1843, Record Unit 7001, Joseph Henry Collection, 1808, 1825–1878, and related papers to circa 1903, Collection Division 6, Incoming and Outgoing Correspondence, 1800–1878 and undated, Smithsonian Institution Archives, in Parshall, *JJS: Life and Work in Letters*, 11 and *Henry Papers*, 5:359 (see note 2).

131. Ibid. (his emphasis).

132. J. J. Sylvester to Joseph Henry, 13 July 1843, Record Unit 7001, Joseph Henry Collection, 1808, 1825–1878, and related papers to circa 1903, Collection Division 6, Incoming and Outgoing Correspondence, 1800–1878 and undated, Smithsonian Institution Archives, in Parshall, *JJS: Life and Work in Letters*, 14, *Henry Papers*, 5:369–370.

133. Minutes of the Faculty, University of Virginia, vols. 6–7, 30 April 1842–26 September 1856.

134. Sylvester to Henry, 13 July 1843, in Parshall, *JJS: Life and Work in Letters*, 14. Predominantly Episcopalian, Columbia officially had a policy against religious discrimination. In practice, however, religion played a negative role both in Sylvester's case in 1843 and in that of the chemist, Oliver Wolcott Gibbs, a decade latter. Gibbs was a Unitarian. Compare Bruce, *The Launching of Modern American Science*, 228–230.

135. Sylvester to Henry, 13 July 1843, in Parshall, *JJS: Life and Work in Letters*, 14.

136. J. J. Sylvester to Joseph Henry, 12 April 1846, in Parshall, *JJS: Life and Work in Letters*, 15, and *Henry Papers*, 6:407–410.

Chapter 4

Epigraph (1): Record Unit 7001, Joseph Henry Collection, 1808, 1825–1878, and related papers to circa 1903, Collection Division 6, Incoming and Outgoing Correspondence, 1800–1878 and undated, Smithsonian Institution Archives, in Parshall, *JJS: Life and Work in Letters*, 15, and *Henry Papers*, 6:407–410.

Epigraph (2): "On the Relation between the Minor Determinants of Linearly Equivalent Quadratic Functions" (1851), or *Math. Papers JJS*, 1:246.

1. Sylvester, "Elementary Researches in the Analysis of Combinatorial Aggregation" (1844), or *Math. Papers JJS*, 1:91–102.

2. Ibid., in *Math. Papers JJS*, 1:92 (his emphasis).

3. J. J. Sylvester to Charles Babbage, 11 April 1844, British Library, 48–49v, 32303. The Graves was most likely Sylvester's former colleague at University College London, John Graves.

4. Augustus De Morgan to J. J. Sylvester, 3 August 1844, JJS Papers, SJC, Box 2.

5. Sylvester, "On the Double Square Representation of Prime and Composite Numbers" (1844), or *Math. Papers JJS*, 2:1.

6. Sylvester, "On the Existence of Absolute Criteria for Determining the Roots of Numerical Equations" (1844), or *Math. Papers JJS*, 1:103–106.

7. Hugh Stewart uncovered a prospectus for the company dated October 1844 in which Sylvester was already mentioned by name as actuary. See Stewart, "Founding Vice-President," 2:485. On Sylvester's salary, see p. 486.

8. Ogborn, *Equitable Assurances*, 26–49.

9. For the statistics, see Ogborn, *Equitable Assurances*, 230. For the quote, see Alborn, "A Calculating Profession," 441.

10. Alborn, "A Calculating Profession," 442, and Walford, *Insurance Cyclopædia*, 1:53.

John Francis of the Bank of England commented on the critical role the mathematically accomplished actuary played in the life insurance industry in his book, *Annals, Anecdotes and Legends: A Chronicle of Life Assurance*. As he put it, "the actuaries, who are intelligent and accomplished gentlemen, must be propitiated, for they are in possession of a somewhat occult science, having justly the ear, the confidence, and the respect of their directors. . . . There is no profession in which subordinates are so respectfully regarded, for the actuary is master of a science in which the director is generally deficient; and knowledge, in this case, as in others, is essentially power" (272–273).

11. Alborn, "A Calculating Profession," 442. See also Porter, "On Some Points Connected with the Education of an Actuary," on the qualities the actuary should possess.

12. Ogborn, *Equitable Assurances*, 230.

13. Alborn, "A Calculating Profession," 446.

14. Ogborn, *Equitable Assurances*, 232–235.

15. Alborn, "A Calculating Profession," 446–448.

16. Quoted in Ogborn, *Equitable Assurances*, 236, as well as in Supple, *Royal Exchange Assurance*, 139.

17. Ogborn, *Equitable Assurances*, 236.

18. Supple, *Royal Exchange Assurance*, 140–143.

19. For the history of the Institute of Actuaries, see Simmonds, *Institute of Actuaries 1848–1948*. For a brief overview of the actuarial profession, its establishment, and its concerns in mid-nineteenth-century Britain, see Porter, *Trust in Numbers*, 101–113.

20. Archibald gives the actual date as 9 December in "Unpublished Letters of James Joseph Sylvester," 87.

21. Walford, *Insurance Cyclopædia*, 3:24.

22. Stewart, "Founding Vice President," 485.

23. Walford, *Insurance Cyclopædia*, 3:70.

24. Stewart, "Founding Vice President," 485.

25. Sylvester to Henry, 12 April 1846, in Parshall, *JJS: Life and Work in Letters*, 16.

26. Ibid. Sylvester only held the dual post through the end of 1847, and it presumably brought with it a commensurate increase in salary. On 1 January 1848, he once again became the Equity and Law's actuary without the additional administrative burdens of its secretary.

27. One of his neighbors and acqaintances across the square was the noted comparative anatomist, Richard Owen, who had set up a medical practice there as early as 1826. Sylvester once described himself and Owen as "doves nesting among hawks" in Lincoln's Inn Fields. See Sylvester, "Presidential Address to Section 'A' of the British Association" (1869), in *Math. Papers JJS*, 2:652 (note *).

28. *Equity & Law 150 Anniversary 1844–1994*, 6.

29. The Equity and Law used Sylvester's mortality tables for some fifty years. In 1864, its directors credited them for the "very favourable mortality" of those insured by the company and for the firm's consistent profitability. Ibid.

30. Compare Ogborn, *Equitable Assurances*, 57–58, 81, 227. See also *Equity & Law 150 Anniversary*, 4. In addition to Sylvester and the bookkeeper, the firm also employed a messenger and a housekeeper and had a physician and a surgeon on retainer.

31. Walford, *Insurance Cyclopædia*, 3:26.

32. On the development of mathematical journals in Britain and on the evolution of the *Cambridge Mathematical Journal*, in particular, see Despeaux, "Development of a Publication Community," 162–184. See also Crilly, "The *Cambridge Mathematical Journal* and Its Descendents." On Thomson's editorship of the journal, in particular, compare Smith and Wise, *Energy and Empire*, 176–190.

33. J. J. Sylvester to William Thomson, 18 November 1845, Kelvin Papers, Add 7342, S594, Cambridge University Library. The quotes that follow in this paragraph are also from this letter.

34. William Thomson to J. J. Sylvester, 19 November 1845, Kelvin Papers, Add 7342, S595, Cambridge University Library (his emphasis). The following quote (with his emphasis) is also from this letter.

35. Walford, *Insurance Cyclopædia*, 3:26.

36. Sylvester to Henry, 12 April 1846, in Parshall, *JJS: Life and Work in Letters*, 16. The quote that follows in this paragraph is from this page (with his emphasis).

37. No religious oaths were required for a call to the English Bar. The first professing Jew to be called, however, was Francis Goldsmid, son of Jewish activist, Isaac Goldsmid. The younger Goldsmid became a barrister of Lincoln's Inn only in 1833. See Finestein, *Jewish Society in Victorian England*, 253–254, and compare Lachs, "A Study of a Professional Elite."

38. For the rules and regulations of the Inner Temple during Sylvester's time there as described here and in the next paragraph, see Pearce, *History of the Inns of Court and Chancery*, 384–401. The quotes are on p. 385. On legal education, in general, at this time, see Abel-Smith and Stevens, *Lawyers and the Courts*, 53–76.

39. Charles Graves to J. J. Sylvester, 26 November 1845, Call Papers, Michaelmas Term 1850, BAR/6/2, Inner Temple Archives.

40. See Admission Papers, ADM/6/13, Inner Temple Archives.

41. Pearce, *History of the Inns of Court and Chancery*, 387, 392, 397. The following quote is from p. 392.

Abel-Smith and Stevens cite figures for admission (£5–£10 depending on the Inn [p. 66]) different from that given in Pearce specifically for the Inner Temple, namely, £29.3.2 (p. 385). Pearce's figures are those recorded on Sylvester's admission papers.

42. Abel-Smith and Stevens, *Lawyers and the Courts*, 65. This sentiment is reinforced in a variety of reminiscences and accounts of those who entered the Inns of Court around midcentury. In describing the experiences at the Inner Temple in the early 1850s of his brother, Sir James Fitzjames Stephen, for example, Leslie Stephen recounted: "His legal education, he says, was very bad. He was for a time in the chambers of Mr. (now Lord) Field, then the leading junior in the Midland Circuit, but it was on the distinct understanding that he was to receive no direct instruction from his tutor." See Stephen, *The Life of Sir James Fitzjames Stephen*, 118–119.

43. Quoted in Abel-Smith and Stevens, *Lawyers and the Courts*, 66 (note 3). A compulsory examination for admission to the Bar was only instituted in 1872. For more on the

reforms of legal education from the 1840s through the 1870s, see the introductory essay in Roxburgh, *The Records of the Honorable Society of Lincoln's Inn*, li–lviii.

44. Pearce, *History of the Inns of Court and Chancery*, 421.

45. Abel-Smith and Stevens, *Lawyers and the Courts*, 169.

46. In a letter to Henry Brougham dated June 1852, Sylvester mentioned in passing that he actually had been "a pupil with Mr. Dugmore and Mr. Francis Turner of *Lincoln's Inn*." See J. J. Sylvester to Lord Brougham, June 1852, UCL Archives, Brougham Correspondence, Sylvester, J. J.:20,393 (my emphasis). It was not common but not unheard of for a member of one Inn to study with members of another Inn. Dugmore and Turner were both conveyancing lawyers. Dugmore was called to the Bar at Lincoln's Inn, while Turner was called at the Inner Temple. Both had offices in Lincoln's Inn. See *The Law List 1853*, Inner Temple Library.

47. Stewart, "Founding Vice President," 486.

48. J. J. Sylvester to William Thomson, 8 July 1846, Kelvin Papers, Add 7342, S596, Cambridge University Library.

49. See Sylvester, "An Account of a Discovery in the Theory of Numbers Relative to the Equation $Ax^3 + By^3 + Cz^3 = Dxyz$" (1847), or *Math. Papers JJS*, 1:107–109; "On the Equation in Numbers $Ax^3 + By^3 + Cz^3 = Dxyz$, and Its Associate System of Equations" (1847), or *Math. Papers JJS*, 1:110–113; and "On the *General* Solution (in Certain Cases) of the Equation $x^3 + y^3 + Az^3 = Mxyz$, &c" (1847), or *Math. Papers JJS*, 1:114–118.

50. On Thomson's attitudes, see Smith and Wise, *Energy and Empire*, pp. 187–190.

51. Sylvester, "An Account of a Discovery in the Theory of Numbers" (1847), in *Math. Papers JJS*, 1:109. The quotes that follow in the next paragraph are also from this paper and page.

52. For an historical analysis of this early work, see Lavrinenko, "Solving an Indeterminate Third Degree Equation in Rational Numbers," 80–83. For Sylvester's later work in this vein, see chapter 10.

53. Both had been present at the annual meeting of the British Association for the Advancement of Science in Cambridge in 1845, but there is no evidence that they actually met there.

54. J. J. Sylvester to Arthur Cayley, 24 November 1847, JJS Papers, SJC, Box 10, or Parshall, *JJS: Life and Work in Letters*, 19 (where there is a typographical error; the letter reads "able hands"). This appears to be the first extant letter in what would become the lifelong correspondence between Sylvester and Cayley.

55. Walford, *Insurance Cyclopædia*, 3:26.

56. Simmonds, *Institute of Actuaries*, 8. Ultimately, these deliberations would result in a rift within the community of actuaries with some of those from the older firms splitting off from the initiative to form an Institute of Actuaries and creating their own, more informal Actuaries' Club. The Actuaries' Club was formed in November 1848. The two groups did not unite until 1884 when the Actuaries' Club was disbanded and the Institute of Actuaries received its royal charter. See ibid., 38, 106–121.

57. For the full text of the document creating the Institute of Actuaries, see Simmonds, *Institute of Actuaries*, 20–22. The quote is on p. 20.

58. On this committee and its work, see Simmonds, *Institute of Actuaries*, 23–29. The details that follow in this paragraph may be found there.

59. *Equity & Law 150 Anniversary*, 8.

60. J. J. Sylvester to Arthur Cayley, 18 April 1849, JJS Papers, SJC, Box 9. For the calculus of operations in which Sylvester was engaged in this letter, see chapter 5.

61. Walford, *Insurance Cyclopædia*, 3:26–27. For a fuller report (presumably written by Sylvester), see "Equity and Law Life Assurance Society" (1851).

62. J. J. Sylvester to Arthur Cayley, 28 November 1849, JJS Papers, SJC, Box 9. The quotes that follow in this paragraph are also from this letter.

63. See John Forshaw to A. Theodore Brown, 22 April 1897 in Brown, *Royal Institution School Liverpool*, 49–50. Recall the discussion of Sylvester's pamphlet in chapter 2. Forshaw, writing decades after the fact, had his facts somewhat wrong, stating that Brougham was in Liverpool for the laying of the Institute's foundation stone in 1837, the year that Sylvester sat the Tripos. Brougham had been present for the cornerstone laying in 1835, but he was also on hand in 1837 for the Institute's actual opening. Compare the *Liverpool Mercury*, 24 July 1835, and "The Liverpool Institute: The History and Development of an Idea," 405.

64. See J. J. Sylvester to Lord Brougham, University College London Archives, Brougham Correspondence, Sylvester, J. J.:20,224, in Parshall, *JJS: Life and Work in Letters*, 20–22. For the quote that follows, see p. 22.

65. J. J. Sylvester to Arthur Cayley, 8 January 1850, JJS Papers, SJC, Box 9.

66. Cayley, "Sur le problème des contacts," or Cayley and Forsyth, *The Collected Mathematical Papers of Arthur Cayley* (hereinafter cited as *Math. Papers AC*), 1:522–531.

67. See J. J. Sylvester to Arthur Cayley, 3 February 1850, JJS Papers, SJC, Box 9.

68. Stewart, "Founding Vice President," 487.

69. "The Royal Society," *Times* (London), 20 May 1850.

70. Despeaux, "Development of a Publication Community," 22–23.

71. On the development of the calculus of operations in Britain and on the numerous contributors to it, see Koppelman, "The Calculus of Operations," 155–157, 187–189. Recall that the *Cambridge Mathematical Journal* became the *Cambridge and Dublin Mathematical Journal* in 1845 under William Thomson's editorial care.

72. On Bronwin and his work in the calculus of operations, see Koppelman, "The Calculus of Operations," 200–201.

73. J. J. Sylvester, 3 July 1850, Referee's Reports, RR.2.39, Royal Society of London. The following quote is also from this report.

74. Bronwin, "On the Solution of Linear Difference Equations."

75. Sylvester, "On the Intersections" 5 (1850), or *Math. Papers JJS*, 1·119–137.

76. Compare the discussion of this work in Parshall, *JJS: Life and Work in Letters*, 28–29. I have adapted Sylvester's notation here, using the variables x, y, z instead of his ξ, η, ζ. In the paper, Sylvester also considered not merely $\Box(U + \lambda V)$ but also the somewhat more general $\Box(\lambda U + \mu V)$.

77. See J. J. Sylvester to Arthur Cayley, 2, 3, 21, and 27 March; 5 and 20 April; 5, 24, and 29 May; 18 and 29 June; and 3 July 1850, JJS Papers, SJC, Box 9.

78. Sylvester, "On the Intersections" (1850), in *Math. Papers JJS*, 1:119.

79. Boole, "Exposition of a General Theory of Linear Transformations."

80. See Despeaux, "'Very Full of Symbols,'" forthcoming.

81. See "Establishing a Mathematical Routine." Of course, the phenomenon of invariance had long been noted in the special case of the discriminant of the binary quadratic. See, for example, Gauss's *Disquistiones arithmeticæ*.

Here, Boole's conception of the linear transformation has been somewhat simplified. Boole literally replaced the variables x and y with new variables X and Y.

82. Sylvester, "On the Intersections" (1850), in *Math. Papers JJS*, 1:119.

83. See *Math. Papers JJS*, 1:119–161.

84. J. J. Sylvester to Arthur Cayley, 26 June 1850, JJS Papers, SJC, Box 9.

85. Ballantine, *Some Experiences of a Barrister's Life*, 13.

86. For the report (again presumably by Sylvester), see "Equity and Law Life Assurance Society (1844).—Fourth Annual Meeting, Held March, 1851" (1851).

87. J. J. Sylvester to Arthur Cayley, 26 December 1850, JJS Papers, SJC, Box 9, in Parshall, *JJS: Life and Work in Letters*, 30. It was, in fact, not the same as Hesse's technique. Compare the commentary in Parshall, *JJS: Life and Work in Letters*, 30–31.

88. See note 88 and Sylvester, "Sketch of a Memoir on Elimination, Transformation, and Canonical Forms" (1851), or *Math. Papers JJS*, 1:191.

89. Compare Hesse, "Über die Elimination der Variabeln aus drei algebraischen Gleichungen vom zweiten Grade mit zwei Variabeln," 90–95. See also the commentary in Parshall, *JJS: Life and Work in Letters*, 30–31.

90. See Sylvester to Brougham, 29 December 1849, UCL Archives, Brougham Correspondence, Sylvester, J. J.:20,224, in Parshall, *JJS: Life and Work in Letters*, 22.

91. J. J. Sylvester to Henry Brougham, 6 January 1851, UCL Archives, Brougham Correspondence, Sylvester, J. J.:20,226.

92. Sylvester, "An Enumeration of the Contacts of Lines and Surfaces of the Second Order" (1851), *Philosophical Magazine*, or *Math. Papers JJS*, 1:219 (note †).

93. Recall the discussion of Sylvester's algebrization of Sturm's theorem in 1839 and 1840 in chapter 3, and compare Sinaceur, *Corps et modèles*, 126, and Parshall, "The Mathematical Legacy of James Joseph Sylvester," 253.

94. See Sylvester, "Addition to the Articles, 'On a New Class of Theorems,' and 'On Pascal's Theorem' " (1850), or *Math. Papers JJS*, 1:150.

95. Sylvester, "On the Relation between the Minor Determinatnts of Linearly Equivalent Quadratic Functions" (1851), in *Math. Papers JJS*, 1:247.

96. J. J. Sylvester to Lord Brougham, 18 June 1861, UCL Archives, Brougham Correspondence, Sylvester, J. J.:20,248, in Parshall, *JJS: Life and Work in Letters*, 103.

97. "The Rotation of the Earth," the *Times* (London), 8 April 1851. The next two quotes are also from this article.

98. J. J. Sylvester, "To the Editor of *The Times*," *Times* (London), 11 April 1851. The quotes that follow in this paragraph are also from this letter.

99. B.A.C., "To the Editor of *The Times*," *Times* (London), 24 April 1851. The quotes that follow in this paragraph are from this letter.

100. J. J. Sylvester, "To the Editor of *The Times*," *Times* (London), 26 April 1851. The next quote is also from this letter.

101. For the quote, see Sylvester, "Sketch of a Memoir on Elimination" (1851), in *Math. Papers JJS*, 1:185. See also Cayley, "On the Theory of Linear Transformations," or *Math. Papers AC*, 1:95–112. In the latter paper, Cayley had, in fact, articulated the main problem of what would become a *theory* of invariants, namely, "to find all the derivatives of any number of functions, which have the property of preserving their form unaltered after any linear transformation of the variables" (95).

102. Sylvester, "Sketch of a Memoir on Elimination" (1851), in *Math. Papers JJS*, 1:187. The following quote is also from this page.

103. He extended these results to homogeneous functions of any odd degree in two unknowns in a privately printed paper entitled, "An Essay on Canonical Forms" (1851), in *Math. Papers JJS*, 1:203–216. Interestingly, Sylvester, who had increasingly withdrawn from the Institute of Actuaries, sent a presentation copy of this paper for its library. See the account of the Institute's eighth sessional meeting of 1850–1851 in *Assurance Magazine* 1(1851): 367.

104. Sylvester, "Sketch of a Memoir on Elimination" (1851), in *Math. Papers JJS*, 1:189. The paper in question is Sylvester, "On a Linear Method of Eliminating between Double, *Treble*, and Other Systems of Algebraic Equations" (1841), or *Math. Papers JJS*, 1:75–85.

105. Sylvester, "Sketch of a Memoir on Elimination" (1851), in *Math. Papers JJS*, 1:189–190 (his emphasis). Sylvester had written on his dialytic method of elimination on several occasions between 1839 and 1842. Recall chapter 3.

106. J. J. Sylvester to Arthur Cayley, 20 May 1851, JJS Papers, SJC, Box 9, in Parshall, *JJS: Life and Work in Letters*, 33 (his emphasis). The following quote is also from this letter and page. For a technical discussion and example of Sylvester's method of compound permutation, see Parshall, *JJS: Life and Work in Letters*, 34.

Chapter 5

Epigraph: *Laws of Verse* (1870), 109 (note ∗) (his emphasis). This passage—in one of the footnotes that Sylvester added to this edition of his Exeter lecture—does not appear in the version reproduced in *Math. Papers JJS*.

1. Stewart, "Founding Vice President," 487.

2. J. J. Sylvester to Arthur Cayley, 25 August 1851, JJS Papers, SJC, Box 9, in Parshall, *JJS: Life and Work in Letters*, 35. For the terms that follow, see p. 36.

3. Sylvester, "On the General Theory of Associated Algebraical Forms" (1851), or *Math. Papers JJS*, 1:200 (note ∗). The following quote is also from this note.

4. Sylvester, "On a Certain Fundamental Theorem of Determinants" (1851), or *Math. Papers JJS*, 1:252.

5. Compare the analysis of Sylvester's efforts in this direction in Parshall and Seneta, "Building an International Reputation."

6. Sylvester, "On Extensions of the Dialytic Method of Elimination" (1851), or *Math. Papers JJS*, 1:256.

7. Sylvester, "On the General Theory of Associated Algebraical Forms" (1851), in *Math. Papers JJS*, 1:201. The quote that follows is also from this page.

8. Sylvester, "On a Remarkable Discovery in the Theory of Canonical Forms and of Hyperdeterminants" (1851), or *Math. Papers JJS*, 1:265. Kung and Rota call this result "Sylvester's Theorem" in their paper, "The Invariant Theory of Binary Forms," 63.

9. See Sylvester, "An Essay on Canonical Forms" (1851), in *Math. Papers JJS*, 1:208–214.

10. See Sylvester, "On a Remarkable Discovery" (1851), in *Math. Papers JJS*, 1:269–270.

11. Ibid. Here, Sylvester has abused his notation, using the letter l in two different senses. The coefficient l of the form need not be the same as the coefficient l in the linear transformation.

12. Ibid., 280 (note †).

13. Sylvester, "Note on a Proposed Addition to the Vocabulary of Ordinary Arithmetic" (1887), or *Math. Papers JJS*, 4:588 (note ∗). Compare the discussion of this in chapter 12.

14. In his biography of Cayley, Tony Crilly has interpreted this naming impulse in the context of Victorian interest in botany and geology. There, the goals were to collect, classify, and describe, goals that also came to characterize the British computational approach to invariant theory. See Crilly, *Arthur Cayley*, 193–198.

15. Sylvester, "On a Remarkable Discovery" (1851), in *Math. Papers JJS*, 1:269 and 277 (note ∗), respectively.

16. Later in his career, Sylvester would become notorious for *not* reading the literature and for appropriating as his own the results of others. In these situations, however, he consistently marveled at the phenomenon of multiple discovery and acknowledged prior

work when it was brought to his attention. See, for example, chapter 11 with its account of Sylvester's rediscovery of Georges Halphen's concept of differential invariants.

17. For an account of the report (again presumably by Sylvester), see "Equity and Law Life Assurance Society.—Fifth Annual General Meeting, 19th March, 1852" (1852).

18. Stewart, "Founding Vice President," 487. By 1853, he would let his membership in the Institute of Actuaries lapse altogether.

19. Arthur Cayley to J. J. Sylvester, 5 December 1851, JJS Papers, SJC, Box 2, in Parshall, *JJS: Life and Work in Letters*, 37. Here, U is an invariant, and a, b, . . . , k are the coefficients of the underlying homogeneous polynomial of degree n in two unknowns.

20. Sylvester, "On the Principles of the Calculus of Forms" (1852), or *Math. Papers JJS*, 1:284.

21. Ibid., in *Math. Papers JJS*, 1:284 (note ∗).

22. Compare ibid., 293–298, and "On a Remarkable Discovery" (1851), in *Math. Papers JJS*, 1:282.

23. Sylvester, "On the Principles of the Calculus of Forms" (1852), in *Math. Papers JJS*, 1:293 and 293 (note †). In fact, an arbitrary binary $2n$-ic form is transformable into the sum of n $2n$ powers of linear expressions in x and y if and only if its catalecticant vanishes. Recall that equation (4) was precisely the invariant that had turned up in "On a Remarkable Discovery" in his algebraic manipulation of equations (1) and (2).

24. Olry Terquem to J. J. Sylvester, 5 October 1853, JJS Papers, SJC, Box 3 (his emphasis; my translation).

25. Compare the discussion of the catalecticant in Parshall, "The Mathematical Legacy of James Joseph Sylvester," 253–255. For reports of Boole's result and Cayley's observation, see Cayley, "On the Theory of Linear Transformations," in *Math. Papers AC*, 1:93–94, and Sylvester, "On the Principles of the Calculus of Forms" (1852), in *Math. Papers JJS*, 1:293, respectively.

26. Salmon, *Treatise on Conic Sections*.

27. Sylvester referred explicitly to one of Salmon's results in his paper, "On Certain General Properties of Homogeneous Functions," in the February number of the *Cambridge and Dublin* in 1851, or *Math. Papers JJS*, 1:177.

28. J. J. Sylvester to George Salmon, dated simply February 1852, JJS Papers, SJC, Box 3. This letter is, unfortunately, incomplete. For the text, see Salmon, *Treatise on the Higher Plane Curves*.

29. George Salmon to J. J. Sylvester, 24 February 1852, JJS Papers, SJC, Box 3. The quote that follows is also from this letter.

30. J. J. Sylvester to George Salmon, 23 March 1852, JJS Papers, SJC, Box 3, and J. J. Sylvester to Arthur Cayley, 23 March 1852, JJS Papers, SJC, Box 9.

31. J. J. Sylvester to Arthur Cayley, 5 February 1852, JJS Papers, SJC, Box 9. For more on orthogonal invariants, compare Parshall, *JJS: Life and Work in Letters*, 91. The following quote is also from this letter.

32. J. J. Sylvester to Arthur Cayley, 13 February 1852, JJS Papers, SJC, Box 9.

33. Compare J. J. Sylvester to Lord Brougham, 14 June 1852, UCL Archives, Brougham Correspondence, Sylvester, J. J.:20,229.

34. *Annual Register . . . of the Year 1851*, 176, and *Annual Register . . . of the Year 1852*, 429, respectively.

35. J. J. Sylvester to Lord Brougham, 23 March 1852, UCL Archives, Brougham Correspondence, Sylvester, J. J.:15,172, 293–298.

36. See J. J. Sylvester to Lord Brougham, [no date] June 1852 and 14 June 1852, UCL Archives, Brougham Correspondence, Sylvester, J. J.:20,393 and 20,229, respectively.

37. Sylvester had referred to Cayley's "bifarious occupation" of lawyer and mathematician in a letter dated 21 December 1850. See J. J. Sylvester to Arthur Cayley, 21 December 1850, JJS Papers, SJC, Box 9, in Parshall, *JJS: Life and Work in Letters*, 28.

38. J. J. Sylvester to Arthur Cayley, 25 March 1852, JJS Papers, SJC, Box 9.

39. J. J. Sylvester to Arthur Cayley, 26 March 1852, JJS Papers, SJC, Box 9.

40. J. J. Sylvester to Arthur Cayley, 29 March 1852, JJS Papers, SJC, Box 9.

41. See J. J. Sylvester to Arthur Cayley, 11 April 1852, JJS Papers, SJC, Box 9.

42. George Salmon to J. J. Sylvester, 7 April 1852, JJS Papers, SJC, Box 3, in Parshall, *JJS: Life and Work in Letters*, 43–44. Other, more specific questions about particular examples followed, but it was this question that lay at the crux of the matter and to which Cayley would only publish a (partially) correct answer in his paper, "A Second Memoir upon Quantics," four years later in 1856. For the mathematical details of the questions Salmon asked in this letter, see the commentary in Parshall, *JJS: Life and Work in Letters*, 42–45. See also Cayley, "A Second Memoir upon Quantics," or *Math. Papers AC*, 2:250–275, and Parshall, "Toward a History of Nineteenth-Century Invariant Theory," 169, 181–182, for a discussion of Cayley's error. Compare chapter 9.

43. J. J. Sylvester to Arthur Cayley, 11 and 15 April 1852, respectively, JJS Papers, SJC, Box 9.

44. George Salmon to J. J. Sylvester, 14 April 1852, JJS Papers, SJC, Box 3, in Parshall, *JJS: Life and Work in Letters*, 45. The quote that follows in this paragraph and those in the following paragraph are also from this letter and appear on p. 46. For more mathematical details on the theorem, see the associated commentary there.

45. Carl Borchardt to J. J. Sylvester (in French), 6 April 1852, JJS Papers, SJC, Box 2, in Parshall, *JJS: Life and Work in Letters*, 38 (my translation). The quote in the following paragraph is also from this letter (pp. 39–40).

46. Sylvester, "On the Principles of the Calculus of Forms" (1852), in *Math. Papers JJS*, 1:289 (note ∗) and pp. 321 and 327.

47. See Eisenstein, "Théorèmes sur les formes cubiques et solution d'une équation de quatrième degré à quatre indéterminés," and "Untersuchungen über die cubischen Formen mit zwei Variabeln." Compare also the discussion in Parshall, "Toward a History of Nineteenth-Century Invariant Theory," 170–171.

48. The papers in question are Hesse, "Über die Elimination der Variabeln aus drei algebraischen Gleichungen von zweiten Grade mit zwei Variablen," and Hesse, "Über die Wendepunkte der Curven dritter Ordnung."

49. Aronhold, "Zur Theorie der homogenen Functionen dritten Grades von drei Variabeln," 140 (my translation). Compare also the discussion of Aronhold's work in Parshall, "Toward a History of Nineteenth-Century Invariant Theory," 172–177.

50. Sylvester, "On the Principles of the Calculus of Forms" (1852), in *Math. Papers JJS*, 1:351–352.

51. Ibid., 352 (note ∗). For an explanation of Aronhold's notation, see Parshall, "Toward a History of Nineteenth-Century Invariant Theory," 173–175.

52. In the 1850s, Cayley did publish with some regularity in Crelle's *Journal* in an effort to bring his research before the German audience. Sylvester would not publish his first paper in a German journal—also Crelle's *Journal*—until 1878. He chose to build his international reputation primarily through the French. See immediately below.

53. On Sylvester's efforts to establish an international reputation, see Parshall and Seneta, "Building an International Reputation."

54. George Salmon to J. J. Sylvester, 27 April, JJS Papers, SJC, Box 3, in Parshall, *JJS: Life and Work in Letters*, 50.

55. Irénée-Jules Bienaymé to J. J. Sylvester, draft of a letter dated 4 April 1852, in the private collection of Arnaud Bienaymé.

56. J. J. Sylvester to Irénée-Jules Bienaymé, 4 June 1852, in the private collection of Arnaud Bienaymé, and in Parshall, *JJS: Life and Work in Letters*, 57.

57. Sylvester, "Sur une propriété nouvelle de l'équation qui sert à déterminer les inégalités séculaires des planètes" (1852), or *Math. Papers JJS*, 1:364–366.

58. Ibid., in *Math. Papers JJS*, 1:366 (my translation). Borchardt had proven his more general result in "Développements sur l'équation à l'aide de laquelle on détermine les inégalités séculaires du mouvement des planètes," 63.

59. Compare J. J. Sylvester to Arthur Cayley, 19 May 1852, JJS Papers, SJC, Box 9, in Parshall, *JJS: Life and Work in Letters*, 54–56.

60. See J. J. Sylvester to Arthur Cayley, 5 June 1852, JJS Papers, SJC, Box 9.

61. See J. J. Sylvester to Arthur Cayley, 21 June 1852, JJS Papers, SJC, Box 9.

62. See J. J. Sylvester to Arthur Cayley, 19 June 1852, JJS Papers, SJC, Box 9.

63. See J. J. Sylvester to George Salmon, 16 July 1852, JJS Papers, SJC, Box 1.

64. See Sylvester, "Observations on a New Theory of Multiplicity" (1852), or *Math. Papers JJS*, 1:370–377, and "Demonstration of the Theorem That Every Homogeneous Quadratic Polynomial Is Reducible by Real Orthogonal Substitutions to the Form of a Sum of Positive and Negative Squares" (1852), or *Math. Papers JJS*, 1:378–395, respectively.

65. Compare the discussion in Parshall, "The Mathematical Legacy of James Joseph Sylvester," 252–253.

66. Cayley, "On the Theory of Linear Transformations," in *Math. Papers AC*, 1:93–94. Cayley and Sylvester would only later distinguish between invariants and covariants and between minimum generating sets of each. Whereas *I* and *J* form a minimum generating set of *in*variants of the binary quartic, a minimum generating set of *co*variants contains five elements. See Elliott, *Introduction to the Algebra of Quantics*, 213–215.

67. Cayley, "On Linear Transformations," or *Math. Papers AC*, 1:95 (his emphasis).

68. J. J. Sylvester to Arthur Cayley, 20 August 1852, JJS Papers, SJC, Box 9. The quotes that follow in this paragraph are also from this letter.

69. See Sylvester's letters to Cayley in September, October, and November 1852 in JJS Papers, SJC, Box 9.

70. J. J. Sylvester to Arthur Cayley, 15 November 1852, JJS Papers, SJC, Box 9.

71. Augustus De Morgan to J. J. Sylvester, 19 November 1852, JJS Papers, SJC, Box 2.

72. Sylvester, "Note on the Calculus of Forms" (1853), or *Math. Papers JJS*, 1:402.

73. Sylvester wrote out the resultant calculation in a letter to Cayley dated 16 October 1852, but he only published it in the November 1853 number of the *Cambridge and Dublin*. See J. J. Sylvester to Arthur Cayley, 16 October 1852, JJS Papers, SJC, Box 9, in Parshall, *JJS: Life and Work in Letters*, 60–62, and Sylvester, "On the Calculus of Forms, Otherwise the Theory of Invariants" (1853), or *Math. Papers JJS*, 1:411–422 and 2:11–27.

74. See Sylvester, "On a Theory of the Syzygetic Relations" (1853); "On the Conditions Necessary and Sufficient To Be Satisfied in Order That a Function of Any Number of Variables May Be Linearly Equivalent To a Function of Any Less Number of Variables," (1853); and "A Proof That All the Invariants to a Cubic Ternary Form Are Rational Functions of Aronhold's Invariants and of a Cognate Theorem for Biquadratic Binary Forms" (1853), or *Math. Papers JJS*, 1:429–586, 587–598, and 599–608, respectively.

75. Lipman, *History of the Jews in Britain since 1858*, 9. Rothschild was finally seated in 1858 after the House of Lords, which had consistently presented the roadblock to this

reform legislation, agreed to allow different oaths in the House of Commons and in the Lords. Compare also Endelman, *Jews of Britain, 1656 to 2000*, 105–108.

76. *Annual Register . . . of the Year 1853*, 33–37.

77. J. J. Sylvester to Lord Brougham, 23 April 1853, UCL Archives, Brougham Correspondence, Sylvester, J. J.:20,230, in Parshall, *JJS: Life and Work in Letters*, 62. The quotes that follow in this and the next paragraph are also from this letter.

78. In fact, Sylvester's name did not actually appear on the list, and then in the *second rang*, until 13 March 1854. He would not become a member until 1863. See chapter 7.

79. Arthur Cayley to C. R. Weld, 16 July 1853, Royal Society of London, Referee's Reports, RR.2.232. William Spottiswoode, the other referee, also reported positively.

80. In the first part of his paper "On Rational Derivation From Equations of Coexistence" of 1839, Sylvester had actually alluded to a work that he hoped would be forthcoming and which he would "present . . . , with all the steps here indicated filled up . . . , as homage to the learned and illustrious society which has lately done me the honour of admitting me to its ranks" (see *Math. Papers JJS*, 1:44). He apparently never forgot that intention. In a sense, "On a Theory of Syzygetic Relations" represents that completed line of thought.

81. Sylvester, "On a Theory of Syzygetic Relations" (1853), in *Math. Papers JJS*, 1:437.

82. Although it was unknown to Sylvester and Cayley at this time, the example given here is that of a minimum generating set of invariants of the binary quintic form. Compare Elliott, *Introduction to the Algebra of Quantics*, 302–303. This example makes abundantly clear that syzygies were not necessarily transparent!

83. See Salmon to Sylvester, 7 and 29 April 1852, in Parshall, *JJS: Life and Work in Letters*, 42–45 and 51–53, respectively.

84. Compare the discussion in Sylvester, "On a Theory of Syzygetic Relations" (1853), in *Math. Papers JJS*, 1:438–440.

85. Sylvester, "A Demonstration of the Theorem That Every Homogeneous Quadratic Polynomial Is Reducible By Real Orthogonal Substitutions to the Form of a Sum of Positive and Negative Squares" (1852), in *Math. Papers JJS*, 1:381.

86. Sylvester, "On a Theory of Syzygetic Relations" (1853), in *Math. Papers JJS*, 1:435 (his emphasis). For the proofs, see ibid., 511–545.

87. Ibid., 548 (his emphasis). For the precise definition and a technical discussion of the Bezoutiant, see the commentary in Parshall, *JJS: Life and Work in Letters*, 61–62. For a discussion in historical context of these aspects of the development of Sturm's Theorem, see Sinaceur, *Corps et modèles*, 129–140.

88. Sylvester, "On a Theory of Syzygetic Relations" (1853), in *Math. Papers JJS*, 1:549 and 436, respectively.

89. Cayley to Weld, 16 July 1853.

90. See the series of short papers published in the *Philosophical Magazine* in *Math. Papers JJS*, 1:620–644.

91. J. J. Sylvester to George Salmon, 6 September 1853, JJS Papers, SJC, Box 1, in Parshall, *JJS: Life and Work in Letters*, 64. Sylvester published this result in "On the Explicit Values of Sturm's Quotients" (1853), or *Math. Papers JJS*, 1:637–640.

92. See Sylvester, "On the Calculus of Forms, Otherwise the Theory of Invariants" (1853), as well as "On the Expressions for the Quotients Which Appear in the Application of Sturm's Method to the Discovery of the Real Roots of an Equation" (1853), or *Math. Papers JJS*, 1:396–398, and "Provisional Report on the Theory of Determinants" (1853), 66–67 (this does not appear in *Math. Papers JJS*), respectively. Compare *Report*

of the Twenty-second Meeting of the British Association for the Advancement of Science (1852), xxxv.

93. In his assessment of the history of invariant theory in the nineteenth century, Wilhelm-Franz Meyer remarked that "the year 1854, in which this first series of works by Sylvester came to an end, must be considered as one of the most important [years] of this first period" in the theory's history. See Meyer, *Sur les progrès de la théorie des invariants projectifs*, 8 (my translation). This was a French translation, abbreviation, and update of Meyer, "Bericht über den gegenwärtigen Stand der Invariantentheorie."

94. See *Math. Papers JJS*, 2:10–49.

95. Stewart, "Founding Vice President," 486.

96. J. J. Sylvester to the Royal Society, 9 March 1854, Royal Society of London, Referee's Reports, RR.2.61. The quotes that follow in this paragraph are from this report.

97. George Boole to J. J. Sylvester, 14 April 1854, JJS Papers, SJC, Box 2, in Parshall, *JJS: Life and Work in Letters*, 67.

98. Cayley, "An Introductory Memoir upon Quantics," or *Math. Papers AC*, 2:221–234.

99. Crilly, *Arthur Cayley*, 179–180, 201–202.

100. Let K be a covariant of a binary n-ic form. It is expressed in terms of the coefficients a_i, for $i = 1, 2, \ldots, n + 1$ and the variables x and y of the underlying binary n-ic form. The order of K is its (constant) degree of homogeneity in x and y, and the degree of K is its (constant) degree of homogeneity in the coefficients a_i. Assigning x weight 1, y weight 0, and a_i weight i, consider a monomial, say, $a_1^2 a_2^3 a_3 x^2 y$ of K. Define the weight of the monomial to be $1 \cdot 2 + 2 \cdot 3 + 3 \cdot 1 + 1 \cdot 2 + 0 \cdot 1$. Each monomial of a covariant K of a binary n-ic form has the same weight, namely, $1/2 \cdot (n\theta + \mu)$. Compare Cayley, "An Introductory Memoir upon Quantics," in *Math. Papers AC*, 2:233, and Parshall, "Toward a History of Nineteenth-Century Invariant Theory," 168. For more on Cayley's early work in invariant theory, see Crilly, "The Rise of Cayley's Invariant Theory (1841–1862)."

101. Arthur Cayley to J. J. Sylvester, 27 May 1854, JJS Papers, SJC, Box 2. The following quote is also from this letter.

102. Hew D. Ross to Sidney Herbert, 7 July 1854, UCL Archives, Brougham Correspondence, Sylvester, J. J.:36,522.

103. Sidney Herbert to Lord Brougham, 10 July 1854, UCL Archives, Brougham Correspondence, Sylvester, J. J.:36,522.

104. Sylvester's collected works at this time amounted to those papers that would ultimately comprise the first volume of *Math. Papers JJS*. See also Brioschi, *La teorica dei determinanti e le sue principali applicazioni*; Sylvester, "Sur une propriété nouvelle de l'équation qui sert à déterminer les inégalités séculaires des planètes" (1852); Sylvester, "Nouvelle méthode pour trouver une limite supérieure et une limite inférieure des racines réelles d'une équation algébrique quelconque" (1853) or *Math. Papers JJS*, 1:424–428; and *Comptes rendus de l'Académie des Sciences de Paris* 38 (1854):514.

105. J. J. Sylvester to Lord Brougham, 11 July 1854, UCL Archives, Brougham Correspondence, Sylvester, J. J.:20,231.

106. Charles Hermite to Peter Lejeune-Dirichlet, 25 July 1854, Dirichlet Nachlaß, Staatsbibliothek Berlin (my translation). I thank Catherine Goldstein for alerting me to this letter and for making it available to me.

107. *DNB*, s.v. "O'Brien, Matthew" by Charles Platts, and compare Rice, "Mathematics in the Metropolis," 392, 402, and Parshall, *JJS: Life and Work in Letters*, 71.

108. J. J. Sylvester to Lord Brougham, 9 August 1854, UCL Archives, Brougham Correspondence, Sylvester, J. J.:20,232, in Parshall, *JJS: Life and Work in Letters*, 71, and 72,

respectively. I inadvertently identified Charles Graves as his brother, John, in ibid., 72 (note 151).

109. As Sylvester pointed out in his letter to Brougham, George Gabriel Stokes, Lucasian Professor of Mathematics at Cambridge and, like himself, a mathematically superior candidate with no prior connections to the Academy, had been unsuccessful as well.

110. Sidney Herbert to Lord Brougham, 7 August 1854, UCL Archives, Brougham Correspondence, Sylvester, J. J.:20,232.

111. J. J. Sylvester to John Lubbock, 24 July 1854, Royal Society of London, Lubbock Letters, LUB.5548, in Parshall, *JJS: Life and Work in Letters*, 69.

112. See, respectively, Arthur Cayley to J. J. Sylvester, 12 October 1854, JJS Papers, SJC, Box 2, in Parshall, *JJS: Life and Work in Letters*, 75, and Arthur Cayley to J. J. Sylvester, undated, JJS Papers, SJC, Box 2, in Parshall, *JJS: Life and Work in Letters*, 77. He would ultimately present a refined version of these results to the Royal Society on 14 April 1855 in his "A Second Memoir upon Quantics."

113. J. J. Sylvester, "On Multiplication by Aid of a Table of Single Entry" (1854–1855). This paper was not included in *Math. Papers JJS*.

114. Walford, *Insurance Cyclopædia*, 3:27.

115. On Smith's life, see Hirsch, *Barbara Leigh Smith Bodichon*, and Herstein, *Mid-Victorian Feminist*.

116. Sylvester to Henry, 12 April 1846, in Parshall, *JJS: Life and Work in Letters*, 17.

Herbert Baker, in his obituary of Sylvester in the fourth volume of *Math. Papers JJS*, remarked that Sylvester had also given private instruction to Florence Nightingale during his years at the Equity and Law (p. xxiii). He based this comment on a statement in the obituary that had appeared just after Sylvester's death in the magazine of St. John's College, Cambridge. See J. W., "James Joseph Sylvester Sc.D.," *The Eagle* 19 (1897):597. I have been unable categorically to substantiate this claim. One of Nightingale's biographers noted that the future heroine of the hospital at Scutari did lobby her father in 1840 to allow her to study mathematics and "after much argument . . . was allowed to study [it] under a tutor." See Small, *Florence Nightingale: Avenging Angel*, 8. If that tutor was Sylvester and if this permission was (reluctantly) given in 1840 or 1841, then Sylvester was still on the faculty at University College London. Nightingale would certainly *not* have been the live-in student to whom Sylvester referred in his letter to Henry.

117. See Sylvester, "On a Simple Geometrical Problem Illustrating a Conjectured Principle in the Theory of Geometrical Method" (1852), or *Math. Papers JJS*, 1:392. Hirsch conjectures that Sylvester had only recently met Barbara Smith in November 1854, although she appears unaware of this earlier connection between Sylvester and Barbara's brother. See Hirsch, *Barbara Leigh Smith Bodichon*, 101.

118. J. J. Sylvester to Barbara Smith, 21 November 1854, 7/BMC, McCrimmon, Bodichon Collection, The Women's Library, London Metropolitan University, in Parshall, *JJS: Life and Work in Letters*, 75. The quote that follows is also from this letter (p. 76). The correspondence between Sylvester and Barbara Smith (later Bodichon) was initially brought to my attention by one of Smith's descendants, Barbara McCrimmon, who subsequently donated the correspondence to what has been newly named the Women's Library.

119. Hirsch, *Barbara Leigh Smith Bodichon*, 101–107, and Herstein, *Mid-Victorian Feminist*, 106–107.

120. Chartres and Vermont, *Brief History of Gresham College 1597–1997*, 6–11.

121. Sylvester, "A Probationary Lecture on Geometry" (1854), in *Math. Papers JJS*, 2:2.

122. MacMahon, "James Joseph Sylvester," *Proceedings of the Royal Society of London*, xvi.

123. J. J. Sylvester to Arthur Cayley, 2 December 1854, JJS Papers, SJC, Box 9, in Parshall, *JJS: Life and Work in Letters*, 76–77 (his emphasis). The quote that follows is also from this letter (77).

Chapter 6

Epigraph: UCL Archives, Brougham Correspondence, Sylvester, J. J.:20,241.

1. *Annual Register . . . of the Year 1855*, 3. The quotes that follow in this paragraph are also from this piece.

2. Norman M. Ferrers to William Thomson, 20 February 1854, Kelvin Papers, Cambridge University Library, Add 7342, as quoted in Despeaux, "Development of a Publication Community," 177.

3. "Address to the Reader" (1855), ii (my emphasis). The other quotes in this paragraph are from this "Address."

4. See Charles Hermite to J. J. Sylvester, 2 January 1855, JJS Papers, SJC, Box 2, and Hermite, "Sur les formes cubiques à deux indéterminées."

5. Sylvester, "On the Change of Systems of Independent Variables" (1855), or *Math. Papers JJS*, 2:65 (note ∗). Note that *Math. Papers JJS* incorrectly lists the date of publication of this paper as 1857. In fact, all six of Sylvester's *Quarterly Journal* papers listed in *Math. Papers JJS* as published in 1857 appeared in 1855. See *Math. Papers JJS*, 2:vii.

6. Sylvester, "On a Discovery in the Partition of Number" (1855), or *Math. Papers JJS*, 2:86–89. For the question, see the *Quarterly Journal of Pure and Applied Mathematics* 1 (1855):79.

7. See the *Quarterly Journal of Pure and Applied Mathematics* 1 (1855):4–6, 7–10, 79, and 85–90, respectively.

8. Spottiswoode, "Note on Axes of Equilibrium" and "On a Theorem in Statics," respectively.

9. De Morgan, "On the Dimensions of the Roots of Equations."

10. Augustus De Morgan to J. J. Sylvester, 14 February 1855, JJS Papers, SJC, Box 2, in Parshall, *JJS: Life and Work in Letters*, 84–85.

11. See *Quarterly Journal of Pure and Applied Mathematics* 1 (1855):57–76.

12. J. J. Sylvester to William Thomson, 13 March 1855, Kelvin Papers, Cambridge University Library, Add 7342, S600 (his emphasis). The quotes that follow in this paragraph are also from this letter.

Despite Sylvester's concern here for regular publication dates, the *Quarterly Journal* essentially never succeeded in actually appearing quarterly during its first two decades of publication, although it did appear with enough regularity to be a viable publication outlet.

13. For the quote, see Augustus De Morgan to J. J. Sylvester, 11 March 1855, JJS Papers, SJC, Box 2. On his new contributions, see Augustus De Morgan to J. J. Sylvester, 16 March and 11 April 1855, JJS Papers, SJC, Box 2.

14. See the *Quarterly Journal of Pure and Applied Mathematics* 1 (1855):91–92.

15. Compare J. J. Sylvester to Enrico Betti, 19 May 1855, Archivio Betti, Scuola Normale Superiore Pisa, lett. 2, V°, 330, 1305–1306. I thank Laura Martini for bringing this and other letters in the Betti Archive to my attention.

16. J. J. Sylvester to the Committee of Papers of the Royal Society, 26 April 1855, Royal Society, Referee Reports, RR.2.226. The quotes that follow in this and the next paragraph are also from this letter.

17. Stewart, "Founding Vice President," 486.

18. On these changes, see Barnard, *Military Schools and Courses of Instruction*, 526.

19. "A Parent," "To the Editor," *Times* (London), 23 June 1855. The quote that follows is also from this letter.

20. J. J. Sylvester to Lord Brougham, 10 July 1855, UCL Archives, Brougham Correspondence, Sylvester, J. J.:43,295.

21. J. J. Sylvester to Lord Brougham, 10 July 1855, UCL Archives, Brougham Correspondence, Sylvester, J. J.:20,236, in Parshall, *JJS: Life and Work in Letters*, 86–87. The quote that follows is also from this letter (87).

22. J. J. Sylvester to Arthur Cayley, 13 August 1855, JJS Papers, SJC, Box 9. The quotes that follow in this parargraph are also from this letter.

23. J. J. Sylvester to Arthur Cayley, 8 September 1855, JJS Papers, SJC, Box 9, in Parshall, *JJS: Life and Work in Letters*, 88 (his emphasis).

24. J. J. Sylvester to Lord Brougham, 16 September 1855, UCL Archives, Brougham Correspondence, Sylvester, J. J.:20,240, in Parshall, *JJS: Life and Work in Letters*, 89. The quotes that follow in this paragraph are also from this letter (90).

25. Compare J. J. Sylvester to Lord Brougham, 21 September 1855, UCL Archives, Brougham Correspondence, Sylvester, J. J.:20,241.

26. Royal Warrant of 30 April 1741, State Papers 44/184 and 41/36, as quoted in Harries-Jenkins, *Army in Victorian Society*, 113.

27. Barnard, *Military Schools and Courses of Instruction*, 526 and 585. The quotes are on the latter page.

28. "Military Education Part II," as abridged in Barnard, *Military Schools and Courses of Instruction*, 526. The quotes that follow come from pp. 526 and 527, respectively.

29. Moseley, *Report on the Examination for Admission to the Royal Military Academy*, 21–23. I thank Paul Garcia for informing me of and providing me with a copy of this report.

For a somewhat quirky overview of mathematics at the Academy, see Johnson, "The Woolwich Professors of Mathematics." Adrian Rice also treated the topic in "Mathematics in the Metropolis," 398–406.

30. See *Annual Register . . . of the Year 1856*, 187.

31. Barnard, *Military Schools and Courses of Instruction*, 530–531, and Harries-Jenkins, *Army in Victorian Society*, 119–121.

32. *Report of the Commissioners Appointed to Consider the Best Mode of Reorganizing the System for Training Officers*, 414, as quoted in Harries-Jenkins, *Army in Victorian Society*, 115 and 117, respectively.

33. In addition to the Royal Military Academy, Woolwich, for the training of members of the artillery and the engineers, there were, among others, the Royal Military College founded at Sandhurst in 1804 for training infantry- and calvarymen and the Honorable East India Company's College established at Addiscombe in 1818 for preparing the Company's artillerymen and engineers. See Barnard, *Military Schools and Courses of Instruction*, 522 and 525–529 for an overview.

34. Fitzgerald, *London City Suburbs As They Are To-day*, 208–216.

35. MacMahon, "James Joseph Sylvester," *Proceedings of the Royal Society of London*, xvi. MacMahon gives Sylvester's starting salary as £550, but an appendix in Guggisberg,

"The Shop," 267, based on data from the 1857 Commission report lists the salary as £500 in 1856.

36. Compare J. J. Sylvester to Arthur Cayley, undated (but from internal mathematical evidence probably from August 1856), JJS Papers, SJC, Box 9.

37. Sylvester, "A Trifle on Projectiles" (1856), and "Letter on Professor Galbraith's Construction for the Range of Projectiles" (1856), or *Math. Papers JJS,* 2:55–58 and 61–62, respectively. For a brief discussion of the first of these papers, see Grattan-Guinness, "The Contributions of J. J. Sylvester, F.R.S., to Mechanics and Mathematical Physics," 258–259.

That the military authorities were also unhappy with Sylvester's job as Professor of Natural Philosophy came through clearly in testimony that the Inspector of Studies at Woolwich, Col. Joseph Portlock, gave before the Yolland Commission. If there were "any deficiency in the staff" at Woolwich, he stated, it lay in Sylvester as Professor of Natural Philosophy with his "hastily prepared and inefficient lectures." As quoted in Hearl, "Military Examinations and the Training of Science, 1857–1870," in *Days of Judgement,* 123.

38. "Dinner at the Mansion-House," *Times* (London), 12 June 1856.

39. Sylvester to Cayley, undated (but probably August 1856), and J. J. Sylvester to Arthur Cayley, 25 August 1856, JJS Papers, SJC, Box 9, respectively.

40. J. J. Sylvester to Arthur Cayley, 16 September 1856, JJS Papers, SJC, Box 9.

41. J. J. Sylvester to Arthur Cayley, 26 September 1856, JJS Papers, SJC, Box 9, in Parshall, *JJS: Life and Work in Letters,* 90–92. The quotes that follow are also from this letter on pp. 91 and 92, respectively. Recall from chapter 3 that Wheatstone and Sylvester had known each other at least since 1838 when both were professors of physics in London.

42. Arthur Cayley to J. J. Sylvester, 2 October 1856, JJS Papers, SJC, Box 9.

43. J. J. Sylvester to Arthur Cayley, 10 October 1856, JJS Papers, SJC, Box 9.

44. J. J. Sylvester to Arthur Cayley, 23 October 1856, JJS Papers, SJC, Box 9, in Parshall, *JJS: Life and Work in Letters,* 92.

45. John Graves to J. J. Sylvester, 24 October 1856, JJS Papers, SJC, Box 6.

46. Arthur Cayley to J. J. Sylvester, 31 October 1856, JJS Papers, SJC, Box 2.

47. Sylvester, "Recherches sur les solutions en nombres entiers positifs ou négatifs de l'équation cubique homogène à trois variables" (1856), or *Math. Papers JJS,* 2:63–64.

48. Hogg, *The Royal Arsenal,* 2:763, 1108.

49. Eaton Hodgkinson to ?, 29 October 1856, UCL Archives, College Correspondence, Hodgkinson, Eaton, 1856: Oct. 29.

50. See Johnson, "Woolwich Professors of Mathematics," 162, and Venn, *Alumni Cantabrigiensis.* The information that follows on these masters also comes from these sources.

51. J. J. Sylvester to Lord Brougham, 10 December 1856, UCL Archives, Brougham Correspondence, Sylvester, J. J.:20,247.

52. Giuseppe Battaglini to Enrico Betti, 13 March 1857, in *Giuseppe Battaglini: Raccolta di lettere (1854–1891) di un matematico al tempo del Risorgimento d'Italia,* ed. Mario Castellana and Franco Palladino, 185 (my translation). I thank Laura Martini for pointing out this letter to me. The quote that follows is also from it (186).

53. On Tortolini and his journal, see Martini, "The Politics of Unification: Barnaba Tortolini and the Publication of Research Mathematics in Italy, 1850–1865," 171–198.

54. J. J. Sylvester to Lord Brougham, 22 January 1857, UCL Archives, Brougham Correspondence, Sylvester, J. J.:24,575. The quotes that follow in this paragraph are from this letter (his emphasis).

55. Timbs, *Clubs and Club Life in London*, 207.

56. Cowell, *The Athenæum: Club and Social Life in London*, 165.

57. J. J. Sylvester to Charles Graves, 26 January 1857, Trinity College Dublin, MS 10047/48/99.

58. J. J. Sylvester to Arthur Cayley, 25 September, JJS Papers, SJC, Box 9, in Parshall, *JJS: Life and Work in Letters*, 94.

59. J. J. Sylvester to Arthur Cayley, 4 January 1858, JJS Papers, SJC, Box 9.

60. On Kirkman's life, see Macfarlane, *Lectures on Ten British Mathematicians*, and Biggs, "T. P. Kirkman, Mathematician."

61. Sylvester to Cayley, 25 August 1856.

62. Thomas Kirkman to J. J. Sylvester, 2 January 1858, JJS Papers, SJC, Box 2. The quote that follows in this paragraph is also from this letter.

63. Barnard, *Military Schools and Courses of Instruction*, 611–612.

64. Thomas Kirkman to J. J. Sylvester, 25 January 1858, JJS Papers, SJC, Box 2.

65. J. J. Sylvester to Arthur Cayley, 25 February 1858, JJS Papers, SJC, Box 9.

66. For the quote, see Timbs, *Clubs and Club Life in London*, 57. Cayley was not a member of the Royal Society Club, yet presumably he could dine there as a guest since Sylvester told him in his letter of 25 February, "if we meet there [I] will be glad to accompany you home at an early hour and stay with you as long as I can."

67. Sylvester regularly participated in the July entrance examinations as part of the duties of his professorship. For the paper, see Sylvester, "Note on the Algebraical Theory of Derivative Points of Curves of the Third Order" (1858), or *Math. Papers JJS*, 2:107–109.

68. Sylvester, "Note on the Equation in Numbers of the First Degree Between Any Number of Variables with Positive Coefficients" (1858), or *Math. Papers JJS*, 2:110. The following quotes are from pp. 110 and 112, respectively.

69. Ibid., in *Math. Papers JJS*, 2:112 (note ∗).

70. See Biggs, "T. P. Kirkman, Mathematician," 118–120 for a list of Kirkman's major mathematical publications.

71. Arthur Cayley to J. J. Sylvester, undated, JJS Papers, SJC, Box 2, in Parshall, *JJS: Life and Work in Letters*, 77–79. For the definitions of degree and order, recall the discussion in the previous chapter. For more on Cayley's ideas and on the problems associated with determining the number of linearly independent covariants, see chapter 9.

72. Compare the commentary in Parshall, *JJS: Life and Work in Letters*, 166–167.

73. Sylvester, "On the Problem of the Virgins, and the General Theory of Compound Partition" (1858), or *Math. Papers JJS*, 2:114. Sylvester would revisit this work almost twenty years later in effecting a key simplification of the British approach to invariant theory. See chapter 9.

74. Harries-Jenkins, *Army in Victorian Society*, 114–117.

75. J. J. Sylvester to Arthur Cayley, 22 December 1858, JJS Papers, SJC, Box 9. The quotes that follow in this paragraph are also from this letter (with his emphasis).

76. Drayson, *Experiences of a Woolwich Professor During Fifteen Years at the Royal Military Academy*, 26. The quotes that follow in this paragraph are also from this source (p. 162).

77. J. J. Sylvester to Arthur Cayley, 26 April 1859, JJS Papers, SJC, Box 9.

78. "Professor Sylvester's Mathematical Lectures," *Times* (London), 11 June 1859. The quote that follows is also from this article.

79. On Noble's collaboration, see MacMahon, "James Joseph Sylvester," *Nature*, 492.

80. The outlines were only published in 1897, the year of Sylvester's death. See Sylvester, "Outlines of Seven Lectures on the Partitions of Numbers" (1859), or *Math. Papers JJS*, 2:119–175.

81. Sylvester, "Note sur certaines séries qui se présentent dans la théorie des nombres" (1860), or *Math. Papers JJS*, 2:178. Sylvester also published one paper on $E(x)$ in his own *Quarterly Journal*. See Sylvester, "On the Equation $P(m) + E\left(\frac{m}{m-1}\right) P(m-1) + E\left(\frac{m}{m-2}\right) P(m-2) + \cdots + E(m) = m\left(\frac{m+1}{2}\right)$" (1860), or *Math. Papers JJS*, 2:225–228.

82. The paper in question is Cauchy, "Mémoire sur les arrangements que l'on peut former avec les lettres données, et sur les permutations ou substitutions à l'aide desquelles on passe d'un arrangement à un autre." Compare also Arthur Cayley to J. J. Sylvester, 18 August 1860, JJS Papers, SJC, Box 2, in which Cayley shared detailed notes and comments on this paper with Sylvester. Cayley had been interested in groups at least since 1854 when he not only gave an abstract definition of a group but also proved that every finite group can be realized as a permutation group, the result now called Cayley's Theorem. See Cayley, "On the Theory of Groups," or *Math. Papers AC*, 2:123–130 and 131–132.

83. Arthur Cayley to J. J. Sylvester, 11 August 1860, JJS Papers, SJC, Box 2.

84. Compare Arthur Cayley to J. J. Sylvester, 16 August 1860, JJS Papers, SJC, Box 2, in Parshall, *JJS: Life and Work in Letters*, 98. The following quote is also from this letter (p. 99). Cayley did not use the modern terminology "semidirect product."

85. Cayley to Sylvester, 18 August 1860.

86. Poncelet, "Sur la valeur approchée des radicaux."

87. See Sylvester, "On a Generalization of Poncelet's Theorems for the Linear Representation of Quadratic Radicals" (1860), or *Math. Papers JJS*, 2:118, and Sylvester, "On Poncelet's Approximate Linear Valuation of Surd Forms" (1860), or *Math. Papers JJS*, 2:181–199.

88. Sylvester, "Meditation on the Idea of Poncelet's Theorem" (1860), or *Math. Papers JJS*, 2:200. The quote that follows is also from this source (p. 207).

89. J. J. Sylvester to Thomas Archer Hirst, 6 October 1860, UCL Archives, London Mathematical Society Papers, Sylvester, J. J, in Parshall, *JJS: Life and Work in Letters*, 101. The next quote is also from this letter.

90. On this stage of Hirst's life, see Gardner and Wilson, "Thomas Archer Hirst." Hirst recorded in his diary having met Sylvester at the Athenæum at least as early as 16 October 1859. Ibid., 828.

91. J. J. Sylvester to Charles Babbage, 7 July 1860, British Library, 87–88, 32308.

92. Sylvester to Hirst, 6 October 1860, in Parshall, *JJS: Life and Work in Letters*, 101.

93. Sylvester, "Notes to the Meditation on Poncelet's Theorem Including a Valuation of the Two New Definite Integrals . . . " (1860), or *Math. Papers JJS*, 2:210 (note *) (his emphasis).

94. On the concept of a "mathematical publication community," see Despeaux, "Development of a Publication Community."

95. Cayley had received a Royal Medal the year before in 1859.

96. Sylvester, "On the Pressure of Earth on Revetment Walls" (1860), or *Math. Papers JJS*, 2:215 (his emphasis). In his study of Sylvester's work in applied mathematics, Grattan-Guinness mistakenly conjectured that this topic was not part of the Woolwich curriculum. See Grattan-Guinness, "The Contributions of J. J. Sylvester, F.R.S., to Mechanics and Mathematical Physics," 259, and compare Moseley, *Report on the Examination for Admission to the Royal Military Academy*, 23.

97. Grattan-Guinness, "The Contributions of J. J. Sylvester, F.R.S., to Mechanics and Mathematical Physics," 259.

98. See *Math. Papers JJS*, 2:229–235. Sylvester also wrote up these results for his English-speaking audience in the February number of the *Philosophical Magazine*. See Sylvester, "Note on the Numbers of Bernoulli and Euler, and a New Theorem Concerning Prime Numbers" (1861), or *Math. Papers JJS*, 2:254–263.

99. J. J. Sylvester to Arthur Cayley, 4 January 1861, JJS Papers, SJC, Box 10.

100. See Möbius, *Lehrbuch der Statik*, and Sylvester, "Sur l'involution des lignes droites dans l'espace considérées comme des axes de rotation" (1861), and "Note sur l'involution de six lignes dans l'espace" (1861), or *Math. Papers JJS*, 2:236–239, 240–241, respectively.

101. Compare Cayley's characterization of Sylvester's findings in Cayley, "On the Six Coordinates of a Line," or *Math. Papers AC*, 7:66. As Cayley showed in this paper, Sylvester's work could be interepreted in terms of the notion of the six coordinates of a line that he had developed as early as 1860 and that Julius Plücker had independently developed several years later. These so-called "Plücker coordinates" would come to play a major role in geometrical developments, in particular, in Germany at the hands of Felix Klein. See the discussion in Parshall and Rowe, *Emergence of the American Mathematical Research Community*, 156–168, and in Rowe, "Klein, Lie, and the 'Erlanger Programm,'" 45–54.

102. J. J. Sylvester to Arthur Cayley, 30 April 1861, JJS Papers, SJC, Box 10 (his emphasis).

103. Arthur Cayley to J. J. Sylvester, 2 May 1861, JJS Papers SJC, Box 2.

104. J. J. Sylvester to Arthur Cayley, 9 May 1861, JJS Papers, SJC, Box 10.

105. J. J. Sylvester to Arthur Cayley, 16 May 1861, JJS Papers, SJC, Box 10.

106. J. J. Sylvester to Arthur Cayley, 26 May 1861, JJS Papers, SJC, Box 10.

Chapter 7

Epigraph: UCL Archives, Brougham Correspondence, Sylvester, J. J.:20,248.

1. Wilford's title, as resident head of the Academy, was actually Lieutenant-Governor; officially, the Governor was the Duke of Cambridge, the head of the entire British Army. Like Sylvester, I sometimes refer to Wilford and his successors as Governor here and below.

2. Unfortunately, it is not known what books Sylvester suggested. It is clear that curricular changes of this sort were slow to occur at the Academy. For example, Charles Hutton's *A Course of Mathematics* had been used there in successive editions for almost fifty years until it was replaced by Samuel Christie's two-volume *An Elementary Course of Mathematics for the Use of the Royal Military Academy*, in the mid-1840s. This text was fairly quickly judged as *too* elementary and had been superceded by 1852. Thus, the text in use in 1861 had been in place for less than ten years. Compare Rice, "Mathematics in the Metropolis," 398–401.

3. J. J. Sylvester to Major General W. F. Forster, 17 June 1861, UCL Archives, Brougham Correspondence, Sylvester, J. J.:20,254. The quotes that follow in this paragraph are also from this letter. A typographical error caused the W. F. Forster here to be inadvertently confused with W. E. Forster in the index of Parshall, *JJS: Life and Work in Letters*.

4. J. J. Sylvester to Lord Brougham, 18 June 1861, UCL Archives, Brougham Correspondence, Sylvester, J. J.:20,248, in Parshall, *JJS: Life and Work in Letters*, 102. The quotes that follow in this paragraph are also from this letter.

5. J. J. Sylvester to Lord Brougham, 20 June 1861, UCL Archives, Brougham Correspondence, Sylvester, J. J.:20,248, in Parshall, *JJS: Life and Work in Letters*, 104 (his emphasis). The quote that follows is also from this letter and page.

6. J. J. Sylvester to Lord Brougham, 25 June 1861, UCL Archives, Brougham Correspondence, Sylvester, J. J.:20,250 (his emphasis).

7. Barnard, *Military Schools and Courses of Instruction*, 537–539.

8. J. J. Sylvester to Lord Brougham, 16 July 1861, UCL Archives, Brougham Correspondence, Sylvester, J. J.:20,251. The next quote is also from this letter.

9. J. J. Sylvester to Arthur Cayley, 26 July 1861, JJS Papers, SJC, Box 10 (his emphasis).

10. J. J. Sylvester to Lord Brougham, 1 August 1861, UCL Archives, Brougham Correspondence, Sylvester, J. J.:20,252. The quote that follows is also from this letter. Herbert, in fact, died on 2 August 1861.

11. J. J. Sylvester to Arthur Cayley, 15 August 1861, JJS Papers, SJC, Box 10.

12. Recall the discussion of his paper, "Elementary Researches in the Analysis of Combinatorial Aggregation" in chapter 4. Compare Sylvester, "Note on the Historical Origin of the Unsymmetrical Six-Valued Function of Six Letters" (1861), or *Math. Papers JJS*, 2:264–271.

13. See Biggs, "T. P. Kirkman, Mathematician," 98–99.

14. Sylvester, "Note on the Historical Origin of the Unsymmetrical Six-Valued Function of Six Letters" (1861), in *Math. Papers JJS*, 2:266. The quote that follows is also from this page.

15. See *Math. Papers JJS*, 2:272–289.

16. J. J. Sylvester to Arthur Cayley, 17 August 1861, JJS Papers, SJC, Box 10, in Parshall, *JJS: Life and Work in Letters*, 106. The quotes that follow in this paragraph are also from this letter on pp. 106 and 106–107, respectively.

In modern terms, the group was the semidirect product of the symmetric group on three letters, S_3, and $V = \mathbb{Z}/3\mathbb{Z} \times \mathbb{Z}/3\mathbb{Z}$. For more technical details, see the commentary in Parshall, *JJS: Life and Work in Letters*, 105–106.

17. Norman M. Ferrers to J. J. Sylvester, 21 August 1861, JJS Papers, SJC, Box 2.

18. Compare the discussion in chapter 9 of Sylvester's editorship of the *American Journal of Mathematics*. From the beginning, he refused to concern himself with the production end of the enterprise, insisting on dealing only with the intellectual content.

19. Compare Sylvester to Cayley, 26 July 1861, and J. J. Sylvester to Arthur Cayley, 30 August 1861, JJS Papers, SJC, Box 10. The quote that follows is from the latter letter.

20. Sylvester, "On the Involution of Axes of Rotation" (1861), or *Math. Papers JJS*, 2:304.

21. Recall the discussion of this work in chapter 6, and see Sylvester, "Généralisation d'un théorème de M. Cauchy" (1861), or *Math. Papers JJS*, 2:244–245. Sylvester essentially reprinted the results in this paper for his British audience in the November issue of the *Philosophical Magazine*. See Sylvester, "On a Generalization of a Theorem of Cauchy on Arrangements" (1861), or *Math. Papers JJS*, 2:290–293.

22. Sylvester, "Addition à la note intitulée: 'Généralisation d'un théorème de M. Cauchy,' et insérée dans le 'Compte rendu' de la séance du 7 octobre dernier" (1861), or *Math. Papers JJS*, 2:247 (my translation).

23. Reminiscences of an "old cadet" as recorded in Guggisberg, "*The Shop*," 98. The quotes that follow in this paragraph are also from this account (p. 99).

24. The "old cadet" who reported on the event to Guggisberg used this term. See ibid. 98–102.

25. See, among many possible examples, "A. Cadet," "To the Editor," *Times* (London), 1 November 1861; "A Father," "To the Editor," *Times* (London), 11 November 1861; and "A Relative of a Woolwich Cadet," "To the Editor," *Times* (London), 22 November 1861.

26. "The Royal Military Academy, Woolwich," *Times* (London), 23 November 1861.

27. See *Math. Papers JJS*, 2:250–253, 305–306, 308–312, and 313–317, respectively.

28. J. J. Sylvester to Arthur Cayley, undated (but from late January or early February 1862), JJS Papers, SJC, Box 10, in Parshall, *JJS: Life and Work in Letters*, 109. The quote that follows in this paragraph is also from this letter (pp. 109–110).

29. J. J. Sylvester to Arthur Cayley, 26 February 1862, JJS Papers, SJC, Box 10.

30. Sylvester to Cayley, undated (but from late January or early February 1862), in Parshall, *JJS: Life and Work in Letters*, 110.

31. In 1862, the political unification of the peninsula was actually not yet complete. The Papal State, for example, would not enter the union until 1871.

32. Sylvester to Cayley, 26 February 1862.

33. J. J. Sylvester to Arthur Cayley, 19 March 1862, JJS Papers, SJC, Box 10. See Novi, *Trattato di algebra superiore*.

34. J. J. Sylvester to Sir George Lewis, 24 June 1862, JJS Papers, SJC, Box 3. The following quote (with his emphasis) is also from this letter.

35. J. J. Sylvester to Arthur Cayley, 19 July 1862, JJS Papers, SJC, Box 10.

36. See Sylvester, "On the Solution of the Linear Equation of Finite Differences in Its Most General Form" (1862), or *Math. Papers JJS*, 2:307, and "On the Integral of the General Equation in Differences" (1862), or *Math. Papers JJS*, 2:318–322.

37. J. J. Sylvester to Thomas Hirst, 23 December 1862, UCL Archives, London Mathematical Society Papers (hereinafter LMS Papers), Sylvester, J. J., in Parshall, *JJS: Life and Work in Letters*, 114 (his emphasis). The quotes that follow in this paragraph are also from this letter and page.

38. J. J. Sylvester to Arthur Cayley, 3 January 1863, JJS Papers, SJC, Box 10.

39. Duming and Guest, *Natal and Zululand from Earliest Times to 1910*, 131–132.

40. J. J. Sylvester to Arthur Cayley, 8 February 1863, JJS Papers, SJC, Box 10, in Parshall, *JJS: Life and Work in Letters*, 117 (his emphasis). The quotes that follow in this paragraph are also from this letter and page.

41. J. J. Sylvester to Thomas Hirst, 8 February 1863, UCL Archives, LMS Papers, Sylvester, J. J. The quotes that follow in this paragraph are also from this letter.

42. J. J. Sylvester to Major General Hamilton, 11 February 1863, JJS Papers, SJC, Box 1, in Parshall, *JJS: Life and Work in Letters*, 117. The quotes that follow in this paragraph are also from this letter (p. 118).

43. Indicative of this heightened activity, Sylvester published three brief notes in the Society's *Proceedings* in 1863, his first publications in that venue. See *Math. Papers JJS*, 2:325–326, 327–328, and 329–330, respectively.

44. See J. J. Sylvester to Thomas Hirst, 19 December 1862, UCL Archives, LMS Papers, Sylvester, J. J., in Parshall, *JJS: Life and Work in Letters*, 113.

45. J. J. Sylvester to Thomas Hirst, 27 February 1863, UCL Archives, LMS Papers, Sylvester, J. J., in Parshall, *JJS: Life and Work in Letters*, 118–119. The following quote is also from this letter (p. 119).

46. J. J. Sylvester to Thomas Hirst, 11 March 1863, UCL Archives, LMS Papers, Sylvester, J. J.

47. J. J. Sylvester to Thomas Hirst, 14 March 1863, UCL Archives, LMS Papers, Sylvester, J. J.

48. J. J. Sylvester to Thomas Hirst, 21 March 1863, UCL Archives, LMS Papers, Sylvester, J. J., in Parshall, *JJS: Life and Work in Letters*, 119 (his emphasis). The quotes that follow in this and the next paragraph are also from this letter (with his emphasis).

49. Interestingly, Monsell and Cave, like Sylvester, were members of the Athenæum.

50. J. J. Sylvester to Thomas Hirst, 29 March 1863, UCL Archives, LMS Papers, Sylvester, J. J. The quotes that follow in this paragraph are from this letter (with his emphasis).

51. J. J. Sylvester to Thomas Hirst, 7 April 1863, UCL Archives, LMS Papers, Sylvester, J. J., in Parshall, *JJS: Life and Work in Letters*, 120. The quote that follows is also from this letter (with his emphasis).

52. J. J. Sylvester to Lord Brougham, 8 May 1863, UCL Archives, Brougham Correspondence, Sylvester, J. J.:20,253, in Parshall, *JJS: Life and Work in Letters*, 121. The quotes that follow in this paragraph are also from this letter. Although it is clear that Sylvester's teaching load did not increase, it is not clear if it was reduced to four attendances, and, if so, what changes may have been made to his salary. Unfortunately, the archival records from this period in the history of the Royal Military Academy no longer exist.

53. J. J. Sylvester to Arthur Cayley, 20 November 1862, JJS Papers, SJC, Box 10.

54. Sylvester, "Sequel to the Theorems Relating to 'Canonic Roots' Given in the March Number of This Magazine" (1863), or *Math. Papers JJS*, 2:331 (his emphasis).

55. Compare the discussion of the canonizant in Parshall, *JJS: Life and Work in Letters*, 111–113.

56. See Sylvester, "An Essay on Canonical Forms" (1851), in *Math. Papers JJS*, 1:208–214.

57. Sylvester to Cayley, 8 February 1863, in Parshall, *JJS: Life and Work in Letters*, 116.

58. Cayley, "Theorems Relating to the Canonic Roots of a Binary Quantic of an Odd Order," or *Math. Papers AC*, 5:103–105.

59. Sylvester, "Sequel to the Theorems Relating to 'Canonic Roots'" (1863), in *Math. Papers JJS*, 2:331.

60. Ibid., 332. The quote that follows in the next paragraph is also from this paper (p. 337).

61. See Crilly, *Arthur Cayley*, 254–263.

62. It is noteworthy, given the volume of correspondence that does survive, that only one known letter from Sylvester to Cayley remains from the period from late April 1863 to the end of January 1866. There are only two known, surviving letters from Cayley to Sylvester during this same period. This suggests—although it does not prove—that Sylvester's contact with Cayley in the years immediately after the latter's move to Cambridge was indeed limited.

63. See Sylvester, "On the Centre of Gravity of a Truncated Triangular Pyramid" (1863), or *Math. Papers JJS*, 2:342–357, and "On the Quantity and Centre of Gravity of Figures Given in Perspective, or Homography" (1863), or *Math. Papers JJS*, 2:323.

64. J. J. Sylvester to Thomas Hirst, 5 October 1863, UCL Archives, LMS Papers, Sylvester, J. J.

65. J. J. Sylvester to Thomas Hirst, 6 October 1863, UCL Archives, LMS Papers, Sylvester, J. J. The quotes that follow immediately are also from this letter with his emphasis.

66. J. J. Sylvester to Thomas Hirst, 27 November 1863, UCL Archives, LMS Papers, Sylvester, J. J. The quote that follows is also from this letter.

67. J. J. Sylvester to the Secretary of the Paris Academy of Sciences, 11 December 1863, Archives de l'Académie des Sciences de Paris, Dossier biographique, Sylvester, J. J., in Parshall, *JJS: Life and Work in Letters*, 122 (my translation). Note that a transcriptional

error crept into the published version. Sylvester wrote "qui sont de son ressort [which are within its purview]" and not the nonsensical "qui sont de son report" as printed.

68. Sylvester, "Théorème sur la limite du nombre des racines réelles d'une classe d'équations algébriques" (1864), or *Math. Papers JJS*, 2:360.

69. J. J. Sylvester to Joseph Bertrand, 12 April 1864, Archives de l'Académie des Sciences de Paris, Pochette de séance, 18 April, 1864, in Parshall, *JJS: Life and Work in Letters*, 122–123 (my translation). The paper Sylvester asked Bertrand to read appeared as Sylvester, "Sur une extension de la théorie des équations algébriques" (1864), or *Math. Papers JJS*, 2:361–362. It contained an announcement, without proof, of a result related to the proof of Newton's Rule. Sylvester presented an actual proof of this result in the massive paper he published in the *Philosophical Transactions*. See immediately below.

70. Sylvester to Bertrand, 2 April 1864, in Parshall, *JJS: Life and Work in Letters*, 123 (my translation).

71. Sylvester, "Algebraical Researches, Containing a Disquisition on Newton's Rule" (1864), or *Math. Papers JJS*, 2:380.

72. J. J. Sylvester to Thomas Hirst, 18 September 1864, UCL Archives, LMS Papers, Sylvester, J. J., in Parshall, *JJS: Life and Work in Letters*, 124 (his emphasis). The "Syzygetic paper" to which Sylvester referred here was "On a Theory of the Syzygetic Relations" (1853). Recall the discussion of this work in chapter 5.

73. Arthur Cayley to the Committee on Papers of the Royal Society, 28 April 1864, Royal Society of London, RR.5.267, Referee's Reports. The following quote is also from this report.

74. Henry J. S. Smith to the Committee on Papers of the Royal Society, 25 May 1864, Royal Society of London, RR.5.268, Referee's Reports.

75. Arthur Cayley to the Committee on Papers of the Royal Society, 2 October 1864, Royal Society of London, RR.5.269, Referee's Reports.

76. Sylvester, "Algebraical Researches, Containing a Disquisition on Newton's Rule" (1864), in *Math. Papers JJS*, 2:380. The following quote is also from this page (note 1).

77. See Cayley, "A Second Memoir upon Quantics," in *Math. Papers AC*, 2:264 and 273.

78. Sylvester, "Algebraical Researches, Containing a Disquisition on Newton's Rule" (1864), in *Math. Papers JJS*, 2:420–423.

79. For the cadet's first-person account of his years at the Academy, see Guggisberg, "'The Shop," 106–107. The quotes that follow in this paragraph are also from this source (p. 107).

80. Ibid., 107. The quotes that follow immediately are also from this source (pp. 107 and 108, respectively).

81. Compare the account in "The Royal Military Academy at Woolwich," *Times* (London), 4 November 1864.

82. "Military and Naval Intelligence," *Times* (London), 24 December 1864.

83. See Sylvester, "Algebraical Researches, Containing a Disquisition on Newton's Rule" (1864), in *Math. Papers JJS*, 2:464.

84. J. J. Sylvester to Thomas Hirst, 28 March 1865, UCL Archives, LMS Papers, Sylvester, J. J.

85. See also Sylvester, "Note sur les conditions nécessaires et suffisantes pour distinguer le cas quand toutes les racines d'une équation du cinquième degré sont réelles" (1865), or *Math. Papers JJS*, 2:482–483.

86. Sylvester to Hirst, 28 March 1865. The quote that follows is also from this letter. As noted in chapter 6, the complete memoir never did appear, although the outlines Noble helped prepare were published by the London Mathematical Society in 1897.

87. J. J. Sylvester to Thomas Hirst, 4 April 1865, UCL Archives, LMS Papers, Sylvester, J. J.

88. See "Sir Andrew Noble," *Times* (London), 23 October 1915.

89. J. J. Sylvester to Thomas Hirst, 11 June 1865, UCL Archives, LMS Papers, Sylvester, J. J. (his emphasis).

90. On the Society's early history, see Rice, Wilson, and Gardner, "From Student Club to National Society."

91. Sylvester, "Sur les limites du nombre des racines réelles des équations algébriques" (1865), or *Math. Papers JJS*, 2:489–490, and "On an Elementary Proof and Generalization of Sir Isaac Newton's Hitherto Undemonstrated Rule for the Discovery of Imaginary Roots" (1865–1866), or *Math. Papers JJS*, 2:498–513. The latter is actually the syllabus of the lecture Sylvester delivered and was the first paper published by the new Society.

92. J. J. Sylvester to Thomas Hirst, 15 June 1865, UCL Archives, LMS Papers, Sylvester, J. J.

93. "A Mathematical Discovery," *Times* (London), 28 June 1865. The quotes that follow in this paragraph (with its emphasis) are also from this article.

94. Collins, *Who Is the Heir?* 3:172. Compare J. J. Sylvester to William Miller, 10 September 1866, David E. Smith Historical Papers, Rare Book and Manuscript Library, Columbia University, in Archibald, "Unpublished Letters of James Joseph Sylvester," 129–130.

95. For a list of Sylvester's questions and solutions, see *Math. Papers JJS*, 4:743–747.

96. J. J. Sylvester to C. E. Hodgson & Son, 16 July 1865, David E. Smith Historical Papers, Rare Book and Manuscript Library, Columbia University, in Archibald, "Unpublished Letters of James Joseph Sylvester," 125.

97. Sylvester, "Solution of Problem 1829," *Mathematical Questions . . . from the Educational Times* 4 (1866):77–79 on p. 78. For an historical account of the work of Sylvester and Crofton as well as the French mathematicians, Joseph-Émile Barbier and Joseph Bertrand, on these issues, see Seneta, Parshall, and Jongmans, "Nineteenth-Century Developments in Geometric Probability."

98. Sylvester, "On a Special Class of Questions on the Theory of Probabilities" (1865), or *Math. Papers JJS*, 2:480. The quote that follows in this paragraph is also from this page. Compare also the discussion in Seneta, Parshall, and Jongmans, "Nineteenth-Century Developments in Geometric Probability," 504–506.

99. Sylvester, "Solution of Problem 1829," 77.

100. J. J. Sylvester to Thomas Hirst, 14 October 1865, UCL Archives, LMS Papers, Sylvester, J. J. (his emphasis). The quotes that follow in this paragraph are also from this letter.

101. Elizabeth Joseph to J. J. Sylvester, undated but sometime before 25 October 1865, JJS Papers, SJC, Box 3.

102. J. J. Sylvester to Thomas Hirst, 25 October 1865, UCL Archives, LMS Papers, Sylvester, J. J. Fanny did die eight years later in 1873 at the age of sixty-six.

103. J. J. Sylvester to Thomas Hirst, 1 November 1865, UCL Archives, LMS Papers, Sylvester, J. J., in Parshall, *JJS: Life and Work in Letters*, 128. The quote that follows is also from this letter and page (with his emphasis). For pertinent references to Chasles's papers, see the commentary in ibid., 128.

104. J. J. Sylvester to Thomas Hirst, 1 November 1865, UCL Archives, LMS Papers, Sylvester, J. J. (his emphasis). This was a separate cover letter also dated 1 November.

105. For a brief overview of Sylvester's applicable work in 1865 and 1866, see Grattan-Guinness, "The Contributions of J. J. Sylvester, F.R.S., to Mechanics and Mathematical Physics," 259.

106. As Sylvester knew, Cayley had considered this theorem in 1862. See Cayley, "On Lambert's Theorem for Elliptic Motion," or *Math. Papers AC*, 3:562–565, where he sketched Lambert's proof.

107. Sylvester, "On Lambert's Theorem for Elliptic Motion" (1865), or *Math. Papers JJS*, 2:496 (my emphasis).

108. For the quote, see Sylvester, "Astronomical Prolusions" (1866), or *Math. Papers JJS*, 2:519. The London Mathematical Society's speakers for 1865 are given in Rice, Wilson, and Gardner, "From Student Club to National Society," table II, p. 412.

109. J. J. Sylvester to Thomas Hirst, 28 November 1865, UCL Archives, LMS Papers, Sylvester, J. J.

110. Rice, Wilson, and Gardner, "From Student Club to National Society," table III, p. 414, and Sylvester, "On an Addition to Poinsot's Ellipsoidal Mode of Representing the Motion of a Rigid Body Turning Freely Round a Fixed Point" (1866), or *Math. Papers JJS*, 2:517–518.

111. Sylvester, "On the Motion of a Rigid Body Acted on by No External Forces" (1866), or *Math. Papers JJS*, 2:577–601.

112. *DSB*, s.v. "Poinsot, Louis" by René Taton.

113. Sylvester, "On an Addition to Poinsot's Ellipsoidal Mode" (1866), in *Math. Papers JJS*, 2:517. The quote that follows is also from this page.

114. William F. Donkin to the Committee on Papers of the Royal Society, 22 June 1866, Royal Society of London, RR.6.274, Referee's Reports. The quote that follows is also from this report.

115. For more on Sylvester's efforts in this direction, see Parshall and Seneta, "Building an International Reputation." For a range of nineteenth-century efforts toward the internationalization of research-level mathematics, see Parshall and Rice, *Mathematics Unbound*.

116. J. J. Sylvester to Thomas Hirst, 14 July 1866, UCL Archives, LMS Papers, Sylvester, J. J. The quote that follows is also from this letter.

117. Gardner and Wilson, "Thomas Archer Hirst," 828. That Sylvester was desperately seeking Hirst's companionship is clear from Hirst's response. There is, however, no evidence to suggest that this was anything but a matter of seeking companionship—a purely homosocial and not a homosexual relationship—typical of the Victorian era and consonant with its conceptions of masculinity. On this issue, see Sussman, *Victorian Masculinities*.

118. Sylvester and Hirst had a serious falling out at some point after this, in which this unanswered invitation may have played a role. Hirst was also a member of the exclusive X Club, a group founded in November 1864 in "devotion to science, pure and free, untrammeled by religious dogmas" which would not shirk from exerting a "concerted effort" for science and its cause. Sylvester clearly shared these ideals. Moreover, the other members of the X Club—medical man and naturalist, George Busk; chemist, Edward Frankland; head of Kew Gardens, Joseph Hooker; arch supporter of Darwin's theory of natural selection, Thomas Huxley; naturalist and Darwinian protégé, John Lubbock; gifted scientific synthesizer, Herbert Spencer; and head of the Royal Institution, John Tyndall—were all well known to Sylvester and among the acquaintances with whom he regularly associated at the Royal Society, at the Athenæum, and elsewhere. Given Sylvester's mood in the late 1860s, his exclusion from this group would have stung. In 1873, he would write that Hirst was "thoroughly insincere" and that he was "probably bound by some engagement to a cabal with whom he is mixed up, being an inveterate plotter." See J. J. Sylvester to William J. C. Miller, 15 February 1873, David E. Smith

Historical Papers, Rare Book and Manuscript Library, Columbia University, in Parshall, *JJS: Life and Work in Letters*, 140–141. The earlier quotes on the X Club are from Huxley's diary. See Gardner and Wilson, "Thomas Archer Hirst," 832–833. On the X Club and its place in Victorian science, see Jones, "The X Club: Fraternity of Victorian Scientists," and Macleod, "The X Club."

119. See J. J. Sylvester to Sir John Herschel, 2 October 1866, Royal Society of London, Herschel Papers, HS.17.163, in Parshall, *JJS: Life and Work in Letters*, 129–131; J. J. Sylvester to Arthur Cayley, [undated] November 1866, JJS Papers, SJC, Box 10; and Sylvester, "Note on the Properties of the Test Operators Which Occur in the Calculus of Invariants" (1866), or *Math. Papers JJS*, 2:567–576.

120. Rice, Wilson, and Gardner, "From Student Club to National Society," table III, p. 414.

121. See J. J. Sylvester to Sir John Herschel, 5 March 1867, Royal Society of London, Herschel Papers, HS.17.164, in Parshall, *JJS: Life and Work in Letters*, 131.

122. See Clebsch, "Ueber die Anwendung der Abel'schen Functionen in der Geometrie."

123. J. J. Sylvester to Arthur Cayley, 27 May 1867, JJS Papers, SJC, Box 10, in Parshall, *JJS: Life and Work in Letters*, 131–132. The quotes that follow are from this letter (p. 132).

Chapter 8

Epigraph: Royal Society of London, Miscellaneous Manuscripts, MM.15.19, in Parshall, *JJS: Life and Work in Letters*, 144 (his emphasis).

1. *Annual Register . . . for the Year 1869*, 148–154.

2. Ibid.

3. *Annual Register . . . for the Year 1870*, 50–73.

4. Recall from chapter 4 that Sylvester had, in fact, first met Brougham on the latter's visit to Liverpool in 1837 for the opening of that city's Mechanics' Institute.

5. Harrison, *Learning and Living*, 64–65. The Society for the Diffusion of Useful Knowledge, another of Brougham's initiatives and founded in 1826, was intended to complement the Mechanics' Institutes by producing accurate but inexpensive educational literature. Compare Kelly, *History of Adult Education in Great Britain*, 165–167.

6. Harrison, *Learning and Living*, 213.

7. Harrison, *History of the Working Men's College*, 21–26, 43, 59.

8. Kelly, *History of Adult Education in Great Britain*, 186.

9. For the Advanced Class, which was specifically for artillery officers, see the section on "The BAAS and Science Education."

10. For Dufferin's objectives for the commission, see Lyall, *The Life of the Marquis of Dufferin and Ava*, 1:145–148.

11. *First Report of the Royal Commission Appointed to Inquire into the Present State of Military Education*, xlvii.

12. Ibid.

13. Ibid.

14. Ibid. (my emphasis).

15. Recall from chapter 6 that the Professor of Surveying and Topographical Drawing had had precisely this same complaint about Sylvester.

16. *First Report of the Royal Commission Appointed to Inquire into the Present State of Military Education*, xlvii. All direct quotes in this paragraph are also from this source and page.

17. Ibid.

18. Sylvester's correspondence from late 1862 through mid-1863 not only chronicles his struggles with the system but also makes clear his personal belief in the nineteenth-century Prussian academic principles of *Lehr-* and *Lernfreiheit.* For the letters, see Parshall, *JJS: Life and Work in Letters,* 114–121, and recall the discussion in chapter 7.

19. *Minutes of Evidence Taken before the Royal Commission Appointed to Inquire into the Present State of Military Education,* 301.

20. Ibid.

21. Ibid.

22. Sylvester, "On the Successive Involutes of a Circle" (1869), or *Math. Papers JJS,* 2:628.

23. See Salmon, *A Treatise on the Higher Plane Curves,* 222–223.

24. Sylvester, "Note on Successive Involutes to a Circle" (1868), or *Math. Papers JJS,* 2:630–640; and "On Successive Involutes to Circles—Second Note" (1868), or *Math. Papers JJS,* 2:641–649. In a letter to Cayley on 28 October 1868, Sylvester had proposed the term "cyclodites," although he had to admit that "it sounds rather *harsh.*" Cayley apparently concurred and suggested "cyclodes," for by 5 November, Sylvester was using that term in bouncing his evolving ideas off of Cayley. See J. J. Sylvester to Arthur Cayley, 28 October 1868 and 5 November 1868, JJS Papers, SJC, Box 11.

25. J. J. Sylvester to Joseph Henry, 26 December 1868, Record Unit 26, Office of the Secretary (Joseph Henry, Spencer Baird), Incoming Correspondence, 1863–1879, Smithsonian Institution Archives, in Parshall, *JJS: Life and Work in Letters,* 133. Henry had asked Sylvester, the former actuary, to referee a paper on computing mortality tables by the American Erastus De Forest. For more on this, see the commentary in Parshall, *JJS: Life and Work in Letters,* 132–134.

26. Sylvester, "Outline Trace of the Theory of Reducible Cyclodes" (1869), or *Math. Papers JJS,* 2:663–700.

27. *Annual Register . . . for the Year 1869,* 149.

28. J. J. Sylvester to Charles Graves, 15 May 1869, British Library, 267-268v, 66520.

29. Ibid.

30. *Annual Register . . . for the Year 1869,* 151.

31. Sylvester had received this invitation sometime before he wrote to Graves on 15 May, because he mentioned there "that the Council of the British Association for the Advancement of Science have invited me to accept the presidency of Section A (that of the Mathematical and Physical Sciences) at the meeting to take place at Exeter in August next. See Sylvester to Cayley, 15 May 1869.

32. See Sylvester, "Presidential Address: Mathematical and Physical Section" (1869). This was reprinted with additional notes and commentary in Sylvester, *Laws of Verse* (1870), 101–130; as "A Plea for the Mathematician" (1869–1870); and as "Presidential Address to Section 'A' of the British Association" (1869), in *Math. Papers JJS,* 2:650–661, 714–719. Sylvester's account of his second thoughts about speaking at Exeter appear on pp. 650–652.

33. On Huxley, see Desmond, *Huxley,* chapter 19, "Eyeing the Prize," which deals specifically with Huxley's mounting fame and popularity in the years from 1868 to 1870.

34. See Huxley, "Scientific Education: Notes of an After-dinner Speech," and Huxley, "The Scientific Aspects of Positivism."

Joan L. Richards discusses the Huxley–Sylvester exchange with an eye to their respective views on geometry in *Mathematical Visions: The Pursuit of Geometry in Victorian England,* 133–136. For a discussion of this exchange in the context of Sylvester's philosophy of science, see Parshall, "Chemistry through Invariant Theory?," 83–87.

35. Huxley, "Scientific Education," 181. The two quotes that follow are also from this page.

36. Ibid., 182 (his emphasis). The next quote in this paragraph and the first quote in the next are also from this page.

37. Huxley, "The Scientific Aspects of Positivism," 666 (his emphasis).

38. Sylvester, "Presidential Address to Section 'A' of the British Association" (1869), in *Math. Papers JJS*, 2:653.

39. Ibid., 656. Recall from chapters 4 and 5 that the German mathematical prodigy, Gotthold Eisenstein, had discovered—in two papers written in December 1843—that what Sylvester would later call the Hessian of the binary cubic form was an invariant. For the references and for a discussion of this work in historical context, see Parshall, "Toward a History of Nineteenth-Century Invariant Theory," especially pp. 170-171.

Paul Belloni Du Chaillu was a noted anthropologist, who traveled and observed extensively in Africa and who published two sensational book-length accounts of his findings in the 1860s.

40. Sylvester, "Presidential Address to Section 'A' of the British Association" (1869), in *Math. Papers JJS*, 2:656.

41. See Ibid., 653, and compare Desmond, *Huxley*, 366–367.

42. Sylvester, "Presidential Address to Section 'A' of the British Association" (1869), in *Math. Papers JJS*, 2:657.

43. Ibid. The quote is from Shakespeare's play, *The Tempest*, Act 5, Scene 1.

44. Ibid., 658; "A putrid excretion of the human mind." Euclid's 47th proposition is the Pythagorean theorem.

45. Sylvester's speech at Exeter reflected his positivistic philosophy of science. He both upheld the ideal of uniting all of human knowledge and viewed mathematics as the key to that unification. Compare the discussion of his poetry in "Poetry Soothes the Soul."

In 1878, he gave his most technical realization of this sort of positivistic synthesis in a paper that linked the atomic theory of chemistry and invariant theory. See chapter 9 for a brief discussion of that work. For a full discussion both of Sylvester's philosophy of science and of his 1878 work, see Parshall, "Chemistry through Invariant Theory?" His philosophy of science *per se* is discussed on pp. 81–87.

46. "Table Showing the Attendance and Receipts of Annual Meetings of the Association," in *Report of the Thirty-ninth Meeting … Held at Exeter in August 1869*, xxxviii–xxxix.

47. See Sylvester, "Presidential Address to Section 'A' of the British Association" (1869), in *Math. Papers JJS*, 2:661.

48. For Sylvester's listing of the foreign participants, see "Presidential Address to Section 'A' of the British Association" (1869), in *Math. Papers JJS* 2:661 (note *).

49. See the various annual *Reports* of the BAAS.

50. For the 1869 report, see *Report of the Thirty-ninth Meeting … Held at Exeter in August 1869*, 383–403. Subsequent reports also appear in the annual BAAS proceedings.

51. See "Applications for Reports and Researches Not Involving Grants of Money," in *Report of the Thirty-ninth Meeting … Held at Exeter in August 1869*, lxxvii, for the record of the committee's appointment. In her book, *Mathematical Visions*, Joan Richards mistakenly dates this committee's formation as 1871. See p. 170.

52. "Report of the Committee Appointed to Consider the Possibility of Improving the Methods of Instruction in Elementary Geometry," in *Report of the Forty-third Meeting Held at Bradford in September 1873*, 459–460.

53. For a brief discussion of the AIGT, see Richards, *Mathematical Visions*, 170–174. Michael H. Price gives a history of the organization in *Mathematics for the Multitude*. Compare also chapter 12 below for Sylvester's later involvement in AIGT affairs.

54. "Resolutions Referred to Council by the General Committee at Exeter," in *Report of the Thirty-ninth Meeting . . . Held at Exeter in August 1869*, lxxix.

55. "Scientific Instruction," *Times* (London), 5 February 1870. The following quote as well as those in the next four paragraphs are also from this account.

56. For a brief history of the Department of Science and Art, see Butterworth, "The Science and Art Department Examinations: Origins and Achievements," 27–44.

57. The BAAS's initiative here resulted almost immediately in the formation of the Devonshire Commission, which issued its findings in a series of reports between 1872 and 1875. The commission marked both a key success for the BAAS as a shaper of British science policy and a major step in the reform of British university education that would take place in the closing decades of the nineteenth century. Huxley and Stokes—but not Sylvester—were among the scientists on the commission.

58. The testimony of Brevet-Lieutenant-Colonel C. F. Young, the Director of Artillery Studies and Superintendent of the Advanced Class, in the *Minutes of Evidence Taken Before the Royal Commission on Military Education*, 326–327, describes the Advanced Class in some detail. The quote is on p. 327.

59. Strange had spent his active military career working on the survey of India. Retiring to England in 1861 at the rank of Lieutenant-Colonel, he became involved in the activities of the Royal Geographical and Royal Astronomical Societies as well as in the Royal Society and the BAAS. Sykes had also served in India and had retired from active duty in 1833 at the rank of Colonel. He became a fellow of the Royal Society in 1834 and pursued a variety of scientitic interests. See *DNB*, s.v. "Strange, Alexander" by Coutis Trotter and "Sykes, William Henry" by Bernard Barham Woodward.

60. "Armed Science," "To the Editor," *Times* (London), 10 February 1870.

61. Ibid.

62. J. J. Sylvester, "To the Editor," *Times* (London), 11 February 1870.

63. Ibid.

64. Ibid. (Sylvester's emphasis).

65. "Armed Science," "To the Editor," *Times* (London), 12 February 1870.

66. J. J. Sylvester, "To the Editor," *Times* (London), 14 February 1870 (his emphasis). Brevet-Lieutenant-Colonel C. F. Young's testimony before the Royal Commission fully substantiates the facts Sylvester presented in his exchange with "Armed Science" and even supports Sylvester's view that the Advanced Class was, indeed, expensive. See the *Minutes of Evidence Taken Before the Royal Commission on Military Education*, 326–329.

67. Lyall, *The Life of the Marquis of Dufferin and Ava*, 1:145.

68. See *First Report of the Royal Commission Appointed to Inquire into the Present State of Military Education*, 40–41. Relative to Woolwich, the Commission also recommended that the course of study be reduced from two-and-a-half to two years, that mathematics play a less prominent role in the curriculum, and that the administration be streamlined by abolishing the Council on Military Education and concentrating its former duties in the single, new post of Director-General of Military Education. For these recommendations, see ibid., 18–21, 31.

69. Fitzgerald, *London City Suburbs*, 154.

70. Sylvester, *Laws of Verse* (1870), 14 and 70.

71. Guthrie, "Matthew Arnold's Diaries," 3:714.

72. Sylvester, *Laws of Verse* (1870), 5.

73. Ibid., 9. The following quote is also from this page.

74. John Conington, *The Satires, Epistles and Art of Poetry of Horace*. Sylvester also took issue with Charles Anthon's *The Works of Horace*, which had come out in numerous nineteenth-century editions, as well as works by Forcellini, Newman, and others.

75. Anonymous, "Sylvester's Laws of Verse," *Times* (London), 12 January 1871.

76. As in poetry, so in mathematics. George Salmon had published his *Lessons Introductory to the Modern Higher Algebra* in 1859 precisely because of the ill-defined terminology and unsystematized techniques that Sylvester and Cayley were employing in their development of invariant theory.

77. Anonymous, "Sylvester's Laws of Verse." The following quote is also from this review.

Not all were as skeptical as the *Times* reviewer of Sylvester's notion of syzygy. The American poet, Sidney Lanier, devoted a short chapter in his book, *The Science of English Verse* (pp. 305–308), to what he called the "happy" term of "Phonetic Syzygy" that Sylvester introduced in his *Laws of Verse* and that Lanier deemed "a genuine contribution to the nomenclature of the science of English verse" (p. 307).

78. Sylvester, "Presidential Address to Section 'A' of the British Association" (1869), in *Math. Papers JJS*, 2:659.

79. Sylvester, *Laws of Verse* (1870), 12. On the interconnectedness specifically of mathematics and music, Sylvester had once waxed lyrical, asking "may not Music be described as the Mathematic of sense, Mathematic the Music of the reason? the soul of each the same!" See Sylvester, "Algebraical Researches Containing a Disquisition on Newton's Rule" (1864), in *Math. Papers JJS*, 2:419 (note *).

80. *Nature*'s first issue had appeared in 1869.

81. Sylvester, "Presidential Address to Section 'A' of the British Association" (1869), in *Math. Papers JJS*, 2:655.

82. Sylvester, "A Plea for the Mathematician" (1869–1870), *Math. Papers JJS*, 2:715. As noted, this was an abridged version of Sylvester's BAAS address commissioned by the journal.

83. Sylvester, *Laws of Verse* (1870), 131 (Lewes's emphasis).

84. In addition to Lewes and Huxley, Clement M. Ingleby, a Shakespearean critic, essayist, and sometime poet, G. Croom Robertson of University College London, and W. H. Stanley Monck of Trinity College Dublin, all joined in the debate.

85. *Times* (London), 17 August 1871, and J. J. Sylvester, "Letter to the Editor," *Times* (London), 24 August 1871.

86. Recall from chapter 4 that this is how Sylvester had described the unsettled years from 1842 through 1845 in a letter to Joseph Henry on 12 April 1846. For the quote, see Parshall, *JJS: Life and Work in Letters*, 15.

87. The text of his commentary may be found in the JJS Papers, SJC, Box 8 under the title, "Prefatory Statement to the Recital of the Ballad of Sir John de Courcy at the Penny Reading Held at the Quebec Club, Tuesday, 11th April 1871," 12 handwritten pages. Sylvester first published the translation under the pseudonym "Syzygeticus" in the *Gentleman's Magazine* (1871). There, he commented that he had first made the translation "upwards of twenty years ago" (p. 316). He also had his translation printed, along with a number of his other poetic flights during the years from 1870 to 1876, in *Fliegende Blätter: Supplement to the Laws of Verse* (1876). See pp. 19–20. This 44-page booklet was dedicated to the poet, Frederick Locker, a fellow Athenæum Club member. Locker's copy is held at St. John's College, Cambridge in the "Sylvester" file.

88. Sylvester, "Prefatory Statement," 9–10.

89. J. J. Sylvester to Benjamin Smith, 20 April 1871, McCrimmon Bodichon Collection, 7/BMC, The Women's Library, London Metropolitan University.

90. Ibid.

91. Sylvester, "Note on the Theory of a Point in Partition" (1871), or *Math. Papers JJS*, 2:701–703. Earlier in 1871, Sylvester had also presented two very brief and inconsequential notes—one relating to Goldbach's conjecture that every even number can be written as the sum of two primes and one on Legendre's theorem that an arithmetical progression containing more than one prime contains infinitely many primes—to the London Mathematical Society for publication in its *Proceedings*. See *Math. Papers JJS*, 2:709–711 and 712–713, respectively.

92. *Times* (London), 17 August 1871. The quote that follows in this paragraph is from this article, a major portion of which was reprinted in *Nature* (24 August 1871), 326.

93. Compare Hermann von Helmholtz to his wife, 24 August 1871 in Thompson, *The Life of William Thomson*, 2:613.

94. *Times* (London), 17 August 1871. Compare J. J. Sylvester, "Letter to the Editor," *Times* (London), 24 August 1871.

95. Ibid.

96. A salary of roughly £600 or more placed one at the lower reaches of the carriage-owning, upper middle class that could afford a semidetached—but not a detached—house. A salary upwards of £400 generally brought with it a semidetached house with fewer frills or an embellished terrace house and a respectable middle-class standing. Best, *Mid-Victorian Britain: 1851–1875*, 19. The precise amount of Sylvester's pension is not quite clear. If he was making £550 (as he was in 1862) at the time of his retirement, then the figure would have been £366. If allowance was made as an "emolument" for the house that he had been provided, then it would have been more.

97. Sylvester, "Science at the London School Board" (1872), 410.

98. Ibid.

99. "A Little Mistake," *Times* (London), 28 March 1872. The *Jewish Chronicle* also picked up on this statement, but it had a different spin on it. It prefaced the quote with "That distinguished journal the *(Church) Record* contains the following pious and liberal remark:" and followed it with the rhetorical question: "Is this the only proof of ignorance on the part of the *Record*? Risum teneatis, amici! [You should hold your laughter, friends!]" See "Toleration," *Jewish Chronicle* (London), 29 March 1872.

100. "The London School Board," *Times* (London), 29 March 1872.

101. J. J. Sylvester to George G. Stokes, 20 April 1872, Stokes Papers, Cambridge University Library, Add 7656, RS849 (Sylvester's emphasis).

102. J. J. Sylvester to Pafnuti Chebyshev, 1 September 1872 in Chebyshev, *Polnoe Sobranie Sochinenii*, 449. I thank Eugene Seneta for pointing out the existence of the Russian versions of the Sylvester–Chebyshev letters to me and for translating them from Russian into English. The letters were originally written by Sylvester in French.

103. J. J. Sylvester to Pafnuti Chebyshev, 27 September 1872 in ibid., 5:449

104. J. J. Sylvester to Contessa Edith M. Gigliucci, 22 November 1872, UCL Archives, MS.ADD.221, in Parshall, *JJS: Life and Work in Letters*, 139. Edith was the daughter of Sylvester's sister, Fanny Joseph Mozley.

105. Ibid.

106. J. J. Sylvester to Pafnuti Chebyshev, 23 December 1872 in Chebyshev, *Polnoe Sobranie Sochinenii*, 5:449.

107. J. J. Sylvester to Walter White, 6 April 1873, Royal Society, Miscellaneous Correspondence, MC 10.235.

108. Thomas, *Michael Faraday and the Royal Institution*, 4.

109. Becker, *Scientific London*, 43–44.

110. Thomas, *Michael Faraday and the Royal Institution*, 222.

111. Becker, *Scientific London*, 45.

112. Sylvester, "On Recent Discoveries in Mechanical Conversion of Motion" (1873–1875), or *Math. Papers JJS*, 3:11 (note ∗).

113. Ibid.

114. Ibid.

115. Ibid., 10 (note ∗) and 15.

116. Ibid., 15. According to Sylvester in a later paper on the subject, the Peaucellier mechanism was, indeed, used in the ventilation machinery in the Houses of Parliament. See Sylvester, "On the Plagiograph *aliter* the Skew Pantigraph" (1875), or *Math. Papers JJS*, 3:33–34.

117. See Sylvester, "Transformation du mouvement circulaire en mouvement rectiligne" (1874–1875), and "On Recent Discoveries in Mechanical Conversion of Motion" (1875). (These two versions do not appear in *Math. Papers JJS*.) Sylvester actually went to France to present the talk again—this time in French. Grattan-Guinness judged these researches on linkworks "Sylvester's most publicly successful forays into applied mathematics" in "The Contributions of J. J. Sylvester, F.R.S., to Mechanics and Mathematical Physics," 260.

118. Penrose's letter was dated 10 March, 1874; Newman's 2 April, 1874. See JJS Papers, SJC, Box 3. Hart wrote at least three letters to Sylvester on the subject, on 3, 9, and 13 July, 1874. See JJS Papers, SJC, Box 2.

119. See Parshall, *JJS: Life and Work in Letters*, 142–144 for one example and for additional commentary. For more letters between Sylvester and Cayley, see the Royal Society, Miscellaneous Manuscripts, MM.15.13-23. Sylvester also corresponded on linkworks with Samuel Roberts, longtime participant in and officer of the London Mathematical Society. See JJS Papers, SJC, Box 3.

120. Sylvester, "On the Plagiograph" (1875) and "On a Lady's Fan, on Parallel Motion, and on an Orthogonal Web of Jointed Rods" (1875), or *Math. Papers JJS*, 3:35–36.

121. Sylvester to Henry, 12 April 1846, in Parshall, *JJS: Life and Work in Letters*, 15.

122. J. J. Sylvester to Charles Graves, British Library, 267–268v, 66520.

123. J. J. Sylvester to Barbara Bodichon, 19 June 1874, McCrimmon Bodichon Collection, 7/BMC, The Women's Library, London Metropolitan University.

124. J. J. Sylvester to Barbara Bodichon, 12 August 1874, McCrimmon Bodichon Collection, 7/BMC, The Women's Library, London Metropolitan University.

125. See Epigraph note. The remark "*living upon charity*" may have referred to the fact that he was living off of his pension and not off of an actual salary.

126. J. J. Sylvester to Arthur Cayley, 19 February 1875, Royal Society, Miscellaneous Manuscripts, MM.15.20.

127. Joseph Henry to Daniel Coit Gilman, 25 August, 1875, DCG Papers, JHU. Gilman did his own homework as well, soliciting opinions on Sylvester from more disinterested parties. These other inquiries revealed that Sylvester could be difficult to get along with. Gilman would need to get a sense of the man for himself. Compare Gilman, *The Launching of a University*, 66.

128. J. J. Sylvester to Barbara Bodichon, 19 September, 1875, McCrimmon Bodichon Collection, 7/BMC, The Women's Library, London Metropolitan University, in Parshall, *JJS: Life and Work in Letters*, 145.

129. For details on the negotiations between Sylvester, Gilman, and the Trustees of the Johns Hopkins University, see Parshall and Rowe, *Emergence of the American Mathematical Research Community*, 72–75.

130. This was a princely salary for an American academic. The highest salaries at Yale College were $3,500, and $4,000 at Harvard. See Daniel Coit Gilman to J. J. Sylvester, 29 November 1875, DCG Papers, JHU, in Parshall, *JJS: Life and Work in Letters*, 149–150. For references to the archival record of Sylvester's negotiations with Hopkins, see Parshall and Rowe, *Emergence of the American Mathematical Research Community*, 75 (notes 70–72).

Chapter 9

Epigraph (1): McCrimmon Bodichon Collection, 7/BMC, The Women's Library, London Metropolitan University, in Parshall, *JJS: Life and Work in Letters*, 155.

Epigraph (2): JJS Papers, SJC, Box 1, in ibid., 166 (his emphasis).

1. *Magee's Illustrated Guide of Philadelphia and the Centennial Exhibition*, and Brown, *Year of the Century: 1876*.

2. The composite of Baltimore and the Johns Hopkins University in 1876 that follows was drawn from these sources: for historical photographs, Jones, *Lost Baltimore: A Portfolio of Vanished Buildings*; Beirne and the Maryland Historical Society, *Baltimore . . . a Picture History, 1858–1958*; and Miller, *Baltimore Transitions: Views of an American City in Flux*; for general history, Hall, *Baltimore: Its History and People*; Olson, *Baltimore: The Building of an American City*; Beirne, *The Amiable Baltimoreans*; and Weishampel, *The Stranger in Baltimore: A New Hand Book*.

3. Weishampel, *The Stranger in Baltimore*, 63.

4. Ammen, "History of Baltimore: 1875–1895," in *Baltimore: Its History and People*, 1:241.

5. Hopkins actually bequeathed a total of $7,000,000 to found both a university and a hospital. At the end of December 1873 when Hopkins died, this was the largest single bequest ever made to higher education in the United States. Although the university opened in 1876, the hospital would not begin its work until 1889. For more on the early history of these institutions, see Hawkins, *Pioneer*.

6. See J. J. Sylvester to Daniel Coit Gilman, 29 May 1876, DCG Papers, Coll #1 Corresp., JHU.

7. J. J. Sylvester to Daniel Coit Gilman, 1 June 1876, DCG Papers, Coll #1 Corresp., JHU. Fortunately, the box finally turned up in Baltimore, but only later in the summer. See J. J. Sylvester to Barbara Bodichon, 27 August 1876, McCrimmon Bodichon Collection, 7/BMC, The Women's Library, London Metropolitan University.

8. Daniel Coit Gilman to J. J. Sylvester, 27 April 1876, DCG Papers, Letter Book 2, Ser. 4, Box 4.1, JHU.

9. Sylvester had, in fact, made recommendations for faculty as early as March, suggesting both Matthew Arnold and his brother, Thomas, to Gilman for the chair of English. Throughout his association with Hopkins, Sylvester worked with Gilman to secure faculty in a range of areas. He was a trusted adviser and a real team player.

10. J. J. Sylvester to Daniel Coit Gilman, 12 June 1876, DCG Papers, Coll #1 Corresp., JHU. On the interactions between Gilman and Story, see Hawkins, *Pioneer*, 44–45. For more on putting together the Hopkins program in mathematics, see Parshall and Rowe, *Emergence of the American Mathematical Research Community*, 75–87.

11. J. J. Sylvester to Barbara Bodichon, 13 June 1876, McCrimmon Bodichon Collection, 7/BMC, The Women's Library, London Metropolitan University.

12. J. J. Sylvester to Daniel Coit Gilman, 21 July 1876, DCG Papers, Coll #1 Corresp., JHU. Joseph Henry had tried to console Sylvester in a letter dated 11 July 1876 by describing the summer's temperatures as "abnormal." For the letter, see JJS Papers, SJC, Box 2.

13. Sylvester, "Note on Spherical Harmonics" (1876), or *Math. Papers JJS*, 3:38.

14. Ibid., in *Math. Papers JJS*, 3:50. Compare also the brief discussion of this paper in Arnold, "Topological Content of the Maxwell Theorem on Multipole Representation of Spherical Functions," 216. I thank the late John Fauvel for pointing out the latter reference to me.

15. Hawkins, *Pioneer*, 69.

16. Desmond, *Huxley*, 470–482.

17. Desmond, *Huxley*, 470–482, and White, *A History of the Warfare of Science with Theology in Christendom*. In 1876, the same year as Huxley's visit, White published a first glimpse at his thoughts on this topic in his short but controversial book, *Warfare of Science*.

18. Hawkins, *Pioneer*, 68–69.

19. See DCG Papers, Letter Book 2, Ser. 4, Box 4.1, JHU for a series of letters Gilman wrote to Craig in the spring of 1876. Gilman was so taken with the young man's earnestness that he even lent him money on a couple of occasions before classes started in the fall. In a letter dated 3 May 1876, Gilman counseled him firmly: "Don't overwork!"

20. See Brown, *Johns Hopkins Half-Century Directory*, and compare J. J. Sylvester to Daniel Coit Gilman, 'Remarks on Candidates for Renewal or Conferral of Fellowships in Johns Hopkins University," DCG Papers, Coll #1 Corresp., JHU.

21. J. J. Sylvester to Barbara Bodichon, 2 November 1876, McCrimmon Bodichon Collection, 7/BMC, The Women's Library, London Metropolitan University, in Parshall, *JJS: Life and Work in Letters*, 162.

22. See Epigraph note 1.

23. Sylvester to Bodichon, 2 November 1876, in Parshall, *JJS: Life and Work in Letters*, 163–164 (his emphasis).

24. Peirce ultimately could not attend, owing to a bad cold, but he sent a testimonial to be read in his absence. See Benjamin Peirce to Daniel Coit Gilman, DCG Papers, Coll #1 Corresp., JHU.

25. Joseph Henry gave a run-down of these events in his diary. See Joseph Henry, desk diary, entry for 3 November 1876, Joseph Henry Papers, Smithsonian Institution Archives. On Gilman's idea of public lectures, see Hawkins, *Pioneer*, 72–73. On Newcomb's place in American science, see Moyer, *A Scientist's Voice in American Culture*.

26. J. J. Sylvester, Simon Newcomb, Henry A. Rowland, and William E. Story to the mathematical scientists of America, 8 November 1876, DCG Papers, Coll #1 Corresp., JIIU, in Parshall, *JJS: Life and Work in Letters*, 164–166.

27. For a list of America's mathematical journals, 1800–1900, and their years of operation, see Parshall and Rowe, *Emergence of the American Mathematical Research Community*, 51. For the early history of the *American Journal of Mathematics*, see ibid., 88–94.

28. Joseph Henry, desk diary, entry for 12 November 1876, Henry Papers, Smithsonian Institution Archives. In fact, when it began coming out in 1878, the *American Journal of Mathematics* did carry a limited number of more expository articles. See "Launching the *American Journal of Mathematics*."

29. See Epigraph note 2.

30. Sylvester to Spottiswoode, 18 November 1876, in Parshall, *JJS: Life and Work in Letters*, 170.

31. Recall from chapter 7 that Sylvester had last engaged seriously in invariant-theoretic research in the context of his work on Newton's Rule in 1864 and 1865. On the British and German schools of invariant theory, see Parshall, "Toward a History of Nineteenth-Century Invariant Theory"; Crilly, "The Rise of Cayley's Invariant Theory (1841–1862)"; and Crilly, "The Decline of Cayley's Invariant Theory (1863–1895)."

32. Cayley, "A Ninth Memoir on Quantics," or *Math. Papers AC*, 7:353. Tony Crilly discusses the "Ninth Memoir" at some length in "The Mathematics of Arthur Cayley with Particular Reference to Linear Algebra," 135–139.

33. Recall the discussion of this work in chapter 6. For more technical details, see the commentary provided in Parshall, *JJS: Life and Work in Letters*, 166–176.

34. Recall Sylvester's critique of Huxley's conception of mathematics as purely deductive, discussed in chapter 8.

35. See Sylvester to Spottiswoode, 18 November 1876, in Parshall, *JJS: Life and Work in Letters*, 167–169 and the associated commentary. For the complete calculation for the quartic, see Elliott, *Introduction to the Algebra of Quantics*, 169–170.

36. Sylvester to Spottiswoode, 18 November 1876, in Parshall, *JJS: Life and Work in Letters*, 173. In fact, no general formula exists for these generating functions.

37. See J. J. Sylvester to William Spottiswoode, 24 and 25 November 1876, JJS Papers, SJC, Box 1.

38. Sylvester to Spottiswoode, 25 November 1876.

39. J. J. Sylvester to William Spottiswoode, 26 and 29 November 1876, JJS Papers, SJC, Box 1.

40. See Sylvester, "Sur les invariants fondamentaux de la forme binaire du huitième degré" (1877), or *Math. Papers JJS*, 3:52–57.

41. J. J. Sylvester to William Spottiswoode, 20 November 1876, JJS Papers, SJC, Box 1.

42. See *Math. Papers JJS*, 3:58–71, 93–100, 105–116, 127–147, and 218–240.

43. Sylvester to Spottiswoode, 25 November 1876.

44. Sylvester to Spottiswoode, 24 November 1876 (his emphasis).

45. Sylvester, "Address on Commemoration Day" (1877), in *Math. Papers JJS*, 3:73.

46. Ibid., 73–75, 78–79. Compare also Storr, *Beginnings of Graduate Education in America*, and Veysey, *Emergence of the American University*.

47. Sylvester, "Address on Commemoration Day" (1877), in *Math. Papers JJS*, 3:80.

48. Coeducation was another matter, however. See chapter 10.

49. Sylvester, "Address on Commemoration Day" (1877), in *Math. Papers JJS*, 3:83.

50. Ibid.

51. Arthur Hathaway to Florian Cajori, as quoted in Cajori, *Teaching and History of Mathematics in the United States*, 265.

52. For more on Sylvester's teaching style, see Parshall, "America's First School of Mathematical Research," and Parshall and Rowe, *Emergence of the American Mathematical Research Community*, 80–83.

53. Sylvester, "Address on Commemoration Day" (1877), in *Math. Papers JJS*, 3:85–87.

54. J. J. Sylvester to William Spottiswoode, 9 April 1877, JJS Papers, SJC, Box 1. Sylvester did get a (renewable) grant of £50 for these calculations but from the British Association for the Advancement of Science. See *Report of the Forty-eighth Meeting of the British Association for the Advancement of Science*, lvii.

55. George Salmon to J. J. Sylvester, 18 April 1877, JJS Papers, SJC, Box 3, in Parshall, *JJS: Life and Work in Letters*, 175.

56. See, for example, Sylvester, "Sur les covariants fondamentaux d'un système cubo-biquadratique binaire" (1878), or *Math. Papers JJS*, 3:127–131, where he acknowledged that "thanks to the intelligent cooperation and the great skill as a calculator of Mr. [F.] Franklin, one of my students in Baltimore, I am in the position to present to the Academy the table of fundamental invariants and covariants," 127.

57. For two such examples, see Parshall, *JJS: Life and Work in Letters*, 178–179 (notes 69 and 72).

58. Sylvester, "Sur les covariants fondamentaux d'un système cubo-biquadratique binaire" (1878), in *Math. Papers JJS*, 3:131.

59. Writing more than 20 years later, Halsted averred, in fact, that "such a course in the creation of modern mathematics, with most precious, elsewhere unattainable, historic indications, will perhaps never be paralleled." See Halsted, "James Joseph Sylvester," 297.

60. Sylvester to Spottiswoode, 9 April 1877.

61. J. J. Sylvester to John Tyndall, 20 April 1877, Tyndall Papers, Sylvester Correspondence, p. 1513, Royal Institution Archives. The announcement appeared as "Sur une méthode algébrique pour obtenir l'ensemble des invariants et des covariants fondamentaux d'une forme binaire et d'une combinaison quelconques de formes binaires" (1877), or *Math. Papers JJS*, 3:58–62.

62. J. J. Sylvester to Arthur Cayley, 23 April 1877, JJS Papers, SJC, Box 11, in Parshall, *JJS: Life and Work in Letters*, 177 (his emphasis). "Grundformen" was the German word for the elements in a minimum generating set.

63. Ibid., 177–178 (his emphasis).

64. Hermite, "Études de M. Sylvester sur la théorie algébrique des formes," 975 (my translation).

65. Camille Jordan to J. J. Sylvester, 13 May 1877, JJS Papers, SJC, Box 2, in Parshall, *JJS: Life and Work in Letters*, 182–184.

66. See Hilbert, "Ueber die vollen Invariantensysteme," and Parshall, "The One-Hundredth Anniversary of the Death of Invariant Theory?"

67. Sylvester to Cayley, 23 April 1877, in Parshall, *JJS: Life and Work in Letters*, 177.

68. Sylvester, "Sur le vrai nombre des covariants élémentaires d'un système de deux formes biquadratiques binaires" (1877), or *Math. Papers JJS*, 3:63–66; and "Théorie pour trouver le nombre des covariants et des contrevariants d'ordre et de degré donnés linéairement indépendants d'un système quelconque de formes simultanées contenant un nombre quelconque de variables" (1877), or *Math. Papers JJS*, 3:67–71.

69. J. J. Sylvester to Daniel Coit Gilman, 8 June 1877, DCG Papers, Coll #1 Corresp., JHU. Preston and Hering also had their fellowships in engineering renewed. They stayed through the 1877–1878 academic year and left without degrees. Clearly, Sylvester's type of mathematics was far removed from their interests. It is, in fact, odd that Gilman appointed fellows in an area that was not formally taught at the University.

70. J. J. Sylvester to William Spottiswoode, 10 April 1877, JJS Papers, SJC, Box 1.

71. See J. J. Sylvester to Arthur Cayley, 26 June 1877, JJS Papers, SJC, Box 11.

72. J. J. Sylvester to Arthur Cayley, 12 October 1877, JJS Papers, SJC, Box 11 (his emphasis).

73. See Sylvester, "Sur les invariants" (1877), or *Math. Papers JJS*, 3:93–100.

74. J. J. Sylvester to Simon Newcomb, 31 October 1877, in Archibald, "Unpublished Letters of James Joseph Sylvester," 133.

75. The elements in a minimum generating set are *algebraically* independent.

76. Thomas Hawkins used this terminology in "Cayley's Counting Problem and the Representation of Lie Algebras."

77. Sylvester consistently referred to Faà di Bruno as Faà *de* Bruno. I adopt the standard Italian version here.

78. Salmon to Sylvester, 18 April 1877, in Parshall, *JJS: Life and Work in Letters*, 176.

79. Compare Sylvester, "Proof of the Hitherto Undemonstrated Fundamental Theorem" (1878), or *Math. Papers JJS*, 3:117 (note ∗).

80. Faà di Bruno, *Théorie des formes binaires*, 150.

81. Sylvester, "Proof of a Hitherto Undemonstrated Fundamental Theorem (1878), in *Math. Papers JJS*, 3:117 (note ∗).

82. J. J. Sylvester to Arthur Cayley, 6 November 1877, JJS Papers, SJC, Box 11, in Parshall, *JJS: Life and Work in Letters*, 184 (his emphasis).

83. J. J. Sylvester to Arthur Cayley, 7 November 1877, JJS Papers, SJC, Box 11.

84. DCG Papers, Coll #1 Corresp., JHU. For the technical details of Sylvester's proof, see the commentary in Parshall, *JJS: Life and Work in Letters*, 184–187. This proof, one of Sylvester's most "modern," was essentially an explicit analysis of what we would now call the weight spaces associated with the finite-dimensional representation theory of the simple Lie algebra $\mathfrak{sl}_2$.

85. In fact, this line of work defined his courses through the end of the 1878–1879 academic year, with papers appearing abroad in the *Comptes rendus de l'Académie des Sciences de Paris*, in Crelle's *Journal für die reine und angewandte Mathematik*, and in the *Messenger of Mathematics*.

86. J. J. Sylvester to Daniel Coit Gilman, 9 May 1877, DCG Papers, Coll #1 Corresp., JHU, in Parshall, *JJS: Life and Work in Letters*, 181–182 (his emphasis).

87. Ibid., 181. Sylvester was still, at least nominally, the co-editor (with Norman Ferrers) of the *Quarterly Journal*. His name would remain on the journal's title page as editor through 1878.

88. J. J. Sylvester to Daniel Coit Gilman, 14 May 1877, DCG Papers, Coll #1 Corresp., JHU.

89. The Trustees paid $500.00 per volume or roughly 20% of the cost of publication. See Hawkins, *Pioneer*, 75.

90. See "Notice to the Reader," *American Journal of Mathematics*, 1 (1878):iii.

91. Sylvester, "On an Application of the New Atomic Theory to the Graphical Representation of the Invariants and Covariants of Binary Quantics" (1878), or *Math. Papers JJS*, 3:148.

92. Ibid., in *Math. Papers JJS*, 3:149.

93. J. J. Sylvester to Simon Newcomb, 14 January 1878, in Archibald, "Unpublished Letters of James Joseph Sylvester," 135.

94. Ibid., 134. For a fuller discussion of Sylvester's chemistrization of invariant theory with explicit examples and a discussion of the earlier and related work of Cayley and Clifford, see Parshall, "Chemistry through Invariant Theory?"

95. Frankland, "Remarks on Chemico-graphs," and Clifford, "Remarks on the Chemico-Algebraical Theory."

96. Weichold, "Historical Data Concerning the Discovery of the Law of Valence," 282, and Mallet, "Some Remarks on a Passage in Professor Sylvester's Paper as to the Atomic Theory," 277.

97. Franklin, "On a Problem of Isomerism," 365–369.

98. In addition to Franklin, Craig, Halsted, and Story each had papers in the journal's first volume.

99. Joseph Henry, desk diary, entry from 3 November 1876, Joseph Henry Papers, Smithsonian Institution Archives.

100. Sylvester, "A Synoptical Table of the Irreducible Invariants and Covariants to a Binary Quintic" (1878), or *Math. Papers JJS*, 3:210–217.

101. Sylvester (assisted by Fabian Franklin), "Tables of the Generating Functions and Groundforms for the Binary Quantics of the First Ten Orders" (1879), or *Math. Papers JJS*, 3:283–311. In 1986 and 1988 Jacques Dixmier and Daniel Lazard showed that Sylvester and Franklin had made a mistake in their calculations regarding the binary septic form. See Dixmier and Lazard, "Le nombre minimum d'invariants fondamentaux pour les formes binaires de degré 7," and "Minimum Number of Fundamental Invariants for the Binary Form of Degree 7." For a modern synthetic treatment of these issues, see Olver, *Classical Invariant Theory*.

102. See Sylvester, "Sur les actions mutuelles des formes invariantives dérivées" (1878), or *Math. Papers JJS*, 3:218–240, and, for example, Sylvester, "Table de nombres de dérivées invariantives d'ordre et de degré donnés, appartenant à la forme binaire du dixième ordre" (1879), or *Math. Papers JJS*, 3:256. In the latter paper, however, Sylvester did take the opportunity to advertise the *American Journal* as the place where the full tables up to and including the binary decimic would appear. See *Math. Papers JJS*, 3:256.

103. See Sylvester (assisted by Fabian Franklin), "Tables of the Generating Functions and Groundforms for Simultaneous Binary Quantics of the First Four Orders, Taken Two and Two Together" (1879), or *Math. Papers JJS*, 3:392–410; Sylvester, "Tables of Generating Functions and Groundforms of the Binary Duodecimic, with Some General Remarks, and Tables of the Irreducible Syzygies of Certain Quantics" (1881), or *Math. Papers JJS*, 3:489–508; and "Tables of Generating Functions, Reduced and Representative, for Certain Ternary Systems of Binary Forms" (1882), or *Math. Papers JJS*, 3:623–632, respectively. On the work in 1881 and 1882, see chapter 10.

104. J. J. Sylvester to Daniel Coit Gilman, 7 June 1878, DCG Papers, Coll #1 Corresp., JHU.

105. J. J. Sylvester to Daniel Coit Gilman, 7 September 1878, DCG Papers, Coll #1 Corresp., JHU, in Parshall, *JJS: Life and Work in Letters*, 191.

106. Clifford died in 1879 at the young age of thirty-three.

107. Sylvester had an early version of the poem privately printed in *Fliegende Blätter* (1876), 5–15. The Rosalind poem was only one of several "studies in monochrome" he published in this volume.

108. Franklin, "An Address Commemorative of Professor James Joseph Sylvester," 8, in DCG Papers, Coll #1 Corresp., JHU.

109. Bond, *When the Hopkins Came to Baltimore*, 37.

110. Franklin, "An Address Commemorative of Professor James Joseph Sylvester," 8.

111. See J. J. Sylvester to Simon Newcomb, 23 March 1879, in Archibald, "Unpublished Letters of James Joseph Sylvester," 143–144. Sylvester had this poem privately printed early in 1880. For a copy, see Sylvester, *Spring's Début*, in DCG Papers, Coll #1 Corresp., JHU.

112. Sylvester, *Spring's Début*, 12 (note 17).

113. J. J. Sylvester to Daniel Coit Gilman, 6 July 1879, DCG Papers, Coll #1 Corresp., JHU.

114. See Kempe, "On the Geographical Problem of the Four Colours," and Story, "Note on the Preceding Paper." As is well-known, Kempe's proof—even with Story's addition—is incomplete. The theorem was only proved in 1976 by Kenneth Appel and Wolfgang Hacken. For the additional references, see Parshall, *JJS: Life and Work in Letters*, 195–196 (note 136).

115. J. J. Sylvester to Daniel Coit Gilman, 22 July 1880, DCG Papers, Coll #1 Corresp., JHU, in Parshall, *JJS: Life and Work in Letters*, 195 (his emphasis).

116. J. J. Sylvester to Sarah Hunt Mills (Mrs. Benjamin) Peirce, 25 March 1880, in Archibald, "Unpublished Letters of James Joseph Sylvester," 144.

117. Sylvester to Gilman, 22 July 1880, in Parshall, *JJS: Life and Work in Letters*, 194–195.

118. Ibid., 195. The next quote is also from this page (his emphasis).

119. Daniel Coit Gilman to William Story, 7 August 1880, DCG Papers, Coll #1 Corresp. (American Journal of Mathematics file), JHU.

120. Daniel Coit Gilman to J. J. Sylvester, 19 February 1881, DCG Papers, Letter Book 4, Ser. 4, Box 4.2, JHU, in Parshall, *JJS: Life and Work in Letters*, 199.

121. See J. J. Sylvester to Arthur Cayley, 21 June 1881, JJS Papers, SJC, Box 11 (his emphasis).

Chapter 10

Epigraph: Archibald, "Unpublished Letters of James Joseph Sylvester," 144.

1. George Salmon to J. J. Sylvester, 18 April 1877, JJS Papers, SJC, Box 3, in Parshall, *JJS: Life and Work in Letters*, 175.

2. Unfortunately, none of the early Hopkins dissertations in mathematics survives, although versions of some of them were published in the *American Journal of Mathematics*, for example, Franklin, "Bipunctual Coordinates," and Stringham, "Regular Figures in n-Dimensional Space." If the original manuscripts did exist, it might be possible to draw even more conclusions about the extent of Sylvester's involvement with graduate students in their research endeavors in the program's early years.

3. Recall the discussion of this work in chapter 9.

4. See Franklin, "On a Problem of Isomerism," and Franklin, "On Partitions," respectively.

5. Craig and Franklin were both promoted to full professor in 1892.

6. Franklin, "On the Calculation of the Generating Functions and Tables of Groundforms for Binary Quantics."

7. Craig, "On the Motion of an Ellipsoid in a Fluid," "Some Elliptic Function Formulæ," "On Quadruple Theta-Functions," in the *American Journal of Mathematics*.

8. George Bruce Halsted to Daniel Coit Gilman, 10 April 1876, DCG Papers, Coll #1 Corresp., JHU. The quotes that follow in this paragraph are also from this source.

9. Daniel Coit Gilman to Simon Newcomb, 12 April 1876, DCG Papers, Letter Book 1, Ser. 4, Box 4.1, JHU.

10. On the immediate intellectual circle of the Peirces, see Menand, *The Metaphysical Club*.

11. Late in 1877, Benjamin Peirce had written to Gilman recommending Charles for the professorship of physics at Hopkins. That position went to Henry Rowland, but in a letter to Peirce on 23 January 1878, Gilman did offer Peirce a lectureship. Peirce, disappointed by the offer, declined in writing on 12 March, proposing instead to keep his job at the U.S. Coast Survey and to serve the University half-time as Professor of Logic. Gilman only received Peirce's proposal on the twentieth, and by that time, the Trustees had decided that, owing to a financial downturn, the University would have to curtail all hiring. Gilman offered nevertheless to take Peirce's suggestion before the Trustees, but Peirce instructed him not to in a letter on the twenty-seventh. See Brent, *Charles Sanders Peirce: A Life*, 120–122, and Fisch and Cope, "Peirce at The Johns Hopkins University." The latter article was reprinted in Ketner and Kloesel, *Peirce, Semeiotic, and Pragmatism: Essays of Max Fisch*, 35–78; subsequent references will be to the reprint edition.

12. George Bruce Halsted to Charles S. Peirce, 14 March 1878, DCG Papers, Coll #1 Corresp., JHU.

13. Charles S. Peirce to J. J. Sylvester, 19 March 1878, DCG Papers, Coll #1 Corresp., JHU (Peirce's emphasis). Passages from these letters were also quoted in Fisch and Cope, in *Peirce, Semeiotic, and Pragmatism*, 54–55.

14. Daniel Coit Gilman to Charles S. Peirce, 20 March 1878, DCG Papers, Letter Book 3, Ser. 4, Box 4.1, JHU. Unfortunately, this was the same letter in which Gilman informed Peirce that the University could do no additional hiring.

15. J. J. Sylvester to Daniel Coit Gilman, 29 March 1878, DCG Papers, Coll #1 Corresp., JHU.

16. Fisch and Cope as well as Brent have given these archives a more pessimistic reading relative to the immediate impact of the Halsted incident to the relationship between Peirce and Sylvester. The difference in our readings seems to hinge on the letter Sylvester wrote to Gilman on the twenty-ninth, informing him of his impending visit to Peirce. Neither Fisch and Cope nor Brent seem aware of this visit. That Peirce and Sylvester *did* have a major falling out in 1882 and into 1883 is uncontested. See "From Number Theory to 'Universal Algebra.'"

17. J. J. Sylvester to Daniel Coit Gilman, 7 June 1878, DCG Papers, Coll #1 Corresp., JHU.

18. George Bruce Halsted to Daniel Coit Gilman, 3 September 1878, DCG Papers, Coll #1 Corresp., JHU.

19. George Bruce Halsted to Daniel Coit Gilman, 16 October [1878], DCG Papers, Coll #1 Corresp., JHU.

20. If Halsted and Sylvester had parted ways in 1878 on less than cordial terms, the relationship was repaired. The two remained in touch, with Sylvester even giving Halsted a valuable selection of letters from his correspondence. See George Bruce Halsted to David Eugene Smith, 10 December 1912, D. E. Smith Professional Papers, Rare Book and Manuscript Library, Columbia University. For his part, Halsted spoke glowingly of Sylvester and his achievement at the Johns Hopkins in a letter to Florian Cajori late in 1888. He summed up with the comment: "That the presence of such a man in America was epoch-making is not to be wondered at. His loss to us was a national misfortune." George Bruce Halsted to Florian Cajori, 25 December 1888 as quoted in Cajori, *The Teaching and History of Mathematics in the United States*, 265.

21. W. Irving Stringham to Daniel Coit Gilman, 20 May 1880, DCG Papers, Coll #1 Corresp., JHU.

22. Parshall and Rowe, *Emergence of the American Mathematical Research Community*, 191–192 and 255.

23. W. Irving Stringham to Daniel Coit Gilman, 28 April 1882, DCG Papers, Coll #1 Corresp., JHU. Actually, the summer of 1882 was the only summer that Sylvester did not spend in England during his tenure at Hopkins. See "A Troubled Transition to the Theory of Partitions."

24. See Epigraph note.

25. This information on the graduate program was gleaned from Brown, *Johns Hopkins Half-Century Directory*. Davis took several of Sylvester's classes after he joined the department in the fall of 1881, but he was more directly a student of Story's and officially earned his Ph.D. after Sylvester's departure.

26. For a brief account of Ladd's life and work, see Green, "Christine Ladd-Franklin (1847–1930)," in *Women of Mathematics*, 121–128.

27. Christine Ladd to J. J. Sylvester, 27 March 1878, DCG Papers, Coll #1 Corresp., JHU, in Parshall, *JJS: Life and Work in Letters*, 188.

28. Bodichon, chief benefactor of the women's college, Girton, at Cambridge (f. 1872), became personally interested in the case of the mathematically gifted, Jewish-born Sara Marks. Bodichon, who came to view Marks as an adopted daughter, funded her studies at Girton from 1876 to 1880. Sylvester had come to know of and take an interest in Marks as early as 1875. See Sylvester to Bodichon, 19 September 1875 and 2 November 1876, in Parshall, *JJS: Life and Work in Letters*, 145 and 163, respectively. For Sara Marks (who later changed her name to Hertha and married William Ayrton), see Tattersall and McMurran, "Hertha Ayrton: A Persistent Experimenter," and Hirsch, *Barbara Leigh Smith Bodichon*, chapters 15 and 16.

29. J. J. Sylvester to Daniel Coit Gilman, 2 April 1878, DCG Papers, Coll #1 Corresp., JHU, in Parshall, *JJS: Life and Work in Letters*, 189.

30. Ibid.

31. Daniel Coit Gilman to Christine Ladd, 26 April 1878, DCG Papers, Coll #1 Corresp., JHU.

32. On the concept of women as "invisible students" in American institutions of higher education, see Rossiter, *Women Scientists in America*, 29–33. On Ladd, the Hopkins Department of Mathematics, and the issue of coeducation, compare Parshall and Rowe, *Emergence of the American Mathematical Research Community*, 84–86.

33. Mitchell is perhaps best remembered, however, for the work in logic that he did in Peirce's course. See Mitchell, "A New Algebra of Logic."

34. *Johns Hopkins University Circulars*, 1 (December 1879):2.

35. Recall the discussion of these results in chapter 4.

36. See Sylvester, "Recherches sur les solutions en nombres entiers positifs ou négatifs de l'équation cubique homogène à trois variables" (1856).

37. Lucas, "Sur l'analyse indéterminée du troisième degré.—Démonstration de pluiseurs théorèmes de M. Sylvester," 182 (my translation).

38. Sylvester, "On Certain Ternary Cubic-Form Equations" (1879–1880), or *Math. Papers JJS*, 3:313–391. For an exhaustive historical and mathematical treatment of this work of Sylvester and Lucas in Diophantine analysis, see Lavrinenko, "Solving an Indeterminate Third Degree Equation in Rational Numbers."

39. *Johns Hopkins University Circulars*, 1 (February 1880), 24.

40. Ibid. For the works, see Lejeune Dirichlet, *Vorlesungen über Zahlentheorie*, 2d ed.; Legendre, *Théorie des nombres*; and Bachmann, *Die Lehre von der Kreistheilung und ihre Beziehungen zur Zahlentheorie*.

41. Mitchell, "On Binomial Congruences; Comprising an Extension of Fermat's and Wilson's Theorems, and a Theorem of Which Both Are Special Cases."

42. As one instance of this influence, compare ibid., 295, and Sylvester, "On Certain Ternary Cubic-Form Equations" (1879–1880), in *Math. Papers JJS*, 3:337 for joint use of Sylvester's term "totitives" for the primes less than a given number. The "totient" was the total number of totitives, or what would today be called the value of the Euler ϕ-function.

43. Mitchell, "Some Theorems in Numbers."

44. Ladd, "On the Algebra of Logic."

45. Hopkins only awarded Ladd her degree in 1926 on the occasion of the fiftieth anniversary of the University's founding. The first Hopkins Ph.D. granted to a woman went to Florence Bascom in 1893 for her work in geology. Compare Rossiter, *Women Scientists in America*, 45.

46. Ely, "Bibliography of Bernoulli Numbers."

47. Ely, "Some Notes on the Numbers of Bernoulli and Euler," 337.

48. Ely remained in his professorship for just one year, leaving it to take the examinership at the United States Patent Office he would hold until his death in 1917.

49. See Sylvester, "On Certain Ternary Cubic-Form Equations" (1879–1880), in *Math. Papers JJS*, 3:345 for Franklin's proof. Franklin revisited cubic curves several years later. See Franklin, "On Cubic Curves," for an exposition of known results—some of them from Sylvester's paper—from a new point of view.

50. The discussion that follows draws from Parshall, "America's First School of Mathematical Research," and Parshall and Rowe, *Emergence of the American Mathematical Research Community*, 118–123.

51. Legendre, *Théorie des nombres*, 2:128–133.

52. Franklin, "Sur le développement du produit infini $(1 - x)(1 - x^2)(1 - x^3)$ $(1 - x^4)\ldots$"

53. Ferrers actually never published his result or its proof. Rather, he entrusted it to Sylvester for publication. It appeared in Sylvester, "On Mr. Cayley's Impromptu Demonstration of the Rule for Determining at Sight the Degree of Any Symmetrical Function of the Roots of an Equation Expressed in Terms of the Coefficients" (1853), or *Math. Papers JJS*, 1:597.

54. For the mathematical details, see Parshall and Rowe, *Emergence of the American Mathematical Research Community*, 120–123.

55. See Story, "On the Theory of Rational Derivation on a Cubic Curve." Lavrinenko discusses Story's work at length and indicates how the work of both Story and Sylvester formed part of the theory of elliptic curves.

56. J. J. Sylvester to Arthur Cayley, 19 January 1881, JJS Papers, SJC, Box 11. The first three quotes in the next paragraph are also from this letter (his emphasis).

57. King, *Benjamin Peirce: A Memorial Collection*, 14.

58. Sylvester had also tried to convince another mathematical confidant, Hermite, to come to Baltimore, for in a letter received on 3 January 1881 Hermite had confessed to Sylvester of being "so touched by [his] invitation" to join him in Baltimore. Nevertheless, he had to decline because it would be absolutely beyond his powers to lecture in English. See Charles Hermite to J. J. Sylvester, 3 January 1881, JJS Papers, SJC, Box 2 (my translation).

59. J. J. Sylvester to Simon Newcomb, 13 January 1881, in Archibald, "Unpublished Letters of James Joseph Sylvester," 145.

60. J. J. Sylvester to Charles E. Norton, 20 February 1881, in ibid., 146.

61. For the quotes, see Williams, *Diary and Letters of Rutherford Birchard Hayes*, 3:644. Sylvester had accepted Gilman's invitation to the event in a letter dated 9 February 1881. See DCG Papers, Coll #1 Corresp., JHU.

62. See J. J. Sylvester to Arthur Cayley, 23 March 1881, JJS Papers, SJC, Box 11.

63. J. J. Sylvester to Arthur Cayley, 26 March 1881, JJS Papers, SJC, Box 11 (his emphasis).

64. Chebyshev, "Mémoire sur les nombres premiers." Morris Kline gave this more modern interpretation of Chebyshev's result in *Mathematical Thought from Ancient to Modern Times*, 830.

65. See Sylvester, "On Tchebycheff's Theory of the Totality of the Prime Numbers Comprised within Given Limits" (1881), or *Math. Papers JJS*, 3:530–545, and compare the discussion of this work in the context of Sylvester's later researches in matrix algebra in Parshall, "Joseph H. M. Wedderburn and the Structure Theory of Algebras," 243–250.

66. William P. Durfee to Florian Cajori, as reported in Cajori, *Teaching and History of Mathematics in the United States*, 267. The quotes that follow in this paragraph are

also from this page. This passage was also quoted in Parshall and Rowe, *Emergence of the American Mathematical Research Community*, 80.

67. See Sylvester, "Lectures on the Principles of Universal Algebra" (1884), or *Math. Papers JJS*, 4:209.

68. Peirce, "Linear Associative Algebra."

69. Much has been written on Peirce's work and its place in the history of mathematics. See, for example, Pycior, "Benjamin Peirce's 'Linear Associative Algebra,'" and Parshall, "Joseph H. M. Wedderburn and the Structure Theory of Algebras," 250–258.

70. On Charles Peirce's contributions to his father's work on the theory of algebras, see, for example, Hawkins, "Hypercomplex Numbers, Lie Groups, and the Creation of Group Representation Theory," 246, and Parshall, "Joseph H. M. Wedderburn and the Structure Theory of Algebras," 258–261.

71. Charles S. Peirce to J. J. Sylvester, 4 January 1882, JJS Papers, SJC, Box 3.

72. Charles S. Peirce to J. J. Sylvester, 5 January 1882, JJS Papers, SJC, Box 3, in Parshall, *JJS: Life and Work in Letters*, 205. The two quotes that follow are also from this letter, pp. 206 and 206–207, respectively.

73. Charles S. Peirce to J. J. Sylvester, 6 January 1882, JJS Papers, SJC, Box 3 (Peirce's emphasis).

74. See, for example, Sylvester, "Sur les puissances et les racines de substitutions linéaires" (1882), or *Math. Papers JJS*, 3:562–567.

75. Sylvester, "Sur les racines des matrices unitaires" (1882), or *Math. Papers JJS*, 3:565–567. Note that Sylvester's sense of the word "unitary" here is not the modern sense.

76. Ibid., 567 (my translation).

77. For more on Sylvester's work in this direction, see Parshall, "Joseph H. M. Wedderburn and the Structure Theory of Algebras," 243–250, especially on pp. 243–245.

78. See Cayley, "A Memoir on the Theory of Matrices," or *Math. Papers AC*, 2:475–496. In fact, Cayley had written personally to Sylvester in November 1857 to convey the paper's key result, the so-called Hamilton-Cayley Theorem. (Arthur Cayley to J. J. Sylvester, 19 November 1857, JJS Papers, SJC, Box 2, in Parshall, *JJS: Life and Work in Letters*, 95–96.) For more on the historical context of this work, see Hawkins, "Another Look at Cayley and the Theory of Matrices," and Crilly, "Cayley's Anticipation of a Generalised Cayley-Hamilton Theorem."

79. J. J. Sylvester (writing in French) to Charles Hermite, 15 January 1882, Archives de l'Académie des Sciences de Paris, Pochette de séance, 13 février 1882 (my translation). The second paper to which Sylvester referred is Cayley, "A Supplementary Memoir on the Theory of Matrices," or *Math. Papers AC*, 5:438–448.

80. Arthur Cayley to Thomas Hirst, 31 March 1882, UCL Archives, LMS Papers, Cayley, A. C.

81. William Pitt Durfee to Florian Cajori, as quoted in Cajori, *Teaching and History of Mathematics in the United States*, 268.

82. J. J. Sylvester to Arthur Cayley, 1 February 1883, JJS Papers, SJC, Box 11.

83. Houser, "Introduction," *Writings of Charles S. Peirce*, 4:liii. See ibid., 334–335, for the text of Peirce's result.

84. Sylvester, "A Word on Nonions" (1882), or *Math. Papers JJS*, 3:647–650. For discussions of this and other of Sylvester's matrix-theoretic work in historical context, see Parshall, "Joseph H. M. Wedderburn and the Structure Theory of Algebra," 243–248, and the commentary in Parshall, *JJS: Life and Work in Letters*, 205–207.

85. Sylvester, "Erratum" (1883), 46.

86. Sylvester, "A Word on Nonions" (1882), in *Math. Papers JJS*, 3:649.

87. Sylvester, "Erratum" (1883) (his emphasis).

88. Charles S. Peirce to Daniel Coit Gilman, 27 March 1883, DCG Papers, Coll #1 Corresp., JHU. The next quote is also from this letter.

89. Peirce, "A Communication from Mr. Peirce."

90. Sylvester, "A Note from Professor Sylvester." For even more details on this nonionic blow-up, see Fisch and Cope, in *Peirce, Semeiotic, and Pragmatism*, 57–62, and Houser, "Introduction," in *Writings of Charles S. Peirce*, 4:lvi–lviii.

91. The same cannot be said for Peirce, who, while as prickly as Sylvester, could long hold a grudge. Peirce had still not forgotten the incident many years later. See the passage quoted in Houser, "Introduction," in *Writings of Charles S. Peirce*, 4:lviii.

92. J. J. Sylvester to Arthur Cayley, 12 May 1881, JJS Papers, SJC, Box 11.

93. J. J. Sylvester to Daniel Coit Gilman, 4 July 1882, DCG Papers, Coll #1 Corresp., JHU. The following two quotes in this paragraph are also from this letter.

94. J. J. Sylvester to Arthur Cayley, 3 August 1882, JJS Papers, SJC, Box 11 (his emphasis). For the report, see *Report of the Fifty-second Meeting of the British Association for the Advancement of Science*, 37–38. Indicative of Sylvester's mental state, in the report, he credited "Messers. Healy and Durfee (fellows), under the able superintendence of Dr. F. Franklin (associate) of the Johns Hopkins University" (p. 37). The Mr. Healy was, in fact, George Ely.

95. J. J. Sylvester to Daniel Coit Gilman, 10 August 1882, DCG Papers, Coll #1 Corresp., JHU, in Parshall, *JJS: Life and Work in Letters*, 209.

96. See Daniel Coit Gilman to Sir William Thomson, 13 August 1882, as reprinted in Thompson, *The Life of William Thomson*, 2:811.

97. Daniel Coit Gilman to J. J. Sylvester, 13 August 1882, JJS Papers, SJC, Box 2.

98. Compare Franklin, *Life of Daniel Coit Gilman*, 266. There, Franklin writes: "It was plain in many ways to their contemporaries at the University that Mr. Gilman felt a full and genuine sympathy with Sylvester's intellectual ardor and a true appreciation of the character of his achievements, though in a field so remote from the apprenhension of any except advanced mathematicians. It required something more than tact to maintain unimpaired the relation of hearty coöperation which existed throughout between the organizing head of the University and the splendid but erratic genius whose presence furnished so much of the inspiration of its early years."

99. Daniel Coit Gilman to J. J. Sylvester, 16 August 1882, JJS Papers, SJC, Box 2, in Parshall, *JJS: Life and Work in Letters*, 210.

100. J. J. Sylvester to Daniel Coit Gilman, 23 August 1882, DCG Papers, Coll #1 Corresp., JHU, in Parshall, *JJS: Life and Work in Letters*, 211 (his emphasis).

101. Ibid., 212 (his emphasis). Indeed, the 1882 fifth volume of the journal was so late in coming out that it was followed only in 1884 by the sixth volume.

102. Daniel Coit Gilman to J. J. Sylvester, 1 September 1882, DCG Papers, Coll #1 Corresp., JHU.

103. See J. J. Sylvester to Arthur Cayley, 23 March, 1881, JJS Papers, SJC, Box 11.

104. J. J. Sylvester to Arthur Cayley, 6 September 1882, JJS Papers, SJC, Box 11, in Parshall, *JJS: Life and Work in Letters*, 215. The paper in question ultimately appeared as Sylvester, "On Subinvariants, That Is, Semi-Invariants to Binary Quantics of an Unlimited Order" (1882), or *Math. Papers JJS*, 3:568–622. The mathematical details of this work are given in the commentary in Parshall, *JJS: Life and Work in Letters*, 213–216.

105. J. J. Sylvester to Arthur Cayley, 6 October 1882, JJS Papers, SJC, Box 11, in Parshall, *JJS: Life and Work in Letters*, 217–218 (his emphasis). For more on Sylvester's repeated

attempts to prove Gordan's theorem, see Parshall, "Toward a History of Nineteenth-Century Invariant Theory," and compare the commentary in Parshall, *JJS: Life and Work in Letters*, 213–219. Neither Sylvester nor any other adherent of the British approach to invariant theory ever succeeded in obtaining this desired proof.

106. Sylvester to Cayley, 6 October 1882, in Parshall, *JJS: Life and Work in Letters*, 218.

107. Ibid., 219. Spencer also mentioned the visit to Baltimore and seeing the University in Sylvester's company in Spencer, *An Autobiography*, 2:471.

108. Sylvester to Cayley, 6 October 1882, in Parshall, *JJS: Life and Work in Letters*, 218. The book in question was Netto, *Substitutionentheorie und ihre Anwendungen auf die Algebra*.

109. Ellery W. Davis to Florian Cajori, 25 December 1888, quoted in Cajori, *Teaching and History of Mathematics in the United States*, 265.

110. J. J. Sylvester to Arthur Cayley, 22 October 1882, JJS Papers, SJC, Box 11.

111. See Sylvester, "On the Fundamental Theorem in the New Method of Partitions" (1882), or *Math. Papers JJS*, 3:658–660.

112. Sylvester, "On Subinvariants" (1882), in *Math. Papers JJS*, 3:605–622.

113. J. J. Sylvester to Arthur Cayley, 25 December 1882, JJS Papers, SJC, Box 11.

114. Sylvester, "A Constructive Theory of Partitions" (1882), or *Math. Papers JJS*, 4:1–83 on p. 1 (his emphasis).

115. For a technical discussion of some of the results contained in this paper, see Parshall, "America's First School of Mathematical Research," 184–188. The discussion in Parshall and Rowe, *Emergence of the American Mathematical Research Community*, 125–128, followed this source. Additional technical discussion may be found in Parshall, *JJS: Life and Work in Letters*, in the commentary on pp. 223–225.

116. J. J. Sylvester to Arthur Cayley, 16 March 1883, JJS Papers, SJC, Box 11, in Parshall, *JJS: Life and Work in Letters*, 223 (his emphasis). Arthur S. Hathaway had joined the program in the fall of 1879 and won one of the Hopkins fellowships for 1882–1883 and then again for 1883–1884, but ultimately left the program without a degree in 1885 to assume an instructorship in mathematics at Cornell. The question mark after Durfee's name may suggest Sylvester's doubts as to Durfee's engagement in this particular issue.

117. Arthur Cayley to J. J. Sylvester, 12 February 1883, JJS Papers, SJC, Box 2.

118. Recall the discussion in chapter 8.

119. J. J. Sylvester to Arthur Cayley, 16 March 1883, JJS Papers, SJC, Box 11, in Parshall, *JJS: Life and Work in Letters*, 222. The quotations that follow in this paragraph are all from this letter.

120. For the Oxford salary, see the *Oxford University Gazette*, 14 (12 October 1883), 26. The Oxford salary was roughly comparable to what professors at Harvard were making.

121. J. J. Sylvester to Arthur Cayley, 10 April 1883, JJS Papers, SJC, Box 11, in Parshall, *JJS: Life and Work in Letters*, 225–227.

122. Simon Newcomb had been at work behind the scenes on recognition for Sylvester by the Academy at least as early as 1878, but questions surrounding Sylvester's intentions regarding citizenship had presented obstacles. See Simon Newcomb to Daniel Coit Gilman, 11 April 1879, DCG Papers, Coll #1 Corresp., JHU.

123. J. J. Sylvester to Daniel Coit Gilman, 20 May 1883, DCG Papers, Coll #1 Corresp., JHU.

124. J. J. Sylvester to Daniel Coit Gilman, 23 May 1883, DCG Papers, Coll #1 Corresp., JHU.

125. J. J. Sylvester to Daniel Coit Gilman, 19 June 1883, DCG Papers, Coll #1 Corresp., JHU.

126. J. J. Sylvester to Arthur Cayley, 10 July 1883, JJS Papers, SJC, Box 11.

127. J. J. Sylvester to Simon Newcomb, 26 July 1883, in Archibald, "Unpublished Letters of James Joseph Sylvester," 150 (Sylvester's emphasis).

128. J. J. Sylvester to Daniel Coit Gilman, 8 August 1883, DCG Papers, Coll #1 Corresp., JHU.

129. J. J. Sylvester to Arthur Cayley, 22 August 1883, JJS Papers, SJC, Box 11.

130. J. J. Sylvester to Arthur Cayley, 22 September 1883, JJS Papers, SJC, Box 11.

131. For much more on this, see Parshall and Rowe, *Emergence of the American Mathematical Research Community*, 138–143.

132. In fact, with this appointment, Sylvester became the first professing Jew to hold a professorship at Oxbridge. See Roth, "The Jews in English Universities," 114.

133. See J. J. Sylvester to Mrs. Daniel Coit Gilman, 7 December 1883, DCG Papers, Coll #1 Corresp., JHU.

134. Trustees of The Johns Hopkins University to J. J. Sylvester, 3 December 1883, DCG Papers, Coll #1 Corresp., JHU.

135. J. J. Sylvester to Arthur Cayley, 13 December 1883, JJS Papers, SJC, Box 11, in Parshall, *JJS: Life and Work in Letters*, 233.

136. These appeared as Sylvester, "Lectures on the Principles of Universal Algebra" (1884). For a discussion in historical context of this paper and some of Sylvester's other work on matrices in the fall of 1883, see Parshall, "Joseph H. M. Wedderburn and the Structure Theory of Algebras," 247–249.

137. "Remarks of Prof. Sylvester, at a Farewell Reception Tendered to Him by the Johns Hopkins University, Dec. 20, 1883 (Reported by Arthur S. Hathaway)," 24 typescript pages, DCG Papers, Coll #1 Corresp., JHU. The quote appears on p. 21.

Chapter 11

Epigraph (1): JJS Papers, SJC, Box 12, in Parshall, *JJS: Life and Work in Letters*, 256.

Epigraph (2): DCG Papers, Coll #1 Corresp., JHU, in Parshall, *JJS: Life and Work in Letters*, 264.

1. "Remarks of Prof. Sylvester, at a Farewell Reception Tendered to Him by the Johns Hopkins University, December 20, 1883 (Reported by Arthur S. Hathaway)," 17. For a brief overview of Sylvester's time in the Savilian chair, see Fauvel, "Sylvester at Oxford," in *Oxford Figures*, 219–239.

2. Engel, "Emerging Concepts of the Academic Profession at Oxford 1800–1854," 328–335.

3. Engel, *From Clergyman to Don*, 56–70.

4. Engel, "Emerging Concepts of the Academic Profession at Oxford," 346–347.

5. Ibid., 348–349. For more details on the Commission of 1877, see Engel, *From Clergyman to Don*, 156–201.

6. Ryan, "Transformation, 1850–1914," 77.

7. Hayter, *Spooner: A Biography*, 28; and Ryan, "Transformation, 1850–1914," 82–84.

8. Ryan, "Transformation, 1850–1914," 84–85.

9. Ibid., 91, and Jackson-Stops, "Restoration and Expansion: The Buildings since 1750," 245–247.

10. Ryan, "Transformation, 1850–1914," 91.

11. *DNB*, 2d supp., s.v. "George, Hereford Brooke" by Robert S. Rait, and Ryan, "Transformation, 1850–1914," 75.

12. Ryan, "Transformation, 1850–1914," 74.

13. Minutes of the College Meetings (hereinafter cited as Minutes NCO) from 23 January 1878 to 31 January 1889 (item 3499) and from 7 January 1890 ff (item 3507), Archives, New College Oxford Library.

14. Hayter, *Spooner: A Biography*, 11–12.

15. Fisher, *An Unfinished Autobiography*, 48.

16. Ryan, "Transformation, 1850–1914," 87, and Minutes NCO, items 3499 and 3507, Archives, New College Oxford Library. Spooner is best remembered today for the linguistic snafu—the spoonerism—to which he so often and amusingly fell victim.

17. Oman, *Memories of Victorian Oxford*, 91, and Fisher, *An Unfinished Autobiography*, 46.

18. Ryan, "Transformation, 1850–1914," 86.

19. Hayter, *Spooner: A Biography*, 41.

20. Ryan, "Transformation, 1850–1914," 86.

21. J. J. Sylvester to Daniel Coit Gilman, 3 January 1884, DCG Papers, Coll #1 Corresp., JHU.

22. The Oxford academic year was divided into three terms: Michaelmas, or the fall term; Hilary, or the winter term; and Trinity, or the spring term. The Trinity term was further subdivided into two consecutive, three-week-long terms called Easter and Acts.

23. J. J. Sylvester to his niece, Edith Gigliucci, 21 January 1884, UCL Archives, MS.ADD.221.

24. J. J. Sylvester to Charles Taylor, 20 January 1884, SJC Archives, W.1.

25. J. J. Sylvester to Arthur Cayley, 22 January 1884, JJS Papers, SJC, Box 12, in Parshall, *JJS: Life and Work in Letters*, 240.

26. Ibid., 241 (Sylvester's emphasis).

27. J. J. Sylvester to Arthur Cayley, 24 January 1884, JJS Papers, SJC, Box 12 (his emphasis), in Parshall, *JJS: Life and Work in Letters*, 241.

28. J. J. Sylvester to Arthur Cayley, 20 January 1884, JJS Papers, SJC, Cambridge, Box 12.

29. For more on these efforts to secure Klein for Hopkins, see Parshall and Rowe, *Emergence of the American Mathematical Research Community*, 138–144.

30. J. J. Sylvester to Arthur Cayley, 3 February 1884, JJS Papers, SJC, Box 12, in Parshall, *JJS: Life and Work in Letters*, 243.

31. Ibid., 244 (his emphasis). Clearly, Sylvester knew so little about Lindemann that he did not even know how properly to spell his name! He did, however, continue to make inquiries about him for Gilman throughout the spring and into the summer of 1884.

32. Daniel Coit Gilman to J. J. Sylvester, 7 February 1884, DCG Papers, Letter Book 4, Ser. 4, Box 4.2, JHU.

33. J. J. Sylvester to Daniel Coit Gilman, 14 March 1884, DCG Papers, Coll #1 Corresp., JHU.

34. Ultimately, Newcomb did succeed Sylvester at Hopkins, but as Professor of Mathematics *and* Astronomy. Newcomb not only divided the position, but he also divided his time between Baltimore and Washington, where he continued to head the U.S. Nautical Almanac Office. It was not a happy choice in the end for the Hopkins mathematics program. Compare Parshall and Rowe, *Emergence of the American Mathematical Research Community*, 143–145.

35. Sylvester, "On the Three Laws of Motion in the World of Universal Algebra" (1884), or *Math. Papers JJS*, 4:146–151. For more specifics on what Sylvester had in mind, see the commentary in Parshall, *JJS: Life and Work in Letters*, 232.

36. Sylvester to Gilman, 3 January 1884. At least one person did follow up on Sylvester's ideas here. Arthur Buchheim, a student of Henry Smith at Oxford and of Felix Klein in Leipzig, corresponded with Sylvester about his matrix-theoretic ideas in 1884. He published a "Proof of Prof. Sylvester's 'Third Law of Motion,'" in the *Philosophical Magazine*. For the correspondence, see JJS Papers, SJC, Box 2.

37. Sylvester to Taylor, 20 January 1884 (his emphasis).

38. J. J. Sylvester to Arthur Cayley, 30 January 1884, JJS Papers, SJC, Box 12, and Sylvester, "Sur les quantités formant un groupe de nonions analogues aux quaternions de Hamilton" (1884), or *Math. Papers JJS*, 4:154–159.

39. See the six papers published in the *Comptes rendus*, in *Math. Papers JJS*, 4:176–207, and the two papers published in the *Philosophical Magazine*, in *Math. Papers JJS*, 4:225–235. Compare the discussion in Parshall, "Joseph H. M. Wedderburn and the Structure Theory of Algebras," 246–249. For the definition, recall chapter 10.

40. Sylvester, "Lectures on the Principles of Universal Algebra" (1884), in *Math. Papers JJS*, 4:224.

41. MacMahon, "Seminvariants and Symmetric Functions," or MacMahon, *Collected Papers*, 2:931–939 (hereinafter cited as *Collected Papers PAM*).

42. Sylvester, "On Subinvariants" (1882).

43. See the commentary in Parshall, *JJS: Life and Work in Letters*, 214–215 and 248–249, for further technical details.

44. J. J. Sylvester to Arthur Cayley, 2 March 1884, JJS Papers, SJC, Box 12.

45. J. J. Sylvester to Arthur Cayley, 30 March 1884, JJS Papers, SJC, Box 12, in Parshall, *JJS: Life and Work in Letters*, 249. The paper to which he alluded is the 1882 paper "On Subinvariants."

46. Sylvester, "Sur la correspondance entre deux espèces différentes de fonctions de deux systèmes de quantités" (1884), and "Note on Captain MacMahon's Transformation of the Theory of Invariants" (1884), or *Math. Papers JJS*, 4:163–165 and 236–237, respectively.

47. Sylvester, "Sur la correspondance entre deux espèces différentes de fonctions de deux systèmes de quantités" (1884), in *Math. Papers JJS*, 4:164 (my translation). Compare also Parshall, *JJS: Life and Work in Letters*, 249. This concentration on invariant theory also brought with it yet another (ultimately failed) claim of a British-style proof of Gordan's finiteness theorem, this time using the Correspondence Theorem. See J. J. Sylvester to Arthur Cayley, 31 March 1884, JJS Papers, SJC, Box 12.

48. J. J. Sylvester to Arthur Cayley, 29 January 1884, JJS Papers, SJC, Box 12, in Parshall, *JJS: Life and Work in Letters*, 242 (his emphasis).

49. J. J. Sylvester to Arthur Cayley, 22 January 1884, JJS Papers, SJC, Box 12. The "Townsend" to which he referred was Richard Townsend, *Chapters on the Modern Geometry of the Point, Line, and Circle*.

50. J. J. Sylvester to Arthur Cayley, 5 February 1884, JJS Papers, SJC, Box 12.

51. Cremona's book had originally appeared in Italian in 1873. Its French translation, *Éléments de géométrie projective*, was published in Paris by Gauthier-Villars two years later.

52. J. J. Sylvester to Arthur Cayley, 20 May 1884, JJS Papers, SJC, Box 12. Recall that drawing on the board had been one of Sylvester's problems in the chair of natural philosophy at University College London. His deficiency had prompted him then to take drawing lessons.

53. J. J. Sylvester to Arthur Cayley, 27 May 1884, JJS Papers, SJC, Box 12. During Cremona's visit, in fact, Sylvester was instrumental in laying the groundwork for the

English translation of Cremona's book that Charles Leudesdorf, mathematics tutor and fellow of Pembroke College, Oxford, would publish in 1885. See Cremona, *Elements of Projective Geometry.*

54. J. J. Sylvester to "President Gilman, and the other signatories of the letter of April 28, 1884," 18 June, 1884, DCG Papers, Coll #1 Corresp., JHU.

55. J. J. Sylvester to Arthur Cayley, 5 August 1884, JJS Papers, SJC, Box 12. The next quote is also from this letter.

56. J. J. Sylvester to Arthur Cayley, 4 and 7 September 1884, JJS Papers, SJC, Box 12.

57. J. J. Sylvester to Arthur Cayley, 20 January 1884, JJS Papers, SJC, Box 12.

58. J. J. Sylvester to Arthur Cayley, 20 May 1884, JJS Papers, SJC, Box 12, in Parshall, *JJS: Life and Work in Letters,* 252–253.

59. J. J. Sylvester to Arthur Cayley, 14 July 1884, JJS Papers, SJC, Box 12.

60. Sylvester to Cayley, 4 September 1884.

61. Sylvester, "The Genesis of an Idea" (1884), 35. This paper was omitted from *Math. Papers JJS.*

62. J. J. Sylvester to Arthur Cayley, 30 September 1884, JJS Papers, SJC, Box 12 (his emphasis). The next quote is also from this letter.

63. Sylvester, "Sur la solution explicite de l'équation quadratique de Hamilton en quaternions ou en matrices du second ordre" (1884), or *Math. Papers JJS,* 4:188–198; "The Genesis of an Idea"; and Poincaré, "Sur les nombres complexes." For a discussion of the import especially of Poincaré's realization, see Hawkins, "Hypercomplex Numbers, Lie Groups, and the Creation of Group Representation Theory," 249, and Parshall, "Joseph H. M. Wedderburn and the Structure Theory of Algebras," 261–263.

64. Sylvester to Arthur Cayley, 2 November 1884, JJS Papers, SJC, Box 12, in Parshall, *JJS: Life and Work in Letters,* 254. The text was Wilhelm Fiedler's German translation and adaptation of George Salmon's *A Treatise on Conic Sections* originally published as *Analytische Geometrie der Kegelschnitte von George Salmon.*

65. Sylvester to Cayley, 2 November 1884.

66. J. J. Sylvester to Arthur Cayley, 8 November 1884, JJS Papers, SJC, Box 12, in Parshall, *JJS: Life and Work in Letters,* 255–256. The quotes that follow in this paragraph are all from this letter. Compare also Epigraph note 1.

67. J. J. Sylvester to Arthur Cayley, 16 November 1884, JJS Papers, SJC, Box 12.

68. J. J. Sylvester to Daniel Coit Gilman, 2 January 1885, DCG Papers, Coll #1 Corresp., JHU.

69. Pamphlet of four poems addressed "To the Warden of New College with the author's kind regards" in the Archives, New College Library, Oxford.

70. Ibid.

71. J. J. Sylvester to Edith Gigliucci, 12 January 1885, UCL Archives, MS.ADD.221.

72. J. J. Sylvester to Arthur Cayley, 23 February and 22 April 1885, JJS Papers, SJC, Box 12.

73. Ibid. The quote is in the 22 April letter.

74. J. J. Sylvester to Arthur Cayley, 14 April 1885, JJS Papers, SJC, Box 12. The research in question is MacMahon, "Memoir on Semivariants," or *Collected Papers PAM,* 2:566–583.

75. Hammond, "On the Solution of the Differential Equations of Sources," 225–226.

76. J. J. Sylvester to Arthur Cayley, 11 May 1885, JJS Papers, SJC, Box 12 (his emphasis). For further technical details, compare Parshall, *JJS: Life and Work in Letters,* 177–178, 214–215.

77. Sylvester withdrew MacMahon's name in the fall when Stokes proposed his former student, Peter Guthrie Tait. Tait won the medal in 1886.

78. J. J. Sylvester to Daniel Coit Gilman, 24 August 1885, DCG Papers, Coll #1 Corresp., JHU, in Parshall, *JJS: Life and Work in Letters*, 256–258. The quotes that follow in this paragraph are all from this letter.

79. J. J. Sylvester to Arthur Cayley, 4 July 1885, JJS Papers, SJC, Box 12. The quotes that follow in this paragraph are all from this letter.

80. Sylvester finally did get the report written up and submitted, but not until 17 November 1885. See Referee's Reports, RR.9.288, Royal Society of London. His report, like Cayley's, was favorable, and Kempe's paper appeared as "A Memoir on the Theory of Mathematical Form."

81. See Sylvester, "Inaugural Lecture at Oxford" (1885), or *Math. Papers JJS*, 4:299, and Forsyth, *Treatise on Differential Equations*.

82. J. J. Sylvester to Arthur Cayley, 9 November 1885, JJS Papers, SJC, Box 12.

83. Archives de l'Académie des Sciences de Paris, Pochette de séance, 23 novembre 1885. The paper appeared as Sylvester, "Sur une nouvelle théorie de formes algébriques" (1885), or *Math. Papers JJS*, 4:242–251. For the quote, see J. J. Sylvester to Daniel Coit Gilman, 28 November 1885, DCG Papers, Coll #1 Corresp., JHU, in Parshall, *JJS: Life and Work in Letters*, 259 (his emphasis).

Actually, the "new theory" was only new to Sylvester. He had discovered it independently of results Georges Halphen had presented in his doctoral dissertation to the Paris Faculty of Science in 1878. On learning from Halphen of his earlier work, Sylvester blamed his "long exile in America" for the oversight, at the same time that he declared that "there is enough difference between the goal and the course of my results in this field and those of M. Halphen to justify" his ongoing efforts. Sylvester, "Sur une nouvelle théorie de formes algébriques" (1885), in *Math. Papers JJS*, 4:245 (my translation). For the pertinent references to Halphen's work, see Parshall, *JJS: Life and Work in Letters*, 259–260.

84. It is not clear whether Cayley was able to make the trip. Sylvester had most definitely invited him and was frustrated on 28 November when he wrote that "you do not say . . . whether there is *any chance* of my having the pleasure of seeing you here for the lecture." J. J. Sylvester to Arthur Cayley, 28 November 1885, JJS Papers, SJC, Box 12 (his emphasis).

85. Sylvester, "Inaugural Lecture at Oxford" (1885), in *Math. Papers JJS*, 4:280.

86. Ibid., 281. The next quote is also from this page.

87. Ibid., 284. The next quote is also from this page.

88. Compare the commentary in Parshall, *JJS: Life and Work in Letters*, 259. Sylvester named the reciprocant in (1) the Schwarzian after the contemporary German mathematician, Hermann Amandus Schwarz, who had first studied it in the context of his work on the theory of hypergeometric series.

89. Sylvester, "Inaugural Lecture at Oxford" (1885), in *Math. Papers JJS*, 4:293.

90. Ibid., 298. The next quote is also from this page.

91. J. J. Sylvester to Arthur Cayley, 13 January 1886, JJS Papers, SJC, Box 12.

92. J. J. Sylvester to Arthur Cayley, 18 February 1886, JJS Papers, SJC, Box 12. The quotes that follow in this paragraph are all from this letter.

93. J. J. Sylvester to Arthur Cayley, 1 February 1886, JJS Papers, SJC, Box 12, in Parshall, *JJS: Life and Work in Letters*, 261–262.

94. Sylvester, "Lectures on the Theory of Reciprocants" (1886–1888), or *Math. Papers JJS*, 4:303–513.

95. Ibid., in *Math. Papers JJS*, 4:303.

96. Ibid. For the references to the various papers, see Parshall, *JJS: Life and Work in Letters*, 265.

97. J. J. Sylvester to Daniel Coit Gilman, 7 July 1886, DCG Papers, Coll #1 Corresp., JHU.

98. On the journal and its importance for the internationalization of mathematics in the nineteenth century, see Barrow-Green, "Gösta Mittag-Leffler and the Foundation and Administration of *Acta Mathematica*."

99. J. J. Sylvester to Gösta Mittag-Leffler, 11 October 1886, Letter 4, 1886, Institut Mittag-Leffler, Djursholm, Sweden. I thank Holly Carley for securing for me copies of the Sylvester correspondence held at the Mittag-Leffler Institute.

100. "University Intelligence," *Times* (London), 18 October 1886.

101. Sylvester, "Music and Mathematics" (1886); J. J. Sylvester to Gösta Mittag-Leffler, 18 November 1886, Letter 6, 1886, Institut Mittag-Leffler, Djursholm, Sweden; J. J. Sylvester to John Tyndall, 25 December 1886, Tyndall Papers, Sylvester Correspondence, 1521, Royal Institution Archives; and the reprint Sylvester hand-corrected and sent to Galton of the poem from *Nature* in Galton Papers 188/1, JJS 1887: Jan. 1, UCL Archives.

102. Ludwig Brill to J. J. Sylvester, 26 November 1886, JJS Papers, SJC, Box 2. On the Brill models, see Fischer, *Mathematische Modelle*.

103. J. J. Sylvester to Sir Joseph Larmor, 18 February 1887, SJC Archives, W.1.

104. J. J. Sylvester to Daniel Coit Gilman, 11 March 1887, DCG Papers, Coll #1 Corresp., JHU, in Parshall, *JJS: Life and Work in Letters*, 263. The quotes that follow in this paragraph are from this letter with Sylvester's emphasis. Compare also Epigraph note 2.

Chapter 12

Epigraph: SJC Archives, 103.163, in Parshall, *JJS: Life and Work in Letters*, 270 (Sylvester's emphasis).

1. "Anglo-Jewish Association," *Times* (London), 4 March 1887.

2. While still in Baltimore in 1882, he had donated $50.00 to aid in the resettlement in Charles County of Jewish refugees fleeing the Russian pogroms. See Mendes Cohen to J. J. Sylvester, 17 November 1882, JJS Papers, SJC, Box 2.

3. Minutes NCO, 257ff and 306, Archives, New College Oxford Library.

4. Reginald's mother, Miriam Josephine Mozley Enthoven, as her sister, Edith Gigliucci, had long been regular recipients of Sylvester's poetic flights of fancy. See Alfred Robinson to J. J. Sylvester, 20 July 1887, Letter Book, p. 329, New College MS. 434, New College Library, Oxford, where Robinson informs him of "a few vacancies [that] were reserved for India Civil Service Candidates."

5. J. J. Sylvester to Gösta Mittag-Leffler, 6 June 1887, Letter 7, 1886, Institut Mittag-Leffler, Djursholm, Sweden.

6. Sylvester and Hammond, "On Hamilton's Numbers" (1887), or *Math. Papers JJS*, 4:553–578. See also Sylvester, "On the So-called Tschirnhausen Transformation" (1887), or *Math. Papers JJS*, 4:531–549, and "Sur une découverte de M. James Hammond relative à une certaine série de nombres qui figurent dans la théorie de la transformation Tschirnhausen" (1887), or *Math. Papers JJS*, 4:550–552.

7. Sylvester, "Sur les nombres dits de Hamilton" (1887), or *Math. Papers JJS*, 4:584–587.

8. See LMS Council Minutes, vol. 2, folios 62–65 on Cayley's selection, and ibid., vol. 3, folios 11–13 on Sylvester's selection. I thank Adrian Rice for sharing with me these particular fruits of his extensive scouring of the LMS archives.

9. Sylvester, "Note on a Proposed Addition to the Vocabulary of Ordinary Arithmetic" (1887), or *Math. Papers JJS*, 4:588 (note ∗).

10. Ibid., in *Math. Papers JJS*, 4:588.

11. See the seven short papers in *Math. Papers JJS*, 4:592–629.

12. "Report of the Annual General Meeting," *Proceedings of the London Mathematical Society* 19 (1887–1888): 1-2 on p. 1.

13. Arthur Cayley to J. J. Sylvester, 16 November 1887, JJS Papers, SJC, Box 2.

14. On the poetry, see his letter to Francis Galton of 12 January 1888, UCL Archives, Galton Papers, 191, James Joseph Sylvester 1888: Jan. 12, in Parshall, *JJS: Life and Work in Letters*, 266–267. He sent some lines he had emended and asked Galton to make the changes in the earlier versions. See below as well for more on Sylvester and the ongoing revision of his verses.

15. J. J. Sylvester to Daniel Coit Gilman, 26 February 1888, DCG Papers, Coll #1 Corresp., JHU, in Parshall, *JJS: Life and Work in Letters*, 267–269 on p. 268.

16. Sylvester, "Preuve élémentaire du théorème de Dirichlet sur les progressions arithmétiques dans les cas où la *raison* est 8 ou 12" (1888), or *Math. Papers JJS*, 4:620–624; he sketched his remarks in Oran on pp. 623–624. Compare also the commentary in Parshall, *JJS: Life and Work in Letters*, 281.

17. J. J. Sylvester to Gösta Mittag-Leffler, 14 April 1888, Letter 11, 1888, Institut Mittag-Leffler, Djursholm, Sweden. The next quote is also from this letter.

18. J. J. Sylvester to Lord Rayleigh, 28 April 1888, Referee's Reports, RR.10.159, Royal Society of London (his emphasis). The quotes which follow in this paragraph are from this letter (with his emphases).

19. J. J. Sylvester to Lord Rayleigh, 29 April 1888, Referee's Reports, RR.10.160, Royal Society of London. The quotes which follow in this paragraph are all from this letter.

20. J. J. Sylvester to Lord Rayleigh, 12 June 1888, Referee's Reports, RR.10.164, Royal Society of London. The quotes which follow in this paragraph are also from this letter.

21. Forsyth, "A Class of Functional Invariants."

22. Elliott, "Why and How the Society Began and Kept Going," lecture at the 200th meeting of the Oxford Mathematical Society, 16 May 1925. The account of the formation and early years of the Society that follows in this and the next paragraph comes from this source.

23. This and several other of Sylvester's poems may be found in JJS Papers, SJC, Box 1.

24. J. J. Sylvester to Frederick Locker, 28 June 1888, SJC Archives, W.1. The other quotes in this paragraph are also from this letter.

25. No such collection ever appeared, although Sylvester did continue to have various of his verses printed for private distribution (see "A Steady Decline" and "The Final Years"). Clearly, the version of the present poem, as preserved in JJS Papers, SJC, differs from the one he had originally sent to Locker.

26. J. J. Sylvester to Robert Forsyth Scott, 3 February 1889, SJC Archives, 103.155.

27. Robert Forsyth Scott to J. J. Sylvester, 31 January 1889, SJC Archives, 103.153.

28. J. J. Sylvester to Robert Forsyth Scott, 30 January [1890], SJC Archives, 103.152. Sylvester inadvertently misdated this letter as 30 January 1889.

29. Scott to Sylvester, 31 January 1889, and Alfred Emslie to Robert Forsyth Scott, 2 February 1889, SJC Archives, 103.154, respectively. Known as a portraitist especially of male subjects, Emslie had a reasonably successful career. His 1890 portrait of Gladstone, for example, is part of the collection of London's National Portrait Gallery.

30. J. J. Sylvester to Robert Forsyth Scott, 28 February 1889, SJC Archives, 103.159.

31. J. J. Sylvester to Robert Forsyth Scott, 11 April 1889, SJC Archives, 103.160. The portrait is the one that appears as the frontispiece of *Math. Papers JJS* and does capture a rather wearied and introspective Sylvester.

32. See Epigraph note.

33. Georges-Louis Leclerc, Comte de Buffon, *Œuvres philosophiques de Buffon*, 473, as quoted and translated in Seneta, Parshall, and Jongmans, "Nineteenth-Century Developments in Geometric Probability," 501. The latter paper provides a fuller history and technical discussion of Sylvester's work in this area.

34. See Seneta, Parshall, and Jongmans, "Nineteenth-Century Developments in Geometric Probability," 506–517, and recall the discussion in chapter 7.

35. J. J. Sylvester to Gösta Mittag-Leffler, 3 May 1889, Letter 12, 1889, Institut Mittag-Leffler, Djursholm, Sweden. The results did appear in *Acta* as Sylvester, "On a Funicular Solution of Buffon's 'Problem of the Needle' in Its Most General Form." (1890–1891), or *Math. Papers JJS*, 4:636–679.

36. *Nature* 40 (1889): 191.

37. J. J. Sylvester to Daniel Coit Gilman, 30 July 1889, DCG Papers, Coll #1 Corresp., JHU, in Parshall, *JJS: Life and Work in Letters*, 271. The following quote is also from this letter (p. 272).

38. J. J. Sylvester to Gösta Mittag-Leffler, 29 August 1889, Letter 13, 1889, Institut Mittag-Leffler, Djursholm, Sweden.

39. Ibid. The next quote is also from this letter.

40. J. J. Sylvester to Gösta Mittag-Leffler, 27 September 1889, Letter 14, 1889, Institut Mittag-Leffler, Djursholm, Sweden (his emphasis).

41. J. J. Sylvester to Julius Petersen, 23 September 1889, Ny kgl. Saml. 3259 4°, II(O–Z), Breve til Julius Petersen, Manuscript Collection, Kongelige Bibliotek, Copenhagen (hereinafter abbreviated Letters to Petersen, KB, Copenhagen), in Sabidussi, "Correspondence," 106.

42. Cayley, "On the Finite Number of Covariants of a Binary Quantic" (not in *Math. Papers AC*), and Hilbert, "Über die Endlichkeit des Invariantensystems für binäre Grundformen."

43. Felix Klein to David Hilbert, 24 February 1889, Hilbert Nachlaß, Niedersächsische Staats- und Universitätsbibliothek, Göttingen (hereinafter abbreviated NSUB, Göttingen), as translated in Sabidussi, "Correspondence," 103. The following quote is also from this letter (p. 103).

44. David Hilbert to Felix Klein, 27 February 1889, Klein Nachlaß, NSUB, Göttingen, as translated in Sabidussi, "Correspondence," 103.

45. Elliott, "How and Why the Society Began," 14–15.

46. Julius Petersen to Felix Klein, 20 [September], 1889, Klein Nachlaß, NSUB, Göttingen, as translated in Sabidussi, "Correspondence," 104. As Sabidussi convincingly argued, this letter, actually dated by Petersen 20 October 1889, was most likely mistakenly dated and actually written one month earlier.

47. It appeared as Petersen, "Ueber die Endlichkeit des Formensystems einer binären Grundform."

48. J. J. Sylvester to Julius Petersen, 3 October 1889, Letters to Petersen, KB, Copenhagen, in Sabidussi, "Correspondence," 107.

49. Compare the commentary in Sabidussi, "Correspondence," 100, 151–152.

50. J. J. Sylvester to Julius Petersen, 1 October 1889, Letters to Petersen, KB, Copenhagen, in Sabidussi, "Correspondence," 106.

51. J. J. Sylvester to Julius Petersen, 11 October 1889, Letters to Petersen, KB, Copenhagen, in Sabidussi, "Correspondence," 107.

52. J. J. Sylvester to Julius Petersen, 12 and 14 October 1889, Letters to Petersen, KB, Copenhagen, in Sabidussi, "Correspondence," 108–110.

53. J. J. Sylvester to Felix Klein, 15 October 1889, Klein Nachlaß, NSUB, Göttingen, in Sabidussi, "Correspondence," 111. The next quote is also from this letter.

54. Recall the discussion of this paper in chapter 9.

55. See Sabidussi, "Correspondence," for the letters and for a discussion of the mathematical details.

56. J. J. Sylvester to Felix Klein, 24 November 1889, Klein Nachlaß, NSUB, Göttingen, in Sabidussi, "Correspondence," 126, and Parshall, *JJS: Life and Work in Letters*, 273–274.

57. For the letters, see Sabidussi, "Correspondence," 130–135. The letter to Petersen is on pp. 133–135. The quotes that follow in this paragraph are also from this source (p. 134).

58. Felix Klein to David Hilbert, 25 December 1889, Hilbert Nachlaß, NSUB, Göttingen, as translated in Sabidussi, "Correspondence," 136.

59. David Hilbert to Felix Klein, 29 December 1889, Klein Nachlaß, NSUB, Göttingen, as translated in Sabidussi, "Correspondence," 136–137.

60. J. J. Sylvester to Felix Klein, 4 January 1890, Klein Nachlaß, NSUB. Göttingen, in Sabidussi, "Correspondence," 137.

61. J. J. Sylvester to Felix Klein, 19 January 1890, Klein Nachlaß, NSUB. Göttingen, in Sabidussi, "Correspondence," 139. The next quote is also from this letter (p. 140).

62. J. J. Sylvester to John Couch Adams, 25 January [1890], Adams Papers, SJC, Box 5, in Parshall, *JJS: Life and Work in Letters*, 276–277. Sylvester mistakenly dated this letter 1889.

63. Julius Petersen to Felix Klein, 26 January 1890, Klein Nachlaß, NSUB, Göttingen, in Sabidussi, "Correspondence," 141–142. The following quote is also from this letter (p. 142).

64. Petersen, "Die Theorie der regulären Graphs." For the assessment of this paper's importance in the history of graph theory, see Sabidussi, "Correspondence," 100.

65. J. J. Sylvester to Julius Petersen, 15 June 1890, Letters to Petersen, KB, Copenhagen, in Sabidussi, "Correspondence," 143. The following quote is also from this letter.

66. See "A Pair of Sonnets," printed "with important alterations" on 5 May 1890, in New College Archives. Gladstone had visited Oxford on 5 February amid much fanfare. Sylvester commented that Gladstone, to whom he referred as "the G. O. M." or "Grand Old Man," "is the object of almost a religious cult to young Oxford—whatever their political opinions might be, they look upon him quite apart from these, as the Typical Oxford Man and scholar." See J. J. Sylvester to Robert Graves, 22 March 1890, Trinity College Dublin, MS 10047/47/148, 150.

67. William Spooner, "Fifty Years in an Oxford College," unpublished, handwritten draft manuscript, p. 53, New College MS. 14,357, New College Oxford Library.

68. Charles Hermite to J. J. Sylvester, 9 March 1890, JJS Papers, SJC, Box 2, in Parshall, *JJS: Life and Work in Letters*, 277 (my translation).

69. Charles Hermite to J. J. Sylvester, 19 March 1890, JJS Papers, SJC, Box 2, in Parshall, *JJS: Life and Work in Letters*, 279 (my translation).

70. J. J. Sylvester to Robert Graves, 24 June 1890, Trinity College Dublin, MS 10047/47/150. On Sylvester's first meeting with Robert Graves, recall chapter 3.

71. Sylvester to Graves, 22 March 1890. The quote that follows is also from this letter.

72. Franklin, *Life of Daniel Coit Gilman*, 260.

73. J. J. Sylvester to Daniel Coit Gilman, 14 May 1890, DCG Papers, Coll #1 Corresp., JHU. The quotes that follow in this paragraph are from this letter.

74. Sylvester had been at work revising some of his earlier poems. On 21 May 1890, for example, he had had 25 copies of a new, improved, and expanded version of "The Lily Fair of Jasmin Dene" privately printed. One of his "studies in monochrone" in which

all 100 lines rhyme with "ean," this poem had been published in a different form in December of 1888 in *The Eagle*, the magazine of St. John's College, Cambridge. A copy of the 1890 version is held in the archives at New College Library, Oxford.

75. *Nature* 42 (29 May 1890): 107.

76. Minutes NCO, item 3507, p. 50, Archives, New College Oxford Library.

77. For the Latin text and an English translation of the public oration that was delivered on this occasion, see "Professor Sylvester," *The Jewish Chronicle* (London), 13 June, 1890. The oration began with explicit reference to Sylvester's Jewish heritage: "More than 53 years ago there wandered among the groves of our Academy a youth, sprung from the ancient stock of the sacred race, whose ancestors, first in the Chaldean plains, next upon the Palestinian hills, looked up to the innumerable stars of heaven, with a sort of reverence as the type of their children yet to be in multitude."

78. Minutes NCO, item 3499, pp. 305–307, and item 3507, pp. 125–126, Archives, New College Oxford Library.

79. Spooner, "Fifty Years in an Oxford College," 52.

80. J. J. Sylvester to Simon Newcomb, 8 January 1891, in Archibald, "Unpublished Letters of James Joseph Sylvester," 151.

81. Sylvester, "Preuve que π ne peut pas être racine d'une équation algébrique à coefficients entiers" (1890), or *Math. Papers JJS*, 4:682–686, and *Oxford University Gazette* (9 December 1890).

82. *Oxford University Gazette* (29 March 1892).

83. Recall the discussion of this work in chapter 10.

84. J. J. Sylvester to Arthur Cayley, 17 April 1891, JJS Papers, SJC, Box 12 (his emphasis).

85. J. J. Sylvester to Arthur Cayley, 19 April 1891, JJS Papers, SJC, Box 12.

86. Sylvester, "On Arithmetical Series" (1891–1892), or *Math. Papers JJS*, 4:702–703. The quotes that follow in this paragraph are from this paper (p. 703).

87. Sylvester to Newcomb, 8 January 1891, and J. J. Sylvester to Arthur Cayley, 11 June 1891, JJS Papers, SJC, Box 12, in Parshall, *JJS: Life and Work in Letters*, 280, respectively.

88. On the history of this work, see Barrow-Green, *Poincaré and the Three Body Problem*.

89. "The Royal Society—Ladies Conversazione," *Times* (London), 18 June 1891.

90. Sylvester to Cayley, 11 June 1891, in Parshall, *JJS: Life and Work in Letters*, 280.

91. J. J. Sylvester to Arthur Cayley, 27 June 1891, JJS Papers, SJC, Box 12.

92. J. J. Sylvester to Arthur Cayley, 27 July 1891, JJS Papers, SJC, Box 12 (his emphasis).

93. Sylvester's collected works were, in fact, assembled posthumously by Henry F. Baker of St. John's College, Cambridge, and published in four quarto volumes by the Cambridge University Press between 1904 and 1912. They were reprinted by the Chelsea Publishing Company in 1973.

94. J. J. Sylvester to Arthur Cayley, 27 May 1891, JJS Papers, SJC, Box 12.

95. Library Discharge and Return Records, item 662, Archives, New College Oxford Library. Sylvester had been regularly checking out issues of these two journals since 1889.

96. Compare Price, *Mathematics for the Multitude*, 33–41, on this expansion especially in the 1880s.

97. J. J. Sylvester to E. M. Langley, 29 September 1891, in Archibald, "Unpublished Letters of James Joseph Sylvester," 152.

98. J. J. Sylvester to Robert Forsyth Scott, 18 February 1892, JJS Papers, SJC, Box 1. The quotes that follow in this paragraph are also from this letter (his emphasis).

99. Herbert A. L. Fisher to J. J. Sylvester, 30 June 1892, JJS Papers, SJC, Box 2. The quotes that follow in this paragraph are from this letter. Fisher later went on to a distinguished career in government, as an historian, and as Warden of New College. See his *An Unfinished Autobiography*.

100. That he did go to Brighton in 1892 seems fairly clear from a later letter from Fisher in which he hopes Sylvester "will be able soon to rejoin us in Brighton." See Herbert A. L. Fisher to J. J. Sylvester, 8 April 1894, JJS Papers, SJC, Box 2.

101. J. J. Sylvester to Daniel Coit Gilman, 17 August 1892, DCG Papers, Series 4, Official Papers, Box 4.6, JHU.

102. Daniel Coit Gilman to J. J. Sylvester, 16 August 1882, JJS Papers, SJC, Box 2, in Parshall, *JJS: Life and Work in Letters*, 210.

103. This book, first published in Dublin in 1848, went into a sixth edition in 1879 published in London by Longmans, Green and Co. and a tenth by 1896.

104. J. J. Sylvester, "Lecture I, October 21st 1892, Invariants and Covariants of Systems of Quantics," handwritten draft, p. 1, JJS Papers, SJC, Box 8. This is one of the few extant drafts of one of Sylvester's classroom lectures.

105. Minutes NCO, item 3507, p. 53, Archives, New College Oxford Library. Hayes was named on 21 February 1894. The same William Esson who had stood in for Sylvester in the winter of 1888 was named his permanent deputy and became the Savilian Professor of Geometry after Sylvester's death.

106. J. J. Sylvester to Arthur Cayley, 30 May 1893, JJS Papers, SJC, Box 12 (his emphasis).

107. Sylvester, "Note on a Nine Schoolgirls Problem" (1893), or *Math. Papers JJS*, 4:732–733.

108. See the French citation in JJS Papers, SJC, Box 3 (my translation). Aleksandr Vasilev, a professor of mathematics at Kasan, had written to him on 11 June with the news.

109. *The Jewish Chronicle* (London), 2 February 1894.

110. Daniel Coit Gilman to J. J. Sylvester, 24 February 1894, JJS Papers, SJC, Box 2, in Parshall, *JJS: Life and Work in Letters*, 281. The quote that follows is also from this letter and page.

111. See C. Limerick to J. J. Sylvester, 21 March 1894, and Herbert A. L. Fisher to Sylvester, 8 April 1894, both in JJS Papers, SJC, Box 2.

112. *The Jewish Chronicle* (London), 1 June 1894. Geoffrey Cantor analyzed the implications of Sylvester's honorary membership in the Maccabæans in "Creating the Royal Society's Sylvester Medal." See also the Epilogue.

113. MacMahon, "James Joseph Sylvester," *Proceedings of the Royal Society of London*, xxiv.

114. William Spooner, Diaries, item 14,355, vol. 2, p. 116, Archives, New College Oxford Library.

115. From an account of the noted geologist and Athenæum member, Sir Archibald Geikie, as quoted in Blondheim, "A Brilliant and Eccentric Mathematician," 140.

116. Spooner, "Fifty Years in an Oxford College," 53. The next quote is also from this source and page.

117. Two printed versions—one dated 19 December 1895 and the other dated 26 February 1896—form part of item 14,214, Archives, New College Oxford Library. The verses quoted below are from the later, February version.

118. J. J. Sylvester to Edith Gigliucci, 31 August 1895, UCL Archives, MS.ADD.221, in Parshall, *JJS: Life and Work in Letters*, 282. The quotes describing Sylvester's trip that follow in this and the next paragraph are all from this letter.

119. Item 14,214, Archives, New College Oxford Library (his emphasis).

120. Sylvester to Gigliucci, 31 August 1895, in Parshall, *JJS: Life and Work in Letters*, 283.

121. MacMahon, "James Joseph Sylvester," *Nature*, 493.

122. See J. J. Sylvester (signing anonymously as F.R.S.), "To the Editor of 'The National Observer,' " JJS Papers, SJC, Box 1, in Parshall, *JJS: Life and Work in Letters*, 284–285, and J. J. Sylvester to Edith Gigliucci, 18 April 1896, UCL Archives, MS.ADD.221, respectively.

123. MacMahon, "James Joseph Sylvester," *Nature*, 493.

124. Sylvester, "On the Goldbach-Euler Theorem Regarding Prime Numbers" (1897), or *Math. Papers JJS*, 4:734.

125. Ibid., in *Math. Papers JJS*, 4:736.

126. David Hilbert to Felix Klein, 20 October 1889, Klein Nachlaß, NSUB, Göttingen, as translated in Sabidussi, "Correspondence," 113.

127. The cemetery, now a part of the National Trust, is located in King Henry's Walk off Ball's Pond Road in Islington.

128. See MacMahon, "James Joseph Sylvester," *Nature*, 494 as well as *Times* (London), 20 March 1897, for lists of some of those in attendance.

129. MacMahon, "James Joseph Sylvester," *Nature*, 494.

Epilogue

Epigraph: Benjamin Peirce, "The Lifework of a Great Mathematician," *Boston Herald*, 14 March 1880. I thank Deborah Kent whose combing of the Peirce archives at Harvard uncovered this piece.

1. O[swald] J. S[imon], "Death of Professor Sylvester," *The Jewish Chronicle* (London), 19 March 1897.

2. "The Death of Professor Sylvester," *The Jewish Chronicle* (London), 19 March 1897.

3. As quoted in "The Anti-Jewish Press and Lord John Russell's Bill," *The Jewish Chronicle* (London), 2 March 1849. The quotes that follow in this paragraph are also from this piece.

4. See, among many other citations and mentions of Sylvester in *The Jewish Chronicle* (London): 21 September 1855; 11 December 1863; 30 June 1865; 18 August 1871; 11 June 1880; 13 June 1890; 2 February 1894, respectively.

5. "Cambridge: Admission of a Jew to a Degree," *The Jewish Chronicle* (London), 24 July 1857. Recall from chapter 8 that the Cambridge University Act was only passed in 1856, so Cohen, like Sylvester, had been illegible to take his degree in the year of his Tripos. Cohen, however, took the first opportunity to rectify the situation.

6. "The Senior Wranglership," *The Jewish Chronicle* (London), 12 February 1869.

7. "Professor Sylvester," *The Jewish Chronicle*, 7 December 1883 and 21 December 1883. Compare Roth, "The Jews in English Universities," 114.

8. *The Jewish Chronicle*, 2 February 1894.

9. The argument that follows is substantially that presented by Geoffrey Cantor in his article, "Creating the Royal Society's Sylvester Medal."

10. *The Jewish Year Book*, 1897, 60–61, as quoted in ibid., 76.

11. Cantor, "Creating the Royal Society's Sylvester Medal," 81.

12. Royal Society Council Minuntes, 20 May 1897, Royal Society Library, MC.17.19, as quoted in ibid., 82.

13. "Proposed Sylvester Memorial—An Appeal to the Jewish Community," *The Jewish Chronicle* (London), 4 June 1897.

14. "The Sylvester Memorial," *The Jewish Chronicle* (London), 4 June 1897.

15. Cantor, "Creating the Royal Society's Sylvester Medal," 84.

16. Recall chapter 12. For a list of the medalists up to 1949, see Ivor Grattan-Guinness, "The Sylvester Medal: Origins and Recipients 1901–1949," 105–108.

17. Franklin, "James Joseph Sylvester," 299–301.

18. See Charles Hermite to J. J. Sylvester, 29 April 1882, DCG Papers, Coll #1 Corresp., JHU, in Parshall, *JJS: Life and Work in Letters*, 201.

19. Compare Charles Hermite to J. J. Sylvester, 19 February 1883, JJS Papers, SJC, Box 2, in Parshall, *JJS: Life and Work in Letters*, 220–221.

20. Noether, "James Joseph Sylvester," 156 (my translation; his emphasis).

21. For just three examples, see Sylvester, "On a Certain Fundamental Theorem of Determinants" (1851), in *Math. Papers JJS*, 1:252; Sylvester, "Presidential Address to Section 'A' of the British Association" (1869), in *Math. Papers JJS*, 2:656; and Sylvester to Spottiswoode, 19 November 1876, in Parshall, *JJS: Life and Work in Letters*, 172.

22. J. J. Sylvester to Arthur Cayley, 6 September 1883, JJS Papers, SJC, Box 11.

23. Sylvester, "On the Relation Between the Minor Determinants of Linearly Equivalent Quadratic Functions" (1851), in *Math. Papers JJS*, 1:247.

24. Sylvester, "Algebraical Researches, Containing a Disquisition on Newton's Rule" (1864), in *Math. Papers JJS*, 2:380 (note 1).

25. On the latter historical developments, see, for example, Corry, *Modern Algebra and the Rise of Mathematical Structures*.

26. See "Remarks of Prof. Sylvester, at a Farewell Reception Tendered to Him by the Johns Hopkins University, Dec. 20, 1883," 5. Compare Sylvester, "Lectures on the Principles of Universal Algebra" (1884), in *Math. Papers JJS*, 4:209. Harriot's manuscript, although written in 1631, was only discovered in 1784.

27. Sylvester, "Presidential Address to Section 'A' of the British Association" (1869), in *Math. Papers JJS*, 2:656.

28. Sylvester, "A Constructive Theory of Partitions" (1882), in *Math. Papers JJS*, 4:1 (his emphasis).

29. Sylvester, "Lectures on the Principles of Universal Algebra" (1884), in *Math. Papers JJS*, 4:209.

30. Noether, "James Joseph Sylvester," 156 (my translation).

31. Tony Crilly makes this analogy in *Arthur Cayley*, 193–196. See also Allen, *The Naturalist in Britain*.

32. See Hilbert, "Ueber die Theorie der algebraischen Invarianten," in *Mathematical Papers Read at the International Mathematical Congress Held in Connection with the World's Columbian Exposition, Chicago 1893*. Edited by Eliakim Hastings Moore, Oskar Bolza, Heinrich Maschke, and Henry S. White (New York: Macmillan and Co., 1896), 124. Meyer echoed Hilbert's assessment in *Sur les progrès de la théorie des invariants projectifs*, 1–26. Compare also the various historical and sociological assessments of the theory's development in Parshall, "Toward a History of Nineteenth-Century Invariant Theory"; Fisher, "The Death of a Mathematical Theory: A Study in the Sociology of Knowledge"; Fisher, "The Last Invariant Theorists: A Sociological Study of the Collective Biographies of Mathematical Specialists"; and Parshall, "The One-Hundredth Anniversary of the Death of Invariant Theory?".

33. MacMahon, "James Joseph Sylvester," *Proceedings of the Royal Society of London*, xxiv.

34. Franklin, "James Joseph Sylvester," 309.

35. Sylvester, "An Essay on Canonical Forms" (1851), in *Math. Papers JJS*, 1:203 (note †). Queen Victoria had offered John Couch Adams a knighthood in 1847 in recognition of his discovery of the planet, Neptune, but he declined the honor.

36. On these developments, see Greene, *American Science in the Age of Jefferson*; Daniels, *American Science in the Age of Jackson*; Bruce, *The Launching of Modern American Science*; and Dupree, *Science in the Federal Government: A History of Policies and Activities*.

37. See, for example, Veysey, *The Emergence of the American University*.

38. Parshall, "America's First School of Mathematical Research," and Parshall and Rowe, *Emergence of the American Mathematical Research Community*.

39. On the numerous failed attempts at sustaining specialized, mathematical journals in the United States, see Parshall and Rowe, *Emergence of the American Mathematical Research Community*, 42–45 and Table 1.2 on p. 51.

40. [Sylvester and Ferrers], "Address to the Reader" (1855), i.

41. Despeaux, "Development of a Publication Community," 296. See also Despeaux, "International Mathematical Contributions to British Scientific Journals, 1800–1900."

42. Desmond, *Huxley*, 626–630.

43. Sylvester, "Presidential Address to Section 'A' of the Bristish Association" (1869), in *Math. Papers JJS*, 2:660.

44. Huxley, "The Scientific Aspects of Positivism," 666 (his emphasis).

45. This is a vision of mathematics that Joan Richards dates to about 1900 in her book, *Mathematical Visions*. See 231–244. It is clear, however, that Sylvester and others embraced it at least three decades earlier.

46. See Epigraph note.

47. Sylvester, "Presidential Address to Section 'A' of the British Association" (1869), in *Math. Papers JJS*, 2:652.

References

Archival Sources

Académie des Sciences de Paris. Archives.

Arnauld Bienaymé. Private Collection.

Athenæum (Liverpool). Archives.

British Library. Sylvester Letters.

Cambridge University Library. Kelvin Papers and Stokes Papers.

Columbia University. Rare Book and Manuscript Library. David E. Smith Historical Papers.

Inner Temple. Archives and Library.

Institut Mittag-Leffler. Archives.

Institute of Actuaries. Library.

The Johns Hopkins University. Milton S. Eisenhower Library. Special Collections. Daniel Coit Gilman Papers Ms. 1.

Liverpool Record Office. Liverpool Old Hebrew Congregation Records.

London Metropolitan University. The Women's Library. McCrimmon Bodichon Collection.

London Metropolitan Archives. United Synagogue and Predecessors.

New College Oxford. Archives.

Niedersächsische Staats- und Universitätsbibliothek, Göttingen. David Hilbert Nachlaß.

Royal Institution. Archives. Tyndall Papers.

Royal Society of London. Archives.

St. John's College Cambridge Library. Archives, John Couch Adams Papers, and James Joseph Sylvester Papers.

Scuola Normale Superiore Pisa. Archivio Betti.

Smithsonian Institution Archives. Joseph Henry Collection.

Staatsbibliothek Berlin. Dirichlet Nachlaß.

Trinity College Dublin. Archives.

University College London. Manuscripts and Rare Books Library. John T. Graves Collection.

University College London Archives. Brougham Correspondence, College Correspondence, and London Mathematical Society Papers.

University of Liverpool. Archives of Liverpool Royal Institution and Liverpool Learned Societies at the University of Liverpool.

University of Virginia Library. Special Collections. Cabell Family Papers and Minor Family Papers; University Archives. Minutes of the Rector and Board of Visitors and Minutes of the Faculty of the University of Virginia.

Newspaper Sources
References to individual articles cited are given in the notes.

Baltimore Sun (Baltimore, MD, USA)
Boston Herald (Boston, MA, USA)
The Jewish Chronicle (London, UK)
The Jewish Comment (Baltimore, MD, USA)
The Liverpool Mercury (Liverpool, UK)
The Times (London, UK)
Watchman of the South (Richmond, VA, USA)

Works by James Joseph Sylvester (in Chronological Order by Year)
This is not a complete listing of Sylvester's works. Insofar as possible, I have listed these papers in the order in which Sylvester wrote them. This ordering differs from that in *Math. Papers JJS*.

[Sylvester, J. J.]. *Collection of Examples on the Integral Calculus, In Which Every Operation of Each Example is Completely Effected*. Cambridge, UK: J. & J. J. Deighton, 1835 (not in *Math. Papers JJS*).

Sylvester, J. I. [sic]. "A Supplement to Newton's First Section, Containing a Rigid Demonstration of the Fifth Lemma, and the General Theory of the Equality and Proportion of Linear Magnitudes." Cambridge, UK: J. Hall, 1836 (privately printed memoir not in *Math. Papers JJS*).

Sylvester, J. J. "Analytical Development of Fresnel's Optical Theory of Crystals." *Philosophical Magazine* 11 (1837): 461–469, 537–541; 12 (1838): 73–83, 341–345.

——— . "On the Motion and Rest of Fluids." *Philosophical Magazine* 13 (1838): 449–453.

——— . "On the Motion and Rest of Rigid Bodies." *Philosophical Magazine* 14 (1839): 188–190.

——— . "On Definite Double Integration, Supplementary to a Former Paper on the Motion and Rest of Fluids." *Philosophical Magazine* 14 (1839): 298–300.

——— . "On Rational Derivation from Equations of Coexistence, That Is to Say, a New and Extended Theory of Elimination. Part I." *Philosophical Magazine* 15 (1839): 428–435.

——— . "Part I of a Memoir on the Dialytic Method of Elimination." *Proceedings of the Royal Irish Academy* 2 (1840–1844): 130–139 and *Philosophical Magazine* 21 (1842): 534–539.

——— . "On Derivation of Coexistence. Part II. Being the Theory of Simultaneous Simple Homogeneous Equations." *Philosophical Magazine* 16 (1840): 37–43.

——— . "A Method of Determining by Mere Inspection the Derivatives from Two Equations of Any Degree." *Philosophical Magazine* 16 (1840): 132–135.

——— . "On the Relation of Sturm's Auxiliary Functions to the Roots of an Algebraic Equation." In *Report of the Eleventh Meeting of the British Association for the Advancement of Science Held at Plymouth in July 1841*. London: John Murray, 1842, 23–24.

——— . "On a Linear Method of Eliminating between Double, Treble, and Other Systems of Algebraic Equations." *Philosophical Magazine* 18 (1841): 425–435.

——— . "Elementary Researches in the Analysis of Combinatorial Aggregation." *Philosophical Magazine* 24 (1844): 285–296.

———. "On the Double Square Representation of Prime and Composite Numbers." In *Report of the Fourteenth Meeting of the British Association for the Advancement of Science Held at York in September 1844*. London: John Murray, 1845, 2.

———. "On the Existence of Absolute Criteria for Determining the Roots of Numerical Equations." *Philosophical Magazine* 25 (1844): 442–445.

———. "An Account of a Discovery in the Theory of Numbers Relative to the Equation $Ax^3 + By^3 + Cz^3 = Dxyz$." *Philosophical Magazine* 31 (1847): 189–191.

———. "On the Equation in Numbers $Ax^3 + By^3 + Cz^3 = Dxyz$, and Its Associate System of Equations." *Philosophical Magazine* 31 (1847): 293–296.

———. "On the *General* Solution (in Certain Cases) of the Equation $x^3 + y^3 + Az^3 = Mxyz$, &c." *Philosophical Magazine* 31 (1847): 467–471.

———. "On the Intersections, Contacts, and Other Correlations of Two Conics Expressed by Indeterminate Coordinates." *Cambridge and Dublin Mathematical Journal* 5 (1850): 262–282.

———. "Addition to the Articles, 'On a New Class of Theorems,' and 'On Pascal's Theorem.'" *Philosophical Magazine* 37 (1850): 363–370.

[Sylvester, J. J.]. "Equity and Law Life Assurance Society." *Assurance Magazine* 1 (1) (1851): 100–102 (not in *Math. Papers JJS*).

———. "Equity and Law Life Assurance Society (1844).—Fourth Annual Meeting, Held March, 1851." *The Assurance Magazine* 3 (1853): 354–355 (not in *Math. Papers JJS*).

Sylvester, J. J. "On Certain General Properties of Homogeneous Functions." *Cambridge and Dublin Mathematical Journal* 6 (1851): 1–17.

———. "An Enumeration of the Contacts of Lines and Surfaces of the Second Order." *Philosophical Magazine* 1 (1851): 119–140.

———. "On the Relation between the Minor Determinants of Linearly Equivalent Quadratic Functions." *Philosophical Magazine* 1 (1851): 295–305.

———. "Sketch of a Memoir on Elimination, Transformation, and Canonical Forms." *Cambridge and Dublin Mathematical Journal* 6 (1851): 186–200.

———. "An Essay on Canonical Forms, Supplement to a Sketch of a Memoir on Elimination, Transformation and Canonical Forms." London: George Bell, 1851 (privately printed memoir).

———. "On a Certain Fundamental Theorem of Determinants." *Philosophical Magazine* 2 (1851): 142–145.

———. "On the General Theory of Associated Algebraical Forms." *Cambridge and Dublin Mathematical Journal* 6 (1851): 289–293.

———. "On Extensions of the Dialytic Method of Elimination." *Philosophical Magazine* 2 (1851): 221–230.

———. "On a Remarkable Discovery in the Theory of Canonical Forms and of Hyperdeterminants." *Philosophical Magazine* 2 (1851): 391–410.

———. "On the Principles of the Calculus of Forms." *Cambridge and Dublin Mathematical Journal* 7 (1852): 52–97.

[Sylvester, J. J.]. "Equity and Law Life Assurance Society.—Fifth Annual General Meeting, 19th March, 1852." *The Assurance Magazine* 3 (1853): 355 (not in *Math. Papers JJS*).

Sylvester, J. J. "Sur une propriété nouvelle de l'équation qui sert à déterminer les inégalités séculaires des planètes." *Nouvelles annales de mathématiques* 11 (1852): 438–440.

———. "Observations on a New Theory of Multiplicity." *Philosophical Magazine* 3 (1852): 460–467.

Sylvester, J. J. "A Demonstration of the Theorem That Every Homogeneous Quadratic Polynomial Is Reducible By Real Orthogonal Substitutions to the Form of a Sum of Positive and Negative Squares." *Philosophical Magazine* 4 (1852): 138–142.

———. "On a Simple Geometrical Problem Illustrating a Conjectured Principle in the Theory of Geometrical Method." *Philosophical Magazine* 4 (1852): 366–369.

———. "On the Conditions Necessary and Sufficient To Be Satisfied in Order That a Function of Any Number of Variables May Be Linearly Equivalent To a Function of Any Less Number of Variables." *Philosophical Magazine* 5 (1853): 119–126.

———. "Note on the Calculus of Forms." *Cambridge and Dublin Mathematical Journal* 8 (1853): 62–64.

———. "On Mr. Cayley's Impromptu Demonstration of the Rule for Determining at Sight the Degree of Any Symmetrical Function of the Roots of an Equation Expressed in Terms of the Coefficients." *Philosophical Magazine* 5 (1853): 199–202.

———. "A Proof That All the Invariants to a Cubic Ternary Form Are Rational Functions of Aronhold's Invariants and of a Cognate Theorem for Biquadratic Binary Forms." *Philosophical Magazine* 5 (1853): 299–303, 367–372.

———. "On a Theory of the Syzygetic Relations of Two Rational Integral Functions, Comprising an Application to the Theory of Sturm's Functions, and That of the Greatest Algebraical Common Measure." *Philosophical Transactions of the Royal Society of London* 143 (1853): 407–548.

———. "On the Expressions for the Quotients Which Appear in the Application of Sturm's Method to the Discovery of the Real Roots of an Equation." In *Report of the Twenty-Third Meeting of the British Association for the Advancement of Science Held at Hull in September 1853*. London: John Murray, 1854, 1–3.

———. "Provisional Report on the Theory of Determinants." In *Report of the Twenty-Third Meeting of the British Association for the Advancement of Science Held at Hull in September 1853*. London: John Murray, 1854, 66–67 (not in *Math. Papers JJS*).

———. "On the Explicit Values of Sturm's Quotients." *Philosophical Magazine* 6 (1853): 293–296.

———. "Nouvelle méthode pour trouver une limite supérieure et une limite inférieure des racines réelles d'une équation algébrique quelconque." *Nouvelles annales de mathématiques* 12 (1853): 329–336.

———. "On the Calculus of Forms, Otherwise the Theory of Invariants." *Cambridge and Dublin Mathematical Journal* 8 (1853): 256–269; 9 (1854): 85–103.

———. "On Multiplication by Aid of a Table of Single Entry." *Assurance Magazine* 4 (1854–1855): 236–238 (not in *Math. Papers JJS*).

———. "A Probationary Lecture on Geometry, Delivered before the Gresham Committee and the Members of the Common Council of the City of London, 4 December 1854" (privately printed memoir).

[Sylvester, J. J., and Ferrers, Norman]. "Address to the Reader." *Quarterly Journal of Pure and Applied Mathematics* 1 (1855): i–ii (not in *Math. Papers JJS*).

Sylvester, J. J. "On the Change of Systems of Independent Variables." *Quarterly Journal of Pure and Applied Mathematics* 1 (1855): 42–56.

———. "On a Discovery in the Partition of Number." *Quarterly Journal of Pure and Applied Mathematics* 1 (1855): 81–84.

———. "A Trifle on Projectiles." *Philosophical Magazine* 11 (1856): 450–453.

———. "Letter on Professor Galbraith's Construction for the Range of Projectiles." *Philosophical Magazine* 12 (1856): 112–114.

————. "Recherches sur les solutions en nombres entiers positifs ou négatifs de l'équation cubique homogène à trois variables." *Annali di scienze matematiche e fisiche* 7 (1856): 398–400.

————. "Note on the Algebraical Theory of Derivative Points of Curves of the Third Order." *Philosophical Magazine* 16 (1858): 116–119.

————. "Note on the Equation in Numbers of the First Degree Between Any Number of Variables with Positive Coefficients." *Philosophical Magazine* 16 (1858): 369–371.

————. "On the Problem of the Virgins, and the General Theory of Compound Partition." *Philosophical Magazine* 16 (1858): 371–376.

————. "Outlines of Seven Lectures on the Partitions of Numbers [1859]." *Proceedings of the London Mathematical Society* 28 (1897): 33–96.

————. "Note sur certaines séries qui se présentent dans la théorie des nombres." *Comptes rendus de l'Académie des Sciences de Paris* 50 (1860): 650.

————. "On the Equation $P(m) + E\left(\frac{m}{m-1}\right) P(m-1) + E\left(\frac{m}{m-2}\right) P(m-2) + \cdots + E(m) = m\frac{m+1}{2}$." *Quarterly Journal of Pure and Applied Mathematics* 3 (1860): 186–190.

————. "On a Generalization of Poncelet's Theorems for the Linear Representation of Quadratic Radicals." In *Report of the Thirtieth Meeting of the British Association for the Advancement of Science Held at Oxford in September 1860*. London: John Murray, 1861, 7.

————. "On Poncelet's Approximate Linear Valuation of Surd Forms." *Philosophical Magazine* 20 (1860): 203–222.

————. "Meditation on the Idea of Poncelet's Theorem." *Philosophical Magazine* 20 (1860): 307–316.

————. "Notes to the Meditation on Poncelet's Theorem Including a Valuation of the Two New Definite Integrals . . . ." *Philosophical Magazine* 20 (1860): 525–533.

————. "On the Pressure of Earth on Revetment Walls." *Philosophical Magazine* 20 (1860): 489–499.

————. "Note on the Numbers of Bernoulli and Euler, and a New Theorem Concerning Prime Numbers." *Philosophical Magazine* 21 (1861): 127–136.

————. "Sur l'involution des lignes droites dans l'espace considérées comme des axes de rotation." *Comptes rendus de l'Académie des Sciences de Paris* 52 (1861): 741–745.

————. "Note sur l'involution de six lignes dans l'espace." *Comptes rendus de l'Académie des Sciences de Paris* 52 (1861): 815–817.

————. "Note on the Historical Origin of the Unsymmetrical Six-Valued Function of Six Letters." *Philosophical Magazine* 21 (1861): 369–377.

————. "On the Involution of Axes of Rotation." In *Report of the Thirty-first Meeting of the British Association for the Advancement of Science Held at Manchester in September 1861*. London: John Murray, 1862, 12.

————. "Généralisation d'un théorème de M. Cauchy." *Comptes rendus de l'Académie des Sciences de Paris* 53 (1861): 644–645.

————. "Addition à la note intitulée: 'Généralisation d'un théorème de M. Cauchy,' et insérée dans le 'Compte rendu' de la séance du 7 octobre dernier." *Comptes rendus de l'Académie des Sciences de Paris* 53 (1861): 722–725.

————. "On a Generalization of a Theorem of Cauchy on Arrangements." *Philosophical Magazine* 22 (1861): 378–382.

————. "On the Solution of the Linear Equation of Finite Differences in Its Most General Form." In *Report of the Thirty-second Meeting of the British Association for the Advancement of Science Held at Cambridge in October 1862.* London: John Murray, 1863, 188.

———. "On the Integral of the General Equation in Differences." *Philosophical Magazine* 24 (1862): 436–441.

———. "Sequel to the Theorems Relating to 'Canonic Roots' Given in the March Number of This Magazine." *Philosophical Magazine* 25 (1863): 453–460.

———. "On the Centre of Gravity of a Truncated Triangular Pyramid, and on the Principles of Barycentric Perspective." *Philosophical Magazine* 26 (1863): 167–183.

———. "On the Quantity and Centre of Gravity of Figures Given in Perspective, or Homography." In *Report of the Thirty-third Meeting of the British Association for the Advancement of Science Held at Newcastle-upon-Tyne in August and September 1863.* London: John Murray, 1864, 2.

———. "Théorème sur la limite du nombre des racines réelles d'une classe d'équations algébriques." *Comptes rendus de l'Académie des Sciences de Paris* 58 (1864): 494–495.

———. "Sur une extension de la théorie des équations algébriques." *Comptes rendus de l'Académie des Sciences de Paris* 58 (1864): 698–691.

———. "Algebraical Researches, Containing a Disquisition on Newton's Rule for the Discovery of Imaginary Roots, and an Allied Rule Applicable to a Particular Class of Equations, Together with a Complete Invariantive Determination of the Character of the Roots of the General Equation of the Fifth Degree, &c." *Philosophical Transactions of the Royal Society of London* 154 (1864): 579–666.

———. "Note sur les conditions nécessaires et suffisantes pour distinguer le cas quand toutes les racines d'une équation du cinquième degré sont réelles." *Comptes rendus de l'Académie des Sciences de Paris* 60 (1865): 759–761.

———. "Sur les limites du nombre des racines réelles des équations algébriques." *Comptes rendus de l'Académie des Sciences de Paris* 60 (1865): 1261–1263.

———. "On an Elementary Proof and Generalization of Sir Isaac Newton's Hitherto Undemonstrated Rule for the Discovery of Imaginary Roots." *Proceedings of the London Mathematical Society* 1 (1865–1866): 1–16.

———. "On a Special Class of Questions on the Theory of Probabilities." In *Report of the Thirty-fifth Meeting of the British Association for the Advancement of Science Held at Birmingham in September 1865.* London: John Murray, 1866, 8.

———. "On Lambert's Theorem for Elliptic Motion." *Monthly Notices of the Royal Astronomical Society* 26 (1865): 27–29.

———. "Astronomical Prolusions: Commencing with an Instantaneous Proof of Lambert's and Euler's Theorems, and Modulating through a Construction of the Orbit of a Heavenly Body from Two Heliocentric Distances, the Subtended Chord, and the Periodic Time, and the Focal Theory of Cartesian Ovals, into a Discussion of Motion in a Circle and Its Relation to Planetary Motion." *Philosophical Magazine* 31 (1866): 52–76.

———. "On an Addition to Poinsot's Ellipsoidal Mode of Representing the Motion of a Rigid Body Turning Freely Round a Fixed Point, Whereby the Time May Be Made to Register Itself Mechanically." *Proceedings of the London Mathematical Society* 1 (1866): 3–4.

———. "On the Motion of a Rigid Body Acted on by No External Forces." *Philosophical Transactions of the Royal Society of London* 156 (1866): 757–780.

———. "Note on the Properties of the Test Operators Which Occur in the Calculus of Invariants, Their Derivatives, Analogues, and Laws of Combination; With an Incidental Application to the Development in a Maclaurinian Series of Any Power of the Logarithm of an Augmented Variable." *Philosophical Magazine* 32 (1866): 461–472.

———. "On the Successive Involutes of a Circle." In *Report of the Thirty-eighth Meeting of the British Association for the Advancement of Science Held at Norwich in August 1868.* London: John Murray, 1869, 10–11.

———. "Note on Successive Involutes to a Circle." *Philosophical Magazine* 36 (1868): 295–306.

———. "On Successive Involutes to Circles—Second Note." *Philosophical Magazine* 36 (1868): 459–466.

———. "Outline Trace of the Theory of Reducible Cyclodes, That Is a Particular Family of Successive Involutes to a Circle Whose Determination Depends on the Solution of an Algebraico-Diophantine Equation, and the Number and Classification of the Forms of Such a Family for Any Given Order of Succession." *Proceedings of the London Mathematical Society* 2 (1869): 137–160.

———. "Presidential Address: Mathematical and Physical Section." In *Report of the Thirty-ninth Meeting of the British Association for the Advancement of Science Held at Exeter in August 1869.* London: John Murray, 1870, 1–9.

———. "A Plea for the Mathematician." *Nature* 1 (1869–1870): 237–239, 260–263 (an abbreviated version of the "Presidential Address" not in *Math. Papers JJS*).

———. *The Laws of Verse or Principles of Versification Exemplified in Metrical Translations: Together with an Annotated Reprint of the Inaugural Presidential Address to the Mathematical and Physical Section of the British Association at Exeter.* London: Longmans, Green, and Co., 1870.

[Syzygeticus] "Ballad of Sir John de Courcy." *Gentleman's Magazine* (February 1871): 313–316 (not in *Math. Papers JJS*).

Sylvester, J. J. "Note on the Theory of a Point in Partition." In *Report of the Forty-first Meeting of the British Association for the Advancement of Science Held at Edinburgh in August 1871.* London: John Murray, 1872, 23–25.

———. "Science at the London School Board." *Nature* (21 March 1872): 410 (not in *Math. Papers JJS*).

———. "On Recent Discoveries in Mechanical Conversion of Motion." *Proceedings of the Royal Institution of Great Britain* 7 (1873–1875): 179–198.

———. "Transformation du mouvement circulaire en mouvement rectiligne." *La Revue scientifique* (1874–1875): 490–498 (translated version of "On Recent Discoveries in Mechanical Conversion" not in *Math. Papers JJS*).

———. "On Recent Discoveries in Mechanical Conversion of Motion." *Van Nostrand's Eclectic Engineering Magazine* 12 (1875): 313–321 (another version of "On Recent Discoveries in Mechanical Conversion" not in *Math. Papers JJS*).

———. "On the Plagiograph *aliter* the Skew Pantigraph." *Nature* 12 (1875): 168, 214–216.

———. "On a Lady's Fan, on Parallel Motion, and on an Orthogonal Web of Jointed Rods." *Proceedings of the London Mathematical Society* 6 (1875): 196–197.

———. *Fliegende Blätter: Supplement to the Laws of Verse.* London: Grant & Co., 1876.

———. "Note on Spherical Harmonics." *Philosophical Magazine* 2 (1876): 291–307, 400.

———. "Sur les invariants fondamentaux de la forme binaire du huitième degré." *Comptes rendus de l'Académie des Sciences de Paris* 84 (1877): 240–244, 532–534.

———. "Address on Commemoration Day at Johns Hopkins University, 22 February 1877." In *Math. Papers JJS*, 3: 72–87.

Sylvester, J. J. "Sur une méthode algébrique pour obtenir l'ensemble des invariants et des covariants fondamentaux d'une forme binaire et d'une combinaison quelconques de formes binaires." *Comptes rendus de l'Académie des Sciences de Paris* 84 (1877): 1113–1116, 1211–1213.

———. "Sur le vrai nombre des covariants élémentaires d'un système de deux formes biquadratiques binaires." *Comptes rendus de l'Académie des Sciences de Paris* 84 (1877): 1285–1289.

———. "Théorie pour trouver le nombre des covariants et des contrevariants d'ordre et de degré donnés linéairement indépendants d'un système quelconque de formes simultanées contenant un nombre quelconque de variables." *Comptes rendus de l'Académie des Sciences de Paris* 84 (1877): 1359–1361, 1427–1430.

———. "Sur les invariants." *Comptes rendus de l'Académie des Sciences de Paris* 85 (1877): 992–995, 1035–1037, 1091–1092.

[Sylvester, J. J.] "Notice to the Reader." *American Journal of Mathematics* 1 (1878): iii (not in *Math. Papers JJS*).

Sylvester, J. J. "On an Application of the New Atomic Theory to the Graphical Representation of the Invariants and Covariants of Binary Quantics,—with Three Appendices." *American Journal of Mathematics* 1 (1878): 64–125.

———. "Proof of the Hitherto Undemonstrated Fundamental Theorem of Invariants." *Philosophical Magazine* 5 (1878): 178–188.

———. "Sur les actions mutuelles des formes invariantives dérivées." *Journal für die reine und angewandte Mathematik* 85 (1878): 89–114.

———. "Sur les covariants fondamentaux d'un système cubo-biquadratique binaire." *Comptes rendus de l'Académie des Sciences de Paris* 87 (1878): 242–244, 287–289.

———. "A Synoptical Table of the Irreducible Invariants and Covariants to a Binary Quintic, with a Scholium on a Theorem in Conditional Hyper-Determinants." *American Journal of Mathematics* 1 (1878): 370–378.

———. "Table de nombres de dérivées invariantives d'ordre et de degré donnés, appartenant à la forme binaire du dixième ordre." *Comptes rendus de l'Académie des Sciences de Paris* 89 (1879): 395–396.

———. "Tables of the Generating Functions and Groundforms for the Binary Quantics of the First Ten Orders." *American Journal of Mathematics* 2 (1879): 223–251.

——— (with Fabian Franklin). "Tables of the Generating Functions and Groundforms for Simultaneous Binary Quantics of the First Four Orders, Taken Two and Two Together." *American Journal of Mathematics* 2 (1879): 293–306, 324–329.

———. "Spring's Début: A Town Idyll." Baltimore: John Murphy & Co., 1880 (privately printed pamphlet not in *Math. Papers JJS*).

———. "On Certain Ternary Cubic-Form Equations." *American Journal of Mathematics* 2 (1879): 280–285, 357–393; 3 (1880): 58–88, 179–189.

———. "Tables of Generating Functions and Groundforms of the Binary Duodecimic, with Some General Remarks, and Tables of the Irreducible Syzygies of Certain Quantics." *American Journal of Mathematics* 4 (1881): 41–61.

———. "On Tchebycheff's Theory of the Totality of the Prime Numbers Comprised within Given Limits." *American Journal of Mathematics* 4 (1881): 230–247.

———. "Sur les puissances et les racines de substitutions linéaires." *Comptes rendus de l'Académie des Sciences de Paris* 94 (1882): 55–59.

———. "Sur les racines des matrices unitaires." *Comptes rendus de l'Académie des Sciences de Paris* 94 (1882): 396–399.

————. "On Subinvariants, That Is, Semi-Invariants to Binary Quantics of an Unlimited Order." *American Journal of Mathematics* 5 (1882): 79–136.

————. "Tables of Generating Functions, Reduced and Representative, for Certain Ternary Systems of Binary Forms," *American Journal of Mathematics* 5 (1882): 241–250.

————. "A Word on Nonions." *Johns Hopkins University Circulars* 1 (August 1882): 241–242.

————. "On the Fundamental Theorem in the New Method of Partitions." *Johns Hopkins University Circulars* 2 (December 1882): 22.

————. "Erratum." *Johns Hopkins University Circulars* 2 (February 1883): 46 (not in *Math. Papers JJS*).

————. "A Note from Professor Sylvester." *Johns Hopkins University Circulars* 2 (April 1883): 86 (not in *Math. Papers JJS*).

————. "A Constructive Theory of Partitions, Arranged in Three Acts, an Interact, and an Exodion." *American Journal of Mathematics* 5 (1882): 251–330.

————. "On the Three Laws of Motion in the World of Universal Algebra." *Johns Hopkins University Circulars* 3 (January 1884): 33–34.

————. "Sur les quantités formant un groupe de nonions analogues aux quaternions de Hamilton." *Comptes rendus de l'Académie des Sciences de Paris* 98 (1884): 273–276, 471–475.

————. "Lectures on the Principles of Universal Algebra." *American Journal of Mathematics* 6 (1884): 270–286.

————. "Sur la correspondance entre deux espèces différentes de fonctions de deux systèmes de quantités, corrélatifs et également nombreux." *Comptes rendus de l'Académie des Sciences de Paris* 98 (1884): 779–781.

————. "Note on Captain MacMahon's Transformation of the Theory of Invariants." *Messenger of Mathematics* 13 (1884): 163–165.

————. "Sur la solution explicite de l'équation quadratique de Hamilton en quaternions ou en matrices du second ordre." *Comptes rendus de l'Académie des Sciences de Paris* 99 (1884): 555–558, 621–631.

————. "The Genesis of an Idea, or Story of a Discovery Relating to Equations in Multiple Quantity." *Nature* 31 (13 November 1884): 35–36 (not in *Math. Papers JJS*).

————. "Sur une nouvelle théorie de formes algébriques." *Comptes rendus de l'Académie des Sciences de Paris* 101 (1885): 1042–1046, 1110–1111, 1225–1229, 1461–1464.

————. "Inaugural Lecture at Oxford 12 December, 1885, on the Method of Reciprocants As Containing an Exhaustive Theory of the Singularities of Curves." *Nature* 33 (7 January 1886): 222–231.

————. "Lectures on the Theory of Reciprocants." *American Journal of Mathematics* 8 (1886): 196–260; 9 (1887): 1–37, 113–161, 297–352; 10 (1888): 1–16.

————. "Music and Mathematics." *Nature* 35 (9 December 1886): 132 (not in *Math. Papers JJS*).

————. "Sur une découverte de M. James Hammond relative à une certaine série de nombres qui figurent dans la théorie de la transformation Tschirnhausen." *Comptes rendus de l'Académie des Sciences de Paris* 104 (1887): 1228–1231.

————. "On the So-called Tschirnhausen Transformation." *Journal für die reine und angewandte Mathematik* 100 (1887): 465–486.

Sylvester, J. J., and Hammond, James. "On Hamilton's Numbers." *Philosophical Transactions of the Royal Society of London* 178 (1887): 285–312.

Sylvester, J. J. "Sur les nombres dits de Hamilton." *Compte rendu de l'Association française pour l'avancement des sciences, Toulouse* (1887): 164–168.

―――. "Note on a Proposed Addition to the Vocabulary of Ordinary Arithmetic." *Nature* (15 December 1887): 152–153.

―――. "Preuve élémentaire du théorème de Dirichlet sur les progressions arithmétiques dans les cas où la *raison* est 8 ou 12." *Comptes rendus de l'Académie des Sciences de Paris* 106 (1888): 1278–1281, 1385–1386.

―――. "On a Funicular Solution of Buffon's 'Problem of the Needle' in Its Most General Form." *Acta Mathematica* 14 (1890–1891): 185–205.

―――. "A Pair of Sonnets." 5 May 1890 (privately printed pamphlet not in *Math. Papers JJS*).

―――. "The Lily Fair of Jasmin Dene." 21 May 1890 (privately printed pamphlet not in *Math. Papers JJS*).

―――. "Preuve que π ne peut pas être racine d'une équation algébrique à coefficients entiers." *Comptes rendus de l'Académie des Sciences de Paris* 111 (1890): 866–871.

―――. "On Arithmetical Series." *Messenger of Mathematics* 21 (1891–1892): 1–19, 87–120.

―――. "Note on a Nine Schoolgirls Problem." *Messenger of Mathematics* 22 (1893): 159–160.

―――. "Corolla Versuum." 19 December 1895 and 26 February 1896 (privately printed pamphlets not in *Math. Papers JJS*).

―――. "On the Goldbach-Euler Theorem Regarding Prime Numbers." *Nature* 55 (31 December 1896): 196–197; (21 January 1897): 269.

―――. *The Collected Mathematical Papers of James Joseph Sylvester*. Edited by Henry F. Baker. 4 Vols. Cambridge, UK: Cambridge University Press, 1904–1912; Reprint ed. New York: Chelsea Publishing Co., 1973.

Published Sources

Abel-Smith, Brian, and Stevens, Robert. *Lawyers and the Courts: A Sociological Study of the English Legal System, 1750–1965*. London: Heinemann, 1967.

Airy, Sir George Biddle. *Autobiography*. Edited by Wilfred Airy. Cambridge, UK: Cambridge University Press, 1896.

Alborn, Timothy L. "A Calculating Profession: Victorian Actuaries among the Statisticians." *Science in Context* 7 (1994): 433–468.

Allen, David Elliston. *The Naturalist in Britain*, 2d ed. Princeton, NJ: Princeton University Press, 1994.

Ammen, S. Z. "History of Baltimore: 1875–1895." In *Baltimore: Its History and People*. Edited by Clayton Colman Hall. Vol. 1, *History*. New York: Lewis Historical Publishing Co., 1912, pp. 241–288.

The Annual Register, or a View of the History and Politics of the Year 1851. London: F. & J. Rivington, 1852 (and successive years).

Archibald, Raymond C. "Material Concerning James Joseph Sylvester." In *Studies and Essays in the History of Science and Learning Offered in Homage to George Sarton on the Occasion of His Sixtieth Birthday, 31 August 1944*. New York: Schuman, n.d., pp. 209–217.

―――. "Unpublished Letters of James Joseph Sylvester and Other New Information Concerning His Life and Work." *Osiris* 1 (1936): 85–154.

Arnold, Vladimir. "Topological Content of the Maxwell Theorem on Multipole Representation of Spherical Functions." *Topological Methods in Nonlinear Analysis* 7 (1996): 205–217.

Aronhold, Siegfried. "Zur Theorie der homogenen Functionen dritten Grades von drei Variabeln." *Journal für die reine und angewandte Mathematik* 39 (1849): 140–159.

"Art. IX: The Cambridge Controversy: Admission of Dissenters to Degrees." *Quarterly Review* 52 (1834): 466–487.

"Art. X: English Corporations and Endowments." *Edinburgh Review* 58 (1834): 469–498.

Ashworth, William. "The Calculating Eye: Baily, Herschel, Babbage, and the Business of Astronomy." *British Journal for the History of Science* 27 (1994): 409–441.

———. "Memory, Efficiency, and Symbolic Analysis: Charles Babbage, John Herschel, and the Industrial Mind." *Isis* 87 (1996): 629–653.

Bachmann, Paul. *Die Lehre von der Kreistheilung und ihre Beziehungen zur Zahlentheorie.* Leipzig, Germany: B. G. Teubner Verlag, 1872.

Baker, Henry F. "Biographical Notice." In *Math. Papers JJS*, 4: xv–xxxvii.

Ballantine, William. *Some Experiences of a Barrister's Life.* New York: J. M. Stoddart, 1883.

Barnard, Henry. *Military Schools and Courses of Instruction in the Science and Art of War.* New York: E. Steiger, 1872; Reprint Ed., New York: Greenwood Press Publishers, 1969.

Barrow-Green, June. " 'A Corrective to the Spirit of Too Exclusively Pure Mathematics': Robert Smith (1689–1768) and his Prizes at Cambridge University." *Annals of Science* 56 (1999): 271–316.

———. "Gösta Mittag-Leffler and the Foundation and Administration of *Acta Mathematica*." In *Mathematics Unbound: The Evolution of an International Mathematical Research Community, 1800–1945.* Edited by Karen Hunger Parshall and Adrian C. Rice. HMATH. Vol. 23. Providence, RI: American Mathematical Society; London: London Mathematical Society, 2002, pp. 138–164.

———. *Poincaré and the Three Body Problem.* HMATH. Vol. 11. Providence, RI: American Mathematical Society; London: London Mathematical Society, 1997.

Becher, Harvey W. "Radicals, Whigs, and Conservatives: The Middle and Lower Classes in the Analytical Revolution in Cambridge in the Age of Aristocracy." *British Journal for the History of Science* 28 (1995): 405–426.

———. "William Whewell and Cambridge Mathematics." *Historical Studies in the Physical Sciences* 11 (1980): 1–48.

Becker, Bernard H. *Scientific London.* New York: D. Appleton & Co., 1875.

Beirne, Francis F. *The Amiable Baltimoreans.* Baltimore: The Johns Hopkins University Press, 1951.

Beirne, Francis F., and the Maryland Historical Society. *Baltimore . . . a Picture History, 1858–1958.* New York: Hastings House, 1957.

Bell, Eric Temple. *Men of Mathematics.* New York: Simon and Schuster, 1937.

Bellot, H. Hale. *University College London 1826–1926.* London: University of London Press, Ltd., 1929.

Benas, Baron L. "Records of the Jews in Liverpool." *Transactions of the Historic Society of Lancashire and Cheshire* 51 (1901): 45–83.

Benas, Bertram B. "A Survey of the Jewish Institutional History of Liverpool and District." *Transactions of the Jewish Historical Society of England* 17 (1953): 23–38.

Besant, Sir Walter. *London North of the Thames.* London: Adam & Charles Black, 1911.

Best, Geoffrey. *Mid-Victorian Britain: 1851–1875.* London: Weidenfeld and Nicolson, 1971.

Betti, Enrico. "Extract from [a] Letter of Signor Enrico Betti to Mr. Sylvester." *Quarterly Journal of Pure and Applied Mathematics* 1 (1855): 91–92.

Biggs, Norman L. "T. P. Kirkman, Mathematician." *Bulletin of the London Mathematical Society* 13 (1981): 97–120.

Blondheim, David. "A Brilliant and Eccentric Mathematician." *Johns Hopkins University Alumni Magazine* 9 (1921): 119–140.

Bond, Allen Kerr. *When the Hopkins Came to Baltimore.* Baltimore: The Pegasus Press, 1927.

Boole, George. "Exposition of a General Theory of Linear Transformations." *Cambridge Mathematical Journal* 3 (1841–1842): 1–20, 106–119.

Borchardt, Carl. "Développements sur l'équation à l'aide de laquelle on détermine les inégalités séculaires du mouvement des planètes." *Journal de mathématiques pures et appliquées* 12 (1847): 50–67.

Brent, George. *Charles Sanders Peirce: A Life.* Rev. and Enl. Ed. Bloomington, IN: Indiana University Press, 1998.

Brioschi, Francesco. *La teorica dei determinanti e le sue principali applicazioni.* Pavia, Italy: Tipografia degli eredi Bizzoni, 1854.

Bristed, Charles Astor. *Five Years in an English University.* 3d Ed. New York: G. P. Putnam & Sons, 1874.

Bronwin, Brice. "On the Solution of Linear Difference Equations." *Philosophical Transactions of the Royal Society of London* 141 (1851): 461–482.

Brooke, Richard. *Liverpool As It Was during the Last Quarter of the Eighteenth Century 1775 to 1800.* Liverpool, UK: J. Mawdsley & Son, 1853; Reprint Ed. Liverpool, UK: The University Press of Liverpool, 2003.

Brown, A. Theodore. *Some Account of the Royal Institution School of Liverpool with a Roll of Masters and Boys (1819 to 1892, A. D.).* 2d Ed. Liverpool, UK: The University Press of Liverpool Ltd.; London: Hodder and Stoughton Ltd., 1927.

Brown, Dee. *The Year of the Century: 1876.* New York: Charles Scribner's Sons, 1966.

Brown, W. Norman, Ed. *Johns Hopkins Half-Century Directory: A Catalogue of the Trustees, Faculty, Holders of Honorary Degrees, and Students, Graduates, and Non-Graduates.* Baltimore: The Johns Hopkins University Press, 1926.

Bruce, Philip Alexander. *History of the University of Virginia: 1819–1919.* 5 Vols. New York: Macmillan, 1920.

Bruce, Robert V. *The Launching of Modern American Science: 1846–1876.* New York: Alfred A. Knopf, 1987.

Buchheim, Arthur. "Proof of Prof. Sylvester's 'Third Law of Motion.'" *Philosophical Magazine* 18 (1884): 459–460.

Burnside, William S., and Panton, Arthur W. *The Theory of Equations with an Introduction to the Theory of Binary Algebraic Forms.* 7th Ed. London: Longmans, Green, and Co., 1912; Reprint Ed., New York: Dover Publications, Inc., 1960.

Butterworth, Harry. "The Science and Art Department Examinations: Origins and Achievements." In *Days of Judgement: Science, Examinations and the Organization of Knowledge in Late Victorian England.* Edited by Roy MacLeod. Driffield, UK: Studies in Education Ltd. Nafferton Books, 1982, 27–44.

Buxton, John, and Williams, Penry, Ed. *New College Oxford 1379–1979.* Oxford, UK: Warden and Fellows of New College Oxford, 1979.

C. C. "Obituary: The Rev. H. H. Hughes." *The Eagle* 13 (1885): 208–213.

Cajori, Florian. *The Teaching and History of Mathematics in the United States.* Washington, DC: Government Printing Office, 1890.

The Cambridge University Calendar for the Year 1831. Cambridge, UK: J. & J. J. Deighton, 1831 (and subsequent years).

The Cambridge Guide, Including Historical and Architectural Notices of the Public Buildings, and a Concise Account of the Customs and Ceremonies of the University.

Cambridge, UK: J. & J. J. Deighton, Thomas Stevenson, and Richard Newby, 1837.

Cannell, D. Mary. *George Green, Mathematician and Physicist, 1793–1841: The Background to His Life and Work.* London: The Athlone Press, 1993; 2d Ed. Philadelphia: Society for Industrial and Applied Mathematics, 2001.

Cannon, Susan Faye. *Science in Culture: The Early Victorian Period.* New York: Dawson and Science History Publications, 1978.

Cantor, Geoffrey. "Creating the Royal Society's Sylvester Medal." *British Journal for the History of Science* 37 (2004): 75–92.

Carrick, Neville, and Ashton, Edward L. *The Athenæum Liverpool 1797–1997.* Liverpool, UK: The Athenæum Liverpool, 1997.

Castellana, Mario, and Palladino, Franco, Ed. *Giuseppe Battaglini: Raccolta di lettere (1854–1891) di un matematico al tempo del Risorgimento d'Italia.* Bari, Italy: Levante Editori, 1996.

Catalogue of the Officers and Students of the University of Virginia: Session of 1841–42. Charlottesville, VA: James Alexander, 1842.

Cauchy, Augustin-Louis. "Mémoire sur les arrangements que l'on peut former avec les lettres données, et sur les permutations ou substitutions à l'aide desquelles on passe d'un arrangement à un autre." *Exercices d'analyse et de physique mathématique* 3 (1844–1846): 151–252.

Cayley, Arthur. "An Introductory Memoir upon Quantics." *Philosophical Transactions of the Royal Society of London* 144 (1854): 244–258.

———. "A Memoir on the Theory of Matrices." *Philosophical Transactions of the Royal Society of London* 148 (1858): 17–37.

———. "A Ninth Memoir on Quantics." *Philosophical Transactions of the Royal Society of London* 161 (1871): 17–50.

———. "On the Finite Number of Covariants of a Binary Quantic." *Mathematische Annalen* 34 (1889): 319–320.

———. "On Lambert's Theorem for Elliptic Motion." *Monthly Notices of the Royal Astronomical Society* 22 (1862): 238–242.

———. "On Linear Transformations." *Cambridge and Dublin Mathematical Journal* 1 (1846): 104–122.

———. "On the Six Coordinates of a Line." *Transactions of the Cambridge Philosophical Society* 11 (1869): 290–323.

———. "On the Theory of Groups, As Depending on the Symbolic Equation $\Theta^n = 1$." *Philosophical Magazine* 7 (1854): 40–47.

———. "On the Theory of Linear Transformations." *Cambridge Mathematical Journal* 4 (1845): 193–209.

———. "A Second Memoir upon Quantics." *Philosophical Transactions of the Royal Society of London* 146 (1856): 101–126.

———. "A Supplementary Memoir on the Theory of Matrices." *Philosophical Transactions of the Royal Society of London* 156 (1866): 25–35.

———. "Sur le problème des contacts." *Journal für die reine und angewandte Mathematik* 39 (1850): 4–13.

———. "Theorems Relating to the Canonic Roots of a Binary Quantic of an Odd Order." *Philosophical Magazine* 25 (1863): 206–208.

Cayley, Arthur, and Forsyth, Andrew R., Ed. *The Collected Mathematical Papers of Arthur Cayley.* 14 Vols. Cambridge, UK: Cambridge University Press, 1889–1898.

Chalklin, C. W. *The Provincial Towns of Georgian England: A Study of the Building Process 1740–1820*. London: Edward Arnold, 1974.

Chartres, Richard, and Vermont, David. *A Brief History of Gresham College 1597–1997*. London: Gresham College, 1997.

Chebyshev, Pafnuti. "Mémoire sur les nombres premiers." *Journal de mathématiques pures et appliquées* 17 (1852): 366–390.

———. *Polnoe Sobranie Sochinenii*. Vol. 5. *Prochie Sochinenia Biograficheskie Materialy*. Moscow: AN SSR, 1951.

Clebsch, Alfred. "Ueber die Anwendung der Abel'schen Functionen in der Geometrie." *Journal für die reine und angewandte Mathematik* 63 (1864): 189–243.

Clifford, William. "Remarks on the Chemico-Algebraical Theory." *American Journal of Mathematics* 1 (1878): 126–128.

Collins, Mortimer. *Who Is the Heir?*. 3 Vols. London: John Maxwell and Company, 1865.

Conington, John. *The Satires, Epistles and Art of Poetry of Horace*. London: Bell and Daldy, 1870.

Corry, Leo. *Modern Algebra and the Rise of Mathematical Structures*. Basel, Switzerland: Birkhäuser Verlag, 1996.

Cowell, F. R. *The Athenæum: Club and Social Life in London, 1824–1974*. London: Heinemann, 1975.

Craig, Thomas. "On the Motion of an Ellipsoid in a Fluid." *American Journal of Mathematics* 2 (1879): 260–279.

———. "On Quadruple Theta-Functions." *American Journal of Mathematics* 6 (1884): 14–59.

———. "Some Elliptic Function Formulæ." *American Journal of Mathematics* 5 (1882): 62–75.

Cremona, Luigi. *Éléments de géométrie projective*. Paris: Gauthier-Villars, 1875.

———. *Elements of Projective Geometry*. Trans. Charles Leudesdorf. Oxford, UK: Clarendon Press, 1885.

Crilly, Tony. *Arthur Cayley: Mathematician Laureate of the Victorian Age*. Baltimore: The Johns Hopkins University Press, 2006.

———. "The *Cambridge Mathematical Journal* and Its Descendents: 1830–1870." *Historia Mathematica* 31 (2004): 255–297.

———. "Cayley's Anticipation of a Generalised Cayley-Hamilton Theorem." *Historia Mathematica* 5 (1978): 211–219.

———. "The Decline of Cayley's Invariant Theory (1863–1895)." *Historia Mathematica* 15 (1988): 332–347.

———. "The Mathematics of Arthur Cayley with Particular Reference to Linear Algebra." Unpublished doctoral dissertation. Middlesex Polytechnic Institute, 1981.

———. "The Rise of Cayley's Invariant Theory (1841–1862)." *Historia Mathematica* 13 (1986): 241–254.

Cross, Patrick S. "The Organization of Science in Dublin from 1785 to 1835: The Men and Their Institutions." Unpublished doctoral dissertation. University of Oklahoma, 1996.

Cumming, James. "On the Application of Magnetism as a Measure of Electricity." *Transactions of the Cambridge Philosophical Society* 1 (1821–1822): 281–286.

———. "On the Connexion of Galvanism and Magnetism." *Transactions of the Cambridge Philosophical Society* 1 (1821–1822): 268–279.

Currie, William Wallace, Ed. *Memoir of the Life, Writings, and Correspondence of James Currie, M.D., F.R.S. of Liverpool.* 2 Vols. London: Longman, Rees, Orme, Brown, and Green, 1831.

Daniels, George H. *American Science in the Age of Jackson.* New York: Columbia University Press, 1968.

Davenport, Guy. *A Table of Green Fields.* New York: New Directions Books, 1993.

De Morgan, Augustus. "On the Dimensions of the Roots of Equations." *Quarterly Journal of Pure and Applied Mathematics* 1 (1855): 1–3.

Desmond, Adrian. *Huxley: From Devil's Disciple to Evolution's High Priest.* Reading, MA: Addison-Wesley, 1997.

———. *The Politics of Evolution: Morphology, Medicine, and Reform in Radical London.* Chicago: University of Chicago Press, 1989.

Despeaux, Sloan Evans. "The Development of a Publication Community: Nineteenth-Century Mathematics in British Scientific Journals." Unpublished doctoral dissertation. University of Virginia, 2002.

———. "International Mathematical Contributions to British Scientific Journals, 1800–1900." In *Mathematics Unbound: The Evolution of an International Mathematical Research Community, 1800–1945.* Edited by Karen Hunger Parshall and Adrian C. Rice. HMATH. Vol. 23. Providence, RI: American Mathematical Society; London: London Mathematical Society, 2002, 61–87.

———. "'Very Full of Symbols': Duncan F. Gregory, the Calculus of Operations, and the *Cambridge Mathematical Journal.*" In *Episodes in the History of Modern Algebra.* Edited by Karen Hunger Parshall and Jeremy J. Gray. HMATH. Providence, RI: American Mathematical Society; London: London Mathematical Society. Forthcoming.

Dixmier, Jacques, and Lazard, Daniel. "Minimum Number of Fundamental Invariants for the Binary Form of Degree 7." *Journal of Symbolic Computation* 6 (1988): 113–115.

———. "Le nombre minimum d'invariants fondamentaux pour les formes binaires de degré 7." *Portugaliae Mathematica* 43 (1986): 377–392.

Drayson, Alfred. *Experiences of a Woolwich Professor During Fifteen Years at the Royal Military Academy.* London: Chapman & Hall, 1886.

Duming, Andrew, and Guest, Bill, Ed. *Natal and Zululand from Earliest Times to 1910: A History.* Pietermaritzburg, South Africa: University of Natal Press, 1989.

Dupree, A. Hunter. *Science in the Federal Government: A History of Policies and Activities.* Baltimore: The Johns Hopkins University Press, 1986.

Durand-Richard, Marie-José. "L'École algébrique anglaise: les conditions conceptuelles et institutionnelles d'un calcul symbolique comme fondement de la connaissance." In *L'Europe mathématique—Mathematical Europe.* Edited by Catherine Goldstein, Jeremy J. Gray, and Jim Ritter. Paris: Éditions de la Maison de l'Homme, 1996.

Dyer, George. *Academic Unity.* London: Richard Taylor, 1827.

Eisenstein, Gotthold. "Théorèmes sur les formes cubiques et solution d'une équation de quatrième degré à quatre indéterminés." *Journal für die reine und angewandte Mathematik* 27 (1844): 75–79.

———. "Untersuchungen über die cubischen Formen mit zwei Variabeln." *Journal für die reine und angewandte Mathematik* 27 (1844): 10–25.

Elliott, Edwin Bailey. *An Introduction to the Algebra of Quantics.* 2d Ed. Oxford, UK: Oxford University Press, 1913; Reprint Ed., Bronx, NY: Chelsea Publishing Co., n.d.

———. "Why and How the Society Began and Kept Going." 16 May 1925. Privately printed pamphlet.

Elliot, Maud Howe. *Uncle Sam Ward and His Circle.* New York: The Macmillan Co., 1938.

Ely, George S. "Bibliography of Bernoulli Numbers." *American Journal of Mathematics* 5 (1882): 228–235.

———. "Some Notes on the Numbers of Bernoulli and Euler." *American Journal of Mathematics* 5 (1882): 337–341.

Endelman, Todd. *The Jews of Britain, 1656 to 2000.* Berkeley, CA: University of California Press, 2000.

———. *The Jews of Georgian England 1714–1830: Tradition and Change in a Liberal Society.* N.p.: The Jewish Publication Society of America, 1979.

Engel, Arthur J. "Emerging Concepts of the Academic Profession at Oxford 1800–1854." In *The University in Society.* Edited by Lawrence Stone. 2 Vols. *Oxford and Cambridge from the 14th to the Early 19th Century.* Vol. 1. Princeton, NJ: Princeton University Press, 1974, 305–351.

———. *From Clergyman to Don: The Rise of the Academic Profession in Nineteenth-Century Oxford.* Oxford, UK: Clarendon Press, 1983.

Enros, Philip C. "The Analytical Society: Mathematics at Cambridge in the Early Nineteenth Century." Unpublished doctoral dissertation. University of Toronto, 1979.

———. "The Analytical Society (1812–1813): Precursor of the Revival of Cambridge Mathematics." *Historia Mathematica* 10 (1983): 24–47.

———. "Cambridge University and the Adoption of Analytics in Early Nineteenth-Century England." In *Social History of Nineteenth Century Mathematics.* Edited by Herbert Mehrtens, Henk Bos, and Ivo Schneider. Boston: Birkhäuser Verlag, 1981, 135–148.

Equity & Law 150 Anniversary 1844–1994. N.p.: Communication (Production) AXA Equity & Law, 1994.

Ezell, John Samuel. *Fortune's Merry Wheel: The Lottery in America.* Cambridge, MA: Harvard University Press, 1960.

Faà di Bruno, Francesco. *Théorie des formes binaires.* Turin, Italy: P. Marietti, 1876.

Fauvel, John. "Sylvester at Oxford." In *Oxford Figures: 800 Years of the Mathematical Sciences.* Edited by John Fauvel, Raymond Flood, and Robin Wilson. Oxford, UK: Oxford University Press, 2000, 219–239.

Feingold, Mordecai. "A Jesuit among Protestants: Boscovich in England c. 1745–1820." In *R. J. Boscovich: Vita e Attività scientifica/His Life and Scientific Work.* Edited by Piers Bursill-Hall. Rome: Istituto della Enciclopedia italiana, 1993, 511–526.

Feuer, Lewis S. "America's First Jewish Professor: James Joseph Sylvester at the University of Virginia." *American Jewish Archives* 36 (1984): 151–201.

Fiedler, Wilhelm, Trans. *Analytische Geometrie der Kegelschnitte von George Salmon.* Leipzig, Germany: B. G. Teubner Verlag, 1860.

Finestein, Israel. *Jewish Society in Victorian England: Collected Essays.* London: Vallentine Mitchell, 1993.

First Report of the Royal Commission Appointed to Inquire into the Present State of Military Education and into the Training of Candidates for Commissions into the Army. London: G. E. Eyre and W. Spottiswoode, 1869.

Fisch, Max, and Cope, Jackson I. "Peirce at The Johns Hopkins University." In *Studies in the Philosophy of Charles Saunders Peirce*. Edited by Philip P. Wiener and Frederic H. Young. Cambridge, MA: Harvard University Press, 1952, 277–311, 355–360, 363–374.

Fisch, Menachem. "The Problematic History of Nineteenth-Century British Algebra." *British Journal for the History of Science* 27 (1994): 247–276.

Fischer, Gerd, Ed. *Mathematische Modelle*. 2 Vols. Berlin: Akademie Verlag, 1986.

Fisher, Charles S. "The Death of a Mathematical Theory: A Study in the Sociology of Knowledge." *Archive for History of Exact Sciences* 3 (1966): 137–159.

———. "The Last Invariant Theorists: A Sociological Study of the Collective Biographies of Mathematical Specialists." *Archives européennes de sociologie* 8 (1967): 216–244.

Fisher, Herbert, A. L. *An Unfinished Autobiography*. London: Oxford University Press, 1940.

Fitzgerald, Percy. *London City Suburbs As They Are To-day*. London: The Leadenhall Press, Ltd., 1893; Reprint Ed., London: The Alderman Press, 1984.

Forsyth, Andrew R. "A Class of Functional Invariants." *Philosophical Transactions of the Royal Society of London* 180 (1889): 71–118.

———. "Old Tripos Days at Cambridge." *Mathematical Gazette* 19 (1935): 162–179.

———. *A Treatise on Differential Equations*. London: Macmillan and Co., 1885.

Francis, John. *Annals, Anecdotes and Legends: A Chronicle of Life Assurance*. London: Longman, Brown, Green, and Longmans, 1853.

Frankland, Edward. "Remarks on Chemico-graphs." *American Journal of Mathematics* 1 (1878): 345–349.

Franklin, Fabian. "Bipunctual Coordinates." *American Journal of Mathematics* 1 (1878): 148–173.

———. "James Joseph Sylvester." *Bulletin of the American Mathematical Society* 3 (1897): 299–309.

———. *The Life of Daniel Coit Gilman*. New York: Dodd, Mead and Company, 1910.

———. "On the Calculation of the Generating Functions and Tables of Groundforms for Binary Quantics." *American Journal of Mathematics* 3 (1880): 128–153.

———. "On Cubic Curves." *American Journal of Mathematics* 5 (1882): 212–217.

———. "On Partitions." *American Journal of Mathematics* 2 (1879): 187–190.

———. "On a Problem of Isomerism." *American Journal of Mathematics* 1 (1878): 365–369.

———. "Sur le développement du produit infini $(1 - x)(1 - x^2)(1 - x^3)(1 - x^4) \cdots$." *Comptes rendus de l'Académie des Sciences de Paris* 82 (1881): 448–450.

Gardner, J. Helen, and Wilson, Robin J. "Thomas Archer Hirst—Mathematician Xtravagant, V. London in the 1860s." *American Mathematical Monthly* 100 (1993): 827–834.

Garland, Martha McMackin. *Cambridge before Darwin: The Ideal of a Liberal Education, 1800–1860*. Cambridge, UK: Cambridge University Press, 1980.

Gascoigne, John. "Mathematics and Meritocracy: The Emergence of the Cambridge Mathematical Tripos." *Social Studies of Science* 14 (1984): 547–584.

Gillispie, Charles C., Ed. *Dictionary of Scientific Biography*. 16 Vols. 2 Suppls. New York: Charles Scribner's Sons, 1970–1990.

Gilman, Daniel Coit. *The Launching of a University*. New York: Dodd, Mead, & Co., 1906.

Goldsmid, Francis Henry. *Remarks on the Civil Disabilities of British Jews*. London: Henry Colburn and Richard Bentley, 1830.

Goldstein, Catherine, Gray, Jeremy J., Ritter Jim, Ed. *L'Europe mathématique—Mathematical Europe*. Paris: Éditions de la Maison de l'Homme, 1996.

Gore's Liverpool Directory. Liverpool, UK: J. Gore, 1790 (and subsequent years under several title variations).

Grattan-Guinness, Ivor. "The Contributions of J. J. Sylvester, F.R.S., to Mechanics and Mathematical Physics." *Notes and Records of the Royal Society of London* 55 (2001): 253–265.

————. "The Sylvester Medal: Origins and Recipients 1901–1949." *Notes and Records of the Royal Society of London* 47 (1993): 105–108.

Graves, Robert Perceval. *Life of Sir William Rowan Hamilton*. 3 Vols. Dublin: Hodges, Figgis, & Co. and London: Longmans, Green, and Co., 1882–1889.

Green, Geoffrey. *The Royal Navy and Anglo-Jewry 1740–1820: Traders and Those Who Served*. London: Naval and Maritime Bookshop, 1989.

Green, Judy. "Christine Ladd-Franklin (1847–1930)." In *Women of Mathematics: A Bio-bibliographic Sourcebook*. Edited by Louise Grinstein and Paul J. Campbell. New York: Greenwood Press, 1987, pp. 121–128.

Greene, John C. *American Science in the Age of Jefferson*. Ames, IA: The Iowa State University Press, 1984.

Guggisberg, F. G. *"The Shop": The Story of the Royal Military Academy*. London: Cassell and Company, Limited, 1900.

Guicciardini, Niccolò. *The Development of Newtonian Calculus in Britain, 1700–1800*. Cambridge, UK: Cambridge University Press, 1989.

Guthrie, William Bell. "Matthew Arnold's Diaries: The Unpublished Items: A Transcription and Commentary." 4 Vols. Unpublished doctoral dissertation. University of Virginia, 1957.

Hall, Clayton Colman, Ed. *Baltimore: Its History and People*. Vol. 1, *History*. New York: Lewis Historical Publishing Company, 1912.

Halsted, George Bruce. "James Joseph Sylvester." *American Mathematical Monthly* 1 (1894): 295–298.

————. "Sylvester at Hopkins." *Johns Hopkins University Alumni Magazine* 4 (1916): 178–188.

Hammond, James. "On the Solution of the Differential Equations of Sources." *American Journal of Mathematics* 5 (1882): 218–227.

Harman, Peter M., Ed. *Wranglers and Physicists: Studies on Cambridge Mathematical Physics in the Nineteenth Century*. Manchester, UK: Manchester University Press, 1985.

Harries-Jenkins, Gwyn. *The Army in Victorian Society*. London: Routledge & Kegan Paul, 1977.

Harrison, J. F. C. *A History of the Working Men's College 1854–1954*. London: Routledge & Kegan Paul, 1954.

————. *Learning and Living 1790–1960: A Study in the History of the English Adult Education Movement*. Toronto: University of Toronto Press, 1961.

Harte, Negley, and North, John. *The World of University College London 1828–1978*. London: University College London, [1978?].

Hawkins, Hugh. *Pioneer: A History of the Johns Hopkins University, 1874–1889*. Ithaca, NY: Cornell University Press, 1960.

Hawkins, Thomas. "Another Look at Cayley and the Theory of Matrices." *Archives internationales d'histoire des sciences* 26 (1977): 82–112.

———. "Cayley's Counting Problem and the Representation of Lie Algebras." In *Proceedings of the International Congress of Mathematicians–Berkeley*. 2 Vols. Providence, RI: American Mathematical Society, 1987, 2: 1642–1656.

———. "Hypercomplex Numbers, Lie Groups, and the Creation of Group Representation Theory." *Archive for History of Exact Sciences* 8 (1972): 243–287.

Hays, Jo N. "The Rise and Fall of Dionysius Lardner." *Annals of Science* 38 (1981): 527–542.

Hayter, William. *Spooner: A Biography*. London: W. H. Allen, 1977.

Hearl, Trevor. "Military Examinations and the Training of Science, 1857–1870." In *Days of Judgement: Science, Examinations and the Organization of Knowledge in Late Victorian England*. Edited by Roy MacLeod. Driffield, UK: Studies in Education Ltd. Nafferton Books, 1982, 109–149.

Hermite, Charles. "Études de M. Sylvester sur la théorie algébriques des formes." *Comptes rendus de l'Académie des Sciences de Paris* 84 (1877): 974–975.

———. "Sur les formes cubiques à deux indéterminées." *Quarterly Journal of Pure and Applied Mathematics* 1 (1855): 20–22.

Herstein, Sheila R. *Mid-Victorian Feminist: Barbara Leigh Smith Bodichon*. New Haven, CT: Yale University Press, 1985.

Hesse, Otto. "Über die Elimination der Variabeln aus drei algebraischen Gleichungen vom zweiten Grade mit zwei Variabeln." *Journal für die reine und angewandte Mathematik* 28 (1844): 68–96.

———. "Über die Wendepunkte der Curven dritter Ordnung," *Journal für die reine und angewandte Mathematik* 28 (1844): 97–107.

Heyck, Thomas W. *The Transformation of Intellectual Life in Victorian England*. London: Croom Helm, Ltd., 1982.

Hilbert, David. "Über die Endlichkeit des Invariantensystems für binäre Grundformen." *Mathematische Annalen* 34 (1889): 223–226.

———. "Ueber die Theorie der algebraischen Invarianten." In *Mathematical Papers Read at the International Congress Held in Connection with the World's Columbian Exposition, Chicago 1893*. Edited by Eliakim Hastings Moore, Oskar Bolza, Heinrich Maschke, and Henry S. White. New York: Macmillan and Co., 1896, 116–124.

———. "Ueber die vollen Invariantensysteme." *Mathematische Annalen* 42 (1893): 313–373.

Hirsch, Pam. *Barbara Leigh Smith Bodichon, 1827–1891: Feminist, Artist and Rebel*. London: Chatto & Windus, 1998.

Hogg, O. F. G. *The Royal Arsenal: Its Background, Origin, and Subsequent History*. 2 Vols. London: Oxford University Press, 1963.

Hopkins, William. *Remarks on Certain Proposed Regulations Respecting the Study of the University*. Cambridge, UK: N.p., 1841 (Privately printed pamphlet).

Houser, Nathan. "Introduction." In *Writings of Charles S. Peirce: A Chronological Edition, Vol. 4 (1879–1884)*. Edited by Christian J. W. Kloesel. Bloomington, IN: Indiana University Press, 1982, xix–lxx.

Huxley, Thomas. "The Scientific Aspects of Positivism." *Fortnightly Review* 11 (1869): 653–670.

———. "Scientific Education: Notes of an After-dinner Speech." *Macmillan's Magazine* 20 (1869): 177–184.

Hyman, Leonard. "Hyman Hurwitz: The First Anglo-Jewish Professor." *Transactions of the Jewish Historical Society of England* 21 (1968): 232–242.

Jackson-Stops, Gervase. "Restoration and Expansion: The Buildings since 1750." In *New College Oxford 1379–1979*. Edited by John Buxton and Penry Williams. Oxford, UK: Warden and Fellows of New College Oxford, 1979, 233–264.

Jacobs, Joseph, Ed. *The Jewish Year Book: An Annual Record of Matters Jewish*. London: N.p., 1897.

Jarvis, Adrian. *Liverpool Central Docks, 1799–1905: An Illustrated History*. Gloucestershire, UK: Alan Sutton Publishing Ltd., 1991.

Jenkins, Hester, and Jones, D. Caradog. "Social Class of Cambridge University Alumni of the 18th and 19th Centuries." *The British Journal of Sociology* 1 (1950): 93–116.

Johnson, Thomas Cary. *The Life and Letters of Robert Lewis Dabney*. Richmond, VA: The Presbyterian Committee of Publication, 1903.

Johnson, W. "The Woolwich Professors of Mathematics, 1741–1900." *Journal of Mechanical Working Technology* 18 (1989): 145–194.

Jones, Carlton. *Lost Baltimore: A Portfolio of Vanished Buildings*. Baltimore: The Johns Hopkins University Press, 1993.

Jones, J. Vernon. "The X Club: Fraternity of Victorian Scientists." *British Journal for the History of Science* 5 (1970): 63–70.

J. W., "James Joseph Sylvester Sc.D." *The Eagle* 19 (1897): 596–605.

Kelland, Philip. "On the Transmission of Light in Crystallized Media." *Transactions of the Cambridge Philosophical Society* 6 (1837): 323–352, 353–360.

Kelly, Thomas. *A History of Adult Education in Great Britain*. Liverpool, UK: The University Press of Liverpool, 1992.

Kempe, Alfred. "A Memoir on the Theory of Mathematical Form." *Philosophical Transactions of the Royal Society of London* 177 (1886): 1–70.

———. "On the Geographical Problem of the Four Colours." *American Journal of Mathematics* 2 (1879): 193–200.

Ketner, Kenneth Laine, and Kloesel, Christian J. W. *Peirce, Semeiotic, and Pragmatism: Essays of Max Fisch*. Bloomington, IN: Indiana University Press, 1986.

King, Moses. *Benjamin Peirce: A Memorial Collection*. Cambridge, MA: N.p., 1881.

Kline, Morris. *Mathematical Thought from Ancient to Modern Times*. New York: Oxford University Press, 1972.

Kloesel, Christian J. W., Ed. *Writings of Charles S. Peirce: A Chronological Edition, Vol. 4 (1879–1884)*. Bloomington, IN: Indiana University Press, 1982.

Koppelman, Elaine. "The Calculus of Operations and the Rise of Abstract Algebra." *Archive for History of Exact Sciences* 8 (1971–1972): 155–242.

Kung, Joseph P. S., and Rota, Gian-Carlo. "The Invariant Theory of Binary Forms." *Bulletin of the American Mathematical Society* 10 (1984): 27–85.

Lachs, Phyllis S. "A Study of a Professional Elite: Anglo-Jewish Barristers in the Nineteenth Century." *Jewish Social Studies* 44 (1982): 125–134.

Ladd, Christine. "On the Algebra of Logic." In *Studies in Logic by Members of the Johns Hopkins University*. Edited by Charles S. Peirce. Boston: Little and Co., 1883; Reprint Ed. (with an introduction by Max Fisch and a preface by Achim Eschbach). Amsterdam, The Netherlands: John Benjamins Publishing Company, 1983, 17–71.

Lane, Tony. *Liverpool: Gateway of Empire*. London: Lawrence & Wishart, 1987.

Lanier, Sidney. *The Science of English Verse*. New York: Charles Scribner's Sons, 1880.

Lavrinenko, Tatiana. "Solving an Indeterminate Third Degree Equation in Rational Numbers: Sylvester and Lucas." *Revue d'histoire des mathématiques* 8 (2002): 67–111.

Lawton, Richard, and Lee, Robert. Ed. *Population and Society in Western European Port-Cities, c. 1650–1939*. Liverpool, UK: The University Press of Liverpool, 2002.

Leclerc, Georges-Louis (Comte de Buffon). *Œuvres philosophiques de Buffon*. Edited by Jean Piveteau, Maurice Fréchet, and Charles Bruneau. Paris: Presses universitaires de France, 1954.

Legendre, Adrien-Marie. *Théorie des nombres*. 2 Vols. Paris: Firmin Didot Frères, 1830; Reprint Ed., Paris: Albert Blanchard, 1955.

Lejeune-Dirichlet, Peter. *Vorlesungen über Zahlentheorie*. 2d Ed. Edited by Richard Dedekind. Braunschweig, Germany: F. Vieweg und Sohn, 1871.

Lipman, Vivian D. "The Age of Emancipation 1815–1880." In *Three Centuries of Anglo-Jewish History*. Edited by Vivian D. Lipman. Cambridge, UK: W. Heffer and Sons Limited, 1961, 69–106.

———. *A History of the Jews in Britain since 1858*. Leicester, UK: Leicester University Press, 1990.

———. *Social History of the Jews of England, 1850–1950*. London: Watts & Co., 1954.

———. Ed. *Three Centuries of Anglo-Jewish History*. Cambridge, UK: W. Heffer and Sons Limited, 1961.

"The Liverpool Institute: The History and Development of an Idea." *The Liverpool Review* 3 (October 1928): 405–408.

Lucas, Édouard, "Sur l'analyse indéterminée du troisième degré.—Démonstration de pluiseurs théorèmes de M. Sylvester." *American Journal of Mathematics* 2 (1879): 178–185.

Lützen, Jesper. *Joseph Liouville 1809–1882: Master of Pure and Applied Mathematics*. New York: Springer-Verlag, 1990.

Lyall, Sir Alfred. *The Life of the Marquis of Dufferin and Ava*. 2 Vols. London: John Murray, 1905.

[Macaulay, Thomas Babington]. "Statement of the Civil Disabilities and Privations Affecting Jews in England." *Edinburgh Review* 52 (1831): 363–374.

McDowell, Robert B., and Webb, David A. *Trinity College Dublin 1592–1952: An Academic History*. Cambridge, UK: Cambridge University Press, 1982.

Macfarlane, Alexander. *Lectures on Ten British Mathematicians of the Nineteenth Century*. New York: John Wiley & Sons, 1916.

MacLeod, Roy. "The X Club." *Notes and Records of the Royal Society of London* 24 (1970): 305–322.

———, Ed. *Days of Judgement: Science, Examinations and the Organization of Knowledge in Late Victorian England*. Driffield, UK: Studies in Education Ltd. Nafferton Books, 1982.

MacMahon, Percy Alexander. *Collected Papers*. Edited by George Andrews. 2 Vols. Cambridge, MA: The MIT Press, 1978–1986.

———. "James Joseph Sylvester." *Nature* 55 (March 1897): 492–494.

———. "James Joseph Sylvester." *Proceedings of the Royal Society of London* 63 (1898): ix–xxv.

———. "Memoir on Semivariants." *American Journal of Mathematics* 8 (1885): 1–18.

———. "Seminvariants and Symmetric Functions." *American Journal of Mathematics* 6 (1884): 131–163.

[Magee, Richard]. *Magee's Illustrated Guide of Philadelphia and the Centennial Exhibition*. Philadelphia: Richard Magee & Son, 1876.

Mallet, John W. "Some Remarks on a Passage in Professor Sylvester's Paper as to the Atomic Theory." *American Journal of Mathematics* 1 (1878): 277–281.

Martini, Laura. "The Politics of Unification: Barnaba Tortolini and the Publication of Research Mathematics in Italy, 1850–1865." In *Il Sogno di Galois: Scritti di storia della matematica dedicati a Laura Toti Rigatelli per il suo 60° compleanno*. Edited by Raffaella Franci, Paolo Pagli, and Annalisa Simi. Siena, Italy: Centro Studi della Matematica Medioevale, 2003, 171–198.

Mehrtens, Herbert, Bos, Henk, and Schneider, Ivo, Ed. *Social History of Nineteenth Century Mathematics*. Boston: Birkhäuser Verlag, 1981.

Menand, Louis. *The Metaphysical Club: A Story of Ideas in America*. New York: Farrar, Straus, and Giroux, 2001.

Meyer, Wilhelm-Franz. "Bericht über den gegenwärtigen Stand der Invariantentheorie." *Jahresbericht der Deutschen Mathematiker-Vereinigung* 1 (1892): 79–292.

———. *Sur les progrés de la théorie des invariants projectifs*. Trans. Henri Fehr. Paris: Gauthier-Villars et Fils, 1897.

"Military Education Part II." *Blackwood's Edinburgh Magazine* 82 (November 1857): 575–592.

Miller, Edward. *Portrait of a College: A History of the College of Saint John the Evangelist in Cambridge*. Cambridge, UK: Cambridge University Press, 1961.

Miller, Mark B. *Baltimore Transitions: Views of an American City in Flux*. Baltimore: The Johns Hopkins University Press, 1998.

Minutes of Evidence Taken before the Royal Commission Appointed to Inquire into the Present State of Military Education and into the Training of Candidates for Commissions in the Army. London: G. E. Eyre and W. Spottiswoode, 1870.

Mitchell, Oscar H. "A New Algebra of Logic." In *Studies in Logic by Members of the Johns Hopkins University*. Edited by Charles S. Peirce. Boston: Little and Co., 1883; Reprint Ed. (with an introduction by Max Fisch and a preface by Achim Eschbach). Amsterdam, The Netherlands: John Benjamins Publishing Company, 1983, 72–106.

———. "On Binomial Congruences; Comprising an Extension of Fermat's and Wilson's Theorems, and a Theorem of Which Both Are Special Cases." *American Journal of Mathematics* 3 (1880): 295–315.

———. "Some Theorems in Numbers." *American Journal of Mathematics* 4 (1881): 25–38.

Möbius, August Ferdinand. *Lehrbuch der Statik*. Leipzig, Germany: G. J. Göschen, 1837.

Moore, Eliakim Hastings, Bolza, Oskar, Maschke, Heinrich, and White, Henry S., Ed. *Mathematical Papers Read at the International Congress Held in Connection with the World's Columbian Exposition, Chicago 1893*. New York: Macmillan and Co., 1896.

Moseley, Henry. *Report on the Examination for Admission to the Royal Military Academy at Woolwich*. London: John W. Parker & Son and John Weale, 1858.

Moyer, Albert E. *Joseph Henry: The Rise of an American Scientist*. Washington, DC: Smithsonian Institution Press, 1997.

———. *A Scientist's Voice in American Culture: Simon Newcomb and the Rhetoric of Scientific Method*. Berkeley, CA: University of California Press, 1992.

Netto, Eugen. *Substitutionentheorie und ihre Anwendungen auf die Algebra*. Leipzig, Germany: B. G. Teubner Verlag, 1882.

Newton, Sir Isaac. *Mathematical Principles of Natural Philosophy*. Trans. Andrew Motte with Rev. Trans. by Florian Cajori. 2 Vols. Berkeley, CA: University of California Press, 1934; Reprint Ed., 1962.

Noether, Max. "James Joseph Sylvester." *Mathematische Annalen* 50 (1898): 133–156.

Novi, Giovanni. *Trattato di algebra superiore*. Florence, Italy: Le Monnier, 1863.

Oehser, Paul H. *Sons of Science: The Story of the Smithsonian Institution and Its Leaders.* New York: Henry Schuman, Inc., 1949.

Ogborn, Maurice Edward. *Equitable Assurances: The Story of Life Assurance in the Experience of the Equitable Life Assurance Society, 1762–1962.* London: George Allen and Unwin, Ltd., 1962.

Olson, Sherry H. *Baltimore: The Building of an American City.* Baltimore: The Johns Hopkins University Press, 1997.

Olver, Peter J. *Classical Invariant Theory.* London Mathematical Society Student Texts. Vol. 44. Cambridge, UK: Cambridge University Press, 1999.

Oman, Charles. *Memories of Victorian Oxford and of Some Early Years.* London: Methuen & Co., Ltd., 1941.

Ormerod, Henry A. *The Liverpool Royal Institution: A Record and a Retrospect.* Liverpool, UK: The University Press of Liverpool, 1953.

Parshall, Karen Hunger. "America's First School of Mathematical Research: James Joseph Sylvester at The Johns Hopkins University." *Archive for History of Exact Sciences* 38 (1988): 153–196.

———. "Chemistry through Invariant Theory?: James Joseph Sylvester's Mathematization of the Atomic Theory." In *Experiencing Nature: Proceedings of a Conference in Honor of Allen G. Debus.* Edited by Paul H. Theerman and Karen Hunger Parshall. Dordrecht, The Netherlands: Kluwer Academic Publishers, 1997, 81–111.

———. *James Joseph Sylvester: Life and Work in Letters.* Oxford, UK: Clarendon Press, 1998.

———. "Joseph H. M. Wedderburn and the Structure Theory of Algebras." *Archive for History of Exact Sciences* 32 (1985): 223–349.

———. "The Mathematical Legacy of James Joseph Sylvester." *Nieuw Archief voor Wiskunde.* 4th Ser. 17 (1999): 247–267.

———. "The One-Hundredth Anniversary of the Death of Invariant Theory?." *The Mathematical Intelligencer* 12 (4) (1990): 10–16.

———. "Telling the Life of a Mathematician: The Case of J. J. Sylvester." *Revue d'histoire des mathématiques* 5 (1999): 285–302.

———. "Toward a History of Nineteenth-Century Invariant Theory." In *The History of Modern Mathematics.* Edited by David E. Rowe and John McCleary. 2 Vols. Boston: Academic Press, Inc., 1989, 1: 157–206.

Parshall, Karen Hunger, and Gray, Jeremy J., Ed. *Episodes in the History of Modern Algebra.* Providence, RI: American Mathematical Society; London: London Mathematical Society. Forthcoming.

Parshall, Karen Hunger, and Rice, Adrian C., Ed. *Mathematics Unbound: The Evolution of an International Mathematical Research Community, 1800–1945.* HMATH. Vol. 23. Providence, RI: American Mathematical Society; London: London Mathematical Society, 2002.

Parshall, Karen Hunger, and Rowe, David E. *The Emergence of the American Mathematical Research Community, 1876–1900: J. J. Sylvester, Felix Klein, and E. H. Moore.* Providence, RI: American Mathematical Society; London: London Mathematical Society, 1994.

Parshall, Karen Hunger, and Seneta, Eugene. "Building an International Reputation: The Case of J. J. Sylvester (1814–1897)." *American Mathematical Monthly* 104 (1997): 210–222.

Pearce, Robert R. *A History of the Inns of Court and Chancery.* London: Richard Bentley, 1848.

Peirce, Benjamin. "Linear Associative Algebra." *American Journal of Mathematics* 4 (1881): 97–229.

Peirce, Charles S. "A Communication from Mr. Peirce." *Johns Hopkins University Circulars* 2 (April 1883): 86.

Peirce, Charles S., Ed. *Studies in Logic by Members of the Johns Hopkins University*. Boston: Little and Co., 1883; Reprint Ed. (with an introduction by Max Fisch and a preface by Achim Eschbach). Amsterdam, The Netherlands: John Benjamins Publishing Company, 1983.

Petersen, Julius. "Die Theorie der regulären Graphs." *Acta Mathematica* 15 (1891): 193–220.

———. "Ueber die Endlichkeit des Formensystems einer binären Grundform." *Mathematische Annalen* 35 (1890): 110–112.

Poincaré, Henri. "Sur les nombres complexes." *Comptes rendus de l'Académie des Sciences de Paris* 99 (1884): 740–742.

Poncelet, Jean-Victor. "Sur la valeur approchée des radicaux." *Journal für die reine und angewandte Mathematik* 13 (1834): 277–291.

Porter, H. W. "On Some Points Connected with the Education of an Actuary." *The Assurance Magazine* 4 (1853): 108–118.

Porter, Theodore M. *Trust in Numbers: The Pursuit of Objectivity in Science and Public Life*. Princeton, NJ: Princeton University Press, 1995.

Price, Michael H. *Mathematics for the Multitude*. Leicester, UK: Mathematical Association, 1994.

Pugh, Ralph Bernard, Ed. *The Victoria History of the Counties of England*. Vol. 3. *A History of Cambridgeshire and the Isle of Ely*. Edited by J. P. C. Roach. Oxford, UK: Oxford University Press, 1959.

Pycior, Helena. "Benjamin Peirce's 'Linear Associative Algebra.'" *Isis* 70 (1979): 537–551.

Quekett, William. "Life at St John's in 1821." *The Eagle* 15 (1889): 149–154.

Reader, W. J. *Professional Men: The Rise of the Professional Classes in Nineteenth-Century England*. London: Weidenfeld and Nicolson, 1966.

Reingold, Nathan, Ed. *Science in Nineteenth-Century America: A Documentary History*. Chicago: University of Chicago Press, 1985.

Reingold, Nathan, and Rothenberg, Marc, Ed. *The Papers of Joseph Henry*. 9 Vols. Washington, DC: Smithsonian Institution Press, 1972–.

Report of the Commissioners Appointed to Consider the Best Mode of Reorganizing the System for Training Officers for the Scientific Corps Together with an Account of Foreign and Other Military Education. London: G. E. Eyre and W. Spottiswoode, 1857.

"Report of the Committee Appointed to Consider the Possibility of Improving the Methods of Instruction in Elementary Geometry." In *Report of the Forty-third Meeting of the British Association for the Advancement of Science Held at Bradford in September 1873*. London: John Murray, 1874, 459–460.

Rice, Adrian C. "Augustus De Morgan and the Development of University Level Mathematics in Nineteenth-Century London." Unpublished doctoral dissertation. Middlesex University, 1997.

———. "Inspiration or Desperation? Augustus De Morgan's Appointment to the Chair of Mathematics at London University in 1828." *British Journal for the History of Science* 30 (1997): 257–274.

———. "Mathematics in the Metropolis: A Survey of Victorian London." *Historia Mathematica* 23 (1996): 376–417.

———. "Sylvester and Friendship." Unpublished lecture given at New College Oxford, 15 March 1997.

Rice, Adrian C., Wilson, Robin J., and Gardner, J. Helen. "From Student Club to National Society: The Founding of the London Mathematical Society in 1865." *Historia Mathematica* 22 (1995): 402–421.

Richards, Joan. *Mathematical Visions: The Pursuit of Geometry in Victorian England.* Boston: Academic Press, Inc., 1988.

Rogers, Emma, and Sedgwick, William T., Ed. *Life and Letters of William Barton Rogers.* 2 Vols. Boston: Houghton Mifflin and Co., 1896.

Rossiter, Margaret. *Women Scientists in America: Struggles and Strategies to 1940.* Baltimore: The Johns Hopkins University Press, 1982.

Roth, Cecil. *A History of the Jews in England.* 2d Ed. Oxford, UK: Clarendon Press, 1964.

———. "The Jews in English Universities." In *Miscellanies of the Jewish Historical Society of England.* Pt. 4. London: The Jewish Historical Society of England, 1942, 102–115.

———. "The Portsmouth Community and Its Historical Background." *Transactions of the Jewish Historical Society of England* 13 (1936): 157–187.

———. *The Rise of Provincial Jewry: The Early History of the Jewish Communities in the English Countryside, 1740–1840.* London: The Jewish Monthly, 1950.

Rothblatt, Sheldon. *The Revolution of the Dons: Cambridge and Society in Victorian England.* New York: Basic Books, Inc., 1968.

Rouse Ball, Walter William. *A History of the Study of Mathematics at Cambridge.* Cambridge, UK: Cambridge University Press, 1889.

Rowe, David E. "Klein, Lie, and the 'Erlanger Programm.'" In *1830–1930: A Century of Geometry.* Edited by Luciano Boi, Dominique Flament, and J.-M. Salanski. Berlin: Springer-Verlag, 1992, 45–54.

Roxburgh, Sir Ronald, Ed. *The Records of the Honourable Society of Lincoln's Inn: The Black Books.* Vol. 5 (A.D. 1845–A.D. 1914). London: Lincoln's Inn, 1968.

Ryan, Alan. "Transformation, 1850–1914." In *New College Oxford 1379–1979.* Edited by John Buxton and Penry Williams. Oxford, UK: Warden and Fellows of New College Oxford, 1979, 72–106.

Sabidussi, Gert. "Correspondence between Sylvester, Petersen, Hilbert and Klein on Invariants and the Factorisation of Graphs 1889–1891." *Discrete Mathematics* 100 (1992): 99–155.

Salmon, George. *Lessons Introductory to the Modern Higher Algebra.* Dublin: Hodges & Smith, 1859.

———. *A Treatise on Conic Sections, Containing an Account of Some of the Most Important Modern Algebraic and Geometric Methods.* 1st Ed. Dublin: Hodges & Smith, 1848.

———. *A Treatise on the Higher Plane Curves: Intended as a Sequel to a Treatise on Conic Sections.* Dublin: Hodges & Smith, 1852.

Schaffer, Simon. "Babbage's Intelligence: Calculating Engines and the Factory System." *Critical Inquiry* 21 (1994): 203–227.

Scott, Robert Forsyth. *St. John's College Cambridge.* London: J. M. Dent & Co; and New York: E. P. Dutton & Co., 1907.

Seneta, Eugene, Parshall, Karen Hunger, and Jongmans, François. "Nineteenth-Century Developments in Geometric Probability: J. J. Sylvester, M. W. Crofton, J.-É. Barbier, and J. Bertrand." *Archive for History of Exact Sciences* 55 (2001): 501–524.

Sharples, Joseph. *Liverpool.* Pevsner Architectural Guides. New Haven, CT: Yale University Press, 2004.

Shaw, George T. "The Athenæum and Its Place in Liverpool History." *The Liverpool Review* 3 (May 1928): 185–188, 3 (June 1928): 229–231.

———. *History of the Athenæum, Liverpool, 1798–1898.* Liverpool, UK: Rockliff Bros., Ltd., 1898.

Shortland, Michael, and Yeo, Richard, Ed. *Telling Lives in Science: Essays on Scientific Biography.* Cambridge, UK: Cambridge University Press, 1996.

Simmonds, Reginald Claud. *The Institute of Actuaries 1848–1948.* Cambridge, UK: Cambridge University Press, 1948.

Sinaceur, Hourya. *Corps et modèles: Essai sur l'histoire de l'algèbre réelle.* Paris: Librarie philosophique J. Vrin, 1991.

Singer, Steven. "Jewish Education in the Mid-Nineteenth Century: A Study of the Early Victorian London Community." *The Jewish Quarterly Review* 77 (1986–1987): 163–178.

Sixteenth Annual Report of the President of Harvard University to the Overseers, on the State of the Institution for the Academical Year 1840–41. Cambridge, MA: Harvard University, 1842.

Small, Hugh. *Florence Nightingale: Avenging Angel.* New York: St. Martin's Press, 1999.

Smith, Crosbie, and Wise, M. Norton. *Energy and Empire: A Biographical Study of Lord Kelvin.* Cambridge, UK: Cambridge University Press, 1989.

Spearman, T. D. "Mathematics and Theoretical Physics." In *The Royal Irish Academy: A Bicentennial History 1785–1985.* Edited by T. Ó Raifeartaigh. Dublin: The Royal Irish Academy, 1985, 200–239.

Spencer, Herbert. *An Autobiography.* 2 Vols. New York: D. Appleton & Co., 1904.

Spottiswoode, William. "Note on Axes of Equilibrium." *Quarterly Journal of Pure and Applied Mathematics* 1 (1855): 36–37.

———. "On a Theorem in Statics." *Quarterly Journal of Pure and Applied Mathematics* 1 (1855): 38–41.

Stapleton, Barry. "The Admiralty Connection: Port Development and Demographic Change in Portsmouth, 1650–1900." In *Population and Society in Western European Port-Cities, c. 1650–1939.* Edited by Richard Lawton and Robert Lee. Liverpool, UK: The University Press of Liverpool, 2002, 212–251.

Stephen, Sir Leslie. *The Life of Sir James Fitzjames Stephen.* London: Smith, Elder, & Co., 1895.

Stephen, Sir Leslie and Lee, Sir Sidney, Ed. *The Dictionary of National Biography.* 22 Vols. Oxford, UK: Oxford University Press, 1885–1901; Reprint Ed., 1949–1950.

Stewart, Hugh. "A Founding Vice-President of the Institute of Actuaries: James Joseph Sylvester (1814–1897)." In *Transactions of the 26th International Congress of Actuaries: Birmingham, UK, 7–12 June, 1998.* 8 Vols. London: Institute of Actuaries, 1998, 2: 481–496.

Stone, Lawrence, Ed. *The University in Society.* 2 Vols. *Oxford and Cambridge from the 14ᵗʰ to the Early 19ᵗʰ Century.* Vol. 1. Princeton, NJ: Princeton University Press, 1974.

Storr, Richard J. *The Beginnings of Graduate Education in America.* Chicago: University of Chicago Press, 1953.

Story, William E. "Note on the Preceding Paper." *American Journal of Mathematics* 2 (1879): 201–204.

———. "On the Theory of Rational Derivation on a Cubic Curve." *American Journal of Mathematics* 3 (1880): 356–387.

Stringham, W. Irving. "Regular Figures in *n*-Dimensional Space." *American Journal of Mathematics* 3 (1880): 1–14.

Sturm, Charles-François. "Mémoire sur la résolution des équations algébriques." *Mémoires présentés par divers savants à l'Académie royale de France* 6 (1835): 271–318.

Supple, Barry. *The Royal Exchange Assurance: A History of British Insurance 1720–1970.* Cambridge, UK: Cambridge University Press, 1970.

Susser, Bernard. *The Jews of South-West England: The Rise and Decline of Their Medieval and Modern Communities.* Exeter, UK: University of Exeter Press, 1993.

Sussman, Herbert. *Victorian Masculinities: Manhood and Masculine Poetics in Early Victorian Literature and Art.* Cambridge, UK: Cambridge University Press, 1995.

[Sylvester, Sylvester Joseph]. *Sylvester's Reporter: A Weekly Report of Lotteries, Bank Notes, Broken Banks, Stocks, &c.*

Tanner, Joseph Robson, Ed. *The Historical Register of the University of Cambridge . . . to the Year 1900.* Cambridge, UK: Cambridge University Press, 1917.

Tattersall, James J., and McMurran, Shawnee L. "Hertha Ayrton: A Persistent Experimenter." *Journal of Women's History* 7 (1995): 86–112.

Thomas, John Meurig. *Michael Faraday and the Royal Institution.* Bristol, UK: Adam Hilger, 1991.

Thompson, Sylvanus P. *The Life of William Thomson, Baron Kelvin of Largs.* 2 Vols. London: Macmillan & Co., 1910.

Thomson, David. *England in the Nineteenth Century (1815–1914).* Hammondsworth, UK: Penguin Books, Ltd., 1978.

Thomson, William (Lord Kelvin). "On the Thermo-Elastic and Thermo-Magnetic Properties of Matter." *Quarterly Journal of Pure and Applied Mathematics* 1 (1855): 57–76.

Timbs, John. *Clubs and Club Life in London.* London: Chatto & Windus, 1908.

Townsend, Richard. *Chapters on the Modern Geometry of the Point, Line, and Circle.* 2 Vols. Dublin: Hodges, Smith, & Co., 1863–1865.

Turner, R. Steven. "The Growth of Professional Research in Prussia, 1818–1848—Causes and Content." *Historical Studies in the Physical Sciences* 3 (1971): 137–182.

University College, London, Session 1839–40: Faculty of Arts and Laws. London: Richard and John E. Taylor, 1839.

Venn, John A. *Alumni Cantabrigiensis: A Biographical List of All Known Students, Graduates and Holders of Office at the University of Cambridge, From the Earliest Times to 1900.* Pt. 2 in 6 Vols. Cambridge, UK: Cambridge University Press, 1940–1954.

————. *Early Collegiate Life.* Cambridge, UK: W. Heffer & Sons Ltd., 1913.

Veysey, Laurence R. *The Emergence of the American University.* Chicago: University of Chicago Press, 1965.

Walford, Cornelius. *The Insurance Cyclopædia.* 4 Vols. London: Charles and Edwin Layton, 1871–1876.

Wall, Charles Coleman, Jr. "Students and Student Life at the University of Virginia, 1825 to 1861." Unpublished doctoral dissertation. University of Virginia, 1978.

Waller, Philip J. *Democracy and Sectarianism: A Political and Social History of Liverpool, 1868–1939.* Liverpool, UK: The University Press of Liverpool, 1981.

Warwick, Andrew. *Masters of Theory: Cambridge and the Rise of Mathematical Physics.* Chicago: University of Chicago Press, 2003.

Weichold, Guido. "Historical Data Concerning the Discovery of the Law of Valence." *American Journal of Mathematics* 1 (1878): 282.

Weinreb, Ben, and Hibbert, Christopher, Ed. *The London Encyclopædia*. London: Macmillan, 1983.

Weishampel, Jr., John F. *The Stranger in Baltimore: A New Hand Book, Containing Sketches of the Early History and Present Condition of Baltimore, with a Description of Its Notable Localities, and Other Information Useful to Both Citizens and Strangers*. Baltimore: J. F. Weishampel, Jr. Bookseller & Stationer, 1876.

White, Andrew D. *A History of the Warfare of Science with Theology in Christendom*. 2 Vols. New York: D. Appleton & Co., 1896.

―――. *Warfare of Science*. New York: D. Appleton & Co., 1876.

Wiener, Philip P., and Young, Frederic H., Ed. *Studies in the Philosophy of Charles Saunders Peirce*. Cambridge, MA: Harvard University Press, 1952.

Wilkes, Maurice V. "Herschel, Peacock, Babbage, and the Development of the Cambridge Curriculum." *Notes and Records of the Royal Society of London* 44 (1990): 205–219.

Williams, Charles Richard, Ed. *Diary and Letters of Rutherford Birchard Hayes: Nineteenth President of the United States*. 5 Vols. Columbus, OH: The Ohio State Archæological and Historical Society, 1924.

Wilson, Arline. "The Cultural Identity of Liverpool, 1790–1850: The Early Learned Societies." *Transactions of the Historic Society of Lancashire and Cheshire* 147 (1998): 55–80.

Wilson, David B. "The Educational Matrix: Physics Education at Early-Victorian Cambridge, Edinburgh and Glasgow Universities." In *Wranglers and Physicists: Studies on Cambridge Mathematical Physics in the Nineteenth Century*. Ed. Peter M. Harmon. Manchester, UK: Manchester University Press, 1985, 12–48.

Wilson, Richard. *A System of Plane and Spherical Trigonometry; to Which Is Added a Treatise on Logarithms*. Cambridge, UK: Cambridge University Press, 1831.

Winstanley, Denys Arthur. *Early Victorian Cambridge*. Cambridge, UK: Cambridge University Press, 1940.

Wolfman, Joseph. "Liverpool Jewry in the Eighteenth Century [Part 1]." *Merseyside Jewish Representative Council Year Book* (1986–1987): 37–40.

―――. "Liverpool Jewry in the Eighteenth Century [Part 2]." *Merseyside Jewish Representative Council Year Book* (1987–1988): 25–41.

―――. "Liverpool's First Jewish Mayor." *Merseyside Jewish Representative Council Year Book* 1993–1994: 60–67

Wright, John M. F. *Alma Mater; or, Seven Years at the University of Cambridge*. 2 Vols. London: Black, Young, and Young, 1827.

Yates, R. C. "Sylvester at the University of Virginia." *American Mathematical Monthly* 44 (1937): 194–201.

Index

Aberdeen, Lord, 137–38

Académie des Sciences, Paris: JJS's participation in meetings of, 62, 90, 132, 183, 264, 237; JJS's publications in the *Comptes rendus* of, 156, 159–60, 167, 179–80, 185, 234, 237, 238, 239, 242, 263, 265, 272, 272–73, 276, 287, 290, 293, 297, 306, 319, 381n85, 394n56, 406n38, 409n6, 410n16; names JJS as foreign correspondent of, 3, 179

actuary. *See under* "professional man"

Adams, John Couch, 168–69, 207, 223, 316, 323, 416n35

Advanced Class (Woolwich), 196; JJS's criticism of, 209–11, 387n66

Airy, George Biddell, 28, 49, 59

Alborn, Timothy, 83

American Academy of Arts and Sciences, JJS's election to membership of, 219

American Journal of Mathematics: Cayley as contributor to, 240, 244, 245; first issue of, 240–42; JJS as editor of, 240–48, 267, 269, 271, 275, 276, 295, 392n28, 402n101; JJS's later connection with, 288–89, 300; Johns Hopkins University as sponsor of, 226, 395n89; plans for launching, 231–32, 234, 239–40; reputation of, 242, 243; Story as production editor of, 259–60; student publications in, 250–51, 253, 257–58, 272–73, 395n98

American Philosophical Society, JJS elected member of (1877), 238

Ampère, André-Marie, 45

Anderson, Henry, 79

Anglo-Jewish Association, 304–5

Anglo-Jewry: boys' education at Highgate, 17–18; cultural identity of, 10–11, 14–15, 345n82; efforts to remove civil and religious disabilities of, 124–25, 273, 329, 345n82, 345n3;

history of, 10–11; *Jewish Chronicle* as recorder of, 329–30; JJS as a symbol of, 325, 329–30, 404n132; Liverpool community of, 12–16; London migration patterns of, 324n35; Maccabæans as representative of, 325, 331; memorialization of JJS by, 331–32; representatives of at JJS's funeral, 328

Anthon, Charles, 388n74

anti-Semitism: expressions of at Columbia College, 79–80; expressions of in England, 10, 124–25, 143, 217–18, 273, 278, 329–30, 331, 389n99; expressions of in Virginia, 68–69

Appel, Kenneth, 396n114

applicable mathematics, JJS's work in, 58–59, 77, 147, 157, 159–60, 188–89, 390n117

Archibald, Raymond Clare, 358n126

Armstrong, William George, Baron Armstrong of Cragside, 184, 326

Arnold, Matthew, 193, 212, 391n9

Arnold, Thomas, 391n9

Aronhold, Siegfried, 119–20, 334

Ashkenazim, 11, 11n6, 12

Association for the Improvement of Geometrical Teaching, 207, 302; JJS as president of, 322

Association française pour l'avancement des sciences, 321; JJS's involvement in, 305, 306–7

Athenæum Club (Liverpool), 15, 212

Athenæum Club (London), 151, 323; JJS's election to under Rule II, 151; as JJS's London base, 151, 212, 216, 225, 274–75, 279, 285, 286, 292, 296, 309, 324, 326; JJS's views of, 213

Atkinson, Charles, 52, 57

Babbage, Charles, 28, 45, 49, 62, 65, 82, 97

Baker, Henry Frederick, 348n55, 357n113, 371n116, 413n93

Ballard, William Henry, 72–76, 357n108, 357n111